Principles of Ultraviolet Photoelectron Spectroscopy

Principles of Ultraviolet Photoelectron Spectroscopy

by

J. Wayne Rabalais

Associate Professor of Chemistry
University of Houston

A Wiley-Interscience Publication

JOHN WILEY & SONS

New York · London · Sydney · Toronto

Library of Congress Cataloging in Publication Data:

Rabalais, J   Wayne, 1944–
    Principles of ultraviolet photoelectron spectroscopy.

    (Wiley-Interscience monographs in chemical physics)
    "A Wiley-Interscience publication."
    Bibliography: p.
    Includes index.
    1. Photoelectron spectroscopy. 2. Ultra-violet spectrometry. I. Title.

QC454.P48R3   539'.1   76–28413
ISBN 0-471-70285-4

Printed in the United States of America

10 9 8 7 6 5 4 3 2 1

To Becki

Preface

The development of the technique of ultraviolet photoelectron spectroscopy ranks as one of the more important advances of chemical physics in the last decade. Information derivable from this technique is of vital importance in understanding the electronic structure of atoms, molecules, ions, and solid materials.

The information explosion in the field of ultraviolet photoelectron spectroscopy in the early 1970s is primarily in the form of individual research papers. Two books dealing exclusively with the subject are an extensive collection of spectra with early interpretations by D. W. Turner, C. Baker, A. D. Baker, and C. R. Brundle entitled *Molecular Photoelectron Spectroscopy* and a nonmathematical guide to most of the phenomena observed in photoelectron spectroscopy by J. H. D. Eland entitled *Photoelectron Spectroscopy*. The field has now moved out of adolescence into a stage of maturity. However, there is no text at present that treats the *principles* of photoelectron spectroscopy in terms of modern chemical physics, that is, the phenomenon of photoionization and its resulting consequences by means of modern quantum mechanical methods. This monograph has been prepared in order to meet the obvious need of such a treatment; the physical process of ionization, spectroscopic classification of the ionic states, the subtle vibrational, rotational, vibronic, and spin-orbit structure of the ionization bands, abundances and angular distributions of the ejected electrons, and interpretation of the results in terms of the electronic structure of the molecule and ion are presented and discussed in detail.

The book has been prepared with the aim of serving both a pedagogic need and a research need. The pedagogic function is particularly evident in Chapters 1 to 4, 10, and 11; these chapters are written at the level of senior undergraduates or beginning graduate students in chemical physics. Many spectra and illustrative diagrams are included to exemplify the discussions. The research function is particularly evident in Chapters 5 to 9; these chapters contain quantum mechanical treatments of areas that are at the brink of current research. In order to keep a keen research edge on the book many references have been compiled, and controversial and difficult areas have not been evaded; in fact, Appendix VIII contains some recent spectra that have not been interpreted. The book should be particularly useful as a text for a senior undergraduate or beginning graduate student level course in uv photoelectron spectroscopy and as a reference and handbook for active researchers in the field. Elementary group theory and quantum mechanics are used throughout the text. Students who have had no previous exposure to these topics may wish to consult an introductory text for background in these areas.

The textual material is organized in the following manner: Chapters 1 and 2, introductory and experimental; Chapters 3 to 5, ionization processes and ionic states; Chapters 6 and 7, cross sections and angular distributions; Chapters 8 and 9, spin-orbit coupling and configurational instabilities; Chapters 10 and 11, simplifying approaches and applications. An attempt has been made throughout the text to amalgamate theory and experiment. The approach has been to use quantum mechanical models to interpret experimental results, for I believe that the material can best be grasped and remembered in this manner. I have not concentrated heavily on experimental methodology and the many recent advances in instrumental design; instead, the experimental results are presented, and their interpretation pondered. The emphasis is on photoionization with low-energy photons, that is, ultraviolet photons ($h\nu < 50$ eV), although high-energy photons (X-rays) are considered for completeness in the treatment of cross sections and angular distributions. The detailed derivations of the quantum mechanical models for open-shell molecules (Chapter 5), cross sections and angular distributions (Chapter 6), and spin-orbit coupling (Chapter 8) could be omitted in a first reading. It was felt that these detailed treatments should be included for the sake of students or researchers who wish to explore extensions or modifications of the methods. Appendix IX contains a useful compliation of molecules with references to their published gas-phase uv photoelectron spectra. The references of Appendix IX cover the period 1963 to early 1975.

The writing of this manuscript is a logical, if not inevitable, consequence of my involvement with photoelectron spectroscopy over the last several years. I am deeply indebted to the persons who assisted in my efforts. My teachers, S. P. McGlynn, K. Siegbahn, and C. Nordling, were particularly stimulating. Many of the experimental spectra and computed results stem from the help of my students and postdoctoral associates—J. L. Berkosky, R. J. Colton, T. P. Debies, J. T. J. Huang, A. Katrib, T. H. Lee, and M. G. White; I am deeply obligated to them. It is a pleasure to acknowledge my colleague, Frank O. Ellison, for his collaboration in the development of theoretical models. I am appreciative of Mrs. Nancy Sattler for her skillful typing of the original handwritten manuscript. Richard Colton and Jan-Tsyu Huang read the manuscript and made valuable suggestions and comments. I am grateful to my wife Becki for her assistance, encouragement, and patience throughout the course of this work.

The manuscript was prepared during my employment at the University of Houston and the University of Pittsburgh. Much of the efforts during the preparation were supported by the U. S. Army Research Office, the Petroleum Research Fund, and the Research Corporation. I am grateful to these institutions and granting agencies for their assistance.

I must acknowledge my debt to the editors of *Chemical Physics Letters, Discussions of the Faraday Society, Inorganic Chemica Acta, Inorganic Chemistry, International Journal of Mass Spectrometry and Ion Physics, Journal of the American Chemical Society, Journal of Chemical Physics, Journal of Electron Spectroscopy, Journal of Pharmaceutical Sciences, Journal of Physical Chemistry, Molecular Physics, Philosophical Transations of the Royal Society of London, Physica Scripta, and Zeitschrift fuer Naturforschung* and the Perkin–Elmer Corporation for permission to reproduce various figures, tables, and equations.

Critical comments by the readers will be greatly appreciated, for it is difficult to write a completely up-to-date book in such a rapidly advancing field. It is my humble hope, however, that the book will prove useful to students and researchers of this fascinating and intriguing branch of spectroscopy.

J. WAYNE RABALAIS

Houston, Texas
May 1976

Contents

Principles of Ultraviolet Photoelectron Spectroscopy

1 Introduction

Photoelectron spectroscopy involves application of the photoelectric effect to the study of electronic structures. Photons of energy $h\nu$ interacting with a molecule that has electrons bound by energy E_I results in ejection of electrons with kinetic energy E_k provided that $h\nu \geq E_I$. Conservation of energy between the ion and electron requires that

$$h\nu = E_k + E_I \tag{1.1}$$

where E_I is usually referred to as the ionization energy or potential of the electron in the molecule; the kinetic energy imparted to the ion is negligible (less than the experimental accuracy of the measurement) due to the large disparity of mass between the ion and electron. The use of monoenergetic radiation for ionization purposes and the precise measurement of E_k permit a direct determination of the ionization energies E_I of the photoelectrons. *Photoelectron spectroscopy involves the measurement of kinetic energies E_k of photoelectrons ejected by monoenergetic radiation in order to determine the ionization energies E_I, intensities, and angular distributions of these electrons and to use this information as a probe to elucidate the electronic structure of molecules and ions.*

Photoelectron spectroscopy is distinguished from conventional forms of spectroscopy in that it uses the analysis of electrons rather than photons as its primary source of information. The unique feature of the technique is its ability to eject electrons from any of the occupied energy levels in a molecule and, since each one of these levels has a different E_I, single out only those particular electrons for study. Thus using sufficiently energetic radiation, it is possible to obtain the ionization energies E_I of electrons in all of the occupied atomic or molecular energy levels. According to Koopmans' theorem these E_Is are equal to the negative eigenvalues of the various atomic and molecular orbitals of the system. Through use of this approximation, photoelectron spectroscopy provides an experimental determination of a molecular orbital diagram. In a more subtle manner, the structure of the ionization bands yields information about the vibrational frequencies and geometries of the molecular ions and the bonding and localization properties of the molecular orbitals.

Electrons can be ejected from a material by exciting it with various types of radiation, the two most common in photoelectron spectroscopy being vacuum ultraviolet radiation and X-rays. The energy of ultraviolet rays is sufficient to only eject electrons from the valence orbitals, while the high energy of X-rays allows ejection of electrons from even the core orbitals of heavy atoms. Both types of radiation have their advantages and disadvantages: Although uv radiation limits one to a study of the valence orbitals, the relatively high resolution attainable, ~0.01 eV, allows resolution of vibrational and sometimes rotational structure in the spectral bands. With X-rays it is possible to study the core as well as the valence orbitals, although the resolution attained in most work is no better than ~0.5 eV. The technique of uv radiation is commonly called *molecular photoelectron spectroscopy, photoelectron spectroscopy, and ultraviolet photoelectron spectroscopy (UPS)*. In this book we use the phraseology, PE spectroscopy. The subject matter of this book is devoted to the original and most extensive use of PE spectroscopy, that is, the study of the electronic structures of gas phase molecules. The technique of X-radiation is commonly called *electron spectroscopy, electron spectroscopy for chemical analysis (ESCA), and X-ray photoelectron spectroscopy (XPS)*. The original X-ray studies were on solids, but have since diversified to gases and even liquids. Due to the lack of suitable photon sources in the intermediate region, most of the work has been confined to ultraviolet (<41 eV) and X-ray (~1000–1500 eV) sources.

A closely related type of spectroscopy is Auger spectroscopy. In Auger spectroscopy a beam of high energy radiation is used to eject an electron from a core orbital. The resulting positive ion can relax to a lower energy state by dropping a valence electron into the core hole. The energy released in this process can be (1) emitted as a quantum of X-radiation (X-ray fluorescence) or (2) used to eject a valence electron from the ion (the Auger process). The ejected electron is called an *Auger electron* and possesses kinetic energy characteristic to the atom from which it is emitted. This kinetic energy is determined by the energy released in relaxation and the binding energy of the ejected electron, hence it is independent of the energy of the exciting beam. For this reason, the excitation source for Auger spectroscopy usually consists of a beam of electrons from a high-powered electron gun. Since an impinging electron can loose any portion of its kinetic energy to the bound electrons, the primary electrons (photoelectrons) are ejected with a random distribution of energies while the secondary electrons (Auger electrons) have discrete energies. The resulting spectrum exhibits Auger peaks superimposed on a continuous background of primary electrons; it is hence displayed as the first derivative of the actual spectrum. Auger peaks are commonly

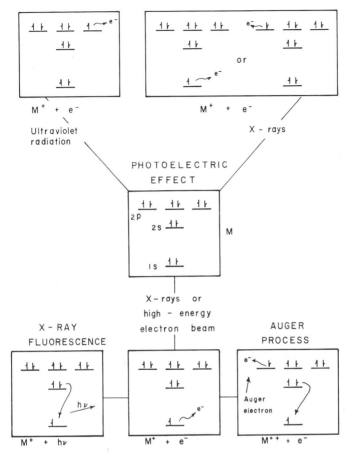

Fig. 1.1. Schematic representation of the photoelectric effect, the Auger process, and X-ray fluorescence. In the photoelectric effect the molecule M is ionized by uv radiation (producing a 'hole' in the valence orbitals) or by X-rays (producing a 'hole' in either the core or valence orbitals). In the Auger process, M is ionized by X-rays or by a high-energy electron beam to produce a 'hole' in the core orbital. The ion relaxes by dropping a valence electron into the core hole and imparting the energy gained to another electron which is in turn ejected (the Auger electron). An alternative and competitive mode of relaxation, X-ray fluorescence, involves emission of an X-ray photon when the valence electron fills the core hole.

observed in ordinary X-ray electron spectroscopy. A schematic representation of PE and Auger spectroscopy is shown in Fig. 1.1.

This book treats the *principles* of PE spectroscopy, that is, the phenomenon of photoionization and its resulting consequences. The physical process of ionization, spectroscopic classification of the ionic states, the subtle vibrational, rotational, vibronic, and spin-orbit structure of the ionization bands, the intensities and angular distributions of the ejected electrons, and interpretation of the results in terms of the electronic structure of the molecule and ion are presented and discussed in detail. Elementary quantum mechanics is used throughout the discussions in order to provide modern interpretations.

1.1. Historical developments

The origins of PE spectroscopy can be traced back to the independent efforts of three research groups. The use of X-rays in photoelectron studies evolved from research in β-ray spectroscopy[1] being carried out by Professor Kai Siegbahn and his group at the University of Uppsala. Their first papers appeared in 1956[2] and 1957[3] and were followed by comprehensive and now classical books covering their instrumentation and experimental and theoretical results on solids in 1967[4] and gases in 1969.[5] They introduced the two most widely used high-energy ionization sources, MgK_α and AlK_α radiation, with energies of 1253.6 and 1486.6 eV, respectively. Most of the early developments and innovations of X-ray excitation are credited to the Swedish group; significant contributions also came from Carlson and Krause at Oak Ridge National Laboratory.[6] The use of ultraviolet radiation in photoelectron studies was developed by Professor A. N. Terenin and his group in Leningrad (1961)[7] and Professor David W. Turner and his group at Imperial College, London University (1963).[8] The Russian group seems to have stopped publishing in the area after their initial exploratory research. Most of the early developments in the ultraviolet area are credited to the English group,[9] who did much of the work prior to 1967, although contributions by Price,[9a] Forst et al.,[10] and Schoen[11] are acknowledged.

The approach of the Russian group[7] was to use the continuum of a hydrogen discharge in conjunction with a vacuum monochromator to produce a monochromatic photon source of variable energy for ionization and a retarding field analyzer for measurement of photoelectron energies. The hydrogen discharge tube was isolated from the target chamber by a LiF window, which has a short wavelength cutoff at 1040 Å, hence limiting the work to $E_I < 11.8$ eV. Besides this, the source was of low intensity and its low energy introduced complications due to autoionization. Turner and his English group[8] introduced the He I resonance line at 584 Å (21.22 eV) using a differentially pumped windowless discharge through helium gas. This source provided an intense and nearly monoenergetic line that was at high enough energy that electrons from most of the valence orbitals could be ionized without serious autoionization interferences. Hence it is the differentially pumped helium lamp used with a retarding field analyzer,[8] followed later by an electrostatic deflection analyzer,[9] that gave birth to modern PE spectroscopy. Later, Professor W. C. Price at King's College in London[12] obtained the He II resonance line at 304 Å (40.81 eV), making ionization from the inner valence orbitals possible.

The tremendous interest generated by these early investigations led to an almost exponential growth in the number of papers after 1968 emanating from laboratories throughout the world with various spectrometer designs and innovations. At present there are several instrument companies[13] marketing X-ray, uv, or both types of spectrometers.

There are now several books and review articles dealing with the subject of PE spectroscopy. The book by Turner and his group[14] provides an extensive collection of spectra with early interpretations. Baker and Betteridge[15] have published a qualitative treatment of PE spectroscopy with emphasis on analytical aspects. The monograph by Eland[16] provides a non-mathematical guide to most of the phenomena observed in PE spectroscopy. A new scientific journal, the *Journal of Electron Spectroscopy and Related Phenomena*, has been established in order to draw together the work of theorists and experimentalists in various aspects of electron spectroscopy. The proceedings of conferences on PE spectroscopy,[17-19] general review articles,[20-38] and analytically oriented review articles[39-42] are available.

1.2. Basic concepts

Molecules introduced into the target chamber of a PE spectrometer are irradiated with a photon beam of specified energy and the ejected electrons are sorted according to their kinetic energies in an electron energy analyzer. The spectrum is obtained by scanning the range of kinetic energies and plotting the number of electrons exhibiting a given kinetic energy, that is, "counts per unit time" versus kinetic energy. Spectra are normally displayed with the ordinate labeled as *count rate* or simply *relative intensity*. The latter designation is sufficient because the absolute count rate is a facet of many complex experimental variables, usually varying from one instrument to another, and is essentially meaningless for any given sample; however, the relative intensities are meaningful. Relative intensities of spectral bands are related to the relative probabilities of photoionization to different states of the molecular ion, that is, the relative differential photoionization cross sections. The abscissa is usually labeled as *ionization energy* in electronvolts as derived from Eq. 1.1 by measurement of the electron kinetic energies. The quantity actually measured is the electrical potential in volts required to focus electrons of a given kinetic energy. Although ionization potentials are measured in volts, which are units of potential, they are traditionally referred to as energies and are quoted in electron volts. Hence the terms *ionization potential* and *ionization energy* are considered synonymous and are referred to as E_I throughout this book.

The recommended direction of the ionization energy scale has only recently been suggested by IUPAC (Section 1.3). Plotting the measured kinetic energy values E_k from left to right requires (Eq. 1.1) that E_I increases from right to left. However, in some spectra the quantity of more direct interest E_I is plotted increasing from left to right. As a result, both representations are found in the literature. Since the spectra presented here are obtained from various instruments operated by different experimentalists throughout the world, both conventions are found in the figures of this book.

1.3. IUPAC recommendations for nomenclature and spectral presentation

The Commission for Molecular Structure and Spectroscopy of the International Union of Pure and Applied Chemistry (IUPAC)[43] has made recommendations on nomenclature and spectral presentation for electron spectroscopy. These recommendations are as follows:

Nomenclature

Examples of existing ambiguities in nomenclature are as follows:

1. The phrase *photoelectron spectroscopy*, commonly abbreviated to PES or UPS, is most commonly used to denote the analysis of the kinetic energies of electrons emitted after excitation by He I or He II photons in the ultraviolet region. But, logically, this description is equally applicable to the main features of spectra produced by X-ray photons (see later discussion).

2. The phrase *electron spectroscopy for chemical analysis*, commonly abbreviated to ESCA or XPS, is widely used in connection with spectra produced by excitation with X-rays. But the spectra produced by ultraviolet irradiation also lead to applications in chemical analysis, particularly the identification of interatomic groupings within molecules.

In some respects, the two methods of excitation are complementary in their applications. For example, the X-ray method is at present mainly applied to the investigation of molecules in the solid state, and the ultraviolet excitation method is mainly applied to small molecules that have the volatility to be studied in the gaseous state, although both techniques find some application to other phases.

A particular problem has arisen, leading to redundant use of language, in connection with acronyms that use 'S' for 'spectroscopy'. For example, a phrase such as 'NMR spectrometer', meaning 'nuclear magnetic resonance spectrometer', is grammatically correct, but, for example, the phrase 'the ESCA spectrum', which literally means 'the electron spectroscopy for chemical analysis spectrum', is incorrect. It is, therefore, recommended that as a general rule abbreviations or acronyms that incorporate 'S' for 'spectroscopy' or 'spectra' should be discouraged. Exceptions are ESCA and PES, which for historical reasons have been widely adopted. It is recommended that these abbreviations be limited to their respective contexts and that in each case care should be taken to see that the acronym or abbreviation is not used in grammatically incorrect situations, such as, for example, in the phrases 'ESCA spectroscopy' or 'PES spectrometer'.

The widely adopted description of spectra excited by ultraviolet photons, such as from He I or He II, as 'photoelectron spectra' is recommended by IUPAC for general use. With this type of excitation, virtually all the electrons which are measured derive from photoionization processes. Higher energy excitation, such as by X-ray photons, commonly leads to the production of electrons derived from Auger processes as well as from photoionization processes. Such complete and uninterpreted spectra are, therefore, logically described by the more general phrase 'electron spectra'. However, the more specific description 'photoelectron spectra' may logically be applied to the appropriately identified features of such spectra. Under such circumstances, the spectra should be reported, as appropriate, in the fashion 'the He I photoelectron spectrum' (spoken: the helium one photoelectron spectrum) or the 'AlK_α photoelectron spectrum' (spoken: the aluminum K alpha photoelectron spectrum).

Energy scale

Three designations of the energy scale—ionization energy, binding energy, or electron kinetic energy—are in common usage in the presentation of spectra derived from X-ray or ultraviolet excitation. The first two of these are alternative names for the same quantity, but the electron kinetic energy, although linearly related to the others (it is the difference between the energy of the exciting photon and the ionization energy), increases in magnitude as the other decreases. Ionization potential (in volts) is also used as an alternative to ionization energy (in electron volts).

The general use of *ionization energy*, E_I (expressed in electron volts, eV, or Joules, J)* is recommended by IUPAC for this scale at the bottom of all spectral diagrams. It is also recommended that the experimentally measured parameter, electron kinetic energy, should be plotted along the top of the spectral diagram in the same units. The latter parameter is meaningful whether the electrons measured derive from photoionization or Auger processes. However, those spectral features derived from Auger processes, such as occur when using high energy excitation by X-rays, should be indicated clearly by asterisks because the ionization energy scale is not applicable to them.

The direction of the ionization energy scale is not well defined in the literature. For X-ray excitation, ionization (binding) energy is usually plotted so as to increase from right to left. For ultraviolet He I or He II excitation the literature shows a limited preference for ionization energy increasing from left to right. For consistency between the two related fields it is recommended by IUPAC that the spectra be plotted with increasing electron kinetic energy to the right, that is, increasing ionization energy to the left.

Other recommendations for the presentation and publication of spectral data

1. The primary ordinate scale of the normal differential curve should be given as (dI/dV), that is, amperes per volt or as counts per second.
2. Information about sweep rate (in volts per second) and/or integration time (in seconds) should be given where appropriate.
3. A statement should be given of the method of calibration and the precision achieved.
4. Because different types of apparatus exhibit different sensitivity as a function of electron kinetic energy, relevant information should be given about the performance of the equipment used.

---

* In this latter case the factor used in conversion from electron volts to Joules (which depends on the accepted value for the electronic charge) should be stated.

1.4. Features of photoelectron spectra

The main features of PE spectroscopy can best be illustrated by considering a practical example. The PE spectrum of water produced by He I radiation is presented in Fig. 1.2. Excitation of water with 21.22 eV photons results in ejection of electrons with kinetic energies in the regions 8.6, 6.4, and 3.0 eV corresponding to binding energies of 12.6, 14.8, and 18.2 eV, respectively. These groups of electrons have kinetic energies which are each spread over approximately a 2 eV range. These groupings correspond to the kinetic energies of electrons ejected in transitions from the ground state of H_2O to various electronic states of the molecular ion H_2O^+, that is, $H_2O + h\nu \rightarrow H_2O^+ + e^-$. The three regions of high intensity in Fig. 1.2 are called *photoelectron bands*. It should be noted that because of the extremely high intensity of the 12.6-eV band, it was necessary to change the ordinate scale at ~12.6 eV and record the remainder of the spectrum on a more sensitive scale.

Each photoelectron band actually gives a group of closely spaced peaks corresponding to the formation of the positive ion H_2O^+ with varying

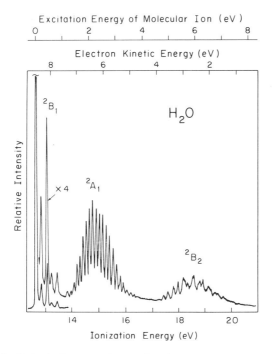

Fig. 1.2.   He I photoelectron spectrum of water.

degrees of vibrational excitation. The vibrational bands correspond to excitation of a definite number of vibrational quanta of H_2O^+. Analysis of these vibrational spacings leads to information about the normal modes of vibration of the molecular ion. Each vibrational band consists of closely spaced rotational lines that can be resolved in some special cases. Analysis of the vibrational structure of photoelectron bands in polyatomic molecules can be complicated due to the possible excitation of more than one of the several normal modes of vibration. Comparison of the vibrational frequencies of the molecule with those of the molecular ion, as measured in PE spectra, provides an approximation to the change in vibrational force constants upon ionization and ultimately provides a clue to the bonding characteristics that the ejected electron posessed in the molecule.

Many PE spectral transitions exhibit broad featureless ionization bands with no evidence of vibrational fine structure. This is usually due to the short lifetime of the ionic state or to a high density of vibrational states that is beyond the resolution capabilities of the technique. Despite this lack of vibrational structure, the spectra still provide useful information as is shown in later chapters. Other features such as spin-orbit coupling, autoionization, coupling between unpaired electrons, and so on, can cause complications in PE spectroscopy.

1.5. Analysis of ionization bands

Analysis of a PE spectrum consists of assigning each spectral band to an electronic state of the molecular ion and identifying the orbital from which electrons are ejected in the molecule. It is extremely useful in such an analysis to have prior knowledge of the energies, character, and ordering of the occupied orbitals in the molecule. This information can be obtained from quantum theory; however, it is not always reliable, requiring that other sources of information be used to assist in assigning spectra. Other features employed for this purpose are analysis of the vibrational structure and shapes of the photoelectron bands, calculating intensities and angular distributions of ionization bands, correlation with spectra of related compounds, using information from other forms of spectroscopy, and a host of other techniques, all of which are discussed in detail in later chapters.

Ejection of electrons from different energy levels of a molecule produces molecular ions in various electronic states. The ionization energies E_I observed in PE spectroscopy are just these energy differences between the ground state of the molecule and the electronic states of the molecular ion. The molecular and ionic states are described in terms of the electron occupancies of the molecular orbitals, that is, the *electron configurations*. Hence ionization energies are interpreted by treating the problem within two levels of approximation: first, a one-electron molecular orbital model is devised, and second, the system is converted to a multielectron state or configuration model. These concepts are now applied to the H_2O spectrum.

One-electron molecular orbital model

Molecular orbital calculations yield eigenvectors ϕ (molecular orbital (MO) wavefunctions) and eigenvalues ε (one-electron MO energies). The term "one-electron" arises from the method of calculation that treats each electron individually as moving in the potential field produced by the fixed nuclei and averaged interactions of all other electrons. The eigenvalues represent the binding energies of electrons in various MOs, that is, the amount of energy required to remove an electron from an MO to a point infinitely far from the molecule, the vacuum level, where it no longer feels the influence of the positive ion. Koopmans' theorem states that the experimental vertical ionization energies are approximately equal to the negative of these eigenvalues. At this level of approximation, all ionization energies are considered to correspond unequivocally to these "negative MO eigenvalues." A one-electron MO diagram for H_2O showing the Koopmans' theorem E_Is as determined from *ab initio* SCF (self-consistent field) calculations[44] and schematic MO wavefunctions are shown on the left side of Fig. 1.3. The MOs are labeled according to group theory[45] using irreducible representations with lower case letters representing their transformation properties in the point group of the molecule. The calculation suggests that the MO structure of H_2O in its ground state is $1a_1^2 2a_1^2 1b_2^2 3a_1^2 1b_1^2$. The $1b_1$ orbital consists of a non-bonding $2p_x$ atomic orbital (AO) that is completely localized on the oxygen atom. The $3a_1$ orbital is a combination of O_{2p} and H_{1s} AOs resulting in a strongly H—H bonding MO. The $1b_2$ orbital is a combination of O_{2p} and H_{1s} AOs in an O—H bonding and H—H anti-bonding configuration. These three largely O_{2p} and H_{1s} combinations constitute the *outer-valence orbitals* of H_2O. The $2a_1$ orbital is an *inner-valence orbital*. It is predominantly composed of an O_{2s} AO, with small admixtures of O_{1s}, O_{2p}, and H_{1s} AOs. Inner valence orbitals generally have only weak bonding character. The $1a_1$ orbital is a *core orbital;* it is an atomic-like orbital in that it is largely composed of an O_{1s} AO that is localized on the oxygen atom, non-bonding, and retained by a massive binding energy.

Comparison of Figs. 1.2 and 1.3 shows that the energies of the three observed ionization bands correspond closely to the eigenvalues of the

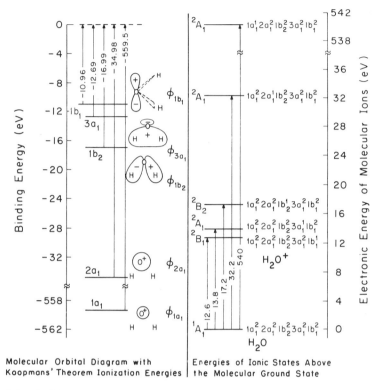

Fig. 1.3. (left) Molecular orbital diagram with Koopmans' theorem ionization energies from *ab initio* SCF calculations[44] and schematic MO wavefunctions. (right) Energies of ionic states of H_2O^+ above the molecular ground state of H_2O as determined from PE spectroscopic measurements. Electron configurations are shown for each state.

$1b_1$, $3a_1$, and $1b_2$ MOs. The two remaining MOs, $2a_1$ and $1a_1$, are more tightly bound, requiring higher energy radiation for ionization. The MO calculation thus leads us to expect only three bands in the He I spectrum of H_2O. In fact, within the Koopmans' theorem approach, a PE spectrum is actually an *MO diagram*. Implicit in this match is the assumption that the MO calculation predicts the "correct" ordering of MOs. For the moment, we tentatively accept this ordering; verifications require analysis of the vibrational structure and shapes of the ionization bands as discussed below.

Multielectron state or configuration model

The electronic structure of a molecule or ion determines its *electronic state*. An electronic state is described by an *electronic configuration* or, sometimes, combinations of several electron configurations. A configuration of electrons is a description of the occupied MOs and the degree of occupancy of each MO. For example, the ground state of H_2O can be described by the electron configuration $1a_1^2 2a_1^2 1b_2^2 3a_1^2 1b_1^2$, where the superscripts describe the number of electrons occupying each MO. This state is labeled $^1A_1$, the superscript corresponding to the multiplicity $2S + 1$ where S is the sum of the spins of individual electrons and the A_1 symbol corresponds to the direct product of the irreducible representations of all the electrons. Capital letters are used to designate states. The ground electronic state of the molecular ion H_2O^+ is obtained by removing the most loosely bound electron from the molecule, that is, a $1b_1$ electron, to produce the configuration $1a_1^2 2a_1^2 1b_2^2 3a_1^2 1b_1^1$ corresponding to a $^2B_1$ term symbol. This configuration provides a good description of the ionic ground state, but any other configurations that give $^2B_1$ symmetry should be considered along with the former in order to provide a complete *configuration interaction* description of the $^2B_1$ state. Actually, in any multielectron system, complete description of an electronic state consists of a combination of all electron configurations corresponding to the symmetry of that state. It is actually the difference in energy between *states*, or specifically, *the energies of the ionic states above that of the molecular ground state* that is measured in PE spectroscopy. The first band observed in the spectrum of H_2O is a result of the $(\cdots 3a_1^2 1b_1^1)^2 B_1 \leftarrow (\cdots 3a_1^2 1b_1^2)\ ^1A_1$ transition. In writing transitions, the upper state is always placed first in accordance with the internationally agreed rule.

The energies of the electronic states of H_2O^+ above the ground state of H_2O, as determined from PE spectroscopy, are shown on the right side of Fig. 1.3. Electron configurations corresponding to each electronic state are indicated. These single configuration descriptions of the states are adequate for these purposes. In a proper configuration interaction wavefunction for each state, the indicated configurations are overwhelmingly predominant. The measured ionization energies corresponding to transitions to ionic states on the right side of Fig. 1.3 are different in magnitude from the calculated binding energies on the left side of the figure. The difference is due to the approximations inherent in Koopmans' theorem and to the non-exact nature of the calculations. However, these differences are rather small, and the one-electron MO calculation certainly provides a reasonable interpretation of the ionization processes in H_2O.

Vibrational structure of the ionization bands

The most intense vibrational peak in a photoelectron band corresponds closely to a transition from the ground state of the molecule to an ionic state in which the nuclear coordinates are unchanged from those of the molecule. The energy of such a transition is called the *vertical ionization energy* because it corresponds to a vertical line on a potential energy diagram. The *adiabatic ionization energy* corresponds to a transition from the rotational, vibrational, and electronic ground state of the molecule to the lowest rotational and vibrational levels of an electronic state of the molecular ion. The adiabatic transition corresponds to the $0 \leftarrow 0$ band of a vibrational progression, which can be equivalent to or very different from the vertical ionization energy. In some cases the adiabatic transition may be of too low intensity to locate. The experimental adiabatic energies are used in the state diagram of Fig. 1.3 (right), whereas the eigenvalues of the MO diagram of Fig. 1.3 (left) represent vertical energies.

The structure in the bands of Fig. 1.1 confirms the assignments from MO calculations. In the first band at 12.6 eV, the adiabatic and vertical ionization energies are equal, indicating that the geometry of the molecule and ion are nearly identical. The outermost MO $1b_1$ is a non-bonding orbital localized on the oxygen atom. Ejection of a $1b_1$ electron should leave the geometry unchanged, in agreement with observations. In the second band the adiabatic and vertical ionization energies are separated by ~ 1 eV, indicating a large change in the geometry upon ionization. A long vibrational progression with a frequency slightly less than that of the molecular ground state bending mode is observed, indicating that the bending mode of vibration is excited in the transition and that the angle between the two hydrogens in the ionic state is very different from that in the molecule. The predicted second MO $3a_1$ is strongly bonding. Ejection of a $3a_1$ electron is expected to produce the large change in angle implied from the vibrational structure. In the third band the adiabatic and vertical ionization energies are separated by more than 1 eV. The vibrational structure is complicated by the presence of progressions in at least two different normal modes. The predicted third MO $1b_2$ is strongly O—H bonding. Ejection of a $1b_2$ electron should produce a large change in geometry as implied from the spectrum. The type of nuclear motions in this case are more complicated than the previous one. The vibrational structure of the H_2O, HDO, and D_2O spectra are discussed in a later chapter.

1.6. Symbiosis of PE spectroscopy and quantum chemistry

PE spectroscopy is at present the best technique available for obtaining a direct measure of atomic and molecular energy levels. The information derivable from these measurements is greatly enhanced if a theoretical model is available for comparison and discussion of the results. As described in the previous section, molecular orbital theory provides a simple, although reasonable, basis for interpretation of PE spectra. More extended calculations of the energies of the molecular and ionic states involved in the photoelectron transitions provide a unique challenge for quantum chemistry. The complementary nature of PE spectroscopic measurements and quantum chemical predictions constitutes a true symbiosis: quantum chemical theories provide necessary models for spectral interpretation, while the spectra serve as both an excellent test for the success of the theories and as a basis for empirical parameterization of the calculations.

Quantum mechanics is becoming increasingly important within the field of chemistry. Its steady growth, application, and influence is evidenced by modern chemical attitudes and approaches as well as the continuous increase in the sophistication of computational techniques. Despite this sophistication, exact solutions of the Schrödinger equation for many-electron molecules are not available. The best solutions to the Schrödinger equation come from experiments; experiments using PE spectroscopy provide direct energy differences between a molecule and its molecular ion, thereby providing "exact" solutions of the Schrödinger equation for those specific systems. Even though exact theoretical solutions to the Schrödinger equation are not yet possible, the approximate MO methods are extremely valuable for understanding spectral data and developing models for interpretation. In this vein, the complementary nature of PE spectroscopy and quantum theory provides an experimental approach for exemplifying the basic concepts of MO theory.

References

1. K. Siegbahn (Ed.), *Alpha, Beta, and Gamma-Ray Spectroscopy*, North Holland, Amsterdam, 1965.
2. K. Siegbahn and K. Edvarson, *Nucl. Phys.*, **1**, 137 (1956).
3. C. Nordling, E. Sokolowski, and K. Siegbahn, *Phys. Rev.*, **105**, 1676 (1957).
4. K. Siegbahn, C. Nordling, A. Fahlman, R. Nordberg, K. Hamrin, J. Hedman, G. Johansson, T. Bergmark, S. E. Karlson, I. Lindgren, and B. Lindberg, *ESCA, Atomic, Molecular, and Solid State Structure Studied by Means of Electron Spectroscopy*, Nova Acta Regiae Soc. Sci. Upsaliensis, Ser. IV, Vol. 20, Uppsala, 1967.
5. K. Siegbahn, C. Nordling, G. Johansson, J. Hedman, P. F. Heden, K. Hamrin, U. Gelius, T. Bergmark, L. O. Werme, R. Manne, and Y. Baer, *ESCA Applied to Free Molecules*, North Holland, Amsterdam, 1969 (reprinted 1971).
6. M. O. Krause, *Phys. Rev. A*, **140**, 1845 (1965); T. A. Carlson, *Phys. Rev.*, **156**, 142 (1967).

7. F. I. Vilesov, B. L. Kurbatov, and A. N. Terenin, *Sov. Phys. Dokl.*, **6**, 490 (1961); ibid., **6**, 883 (1962); M. E. Akopyan, F. I. Vilesov, and A. W. Terenin, *Sov. Phys. Dokl.*, **6**, 890 (1962); B. L. Kurbatov and F. I. Vilesov, *Sov. Phys. Dokl.*, **6**, 1091 (1962).

8. M. I. Al-Joboury and D. W. Turner, *J. Chem. Soc.*, 5154 (1963); D. W. Turner and M. I. Al-Joboury, *J. Chem. Phys.*, **37**, 3007 (1963).

9. M. I. Al-Joboury, D. P. May, and D. W. Turner, *J. Chem. Soc.*, 616, 4434, 6350 (1965); T. N. Radwan and D. W. Turner, *J. Chem. Soc.*, 85 (1966); M. I. Al-Joboury and D. W. Turner, *J. Chem. Soc.*, 373 (1967); D. W. Turner and D. P. May, *J. Chem. Phys.*, **45**, 471 (1966); ibid., **46**, 1156 (1967); D. W. Turner, *Nature (London)*, **213**, 795 (1966); D. W. Turner, *Tetrahedron Lett.*, **35**, 3419 (1967); C. R. Brundle and D. W. Turner, *Chem. Commun.*, 314 (1967); C. Baker and D. W. Turner, *Chem. Commun.*, 797 (1967).
    a. W. C. Price, *Endeavour*, **26**, 78 (1967).

10. D. C. Frost, C. A. McDowell, and D. A. Vroom, *Phys. Rev. Lett.*, **15**, 612 (1965); *Proc. Roy. Soc., Ser. A*, **296**, 516 (1967); *Chem. Phys. Lett.*, **1**, 93 (1967); *J. Chem. Phys.*, **46**, 4255 (1967).

11. R. I. Schoen, *J. Chem. Phys.*, **40**, 1830 (1964); P. H. Doolittle and R. I. Schoen, *Phys. Rev. Lett.*, **14**, 348 (1965).

12. A. W. Potts, H. J. Lempka, D. G. Streets, and W. C. Price, *Phil. Trans. Roy. Soc. London A*, **268**, 59 (1970).

13. Advanced Research Instrument Systems, Inc., Austin, Tex.; AEI Scientific Apparatus LTD, Manchester, England; E. I. duPont de Nemours and Co., Inc., Wilmington, Del.; Hewlett-Packard, Palo Alto, Calif.; Japan Applied Spectroscopy Co.; Leybold-Heraeus GMBH and Co., Koln, Germany; McPherson Inst. Co., Acton, Mass.; Perkin-Elmer LTD, Beaconsfield, England; Physical Electronics Industries, Inc., Edina, Minn.; Vacuum Generators Limited, East Grinstead, England.

14. D. W. Turner, C. Baker, A. D. Baker, and C. R. Bundle, *Molecular Photoelectron Spectroscopy*, Wiley-Interscience, New York, 1970.

15. A. D. Baker and D. Betteridge, *Photoelectron Spectroscopy: Chemical and Analytical Aspects*, Pergamon, Oxford, 1972.

16. J. H. D. Eland, *Photoelectron Spectroscopy; An Introduction to Ultraviolet Photoelectron Spectroscopy in the Gas Phase*, Halsted, New York, 1974.

17. D. A. Shirley (Ed.), *Electron Spectroscopy*, North Holland, Amsterdam, 1972.

18. *The Photoelectron Spectroscopy of Molecules, Faraday Discuss. Chem. Soc. Lond.*, **54**, 1–306, 1972.

19. R. Caudano and J. Verbist (Eds.), *Electron Spectroscopy: Progress in Research and Applications*, Elsevier, Amsterdam, 1974.

20. R. L. DeKock and D. R. Lloyd, in *Advances in Inorganic Radiochemistry*, H. J. Eméleus and A. G. Sharpe (Eds.), Vol. 16, p. 65, 1974, Academic, New York.

21. H. Bock and P. D. Mollére, *J. Chem. Educ.*, **51**, 506 (1974).

22. H. Bock and B. G. Ramsey, *Angew. Chem. (Int. Edit.)*, **12**, 734 (1973).

23. R. E. Ballard, *Appl. Spectrosc. Rev.*, **7**, 183 (1973).

24. J. L. Bahr, *Contemp. Phys.*, **14,** 329 (1973).
25. G. Leonhardt, *Z. Chem.*, **13,** 81 (1973).
26. A. Hamnett and A. F. Orchard, in *Electronic Structure and Magnetic Properties of Inorganic Compounds*, P. Day (Ed.), Specialist Periodical Rep., p. 1, Chem. Soc. London, 1972.
27. A. D. Baker, C. R. Brundle, and M. Thompson, *Chem. Soc. Rev.*, **1,** 355 (1972).
28. T. L. James, *J. Chem. Educ.*, **48,** 712 (1971).
29. S. D. Worley, *Chem. Rev.*, **71,** 295 (1971).
30. N. Knöpfel, Th. Olbricht, and A. Schweig, *Chem. Unserer Zeit*, **3,** 65 (1971).
31. C. R. Brundle, *Appl. Spectrosc.*, **25,** 8 (1971).
32. A. D. Baker, *Account. Chem. Res.*, **3,** 17 (1970).
33. D. W. Turner, *Annu. Rev. Phys. Chem.*, **21,** 107 (1970).
34. R. S. Berry, *Annu. Rev. Phys. Chem.*, **20,** 357 (1969).
35. D. W. Turner, *Chem. Brit.*, **4,** 435 (1968).
36. D. W. Turner, in *Physical Methods in Advanced Inorganic Chemistry*, H. A. O. Hill and P. Day (Eds.), p. 74, Wiley, New York, 1968.
37. D. W. Turner, *Advan. Mass Spectrom.*, **4,** 755 (1968).
38. W. C. Price, in *Molecular Spectroscopy*, P. Hepple (Ed.), p. 221, Institute of Petroleum, London, 1968.
39. D. Betteridge and M. A. Williams, *Anal. Chem.*, **46,** 125R (1974).
40. D. Betteridge, *Anal. Chem.*, **44,** 100R (1972).
41. D. Betteridge, *Int. J. Environ. Anal. Chem.*, **1,** 243 (1972).
42. D. Betteridge and A. D. Baker, *Anal. Chem.*, **42,** 43A (1970).
43. "Recommendations for Nomenclature and Spectral Presentation in Chemical Electron Spectroscopy Resulting from Excitation by Photons," International Union of Pure and Applied Chemistry, Information Bulletin, Appendices on Provisional Nomenclature, Symbols, Units, and Standards, No. 37, Cowley Centre, Oxford OX4 3YF, UK, August 1974.
44. S. Aung, R. M. Pitzer, and S. I. Chan, *J. Chem. Phys.*, **49,** 2071 (1968).
45. F. A. Cotton, *Chemical Applications of Group Theory*, Wiley-Interscience, New York, 1971; L. H. Hall, *Group Theory and Symmetry in Chemistry*, McGraw-Hill, New York, 1969.

2 Experimental Methods

This chapter treats the basic design elements of a PE spectrometer, experimental procedures, and the factors affecting experimental spectra. Many new developments in instrumentation have appeared in recent years and have resulted in several novel spectrometric designs. It is not possible to discuss all of these new developments here. Hence the objective of this chapter is to focus attention on the most commonly used instrumentation and procedures rather than to be encyclopedic.

2.1. General

The basic requirements of a PE spectrometer are shown in Fig. 2.1. The equipment consists of a source of ionizing radiation, a collision chamber where ionization takes place, a sample inlet system for allowing an appropriate amount of sample to enter the collision chamber, an electron kinetic energy analyzer to measure the energy of the ejected electrons, an electron detection system, amplification and recording equipment, and vacuum pumps for maintaining low pressures within the system. In a typical experiment monochromatic photons are directed such that they impinge on sample vapor that is flowing through the target chamber. An electron can be ejected from any energy level, provided that its binding energy is less than that of the incident photon energy and its ionization cross section is substantial. These ejected electrons possess kinetic energies E_k given by Eq. 1.1. Electrons ejected within the solid angle of acceptance of the collision chamber slit enter the analyzer. The trajectories of these electrons within the analyzer are a function of their kinetic energies and the voltages applied to the analyzer plates. Electrons of various kinetic energies are brought into focus on the exit slit at the end of the analyzer by scanning the voltage on the plates or by scanning a voltage applied to the collision chamber slit. Once through the exit slit, the electrons enter a detector (multiplier) where a pulse is generated, amplified, and recorded. The information is displayed as a differential spectrum by plotting the electron current registered by the detector

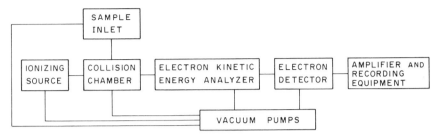

Fig. 2.1. Block diagram of the principal parts of a photoelectron spectrometer.

against the voltage applied to the analyzer plates. The resulting analyzer voltages can be converted by means of suitable calibrants into kinetic energy E_k or ionization energy E_I.

The resolution of a PE spectrometer is the smallest energy difference between two groups of electrons that will result in separate photoelectron bands in the spectrum. It is usually measured as the *fullwidth at half-maximum intensity* of the $^2P_{3/2,1/2}$ lines produced by ejection of Ar $3p$ electrons. The best resolution achieved thus far is ~5 meV (~40 cm^{-1}).

Photoelectron spectra exhibit considerable random fluctuations and statistical noise due to the low electron count rates. The intensities have an uncertainty that is equal to the square root of the total number of electrons counted at a given voltage V. If 100 electrons are counted, the statistical uncertainty is ±10 electrons or ±10% of the signal. If the experimental recording time is increased by a factor of five, 500 electrons are counted, and the uncertainty is ±22.4 electrons or only ±4.5% of the signal. Hence the statistical noise in a PE spectrum can be reduced by increasing the product of the electron count rate and the counting time, that is, by increasing the number of electrons counted at each given voltage V. As a result of these random fluctuations, PE spectra usually contain much more statistical noise than optical spectra.

The individual components in the block diagram of Fig. 2.1 are now considered in detail.

2.2. Ionizing sources

The incident radiation used in PE spectroscopy must be monochromatic so that monoenergetic electrons can be ejected from a given level. Photon sources that inherently produce monochromatic radiation are the most efficient and most intense ionizing sources. For this purpose, resonance lamps or X-ray tubes have proven to be best. The most common source

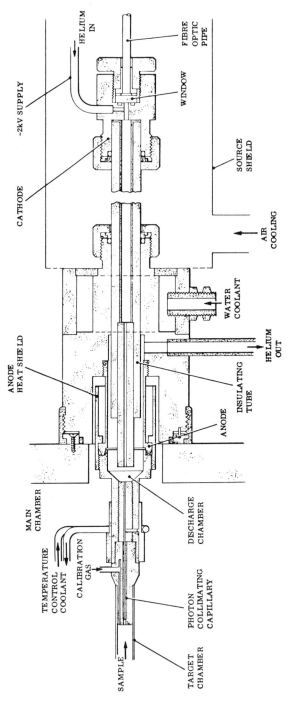

Fig. 2.2. Schematic diagram of a water-cooled dc discharge lamp for producing resonance radiation of rare gases. Reproduced with permission from Ref. 6.

for ionizing low energy electrons has been the intense resonance lines of the rare gases produced by discharge or microwave lamps. The characteristic X-rays produced by X-ray tubes have a large natural width (~1 eV) that places a limit on the resolution attainable. This linewidth can be reduced by using a monochromator on the X-ray source;[1,2] however, even with this monochromator the intensity and monochromacity of the resonance lines cannot be matched. Resonance lines have typical intensities of $\sim 10^{12}$ photons/cm · sec whereas X-ray sources may be two orders of magnitude weaker.

Resonance radiation

Resonance radiation[3] can be produced by a condensed spark, microwave,[4] or direct-current[5] discharge through a rare gas confined in a 1- to 2-mm diameter capillary. A schematic diagram of a water-cooled dc discharge lamp[6] is shown in Fig. 2.2. Normal discharge lamps used in optical studies are usually all-glass constructions containing the electrodes in side arms and using the capillary section only for collimation. These lamps require high voltages and currents (5 kV, 150 mA) and need frequent cleaning and electrode changes. The lamp in Fig. 2.2 operates with only ~2 kV, applied through a ballast resistor bank, and draws only ~30 mA current. The high voltage and rare gas inputs are supplied through a stainless steel pipe that is brazed onto the cathode. The clamping ring that secures the anode to the main chamber serves as the ground connection. Also at ground potential are the collimating capillary holder and cooling block that screw into the anode. The discharge between the anode and cathode takes place through the capillary in the center of the quartz tube and is viewed through the fiber optic pipe. Since the discharge takes place down the capillary, a high flux is generated. Photons pass to the target chamber through the coaxially aligned collimating capillary tube. A pumping port is provided in the intermediate region between the two lengths of capillary for removing the rare gas. This differential pumping of the lamp is necessary to prevent self-absorption and to keep the rare gas from entering the collision chamber, for there are no practical window materials for passing photons of the wavelengths desired. The lamp is cooled by circulating water through a cooling block and by air circulation through use of a small fan. An additional jacket for circulating fluids is placed around the upper capillary, directly beneath the target chamber. This allows heating or cooling of the collision chamber, which is in thermal contact with the upper end of the lamp. The radiation produced by a resonance lamp is usually used without monochromatization in order to preserve its full intensity.

   The principal resonance lines obtained from these direct-current or microwave lamps are listed in Table 2.1. These lines represent resonance fluorescence produced when the rare gas is excited in the discharge and subsequently decays back to its ground state. The Roman numerals placed after the symbols for the elements indicate whether the emitting species is the neutral atom (I), the singly ionized atom (II), or the doubly ionized atom (III). An energy level diagram for helium and singly ionized

Table 2.1. Principal Resonance Lines and Ionization Energies for the Rare Gases, Hydrogen, Oxygen, and Nitrogen

Gas	Resonance Line		Ionization Energy	
	(Å)	(eV)[a]	(Å)	(eV)
He I	584.3340	21.2175 (100)	504.259	24.5868
	537.0296	23.0865 (2)		
	522.2128	23.7415 (0.5)		
He II	303.781	40.8136 (<1)	227.8	54.43
	256.317	48.3702		
	243.027	51.0153		
	237.331	52.2397		
Ne I	743.718	16.6704 (15)	574.938	21.5642
	735.895	16.8474 (100)		
Ar I	1066.659	11.6233 (100)	786.721	15.7592
	1048.219	11.8278 (50)		
Kr I	1235.838	10.0321	885.620	13.9994
	1164.867	10.6434		
Xe I	1469.610	8.4363	1022.140	12.1296
	1295.586	9.5695		
H(Lyman α)	1215.668	10.1986 (100)	911.75	13.598
H(Lyman β)	1025.722	12.0872 (10)		
H(Lyman γ)	972.538	12.7482 (1)		
N I	1134.981	10.9236	852.19	14.5485
	1134.415	10.9291		
	1134.166	10.9315		
	1199.550	10.3356		
	1200.224	10.3298		
	1200.710	10.3256		
O I	1302.168	9.5211	910.44	13.6177
	1304.858	9.5015		
	1306.029	9.4930		

[a] Relative intensities of some of the resonance lines are given in parenthesis for a typical capillary discharge under normal He I operating conditions. The relative intensities of the lines at higher energy can be increased by lowering the gas pressure and increasing the current density as explained in the text. The H, N, and O lines are typical impurity lines that sometimes appear in a discharge.

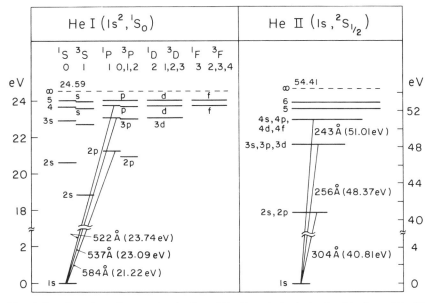

Fig. 2.3. Grotrian diagram for helium (He I) and singly ionized helium (He II) showing the strongest resonance lines. The $^1P \rightarrow {}^1S$ (584 Å) and the $^2P \rightarrow {}^2S$ (304 Å) transitions produce the most intense lines.

helium is shown in Fig. 2.3. The principal resonance lines, that is, transitions between excited states and the ground state of the atom or ion, are indicated on the diagram. The resonance line produced by the transition from the first excited state to the ground state of the atom or ion is usually the most intense line and is called the *raie ultime*. The most highly ionized atoms generally produce the highest energy resonance radiation.

The source most frequently used to date is the He I resonance line at 584 Å because of its high energy, ease of production, and relative spectral purity. Under appropriate conditions ($\sim$2 kV, $\sim$30 mA, $\sim$0.30 torr He), the emission spectrum in the dc discharge consists of a series of lines leading to the ionization limit at 24.59 eV. The He 584-Å line resulting from the $^1P_1 \rightarrow {}^1S_0$ emission accounts for more than 98% of the emission intensity under these conditions. Careful analysis[7,8] of the lines present in a helium discharge have shown that the most intense minority emission lines occur at 537, 522, and 304 Å and can have intensities of the order of 2% of the 584-Å line. Their presence must be kept in mind in order to avoid confusion in spectral identification. Longer wavelength emissions are also produced in the He discharge with the most energetic of these being the 3000-Å line (~4 eV) produced by the $^3P \rightarrow {}^3S$ series. Since all substances investigated thus far have a first ionization energy of at least 5 eV or greater, these low energy lines do not cause interferring ionizations.

The He II line at 304 Å results from the resonance fluorescence of singly ionized He atoms. Relatively intense emission at 304 Å can be produced by employing a much higher current density than that used for the 584-Å line. Typical operating conditions for a He II lamp are ~4 kV and 75 mA. The current and voltage conditions as well as the He gas pressure used for this mode of operation are a function of the lamp construction and must be determined for each individual lamp. Usually higher voltages (up to ~10 kV) and currents and lower He gas pressures produce a higher intensity of He II radiation. These operating conditions generate a substantial amount of heat; in the most successful He II lamps the Pyrex or fused quartz discharge capillaries are replaced by high purity boron nitride with tantalum electrodes. Two practical difficulties are encountered in this mode of operation. First, self-ionization of helium gas usually occurs with a resulting sharp spectral peak at 24.6 eV. Second, and more important, the He I line remains dominant, even under conditions of higher voltages and currents. This He I dominance tends to obscure the ionization region between 30 and 40.8 eV (0–10 eV E_k) due to the superposition of bands arising from ejection of outer electrons by the He I radiation with the bands arising from ejection of inner electrons with He II radiation. Unmonochromatized He II radiation therefore, is, useful only to approximately 30 eV. This is very important, however, because most valence molecular levels lie in the range of approximately 7–30 eV. The He I radiation can be partially filtered by using a thin polystyrene film across the capillary tip or by using a gas cell in the light path. These methods are dependent on the higher absorption coefficients of molecules composed of light atoms at 584 Å than at 304 Å. Using these procedures, He I and He II radiation are all that is required to eject the valence electrons of most molecules.

Traces of hydrogen, nitrogen, or oxygen as impurities in the lamp structure or helium gas produce many emission lines, some with energies greater than 10 eV. It is essential that the gas in the discharge tube be extremely pure in order to avoid ionization by these impurity lines. Trace impurities can be adequately removed by allowing the He to flow through activated charcoal, which is cooled by liquid nitrogen. When operating correctly, a helium lamp appears "peach" colored. If the light source is contaminated with air, it takes on a blue or purple tint due to the hydrogen Balmer series and oxygen and nitrogen visible emission lines.

It is possible to make use of the resonance lines of Ne, Ar, Xe, and Kr by operating the discharge lamp under conditions similar to that used for He I radiation and introducing the appropriate rare gas other than He. The advantage of using these lines for excitation is that they are of low energy and thus produce low kinetic energy electrons. As shown later, higher resolution can be obtained with electrons of low kinetic energy. The disadvantages of these rare gases is that (1) their low energy resonance lines do not allow one to obtain the spectrum of the entire valence region and (2) their emission lines are doublets resulting in doubling of all the features in the spectrum. These sources are most valuable for investigating one ionization band or a portion of a band under high resolution. For example, Ne I has been used to excite the spectra of the outer bands of CH_4,[9] C_2H_6,[10] and benzene[11] when high resolution was required for the analysis. In these spectra the resolved vibrational structure repeats itself every 0.20 eV because of the double excitation line; however, the 736-Å line is considerably more intense than the 744-Å line under certain conditions of pressure and current. By careful adjustment of these parameters it is possible to produce an intensity ratio of 736 Å/744 Å of approximately 10/1.

When H_2, N_2, or O_2 is used in the discharge lamp, the atomic resonance lines are excited as well as the weaker molecular bands. For the case of H_2, it is possible to greatly reduce the intensity of the molecular bands without altering the intensities of the atomic lines by using a mixture of H_2 (25%) and He (75%) gas in the discharge. For such a mixture, the radiation is nearly monochromatic in the Lyman-α line.

Linewidths from resonance radiation

Resonance radiation produced by lamps as described above is not completely monochromatic but actually appears as a distribution of wavelengths over a finite range. The maximum of this distribution is taken as the wavelength of the radiation. The width of the excitation line contributes a spread to the electron kinetic energy such that photoelectron bands cannot be sharper than the exciting line width.

The factors that may contribute to the finite energy width of the ionizing flux are the natural width, self-absorption, pressure (or resonance) effects, Doppler effect, Stark effect, and self-reversal. A brief description of these effects on the He I resonance line is given here. More rigorous treatments are available elsewhere.[12,13]

1. Natural Widths—The natural width $\delta\lambda_n$ of an emission line is a function of the mean lifetime of the excited state τ and is given by

$$\delta\lambda_n = \frac{\lambda^2}{2\pi c \tau} \tag{2.1}$$

where $\delta\lambda_n$ is a measure of the width of the emission line profile at half its maximum height. For He I emission, $\delta\lambda_n \simeq 3.0 \times 10^{-5}$ Å.

2. Self-absorption—Self-absorption is caused by the existence of unexcited helium gas in the region between the capillary discharge and the collision chamber. Resonance radiation is absorbed by this gas, thus decreasing the photon flux entering the target chamber. Self-absorption changes the profile of the emission line by decreasing the peak amplitude, thus resulting in two maxima occurring symmetrically about the center of the profile. This produces a line with an increased half-width. Samson[14] has studied the line profiles from direct-current, microwave, and condensed spark lamps as a function of the helium pressure. Self-absorption is observed for pressures greater than 0.6 torr. A relation between the length of the discharge and half-width of the resonance line is observed; dividing the half-widths obtained from the various light sources by their respective capillary lengths produces a constant value for all sources. This value is less than 0.5 meV (half-width)/cm (capillary length). Samson concludes that the broadening of the 584-Å resonance line depends primarily on the length of the discharge capillary and the helium pressure and not on the type of lamp.

The 584-Å He line can be produced with a half-width of $\sim$1 meV using a microwave discharge at the lowest possible pressure. However, this lamp produces a low photon flux. The best general light source appears to be a dc discharge in helium with a capillary length of $\sim$3 cm designed to have a minimum length of cold gas between the discharge and the collision chamber. The diameter of the capillary does not seem to affect the width of the resonance line.

3. Pressure or Resonance Broadening—Pressure broadening is a result of collisions of excited atoms with unexcited atoms. Its contribution for He I can be expressed as

$$\delta\lambda_r = \left[\frac{3}{2\pi^2}\right]\left[\frac{g_1}{g_2}\right]^{1/2}\left[\frac{e^2 f}{8mc^2\varepsilon_0}\right] \cdot \lambda^3 \cdot N = 7.5 \times 10^{-6} \text{ Å/torr} \quad (2.2)$$

where g_1 and g_2 are the statistical weights of the lower and upper states, respectively, f is the oscillator strength, ε_0 is the dielectric constant of free space, and N is the particle density.

4. Doppler Effect—The thermal velocity of atoms in the discharge is sufficient to cause a Doppler shift in the emitted radiation. This can be expressed as

$$\Delta\nu = \left(\frac{2(\ln 2)\nu^2 kT}{Mc^2}\right)^{1/2} \quad (2.3)$$

where M is the atomic or molecular mass. For a He I lamp at room temperature, the Doppler broadening is $\delta\lambda_D = 3.6 \times 10^{-3}$ Å.

5. Stark Effect—Stark broadening may be caused from the electric field set up by the motion of electrons and ions in the discharge. Samson[14] investigated this effect by observing the half-width of the line upon changing the current in the discharge from 100 to 500 mA at constant voltage across the discharge. No broadening was observed throughout this range, and it was, therefore, concluded that Stark broadening is not a major contributor to the observed linewidth.

6. Self-reversal—Self-reversal is a result of temperature gradients in the discharge capillary. The plasma at the center of the discharge is hotter than that at the sides, resulting in absorption of the radiation emitted from the hotter zone by the cooler atoms. The absorption profile of the cooler atoms is narrower than the emission profile of the hotter atoms, resulting in an overall emission with a minimum at the center of the line profile. It is best to keep the capillary short in order to avoid this absorption effect. Short capillaries have a reduced photon flux, however, so some compromise must be made for the capillary length.

Other ionizing sources

Short-wavelength photons can be generated in many different discharge or X-ray lamps, but most of these lamps produce polychromatic radiation. Grating monochromators can be used to select a narrow range of wavelengths for excitation; however, there is a considerable loss of intensity at the grating. Despite the reduction in intensity, such sources have been used for PE spectroscopy.[15,16]

A potentially excellent source of radiation for PE spectroscopy is the synchrotron.[3] It is well known that an accelerating electron should radiate energy. Synchrotron radiation is the continuum of radiation between ~600 Å and ~4 Å produced by accelerating free electrons in a magnetic field. Use of the synchrotron in conjunction with a monochromator can provide a useful source between the low-energy (uv) and high-energy (X-ray) photoelectron studies.

An electron beam can be used as an exciting source. Electrons emitted from a hot filament have a distribution of energies requiring an electron lens to obtain a monochromatic beam. Polychromatic electron beams have been used mainly for excitation of Auger spectra,[1] since the energies of Auger electrons are independent of the energy of the ionizing radiation. Also, electron sources produce a higher yield of Auger electrons over photon sources.

A process known as "Penning ionization"[17] uses metastable or excited atoms as the ionizing source. In this process, a quantum of radiation bound to an impinging atom is delivered to the target molecules. The reaction

$$A^* + B \rightarrow A + B^+ + e^-$$

passes through an intermediate state [A*B]. The method can lead to excitation of states inaccessible by photoionization through various ionizations and dissociations of the [A*B] state. Because the process may leave A and B^+ in a compound state AB^+, the method can be used to study the levels of compound systems.

2.3.   Collision chambers

The collision chamber serves as an enclosure for the gaseous molecules being ionized. The specific geometry of this chamber is not critical. The most common chamber appears to be of cylindrical construction and fits over the end of the collimating capillary tube from the discharge lamp (Fig. 2.2). In this arrangement the photons enter from the bottom of the chamber, and the target gas is fed in from the top. The sample entrance bore is much wider than the collimated light beam, so that it also serves as a light trap for unabsorbed radiation. A slit in the wall of the collision chamber allows the photoejected electrons to enter the analyzer. Slit widths between 100 and 200 microns with heights of ~ 1 cm have proven most useful in this arrangement. The sample vapor can be differentially pumped through this slit or some other tiny port in the collision chamber. An electrical potential can be applied to the slit of the collision chamber if acceleration or retardation of the ejected electrons is desired before they enter the analyzer chamber.

Resolution and signal intensity are critically dependent on the cleanliness and surface charges of the collision chamber and slit jaws. These areas must be kept smooth and clean in order to minimize surface charges and contact potentials. Shifts in photoelectron lines by as much as several tenths of an electron volt can result from poorly prepared surfaces. Also, the resolution of a clean system can easily be degraded by the introduction of sample gases which have high dipoles and become adsorbed on the working surfaces of the chamber.[18] These effects can be minimized by treatments such as gold plating or coating with colloidal graphite. Coating the surfaces with a thin layer of colloidal graphite appears to be the simplest and most effective method of obtaining good resolution and minimizing surface charges.

Solid samples and liquids with low volatility can be studied in the vapor phase by heating them to a temperature at which they are sufficiently volatile to produce adequate vapor pressure. Materials which produce adequate vapor pressures at temperatures less than $\sim 300°C$ can be studied by placing them in a small vial in the upper part of the collision chamber (just above the slit). The target chamber is then heated and maintained at a temperature that gives a good compromise between emitted electron flux and sample longevity and stability. The heating can be accomplished by circulating a liquid through channels cut in the walls of the collision chamber, resistance heating, or by using the heat produced by the discharge lamp. The latter method can be used most effectively if the collision chamber is in thermal contact with the top of the capillary tube from the lamp. Berkowitz[19] has designed an oven which is capable of reaching temperatures of $\sim 1000°C$ for studying samples of very low volatility. In this design, an extension of the oven protrudes into the collision chamber. The oven is constructed of 0.010-in-thick platinum that is noninductively wound with glass-fiber-insulated nichrome heater wire and covered with a tantalum shield.

2.4. Electron kinetic energy analyzers

The electron energy analyzer is the place where photoelectrons are separated according to their kinetic energies. The ideal analyzer provides high resolution and high sensitivity simultaneously. These are conflicting requirements, for any restrictions placed on the paths of the electrons naturally decrease the transmittance or sensitivity of the analyzer.

The kinetic energy of a beam of electrons can be measured by different techniques,[20-22] which range from very simple electrostatic retarding fields to deflection analysis using electric or magnetic fields. Several different types of analyzers have been applied in photoelectron spectroscopy; some of these are the cylindrical electrostatic deflector,[23-31] hemispherical plates,[32,33] parallel plates,[34-39] retarding fields,[40,41] cylindrical mirrors,[19,42-45] spherical grids,[46] spherical plates,[37] magnetic deflectors,[1] Wien filters,[47] the trochoidal electron monochromator,[48] quadrupole energy filters,[49] and the pill-box spectrometer[50]. The analyzer requirements of high collecting efficiency and resolving power are expressed in terms of Ω, the solid angle in which electrons are collected, and $E_k/\Delta E$, where ΔE is the full width of a peak of energy E_k at half its maximum intensity. The performance of a spectrometer is evaluated from the product $\Omega \times (E_k/\Delta E)$.

The first analyzers used in PE spectroscopy were the simple retarding field type. These analyzers consist of cylindrical or spherical grids which surround the ionization region. The potential on the grids is varied,

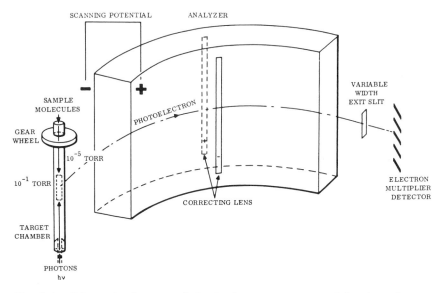

Fig. 2.4. Schematic diagram of the basic components used in photoelectron excitation and analysis. Reproduced with permission from Ref. 6.

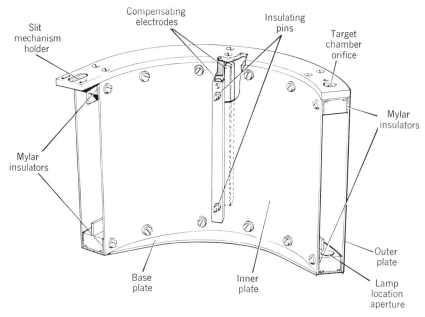

Fig. 2.5. Cylindrical electrostatic deflection analyzer used in electron energy analysis. Reproduced with permission from Ref. 6.

starting from high to low potential, such that only electrons having energies greater than the retarding potential reach the detector. A step in the detector current is recorded for each group of photoelectrons of discrete energy. This produces an integral spectrum in which the steps for low energy photoelectrons are superimposed on the electron current of all higher energy photoelectrons. These analyzers generally have a high collecting efficiency and are approximately equally sensitive for electrons of different kinetic energies. However, the resolution attained with these earlier analyzers does not match that of the deflection analyzers.

Deflection analyzers

Deflection analyzers are used extensively in PE spectroscopy. These analyzers use electric or magnetic fields to focus electrons according to their kinetic energies. Magnetic analyzers are not well suited for focusing low energy electrons because it is difficult to shield the weak magnetic fields from the stray fields of the environment. Electrostatic analyzers are much more convenient to use because they are easier to shield. The many different types of electrostatic analyzers offer various advantages and disadvantages; for a discussion of these as well as a mathematical treatment of their properties the reader is referred to the above-mentioned references. Only the cylindrical electrostatic deflector is described here.

The cylindrical electrostatic deflector ($\pi/2^{1/2}$ sector) analyzer[23–31] is one of the most common analyzers used in PE spectroscopy. The analyzer plates, Figs. 2.4 and 2.5, are concentric brass segments with an angle of 127°17' ($\pi/2^{1/2}$) between the entrance and exit slits. The surfaces are gold plated, and graphite may be deposited on them in order to minimize surface potentials. The segments are insulated from the base and top plate by Mylar (Melinex) spacers. The exit slit of the collision chamber serves as the entrance slit to the analyzer. This analyzer is easily adapted to a capillary discharge lamp that produces a column of photoelectrons in the collision chamber whose width equals the diameter of the collimating capillary. The straight slits of the analyzer are easily aligned with this column. Two vertical electron lenses are inserted at the midpoint of each analyzer plate. These lenses are used for refocusing of electrons whose trajectories are slightly off from the principal orbit of the analyzer. This is accomplished by applying potentials to the lens of opposite sign to those on the analyzer plates. These lenses appear to be most useful when the analyzer plates become contaminated and surface charges accumulate or when the plates have not been machined to close tolerances.

A sweep voltage is applied to the analyzer plates by means of a variable resistor network. This sweep voltage can be controlled by a helipot driven by the pen rider of an *xy* recorder. A constant voltage supplied by a battery pack converts the input to a digital voltmeter to a value corresponding to the ionization potential in electron volts. The analyzer can be operated in two basic modes: (1) Electrons of different kinetic energies can be focused by sweeping the voltage on the analyzer plates. Fixed accelerating or retarding potentials can be applied to the collision chamber electrode in order to accelerate or retard the electrons. In this mode the analyzer discriminates against electrons of different kinetic energy; in particular, the transmission diminishes nearly linearly as electron kinetic energy decreases. This effect can be corrected to a first approximation[21] by dividing the observed band intensities by the corresponding kinetic energy of the ejected electrons. (2) The voltage on the analyzer plates may be set to focus a certain electron energy, and the voltage on the collision chamber electrode may be swept in order to bring the electrons to the proper energy required for transmission and focusing. This improves the transmission properties of low energy electrons; however, it produces a high background of electrons at low energies.

The exit slit is a gap between two specially shaped brass blades held together by a spring clamp. The slit width is controlled by rotation of a threaded rod that depresses a plunger, forcing the blades apart. An angle of 127°17′ is used between the entrance and exit slits because of a refocusing property that occurs at this angle. This refocusing property is such that electrons entering the analyzer at an angle α to the normal (Fig. 2.4) will still be focused correctly on the exit slit provided that they have the correct kinetic energy. In order to focus an electron of energy E_k, potentials must be applied to the analyzer plates according to[24,25]

$$V_2 - V_1 = 2E_k R \log \frac{R_1}{R_2} \tag{2.4}$$

where V_i and R_i are the potential and radius of analyzer plate i and R is the mean radius of the analyzer. The resolving power of the electrostatic analyzer is given by[51]

$$\frac{\Delta E}{E_k} = \frac{W}{2R} \tag{2.5}$$

where R is the radius of the electron trajectory (mean radius of analyzer) and W is the total slit width (entrance plus exit). Considering common spectrometers with a mean radius of 10 cm and entrance and exit slits of 100 microns, $E_k/\Delta E = 1000$. An electron with 5-eV kinetic energy, therefore, has $\Delta E_{theor} = 5$ meV.

2.5. Detection and recording system

After passing through the exit slits of the analyzer, the electrons impinge upon an electron detector. Electron multipliers are used to detect the electron fluxes, which can be as low as 20 electrons per second. Two types of multipliers have been used for detection: (1) The "venetian blind" multiplier tube[52] is the older type of detector. The "blind" consists of Cu–Be dynodes that are sensitive to surface contaminants such as adsorbed chemicals or water vapor. For this reason they should be maintained under vacuum at all times. (2) The electron channel multiplier (channeltron) is a recently developed[53] multiplier that is more satisfactory for PE spectroscopy. These detectors are very resistant to chemical contamination, they do not have to be kept under vacuum when not in operation, and their small size makes them easily adaptable to various instrumental designs. The detectors should be operated at a pressure of less than 10^{-4} torr. The gain eventually decreases due to surface contamination and can be rejuvenated to a certain extent by rinsing with cyclohexane.

The channel electron multiplier is a thin tube coated internally with a layer of high-resistance material. A voltage of ~ 2 kV is applied across the tube by means of electrodes connected at each end. Electrons entering the tube go through repeated cycles of acceleration and collision with the walls with each collision producing a cascade of secondary electrons. The gain, which can be as high as 10^9, usually decreases with age. If the voltage is raised too high (above ~ 3.2 kV) or if the pressure in the instrument rises significantly, arcing may occur across the multiplier, resulting in a further loss of gain.

The signal produced by the multiplier proceeds to the counting equipment. Here the pulses can be counted digitally or the rate in pulses per second can be converted into analogue form by a ratemeter. The spectrum is displayed by recording the count rate on the ordinate of an xy recorder and the analyzer potential or preaccelerating potential on the abscissa. The potential on the abscissa scale can be converted to E_k or E_I. Another method of collecting spectra is to scan the spectrum rapidly many times and accumulate the number of electrons detected at each energy in a multichannel analyzer, the contents of which can be displayed or recorded after sufficient accumulation.

2.6. Magnetic shielding

The trajectory of the ejected electron is a function of the stray magnetic fields as well as the analyzer field. For this reason, the analyzer and many of the instrumental components must be constructed from nonferromagnetic materials such as stainless steel, aluminum, or brass. Magnetic shielding must be used to exclude the inhomogeneous magnetic fields in the laboratory and to insure that the trajectory of the electrons is solely determined by the field of the analyzer. Ideally, the external magnetic field should be reduced below 10^{-4} G. Shielding has taken two forms, Helmholtz coils or magnetic shielding metals that encase the analyzer. Helmholtz coils consist of three pairs of circular coils arranged perpendicular to each other. The three pairs of coils are connected in parallel to a power supply with a variable resistance in series with each coil. For two coils of radius r, separated by a distance r, the field produced at the central position, $r/2$, is given by

$$H(\text{Oersteds}) = \frac{4\pi nIr^2}{10(r^2 + r^2/4)^{3/2}} \qquad (2.6)$$

where I is the current. The value of H is relatively constant over a large area between the coils.

Magnetic shielding materials of high permeability such as "mu metal" alloy or "co-nectic" foil can be used to encase the analyzer, collision chamber, and detector regions. These materials effectively shield the sensitive areas from external magnetic fields and are usually reliable and easy to use. The shielding properties of these materials can be impaired if they receive sharp blows or agitations.

2.7. Vacuum system

It is not necessary to have a very high vacuum in order to do gas-phase PE spectroscopy. The pressure should be maintained below 10^{-4} torr in order to prevent arcing of the channeltron and to provide a collision-free path for the photoelectrons. The mean free path of electrons at 10^{-4} torr is greater than one meter. Most spectrometers for gas-phase work employ Viton O-rings and oil diffusion pumps backed by two-stage rotary pumps. Additional pumps are generally required for differential pumping of the lamp and sample gases. The sample gas pressure in the target chamber required to produce a satisfactory spectrum is $\sim 10^{-1}$ torr.

2.8. Working resolution

The fundamental limitations on resolution and factors contributing to linewidths are important to the design of spectrometers and to the determination of their limitations. As described in Sections 2.2 and 2.4, the broadening due to the linewidth of the excitation source and the analyzer aberrations cannot account for the linewidths actually observed in PE spectra. There are other fundamental physical limitations on the linewidths. The resolution obtained in practice is called "working resolution." These physical limitations can be described as follows:

1. Conservation of Energy—The small fraction of the energy carried off by the ion is equal to the ratio of the electron mass to that of the ion. In the most unfavorable case, hydrogen, this quantity of energy is negligible for ionization by photons.

2. Thermal Motion of Target Molecules—The observed electron energies are modified by a Doppler-type effect due to the thermal motion of the target molecules. The molecules have velocities V and are traveling in random directions. The photoelectrons are ejected with velocity v. The maximum or minimum observed electron energy occurs when the photoelectron is ejected in the same or opposite direction as the velocity of the target molecule. That is, $E_{k,max} = m(v + V)^2/2$ and $E_{k,min} = m(v - V)^2/2$ where m is the mass of the electron. The broadening is then $\Delta E = 2mvV$. We assume that the target molecules have a Maxwell-Boltzmann velocity distribution, giving a gaussian distribution of velocities. The spread of velocities at half the maximum peak height is given by

$$V = \pm \left[\frac{2kT \ln 2}{M} \right]^{1/2} \qquad (2.7)$$

where M is the mass of the target molecule. Using $v = (2E_k/m)^{1/2}$ for the electron velocities, the width of the electron energy profile measured at half its full height is

$$\Delta E = 4 \left[\frac{mE_k kT \ln 2}{M} \right]^{1/2} \qquad (2.8)$$

where k is the Boltzmann constant and T is the absolute temperature. If M is expressed in atomic units and E_k is in electron volts

$$\Delta E = 0.724 \left[\frac{E_k T}{M} \right]^{1/2} \text{ meV} \qquad (2.9)$$

This thermal broadening can, in principle, be reduced to zero by using a molecular beam directed perpendicular to the slit of the instrument. An experiment of this type, performed by Wiess et al.[54] was not carried out under sufficiently high resolution to demonstrate the utility of molecular beams for PE spectroscopy. In Figs. 2.6 and 2.7 this thermal broadening is plotted in parametric form for molecules of various masses as a function of the kinetic energy of the

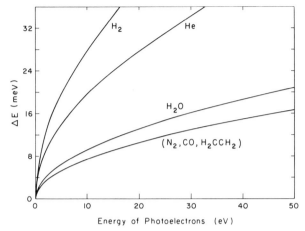

Fig. 2.6. Thermal broadening as a function of kinetic energy of the ejected electrons for light molecules.

ejected electron, the gas being at room temperature (300°K). Figure 2.8 shows the thermal spread plotted as a function of temperature for molecules of various masses. The kinetic energies of the photoelectrons in this plot are those corresponding to ejection of the most loosely bound electron in the respective molecule using He I radiation.

3. Lifetime of the Excited State—Natural processes that shorten the lifetime of the excited state can produce considerable spectral

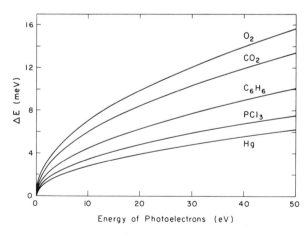

Fig. 2.7. Thermal broadening as a function of kinetic energy of the ejected electrons.

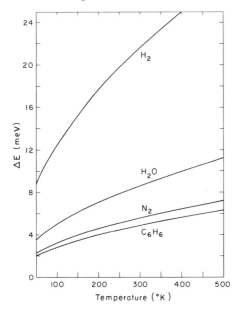

Fig. 2.8. Thermal broadening as a function of temperature for molecules of various masses. The kinetic energies used in these plots are those corresponding to ejection of the most loosely bound electron using He I radiation, that is, H_2 ~5.8 eV, H_2O ~8.6 eV, N_2 ~5.6 eV, and C_6H_6 ~12 eV.

broadening as follows: (a) An ion in an excited state that decays by radiative transitions such as fluorescence has a lifetime of the order of ~10^{-8} sec. This mode of decay will, therefore, cause no significant broadening. (b) If excitation takes place to a repulsive part of the potential energy curve, the ion may dissociate into fragments. Direct dissociation such as this can shorten the lifetime to ~10^{-14} sec (that is, lifetimes comparable to the period of one vibrational amplitude). This process can produce spectral broadening by several tenths of an electron volt. (c) Nonradiative transitions or curve crossings to other ionic states can be fast or slow resulting in broadenings comparable to those from dissociation.

From the preceding discussion it is obvious that two of the most important factors contributing to spectral broadening are the thermal motion of the target molecules and the short lifetimes of some ionic states. The best "working resolution" that has been obtained thus far on molecules such as H_2O is ~6 meV measured as the full width of the Ar lines at half height while Ar and H_2O are flowing through the collision chamber.[18] The argon $^2P_{3/2}$ line, which is used as a standard test of resolution, is thermally broadened by 4.6 meV when ionized by 21.22-eV photons and by 9.9 meV when ionized by 40.81-eV photons at 300°K. The thermal spread is most important for light molecules and high electron energies or high temperatures. Reduction of this effect for high-resolution work is usually effected by reducing the kinetic energies of the photoelectrons as much as possible by employing radiation of the longest usable wavelength. The reduction of the $E_k^{1/2}$ term in Eq. 2.9 produces a significant improvement in resolution. Cooling the gas below room temperature is not practical for most substances and has little effect until very low temperatures are reached. The ideal solution for reducing the thermal spread is to introduce the target molecules into the collision chamber in the form of a molecular beam in which the transverse kinetic energies are very small.

2.9. Photoelectron band intensities

The PE spectrum of a given substance should be an absolute physical characteristic of that substance. The general type of PE spectrum found in the literature is a function of many facets of the experimental apparatus used to measure it, rendering quantitative comparison of spectra from different instruments a dubious endeavour. Photoelectron band intensities are a function of the following parameters: pressure of sample vapor, intensity of the incident radiation, size and direction of the solid angle of collection, type of kinetic energy analyzer, retardation or acceleration voltages, surface charges on the walls of the collision and analyzer chambers and slits, sensitivity of the electron detector, slit widths, and residual magnetic fields in the collision and analyzer chambers. These parameters can be kept more or less constant in any given spectrometer. However, comparison of spectra from different types of experimental arrangements should be made with caution. For a given spectrometer one of the most important factors in distorting the intensities of the photoelectron bands is the discrimination of the analyzer for electrons of different kinetic energy.

It is difficult to measure the energy discriminations of a spectrometer because there are no sources of electrons of various kinetic energies with precisely known intensities. Spectra are usually corrected by assuming the theoretical variations of analyzer sensitivity with kinetic energy. Usually retarding field analyzers and deflection analyzers used with preacceleration have little or no energy discriminations (except for very low energy electrons). Deflection analyzers without preacceleration have severe energy discriminations.

In order to obtain meaningful band intensities or photoionization cross sections from recorded spectra, it is necessary to analyze the transmission and broadening characteristics of the spectrometer. A simplified analysis of these characteristics for a $127°17'$ electrostatic deflection spectrometer is presented in this section.

The measured widths of the photoelectron bands ΔE_{exp} are a result of the natural linewidth ΔE_n and the broadening by instrumental aberrations ΔE_i:

$$\Delta E_{exp} = [(\Delta E_n)^2 + (\Delta E_i)^2]^{1/2} \tag{2.10}$$

ΔE_n is a pseudonatural linewidth that is determined by the lifetime of the ionic state and thermal broadening. Barring any fast dissociative processes that decrease the lifetime of the ionic state or rotational and vibrational broadening, by far the largest contributor to ΔE_n is thermal broadening; this can be calculated from Eq. 2.9. The energy resolution of an electrostatic analyzer ΔE_i is given by Eq. 2.5. From this expression it is obvious that when recording a spectrum with fixed slits W, the energy band width ΔE_i increases in proportion to the energy E_k.

There are two limiting cases[21,55] that should be considered in photoelectron experiments.

1. $\Delta E_n < \Delta E_i$, *that is, the natural linewidth is less than the resolution width of the instrument.* If the exit slit is wider than the entrance slit, all electrons that go through the entrance slit are collected, resulting in flat-topped peaks whose heights are proportional to the line intensities. From Eq. 2.5 a similar peak scanned at higher energy will involve a larger ΔE_{exp} and hence a larger peak area A. The true line intensity will be given by the peak height or by A/E_k. If the exit slit is narrowed, the peaks become narrower and approach a triangular shape at the optimum slit width. As before, the peak heights remain independent of energy and the true line intensities are given as above.

2. $\Delta E_i < \Delta E_n$, *that is, the resolution width of the instrument is less than the natural line width.* Only a fraction of the electrons that go through the entrance slit are collected by the detector. Since this fraction increases proportionally with energy, the peak heights will be dependent upon the kinetic energy of the electrons. The peak width ΔE_{exp} will be approximately equal to ΔE_n. For such bands, the areas are proportional to E_k and hence peak height/E_k and A/E_k are proportional to the true line intensities. In this case the peak widths ΔE_{exp} will be almost constant in contrast to case (1) where $\Delta E_{exp}/E_k$ is approximately constant.

The performance of an electrostatic deflection spectrometer, analyzed by measuring the $^2P_{3/2}$ and $^2P_{1/2}$ lines of Ne, Ar, Kr, and Xe using Ne I, He I, and He II radiation, is illustrated in Fig. 2.9. The rare gases were

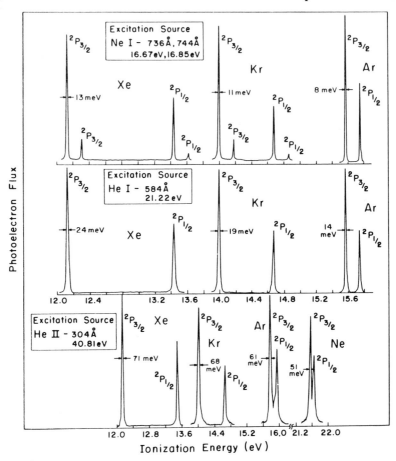

Fig. 2.9.   The $^2P_{3/2}$ and $^2P_{1/2}$ lines of the rare gases produced by Ne I, He I, and He II resonance radiation. The widths indicated are FWHM (full-width at half-maximum intensity) as obtained from an average of three spectra. Reproduced with permission from Ref. *a* of Table 2.2.

chosen for this example because of the large range of kinetic energies of their photoelectrons and the small natural widths of the $^2P_{3/2,1/2}$ lines. Aquadag (collidal graphite) was used to reduce surface potentials in the collision and analyzer chambers. The largest contributor to ΔE_n for these lines is thermal broadening, which, as calculated from Eq. 2.9, ranges from a minimum of 1.8 meV for ionization of Ar with 736-Å radiation to a maximum of 12.2 meV for ionization of Ne with 304-Å radiation. The full widths of the peaks at half height, ΔE_{exp}, are larger than

ΔE_n, indicating that the spectra correspond to case (1) above. Assuming that ΔE_n is equal to the thermal broadening and that the radiation sources contribute no significant broadening, it is possible to calculate ΔE_i for the $^2P_{3/2,1/2}$ peaks using Eq. 2.10. The results are plotted in Fig. 2.10 as a function of kinetic energy of the photoelectrons. The

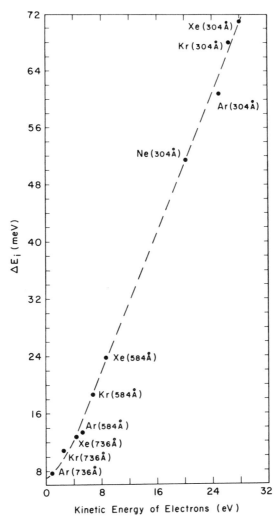

Fig. 2.10. Total energy spread ΔE_i due to instrumental aberrations as a function of electron energy. The points correspond to ionization of the rare gases by the radiation indicated in brackets. Reproduced with permission from Ref. *a* of Table 2.2.

Table 2.2. The Ratio $\sigma_{3/2}/\sigma_{1/2}$ of Specific Photoionization Cross Sections for Producing Rare Gas Ions in Their $^2P_{3/2}$ and $^2P_{1/2}$ States Using Different Excitation Wavelengths

Rare Gas	Rabalais et al[a]			Sampson and Cairns[b]	Turner and May[c]	Frost et al.[d]	Comes and Sälzer[e]
	Ne I	He I	He II				
Ne	—	—	1.83 (19.20 eV)	2.18	—	—	—
Ar	1.88 (0.82 eV)	1.94 (5.37 eV)	1.86 (24.96 eV)	1.98	—	1.96	2.14
Kr	1.82 (2.34 eV)	1.80 (6.89 eV)	1.87 (26.48 eV)	1.79	1.79	1.73	1.69
Xe	1.66 (3.89 eV)	1.58 (8.44 eV)	1.54 (28.03 eV)	1.60	1.69	1.68	1.66

[a] J. W. Rabalais, T. P. Debies, J. L. Berkosky, J. T. J. Huang, and F. O. Ellison, *J. Chem. Phys.*, **61**, 516 (1974). Numbers in parentheses are the average of the kinetic energies for $^2P_{3/2}$ and $^2P_{1/2}$ photoelectrons.
[b] J. A. R. Samson and R. B. Cairns, *Phys. Rev.*, **173**, 80 (1968).
[c] D. W. Turner and D. P. May, *J. Chem. Phys.*, **45**, 471 (1966).
[d] D. C. Frost, C. A. McDowell, and D. A. Vroom, *Proc. Roy. Soc.*, **296**, 566 (1967).
[e] F. J. Comes and H. G. Sälzer, *Z. Naturforsch.*, A, **19**, 1230 (1964).

graph indicates a considerable variation of ΔE_i with kinetic energy with the limit of ΔE_i as $E \to 0$ being $\Delta E \sim 7$ meV. Figure 2.10 gives a good indication of the ultimate resolution to be expected from an electrostatic instrument ($127°17'$ analyzer) when ionizing molecules.

The specific photoionization cross sections σ_i for these peaks are obtained by dividing the integrated peak areas by E_k, that is, A/E_k. The ratios of these specific cross sections $\sigma_{3/2}/\sigma_{1/2}$ are listed in Table 2.2 along with the average kinetic energy of $^2P_{3/2}$ and $^2P_{1/2}$ electrons and similar ratios obtained by other investigators. The values of Sampson and Cairns and Frost et al. are obtained from concentric-sphere analyzers, which are not susceptible to angular and kinetic energy aberrations. The ratios $\sigma_{3/2}/\sigma_{1/2}$ remain fairly constant over the range of kinetic energies investigated. These ratios are all less than the statistical value of $2:1$ expected for the $^2P_{3/2}$ and $^2P_{1/2}$ states. This deviation from the statistical value is a result of different photoionization cross sections per electron for the $^2P_{3/2}$ and $^2P_{1/2}$ levels and variations in the $\sigma_{3/2}/\sigma_{1/2}$ ratio as a function of kinetic energy.

In the spectra of most molecules, the rotational and vibrational broadening are major contributors to ΔE_n, resulting in case (2) above, that is, $\Delta E_n > \Delta E_i$. For such cases the intensities must be determined as

A/E_k, where A is the total area of the band obtained by integrating over all its vibrational components and E_k is the electron kinetic energy at the band maximum. The intensities obtained in this manner are only first-order approximations to the true line intensities for they neglect autoionization phenomena, changes in molecular geometry, and vibronic interaction upon ionization.

2.10. Calibration of spectra

The analyzer voltage at which electrons of a given kinetic energy are brought into focus is a function of the surface charges on the collision and analyzer chambers, cleanliness of the surfaces, sample pressure, and so forth. In order to obtain absolute binding energies it is, therefore, necessary to calibrate each spectrum individually using an internal standard. Calibration is achieved by admitting a small amount of calibration gas into the collision chamber along with the sample gas and simultaneously measuring the positions of the calibrant and sample peaks. The sharp lines corresponding to the $^2P_{3/2}$, $^2P_{1/2} \leftarrow {}^1S$ transitions of the rare gases have been used most effectively for this purpose, although sharp lines from any atom or molecule can, in principle, be used as a calibrant. Listed in Table 2.3 are the sharp lines of the rare gases and numerous molecules that can be used for calibration.[56] The values in Table 2.3 can be used to check the linearity of the spectrometer energy scale. In cases where the

Table 2.3. Calibration Lines

Calibrant	Ionic State	Ionizing Radiation	Electron Kinetic Energy (eV)	Ionization Potential (eV)
CH_3I	$^2E_{3/2}$	He I	11.680	9.538
CH_3I	$^2E_{1/2}$	He I	11.053	10.165
Xe	$^2P_{3/2}$	He I	9.088	12.130
Xe	$^2P_{1/2}$	He I	7.782	13.436
Kr	$^2P_{3/2}$	He I	7.219	13.999
Kr	$^2P_{1/2}$	He I	6.553	14.665
Hg	$^2D_{5/2}$	He I	6.378	14.840
Ar	$^2P_{3/2}$	He I	5.459	15.759
Ar	$^2P_{1/2}$	He I	5.281	15.937
Hg	$^2D_{3/2}$	He I	4.514	16.704
N_2	$^2\Sigma_u^+$	He I	2.467	18.751
Ne	$^2P_{3/2}$	He II	19.249	21.564
He	^2S	He II	16.227	24.587

spectrometer energy scale is not linear, it is necessary to make a calibration curve of observed energy versus true energy.

2.11. Electron-electron coincidence spectroscopy

Although PE spectroscopy is ideal for studying relative transition probabilities to various ionic states of an atom or molecule, data of this type using many different excitation energies are scarce due to the lack of photon sources providing continuous energy variation over a sufficiently wide range. It has recently been shown[57] that a photon source of the required characteristics can be simulated by energy analysis of fast, forward scattered electrons detected in coincidence with ions ("photo" ionization). This method has been extended[58-60] to include the study of ejected electron spectra as a function of energy transfer. The technique is called electron-electron coincidence spectroscopy and involves the detection in coincidence of scattered and ejected electrons by an electron beam source. The incident electron beam energy is sufficiently high for the Born approximation to be valid, and the forward scattering angle can be chosen to be sufficiently small so that the momentum transfer K is close to the optical limit $K = 0$. The ionization source used in this technique is an electron beam emitted by a hot filament. A collision between one of these primary electrons and a sample molecule results in emission of a photoelectron and scattering of the primary electron. Since the primary electron can give up any fraction of its kinetic energy to the molecule (unlike a photon), meaningful information can only be obtained by examining the kinetic energies of the scattered and the "photo" electrons. This can be accomplished by coincidence counting techniques in which one only analyzes "photo" electrons that have accepted a prescribed amount of kinetic energy from the primary electrons. This requires analysis of the scattered and "photo" electrons in different analyzers linked by a coincidence circuit.

The apparatus used by van der Wiel and Brion[60-62] for such a study is shown schematically in Fig. 2.11. An electron beam of 3.5-keV primary energy and 0.3-mm diameter is sent through 2 cm of target gas at a pressure of $\simeq 10^{-3}$ torr. Projectiles scattered through $0 \pm 1.2 \times 10^{-2}$ radians are accepted into a decelerating lens and energy analyzed in a 180° hemispherical analyzer of 50-mm radius. Electrons ejected from the target at 90° to the incident beam pass through a 127°17' analyzer of 30-mm radius with its slit parallel to the electron beam and are detected in coincidence with the forward scattered electrons. The coincidence time resolution (15 nsec) is set to be slightly larger than the actual time spread

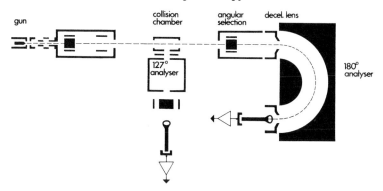

Fig. 2.11. Schematic description of apparatus for electron impact coincidence measurements of scattered and ejected electrons. Reproduced with permission from Ref. 60.

of coincidences in one ejected electron peak in order to allow for the small differences in time-of-flight of different groups of ejected electrons. For data accumulation it is desirable to have a system that provides automatic subtraction of accidental coincidences.[57,63]

The measuring procedure consists of setting the energy loss of the incident electrons at a fixed value and recording the true coincidences as a function of ejected electron energy. Since it is convenient to work at constant resolution, the 127°17′ analyzer can be operated with a constant electric field between the plates. Scanning the energy in this mode involves accelerating or retarding the ejected electrons. The energy-dependent transmission of the electron optics between the collision chamber and entrance slit of the analyzer can be calibrated in the following way.[60] Electrons at the energies required in the calibration (1–100 eV) are produced by 90° elastic scattering of a low-energy beam. The elastic peak is then scanned in the two operating modes: (1) scanning the plate voltages, without electric fields between collision chamber and entrance slit (no energy dependent discrimination); (2) the normal mode of accelerating or decelerating. The peak obtained in mode (1) represents a convolution of the thermal spread in the gun and the theoretical resolution. Removing the effect of the gun spread by taking

$$\frac{\text{peak area}}{\text{theoretical FWHM}}$$

where peak area equals the peak height times the observed FWHM, gives a number that is proportional to the intensity of a "true" scattered signal having zero energy width. This signal is used to calibrate the peak height observed in mode (2) at the same energy, the correction factor being a function of the energy.

References

1. K. Siegbahn, C. Nordling, A. Fahlman, R. Nordberg, K. Hamrin, J. Hedman, G. Johansson, T. Bergmark, S. E. Karlsson, I. Lindgren, and B. Lindberg, *ESCA: Atomic, Molecular, and Solid State Structure Studied by Means of Electron Spectroscopy*, Alinquist and Wiksells, Uppsals, 1967.
2. K. Siegbahn, D. Hammond, H. Fellner-Feldegg, and E. F. Barnett, *Science*, **176,** 245 (1972).
3. J. A. R. Samson, *Techniques of Vacuum Ultraviolet Spectroscopy*, Wiley, New York, 1967.
4. L. Åsbrink, O. Edqvist, E. Lindholm, and L. E. Selin, *Chem. Phys. Letters*, **5,** 192 (1970).
5. D. W. Turner, A. D. Baker, C. Baker, and C. R. Brundle, *Molecular Photoelectron Spectroscopy*, Wiley, New York, 1970.
6. Model PS-18 Operator's Manual, Perkin-Elmer Ltd., Beaconfield, Buckinghamshire, England.
7. C. R. Brundle, *Appl. Spectrosc.* **25,** 8 (1971).
8. R. B. Cairns, H. Harrison, and R. I. Schoen, *Appl. Opt.* **9,** 605 (1970).
9. J. W. Rabalais, T. Bergmark, L. O. Werme, L. Karlsson, and K. Siegbahn, *Phys. Scripta*, **3,** 13 (1971).
10. J. W. Rabalais and A. Katrib, *Mol. Phys.*, **27,** 923 (1974).
11. A. W. Potts, W. C. Price, D. G. Streets, and T. A. Williams, *Faraday Discussions Chem. Soc.*, **54,** 168 (1972).
12. H. R. Griem, *Plasma Spectroscopy*, McGraw-Hill, New York, 1964.
13. A. Mitchell and M. Zemansky, *Resonance Radiation and Excited Atoms*, Cambridge U.P., London, England, 1961.
14. J. A. R. Samson, *Rev. Sci. Instr.*, **40,** 1174 (1969).
15. R. I. Schoer, *J. Chem. Phys.*, **40,** 1830 (1964).
16. F. J. Comes and H. J. Salzer, *Z. Naturforsch.*, **19A,** 1230 (1964).
17. A. A. Kruithoff and F. M. Penning, *Physica*, **4,** 430 (1937); V. Cermak, *J. Chem. Phys.*, **44,** 3781 (1966); V. Cermak, *Collect. Czech. Chem, Commun.*, **33,** 2739 (1968).
18. L. Åsbrink and J. W. Rabalais, *Chem. Phys. Letters*, **12,** 182 (1971).
19. J. Berkowitz, *J. Chem. Phys.*, **56,** 2766 (1972).
20. O. Klemperer and M. E. Barnett, *Electron Optics*, 3rd Edit., Cambridge U.P., London, England, 1971.
21. O. Klemperer, *Rep. Progr. Phys.*, **28,** 77 (1965).
22. H. Z. Sar-El, *Rev. Sci. Instr.*, **41,** 461 (1970).
23. A. L. Hughes and V. Rojansky, *Phys Rev.*, **34,** 284 (1929).
24. A. L. Hughes and McMillan, *Phys Rev.*, **34,** 293 (1929).
25. G. C. Theodoridis and F. R. Paolini, *Rev. Sci. Instr.*, **39,** 326 (1968).

26. P. H. Citrin, R. W. Shaw, Jr., and T. D. Thomas, *Electron Spectroscopy*, D. A. Shirley (Ed.), North-Holland, Amsterdam, 1972, p. 105.

27. B. P. Pullen, T. A. Carlson, W. E. Moddeman, G. K. Schweitzer, W. E. Bull, and F. A. Grimm, *J. Chem. Phys.*, **53**, 768 (1970).

28. R. Herzog, *Z. Phys.*, **97**, 586 (1935).

29. J. J. Leventhal and G. R. North, *Rev. Sci. Instr.*, **42**, 120 (1971).

30. A. D. Johnstone, *Rev. Sci. Instr.*, **43**, 1030 (1972).

31. M. Arnow and D. R. Jones, *Rev. Sci. Instr.*, **43**, 72 (1972).

32. J. A. Simpson, *Rev. Sci. Instr.*, **35**, 1968 (1964).

33. C. E. Kuyatt and J. A. Simpson, *Rev. Sci. Instr.*, **38**, 103 (1967).

34. J. H. D. Eland and C. J. Danby, *J. Sci. Instr.*, **1**, 406 (1968).

35. S. Aksela, M. Karras, M. Pessa, and E. Suoninen, *Rev. Sci. Instr.*, **41**, 351 (1970).

36. D. W. O. Heddle, *J. Phys. E*, **4**, 589 (1971).

37. G. H. Harrower, *Rev. Sci. Instr.*, **26**, 850 (1955).

38. T. S. Green and G. A. Proca, *Rev. Sci. Instr.*, **41**, 1409 (1970).

39. W. Schmitz and W. Melhorn, *J. Phys., E, Sci. Instr.*, **5**, 64 (1972).

40. R. Spohr and E. Von Puttkamer, *Z. Naturforsch., A*, **22**, 409 (1967).

41. D. C. Frost, C. A. McDowell, and D. A. Wroom, *Proc. Roy. Soc., Ser. A*, **296**, 566 (1967).

42. V. Zashkvara, M. I. Korsanskii, and O. S. Kosmachev, *Sov. Phys.—Tech. Phys.*, **11**, 96 (1966).

43. E. Blauth, *Z. Phys.*, **147**, 278 (1957).

44. H. Z. Sar-El, *Rev. Sci. Instr.*, **38**, 1210 (1967); ibid., **39**, 533 (1968).

45. S. Aksela, *Rev. Sci. Instr.*, **42**, 810 (1971).

46. D. A. Huchital and J. D. Rigden, *Electron Spectroscopy*, D. A. Shirley (Ed.), North-Holland, Amsterdam, 1972, p. 79.

47. W. H. J. Anderson and J. B. Lepoole, *J. Phys. E*, **3**, 121 (1970).

48. A. Stamatovic and G. J. Schulz, *Rev. Sci. Instr.*, **41**, 423 (1970).

49. C. Schmidt, *J. Phys. E*, **5**, 1063 (1972).

50. J. D. Allen, J. P. Wolfe, and G. K. Schweitzer, *J. Mass Spectrom. Ion Phys.*, **8**, 81 (1972).

51. P. Marmet and L. Kerwin, *Can J. Phys.*, **38**, 787 (1960).

52. J. S. Allen, *Rev. Sci. Inst.*, **18**, 739 (1947).

53. L. Heroux and H. E. Hinteregger, *Rev. Sci. Instr.*, **31**, 280 (1960).

54. M. J. Wiess, G. M. Lawrence, and R. A. Young, *J. Chem. Phys.*, **52**, 2867 (1970).

55. J. Berkowitz and P. M. Guyon, *Int. J. Mass Spectrom. Ion Phys.*, **6**, 302 (1971).

56. D. R. Loyd, *J. Phys. E*, **3**, 629 (1970).

57. M. J. van der Wiel and G. Wiebes, *Physica*, **53**, 225 (1971); Th.M. El-Sherbini and M. J. van der Wiel, ibid., **59**, 433 (1972).

58. H. Ehrhardt, M. Schulz, T. Tekaat, and K. Willmann, *Phys. Rev. Letters*, **22**, 89 (1969).

59. U. Amaldi, A. Egidi, R. Marconero, and G. Pizzella, *Rev. Sci. Instr.*, **40**, 1001 (1969).

60. M. J. van der Wiel and C. E. Brion, *J. Electron Spectrosc.*, **1,** 309, 439, 443 (1972/1973).
61. G. R. Branton and C. E. Brion, *J. Electron Spectrosc.*, **3,** 123 (1974); ibid., **3,** 129 (1974).
62. W. C. Tam and C. E. Brion, *J. Electron Spectrosc.*, **2,** 111 (1973).
63. P. Ikelaar, M. J. van der Wiel, and W. Tebra, *J. Phys. E*, **4,** 102 (1971).

3 Photoionization Processes

When molecules are irradiated with high-energy photons, processes such as photoexcitation, photoionization, or some combination of the two may occur simultaneously to produce a multitude of final states. In this chapter, the processes occurring when molecules are irradiated are deciphered and those phenomena that are important to PE spectroscopy are discussed. The first section deals with direct photoionization. Sections 2 to 6 are concerned with the shapes of PE bands and interpretation of the vibrational fine structure. Sections 7 to 10 present selection rules pertinent to PE spectroscopy. Configurational mixing and multielectron processes are considered in Sections 11 and 12, and the duration of electron transitions is pondered in Section 13.

In the process of *photoexcitation* a molecule is transferred from its initial state to a bound excited state by absorption of a photon. The energy absorbed is

$$h\nu_{ij} = E_i - E_j \tag{3.1}$$

where ν_{ij} is the frequency of the incident radiation and E_i and E_j are the energies of the states involved in the transition. If the exciting radiation is of sufficiently high energy, the molecule can be excited to a region of continuous states lying above the bound states. In this case the absorption spectrum is continuous because the final state energy in Eq. 3.1 can take on continuous energy values over a significant range. The region of continuous absorption corresponds to dissociation and ionization processes of the molecule. In order to investigate the ionization processes, it is necessary to use high-energy monochromatic radiation for excitation and to measure the kinetic energies of the photoejected electrons. In such a photoionization process the incident photon energy is used to ionize the molecule, thereby releasing an electron; any energy in excess of the ionization energy E_I appears as kinetic energy E_k of the photoelectron according to Eq. 1.1.* Only those ionization processes for which $E_I \leq h\nu$ can occur.

---

* It is not necessary to include the kinetic energy of the positive ion in this expression because the ion is always several thousand times heavier than the electron; in order to conserve momentum, nearly all the kinetic energy must be taken up by the electron.

The ionization energy E_I can be written as

$$E_I = E_j + E_{vib}^+ + E_{rot}^+ \tag{3.2}$$

where E_j is the adiabatic ionization energy for ejection of an electron from level j and E_{vib}^+ and E_{rot}^+ are the vibrational and rotational energies, respectively, of the positive ion. The vibrational structure is often well within the resolving power of the technique. The rotational structure is much more closely spaced than the vibrational structure and, consequently, has been partially resolved only in some ideal cases. Equation 1.1 can now be rearranged with the inclusion of the above expansion for E_I to obtain

$$E_k = h\nu - E_j - E_{vib}^+ - E_{rot}^+ \tag{3.3}$$

It is through studies of E_j, E_{vib}^+, and E_{rot}^+ that we obtain valuable information about molecular ions and their parent molecules.

3.1. Direct photoionization

Direct photoionization is the process whereby an electron is ejected from a molecule $M(X; v'')$ in its ground electronic state X and vibrational level v'' to form the corresponding molecular ion $M^+(x, a, b, c, \cdots; v' = 0, 1, \cdots)$ in either of its electronic states $x, a, b, c, \cdots$ and vibrational levels v' according to the expression

$$M(X; v'') + h\nu \to M^+(x, a, b, c, \cdots; v' = 0, 1, \cdots) + e^- \tag{3.4}$$

The probability of such a transition between the molecular ground state characterized by the eigenfunction Ψ'' and the final state (ion + photoelectron) characterized by Ψ' is determined by the square of the transition moment integral[1]

$$\mathbf{M} = \langle \Psi'' | \sum \mathbf{p} | \Psi' \rangle \tag{3.5}$$

where $\mathbf{p}$ is the dipole moment operator and the sum extends over all electrons i and nuclei j. Ψ'' and Ψ' are functions of the electron r and nuclear R coordinates. The dipole length matrix element of Eq. 3.5 can be expressed as a dipole velocity integral

$$\mathbf{M}_v = \frac{\hbar^2}{mE} \langle \Psi'' | \sum \boldsymbol{\nabla} | \Psi' \rangle \tag{3.6}$$

or as a dipole acceleration integral

$$\mathbf{M}_a = \frac{\hbar^2}{mE^2} \langle \Psi'' | \sum \boldsymbol{\nabla} \mathbf{V} | \Psi' \rangle \tag{3.7}$$

where ∇ is the gradient operator and $\nabla V = Z/r^2$. The dipole length, velocity, and acceleration forms of the matrix elements are formally identical when exact wavefunctions are used. Their values differ when approximate wavefunctions are employed because of their dependence on different parts of the radial wavefunction. Inaccuracies occur because of overemphasis of large radial distances by the dipole length formulation and short radial distances by the dipole acceleration formulation. The dipole velocity approximation is somewhat intermediate between these two extremes.

Using the Born–Oppenheimer approximation, the eigenfunctions can be separated into the product of electronic and nuclear functions

$$\Psi(r; R) = \Psi_e(r; R)\Psi_n(R) \tag{3.8}$$

$\Psi_e(r; R)$ has a parametric dependence upon the instantaneous nuclear configuration, and $\Psi_n(R)$ is dependent upon the particular electronic configuration as well as the nuclear coordinates. Neglecting the interactions between vibrational and rotational motion, $\Psi_n(R)$ can be expressed as a product of vibrational Ψ_v and rotational Ψ_τ wavefunctions

$$\Psi_n(R) = \left(\frac{1}{R}\right)\Psi_v(R)\Psi_\tau(R) \tag{3.9}$$

The dipole operator can also be separated into an electronic and a nuclear dependent part $\sum_{i,j} \mathbf{p} = \sum_i \mathbf{p}_e + \sum_j \mathbf{p}_n$. Substituting these results into Eq. 3.5, we obtain

$$\mathbf{M} = \int\int \Psi_e^{*\prime\prime}(r; R)\left(\frac{1}{R}\right)\Psi_v^{*\prime\prime}(R)\Psi_\tau^{*\prime\prime}(R) \left| \sum_i \mathbf{p}_e + \sum_j \mathbf{p}_n \right| \Psi_e'(r; R)$$
$$\times \left(\frac{1}{R}\right)\Psi_v'(R)\Psi_\tau'(R) \, dr \, dR$$
$$\tag{3.10}$$

This integral can be separated according to the electronic and nuclear dipole operators

$$\mathbf{M} = \int\left(\frac{1}{R}\right)\Psi_v^{*\prime\prime}(R)\Psi_\tau^{*\prime\prime}(R)\left(\frac{1}{R}\right)\Psi_v'(R)\Psi_\tau'(R)\,dr \int \Psi_e^{*\prime\prime}(r; R) \left| \sum_i \mathbf{p}_e \right| \Psi_e'(r; R) \, dr$$
$$+ \int\left(\frac{1}{R}\right)\Psi_v^{*\prime\prime}(R)\Psi_\tau^{*\prime\prime}(R) \left| \sum_j \mathbf{p}_n \right| \left(\frac{1}{R}\right)\Psi_v'(R)\Psi_\tau'(R) \, dR \int \Psi_e^{*\prime\prime}(r; R)\Psi_e'(r; R) \, dr$$
$$\tag{3.11}$$

Since electronic eigenfunctions belonging to different electronic states are orthogonal to one another, the second term in Eq. 3.11 vanishes for electronic transitions.

We now wish to simplify Eq. 3.11 by integrating over rotational coordinates. Consider, for example, the z component of, the dipole operator $\mathbf{p}_{ez} = \mathbf{p} \cos \theta$ and the volume element of configuration space $d\tau = dr \, R^2 \, dR \sin \theta \, d\theta \, d\phi$, where dr is the volume element of the configuration space of the electrons and θ and ϕ are the Euler angles. With these substitutions the first term in Eq. 3.11 becomes

$$\mathbf{M}_z = \int \Psi_\tau^{*\prime\prime}(R) \Psi_\tau^\prime(R) \sin \theta \cos \theta \, d\theta \, d\phi \int \Psi_v^{*\prime\prime}(R) \Psi_v^\prime(R) \, dR$$

$$\times \int \Psi_e^{*\prime\prime}(r; R) \left| \sum_i \mathbf{p}_e \right| \Psi_e^\prime(r; R) \, dr \quad (3.12)$$

The first integral in Eq. 3.12 is a constant for any given combination of rotational levels J'', J' and the second and third integrals are, to a good approximation, independent of J. For this reason, neglect of the rotational part of Eq. 3.12 is acceptable when it is applied to PE spectroscopy, where the rotational structure is unresolved. We, therefore, simplify Eq. 3.12 to

$$M = \int \Psi_v^{*\prime\prime}(R) \Psi_v^\prime(R) \, dR \cdot \int \Psi_e^{*\prime\prime}(r; R) \left| \sum_i \mathbf{p}_e \right| \Psi_e^\prime(r; R) \, dr \quad (3.13)$$

The second integral in Eq. 3.13

$$\mathbf{M}_e(r, R) = \int \Psi_e^{*\prime\prime}(r; R) \left| \sum_i \mathbf{p}_e \right| \Psi_e^\prime(r; R) \, dr \quad (3.14)$$

is the matrix element of the electric dipole moment for a given nuclear configuration (R); in most cases it varies only slightly with R. We may express its functional dependence in a Taylor series expansion about the R value corresponding to the maximum in the product $\Psi_v^{*\prime\prime}\Psi_v^\prime$ (usually the equilibrium ground state configuration R_0) as

$$\mathbf{M}_e(r; R) = \mathbf{M}_e(r; R_0) + \left[\frac{\partial \mathbf{M}_e(r; R)}{\partial R} \right]_{R_0} (R - R_0) + \cdots \quad (3.15)$$

In discussing band envelopes in spectroscopy it is common to terminate this expansion after the first term, for this term is considerably larger than the other members of the series. This first term is always nonzero for one-electron photoionization transitions because the continuum functions $\Psi_e^\prime$ in Eq. 3.14 always have appropriate symmetry elements to make the integral finite (see Section 3.7). Electronic transition moments are usually calculated using only this first term,

$$\mathbf{M}_e(r; R_0) = \int \Psi_e^{*\prime\prime}(r; R_0) \left| \sum_i \mathbf{p}_e \right| \Psi_e'(r; R_0) \, dr \qquad (3.16)$$

with the assumption that the variation of $\mathbf{M}_e$ with R is negligible.

3.2. Adiabatic and vertical ionization energies

Several different ionization energies can be measured, depending on the degree of vibrational excitation of the ions. Two types of ionization energies are usually considered:

1. *Adiabatic ionization energy E_{Ia}—The energy corresponding to the transition*

$$M(X, v'' = 0) + h\nu \rightarrow M^+(x, v' = 0) + e^- \qquad (3.17)$$

   that is, the minimum energy required to eject an electron from a molecule in its ground vibrational state and transform it into a positive ion in the lowest vibrational level of an electronic state x of the ion.

2. *Vertical ionization energy E_{Iv}—The vertical ionization energy corresponds to the transition*

$$M(X, v'' = 0) + h\nu \rightarrow M^+(x, v' = n) + e^- \qquad (3.18)$$

   where the value n of the vibrational quantum number v' corresponds to the vibrational level whose wavefunction gives the largest overlap with the $v'' = 0$ wavefunction. This is the most probable transition and usually corresponds to the vertical transition where the internuclear separations of the ionic state are similar to those of the ground state. Obviously, transitions to each of the ionic states $x, a, b, c, \cdots$ will have individual adiabatic and vertical ionization energies.

3.3. Franck–Condon factors

From the above considerations we may write the photoionization transition probability, to a good approximation, as

$$\mathscr{P} \propto |\mathbf{M}_e(r; R_0)|^2 \, |\langle \Psi_v''(R) | \Psi_v'(R) \rangle|^2 \qquad (3.19)$$

The vibrational overlap integral in Eq. 3.19 is called the *Franck–Condon factor* and is largely responsible for the relative intensities of the vibrational bands in photoionization transitions. This overlap integral does not vanish by orthogonality because Ψ_v' and Ψ_v'' are vibrational functions belonging to different electronic states. Equation 3.19 contains the *wave mechanical statement of the Franck–Condon principle: The intensity of a vibrational band in an electronically allowed transition is proportional to the absolute square of the overlap integral of the vibrational wavefunctions of the initial and final states.* This statement rests on the assumption that the variation of $\mathbf{M}_e$ with R is slow and that $\mathbf{M}_e$ may be replaced by an average value $\bar{\mathbf{M}}_e(r; R_0)$.*

It should be noted that the Franck–Condon principle does not specify a "verticality" of electronic transitions nor the time duration of an electronic transition as determined from the mass disparity between electrons and nuclei. It indicates only that transitions are favored when there is a large overlap between the vibrational wave functions of the initial and final states of the transition. This condition favors, but does not require, transitions in which the relative positions of the nuclei are the same in the initial and final states.

Potential energy curves are shown in Fig. 3.1 for a hypothetical molecule AB in its ground state and the corresponding molecular ion AB^+ in several ionic states. The vibrational eigenfunctions drawn into the upper and lower potential wells are intended to approximate those of an anharmonic oscillator. The wavefunctions for the $v = 0$ levels are bell shaped curves whose maxima lie at the equilibrium internuclear separation for the particular electronic configuration. The eigenfunctions of the

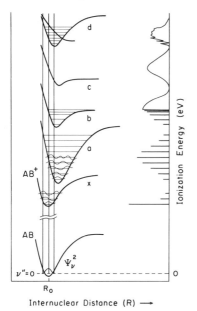

Fig. 3.1 Potential energy curves for the molecule AB in its ground state and the corresponding molecular ion AB^+ in several different ionic states. The Franck-Condon region is between the vertical lines. Photoelectron bands resulting from the various transitions are shown schematically on the right ordinate.

* The variation of M_e with R is discussed in Chapter 7.

higher vibrational levels have broad maxima or minima near the classical turning points of the motion. The maxima and minima between these terminal loops are smaller and narrower than those at the terminal position. The contributions to the overlap integral from these intermediate maxima and minima will roughly cancel one another. For a transition in which $v'' = 0$, the overlap integral has a maximum value for the upper vibrational levels whose eigenfunctions have their broad terminal maxima or minima roughly vertically above the maximum of the eigenfunction of the lower state. Considering the $0 \leftarrow 0$ band, the product of the two eigenfunctions for $v'' = 0$ and $v' = 0$ at each R value, and, therefore, the integral of this product over all R values, is usually greatest when the minima of the two potential curves lie exactly one above the other, that is, vertically. As the minima become separated, the overlap between the $v'' = 0$ and $v' = 0$ eigenfunctions decreases and the intensity of the $0 \leftarrow 0$ band is reduced. When the minima of the two potential curves lie at equal internuclear distances, the overlap integral for the $0 \leftarrow 0$ band is large, but it is obviously small for the $1 \leftarrow 0$ or higher levels since the positive and negative contributions to the integral effectively cancel each other. If the potential curve for the upper state is shifted to higher or lower R values, the integral, and, therefore, the intensity, of the $1 \leftarrow 0$ band increases while that of the $0 \leftarrow 0$ band decreases. This maximum in the vibrational transition shifts to consequently higher v' values as the potential curves are shifted further.

The shapes of photoelectron bands can provide useful information concerning the type of electron ejected. Referring to Fig. 3.1, the ion has the same nuclear configuration in the ground ionic state x as in the molecular ground state. Such a configuration is produced by ejection of a nonbonding electron. The resulting spectrum exhibits a very intense $0 \leftarrow 0$ band followed by a relatively short progression. Figure 3.1 is not meant to imply that the ground state of the ion is always obtained by ejection of a nonbonding electron; indeed, nonbonding orbitals can have higher E_Is than bonding orbitals. Curve a represents a potential surface in which R is increased from its ground state value. In the corresponding photoelectron band, the Franck–Condon maximum appears near the middle of a long vibrational progression. In case b, the photoelectron band consists of a vibrational progression which converges into a continuum. This represents ionization into the region both above and below the dissociation limit of the ion in state b. The threshold of the continuum corresponds to the mass spectrometric appearance potential of the dissociated fragments. Case c represents ionization to a repulsive potential surface resulting in a broad featureless band.

Case d in Fig. 3.1 corresponds to the crossing or close approach of a repulsive potential surface and a bound potential. In such a case, the wavefunctions describing the two states become mixed and the ion is subject to the lifetime limitations of the repulsive state. Such a situation is called *predissociation.* It results in a spectral band with discrete vibrational structure in the region below the crossing, broadened vibrational structure in accordance with the uncertainty principle in the region of the crossing, and finally a continuum in the region above the crossing. Predissociation can only occur if the total wavefunctions of the two interacting states have the same symmetry.

3.4. Vibrational intervals

The vibrational energy spacings observed in a PE spectrum are those of the positive ion. For a diatomic ion with vibrational frequency v, the spacings can be represented as

$$E_{v'} = hv'(v' + \tfrac{1}{2}) + hv'x'(v' + \tfrac{1}{2})^2 + hv'y'(v' + \tfrac{1}{2})^3 + \cdots \tag{3.22}$$

where v' is the vibrational quantum number, x' and y' are anharmonicity constants, and v' is the vibrational frequency of the ion. Classically this frequency is $v' = (k/\mu)^{1/2}/2\pi$, where k is the force constant and μ the reduced mass of the ion.[2] The magnitude of the anharmonicity constants x' and y' is such that $v'x' \ll v'$ and $v'y' \ll v'x'$. If an electron is ejected from a nonbonding orbital, the bond strengths will be altered very little. The internuclear separations of the ion will be very similar to those of the molecule, and the minimum in the ionic potential will lie vertically above that of the ground state (Fig. 3.1). For such a case, the $0 \leftarrow 0$ band will be the most intense transition and the force constant k, and hence v, will be essentially unchanged, giving a vibrational energy in the ionic state that is very similar to that of the ground state molecule. This situation arises in the lowest energy photoionization transition of CO, that is, the $^2\Sigma^+$, $CO^+ \leftarrow {}^1\Sigma^+$, CO transition (Fig. 3.2) for which the $0 \leftarrow 0$ band is by far the strongest, the $1 \leftarrow 0$ and $2 \leftarrow 0$ transitions are extremely weak, and the vibrational frequency in the ionic state (2160 cm^{-1}) is very similar to that of the ground state (2170 cm^{-1}).

When an electron is ejected from a bonding orbital, the bond strength is weakened and the internuclear separation in the corresponding ion is larger than in the neutral parent. As a result, the steeply rising repulsive part of the ionic potential lies directly above the minimum of the molecular ground state potential. This is a region where the number of accessible excited vibrational states is high, so that transitions to a large number of these excited vibrational states occur with monotonically increasing then decreasing Franck–Condon factors. In this case, the

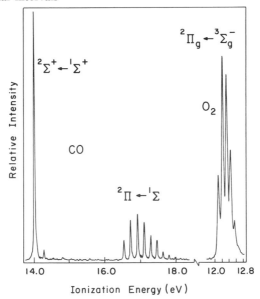

Fig. 3.2. Portions of the He I photoelectron spectra of CO and O_2.

vertical ionization energy may be considerably removed from the adiabatic ionization energy. The decreased bond strength will result in a decreased k'_i and v'_i; hence the vibrational spacings will be reduced from those of the ground state. This situation arises in the second transition (Fig. 3.2) of CO, $^2\Pi$, $CO^+ \leftarrow {}^1\Sigma^+$, CO for which the vibrational frequency of the ionic state is reduced to $1610\ cm^{-1}$ and E_{Iv} is $\sim 0.93\ eV$ higher than E_{Ia}.

If the electron ejected is significantly antibonding, the ion will be smaller than the neutral parent and the slowly sloping attractive region of the ionic potential will lie vertically above the minimum of the molecular potential. Relatively few vibrational levels of the ion are likely to be excited (with respect to the ejection of a bonding electron). There will usually be a slight increase in k'_i and hence v'_i, resulting in larger vibrational intervals in the ion than in the molecule. This effect is clearly apparent in the first ionization band of O_2 (Fig. 3.2) corresponding to the $^2\Pi_g$, $O_2^+ \leftarrow {}^3\Sigma_g^-$, O_2 transition. In this transition the vibrational spacings of the ionic state are $\sim 1800\ cm^{-1}$, a considerable increase over the ground state frequencies of $1,580\ cm^{-1}$.

It is clear that ejection of bonding, nonbonding, or antibonding electrons can alter the bond strengths, thus affecting the equilibrium internuclear distances R_e, v', and k. The bonding characteristics of ejected

Table 3.1 Relation of Bonding Characteristics of Ejected Electrons to
Changes in R_e, k, and ν Upon Ionization

Type of Electron Ejected	R_e	k	ν
Bonding	increased	decreased	decreased
Nonbonding	unchanged	unchanged	unchanged
Antibonding	decreased	increased	increased

electrons and the relation between R_e, k, and ν are summarized in Table 3.1.

These qualitative relationships between electron bonding character and photoelectron band structure must be used with caution for they can be misleading. For example, although the nonbonding electrons of ammonia, amines, phosphines, water, alcohols, ethers, and similar compounds have almost no effect on bond lengths, they are very important in determining the bond angles. Ejection of an electron from these orbitals can produce large changes in geometry and hence photoelectron bands with a large separation between E_{Ia} and E_{Iv}.

Bonding and nonbonding character in ionic compounds

The qualitative relations discussed above are appropriate for molecules that are largely covalent. Recent work by Berkowitz[3] on the PE spectra of high-temperature vapors of compounds that are largely ionic such as TlX and InX (X = Cl, Br, I) has shown that these relations are different in ionic compounds. If the primary contribution to the bonding is ionic, then a compound such as thallium chloride can be considered to be Tl^+Cl^-. Removal of an electron localized on the halogen produces Tl^+Cl, thus effectively destroying the ionic bond. The MO from which the electron was removed is of π-type and is appropriately classified as "bonding," although its electron distribution is very similar to that found in the atomic halogen or π-type orbitals of the halogen acid where it is "nonbonding." The resulting photoelectron band of the ionic compound has the broad characteristic appearance of a bonding-type electron band. A similar reversal of roles is found for the σ MOs of ionic and covalent species. The σ MO electron density is distributed between the two atoms of the diatomic molecule. Removal of a σ electron from a covalent molecule usually weakens the bond, whereas removal of a similar electron from an ionic compound does not appreciably change the positive and negative distribution of charges on the constituent ions. Ejection of σ

electrons from ionic compounds such as Tl^+Cl^- produces sharp bands characteristic of nonbonding electrons. This reversal of the roles of bonding and nonbonding electrons should be kept in mind when analyzing spectra of ionic compounds. Relatively little work[3,3a] has been done on the PE spectra of ionic compounds because of the high temperatures required for volatilization.

Anharmonicities

Comparing the vibrational frequencies of the ion and molecule provides information on the change in geometry upon ionization and thus the bonding characteristics of the electron ejected. It is also possible to obtain some information about the shape of the ionic potential well by studying changes in the vibrational spacings of long progressions and deriving the anharmonicity constant x' from the data. Retaining only the first two terms in Eq. 3.22, the spacings between the vibrational bands can be expressed as

$$\Delta E_{v'} = h\nu + 2hx'\nu(v'+1) \qquad (3.23)$$

where v' is the vibrational quantum number of the lower of the two adjacent levels. The vibrational intervals $\Delta E_{v'}$ are linear functions of the quantum number v' with a slope of $2hx'\nu$. The value of the slope is obtained by plotting $\Delta E_{v'}$ versus v'. If the vibrational intervals converge with increasing v' as they approach the dissociation limit, the slope of the plot is negative, and hence the anharmonicity constant x is negative. Such convergence is considered normal and is called positive anharmonicity. If the vibrational intervals increase with increasing v', the slope of the plot is positive and the anharmonicity constant x is positive. Such divergence is considered abnormal and is called negative anharmonicity. Both positive and negative anharmonicity have been observed in photoionization bands, the latter being more common in progressions of bending vibrations. In such bending vibrations the restoring force usually increases rather than decreases as the amplitude of vibration increases due to steric interactions between the oscillating groups. Examples of negative anharmonicity can be observed in the second ionization bands of the spectra of H_2O, H_2S, and NH_3.

Ionic geometries from vibrational structure

Changes in geometry that occur upon ionization can be determined by calculating the Franck–Condon factors from vibrational wavefunctions. For diatomics, the calculation can be performed for different bond lengths until the best agreement between theoretical and experimental Franck–Condon factors is obtained. For polyatomics, the match between calculated and experimental vibrational band intensities and changes in vibrational frequencies lead to changes in normal coordinates of vibration. These changes in normal coordinates can be transformed into changes in internal coordinates such as bond lengths and bond angles using matrix algebra techniques.[4,5] An example of the calculation of Franck–Condon factors from Morse oscillator wavefunctions is provided in Chapter 7.

For diatomic molecules it is possible to use classical expressions to estimate the change in bond length upon ionization. Considering Fig. 3.1, the maximum Franck–Condon factor occurs when the turning point of the classical vibrational motion in the ionic state occurs at the same nuclear coordinates as the equilibrium position of the ground state molecule. This corresponds to the peak of maximum intensity in a vibrational progression. The vibrational energy of the classical oscillator is proportional to the square of the vibrational amplitude at the turning point. Hence the energy of the oscillator can be set equal to the experimental vibrational energy of the ion at the vertical ionization energy E_{Iv}. For a diatomic molecule with reduced mass μ and vibrational frequency ν', this expression is

$$h\nu'(v'_{max}+\tfrac{1}{2}) = 2\pi^2 \mu\nu'^2 (\Delta R)^2 = E_{Iv} - E_{Ia} \qquad 3.24)$$

where v'_{max} is the vibrational quantum number at the maximum of the progression and ΔR is the equilibrium bond length change between the molecule and ion. The equation can be used for polyatomics, in which case the reduced mass must be chosen for the particular normal mode of vibration under consideration.[6] Application to bending vibrations is also possible if ΔR is replaced by $R(\Delta\theta)$, where R is the bond length and $\Delta\theta$ is the change in equilibrium bond angle between molecule and ion. Expressing ΔR and R in angstroms, $\Delta\theta$ in radians, E_{Iv} and E_{ia} in electron volts, μ in atomic units, and ν' in cm^{-1}, Eq. 3.24 becomes

$$(\Delta R)^2 = R^2(\Delta\theta)^2 = 5.439 \times 10^5 \frac{[E_{Iv} - E_{Ia}]}{\mu\nu'^2} \qquad (3.25)$$

Since the classical and wave mechanical oscillators used in deriving Eq. 3.24 become nearly equivalent only at high vibrational quantum numbers, the expression is valid only for long vibrational progressions where large changes in molecular coordinates are expected. This method does not determine the sign of the changes in bond length or angle; the direction of change is usually determined from the direction of the change in vibrational frequency upon ionization. For example, applying Eq. 3.25 to the $^2\Pi \leftarrow {}^1\Sigma$ transition of CO and the $^2\Pi_g \leftarrow {}^3\Sigma_g^-$ transition of O_2 in Fig.

3.2 yields $\Delta R_{CO} = 0.11\,\text{Å}$ and $\Delta R_{O_2} = -0.07\,\text{Å}$ upon ionization. These values are very close to those obtained from rotational analysis of optical spectra. Although Eq. 3.25 usually yields reliable estimates of changes in nuclear coordinates, it should be used with care because of the simplifying assumptions in its derivation. The most serious assumptions are the neglect of anharmonicity and the neglect of interactions between different normal modes in a polyatomic molecule.

Unresolved photoelectron bands

Most molecules with more than three atoms exhibit PE spectra in which some, or even a majority, of the bands are broad structures with no apparent vibrational structure. The existence of such continuous bands is usually due to one of the following reasons.

1. The resolution of the instrument is insufficient to separate the vibrational peaks. If an instrumental resolution of $10\,\text{meV}$ ($\sim 81\,\text{cm}^{-1}$) can be achieved, one can be reasonably sure that this will not be the case and any existing vibrational structure will be resolved.
2. The vibrational spacings are less than the width of the vibrational peaks, and the band appears to be continuous. The rotational broadening of the vibrational peaks is of the order of kT ($24\,\text{meV} \sim 200\,\text{cm}^{-1}$). Vibrations with such low frequencies are common only in molecules composed of heavy atoms. Molecules composed of light atoms generally have much higher vibrational frequencies, which are easy to resolve in a PE spectrometer. Therefore, continuous bands in the spectra of light molecules can be attributed to overlapping unresolvable vibrations only if it is possible to excite several modes of vibration simultaneously in the transition.
3. Configurational instability such as Jahn–Teller or Renner interactions can produce bands that appear continuous or have only weak vibrational progressions. See Chapter 9 for details.
4. The lifetime of the molecular ions can be so short that broadening results from the uncertainty principle, $\Delta E \Delta t \approx h/2\pi$. The natural lifetimes of most molecular ions ($\sim 10^{-8}\,\text{sec}$) are much longer than the period of a vibration ($\sim 10^{-14}\,\text{sec}$). Broadening only becomes significant when the lifetime of the ion approaches or becomes shorter than the period of a vibration. Molecular ions with such short lifetimes result (1) when the ionic state is repulsive and excitation to any portion of the potential surface leads to direct dissociation or (2) when the ionic state is stable but is crossed by some other unstable state that leads to predissociation. Unstructured photoelectron bands

are common for large molecules because of the many decomposition mechanisms and pathways in polyatomic ions.

Photoelectron bands still provide useful information even though they contain no vibrational structure. Their energies and relative intensities as well as their shapes can yield valuable information about electronic structure. The shapes of the bands are used in the following manner. If the low E_I edge of the band is sharp, it indicates that the adiabatic transition is strong and the change in geometry between molecule and ion is relatively small. Bands that are roughly symmetrical usually have weak adiabatic transitions, indicating that there is a substantial geometry change upon ionization. A continuous band may also exhibit two or more maxima indicating that there are two or more overlapping photoelectron transitions within the band.

3.5. Sequences and hot bands

When molecules have low vibrational frequencies it is possible to have several vibrational levels populated in the ground state. The PE spectra of such molecules may be complicated by hot bands and sequences.[2,7] *Hot bands are produced by transitions from an excited vibrational level of the ground state, $v'' \neq 0$, to various levels of the ionic state.* Examples of hot bands in the PE spectra of 2-bromothiophene[8] and molecular iodine[9] are presented in Figs. 3.3 and 3.4, respectively. The intensities of these hot bands are determined by the Boltzmann distribution of molecules in the low-energy vibrational levels. In the spectrum of I_2 there is a distinct loss in intensity of the components below 9.31 eV at the lower temperature. Considering the 9.31 eV component as E_{Iv}, the intensity change is just that expected from the Boltzmann factors. Plots of the logarithm of intensity versus energy are almost linear up to 9.31 eV, but break sharply at that value. Thus, the first vibrational components of the transition are considered to be hot bands, with $E_{Ia} = 9.31$ eV and $E_{Iv} = 9.34$ eV.

Sequences are transitions involving constant differences between the vibrational quantum numbers of a specific vibration i in the molecular and ionic state, that is, $\Delta v_i = v'_i - v''_i = 0,\ \pm 1,\ \pm 2,\ \cdots$. If the vibrational frequencies of the initial and final state are the same, all bands of a given set of sequences (a given Δv_i) coincide. If the vibrational frequencies are slightly different, then each subsequence of a given Δv_i forms a group of close-lying and approximately equidistant bands of spacing, $\Delta v_i = v'_i - v''_i$. The resolution of photoelectron spectroscopy is generally too low to observe Δv_i sequences when v'_i and v''_i differ very little. If v'_i and v''_i differ considerably, strong $\Delta v_i = 0$ sequences can be resolved with the spacings

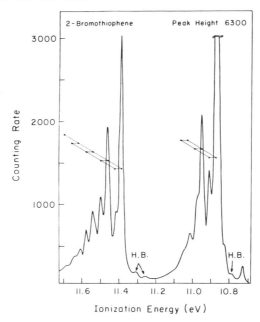

Fig. 3.3. The photoelectron bands resulting from ejection of the nonbonding electrons of bromine in 2-bromothiophene. Hot bands are labeled H.B. Two vibrational progressions of $\simeq36$ meV and $\simeq78$ meV are observed in both transitions. The weak sharp peak at ~10.73 eV is due to a small amount of 3-bromothiophene impurity. Reproduced with permission from Ref. 8.

between the observed bands equal to $\nu_i' - \nu_i''$. Such a sequence has been observed[10] in the PE spectrum of carbon suboxide, $O{=}C{=}C{=}C{=}O$ (Fig. 3.5). The spacings between these bands correspond to $\nu_7' - \nu_7''$ as illustrated in Fig. 3.6. The transition is dominated by a progression in the totally symmetric stretching mode v_1. Each of these bands is accompanied

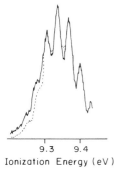

Fig. 3.4. The first photoelectron band in the He I spectrum of I_2. The spectrum was obtained at 30°C (solid line) and −10°C (dotted line). The −10°C spectrum has been normalized so that both spectra have the same intensity for the most intense component. Reproduced with permission from Ref. 9.

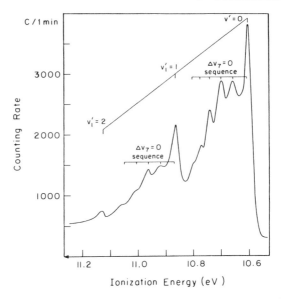

Fig. 3.5. The He I photoelectron spectral band resulting from the $^2\Pi_u(C_3O_2^+) \leftarrow {}^1\Sigma_g^+(C_3O_2)$ transition. Reproduced with permission from Ref. 10.

by $\Delta v_7 = 0$ sequences with $\nu_7' \simeq 400 \text{ cm}^{-1}$, which is a large although not unexpected increase from the ν_7'' value of 63 cm^{-1}. This large increase in ν_7' is caused by the reduction of the high negative charge on the central carbon atom upon ionization with a resulting increase in the bending force constant.

The intensity distribution in sequences is different from that in progressions. It is obtained from the relative populations of the vibrational levels as determined by Boltzmann and statistical weight factors and the overlap integrals of Eq. 3.19. For a $\Delta v_i = 0$ sequence the overlap integral in general does not vary greatly and to a first approximation can be regarded as constant when there is little change in the equilibrium positions of the nuclei. The relative intensities of the C_3O_2 sequences in Fig. 3.5, as determined with constant Franck–Condon factors, are 1.00, 1.48, 1.64, 1.62, 1.49, 1.33, and 1.14 for $v_7'' = 0, 1, \cdots, 6$, respectively. These populations predict the $v_7' = 2 \leftarrow v_7'' = 2$ transition to be most intense, in agreement with the spectrum. The remaining band intensities, however, do not fit these populations exactly, probably because they are superimposed on a background of unknown shape. Also, the breadth of these sequence bands indicates that they may conceal additional complexity, possibly due to $\Delta v_7 = \pm 2$ sequences. Generally such sequences with $\Delta v_i \neq 0$ are considerably weaker than those for which $\Delta v_i = 0$.

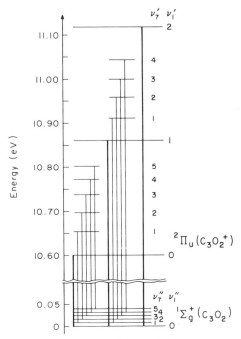

Fig. 3.6. Energy level diagram for the $^2\Pi_u \leftarrow {}^1\Sigma_g^+$ transition of C_3O_2. The v_1 and v_7 energy levels are shown for the molecular and ionic states. The heavy vertical lines represent a progression, and the lighter vertical lines indicate $\Delta v_7 = 0$ sequences. Reproduced with permission from Ref. 10.

3.6. Isotope effects

As in the case of optical spectroscopy, a study of isotope shifts in PE spectra may be an aid in making vibrational assignments. The potential functions of two isotopic species in a given electronic state are very nearly identical. The vibrational levels of the normal ion may be represented by Eq. 3.22, while those of the isotopic species are formulated by replacing x' and v' in Eq. 3.22 by $x^{*'}$ and $v^{*'}$. In all but exceptional cases where the ionic state may be strongly perturbed by some other state, the origins (0–0 transitions) of the band systems in the two isotopic molecules are nearly identical. Since the vibrational force constant k is determined by electronic motion only, it is exactly the same for different isotopic molecules. Therefore using the superscript * to distinguish the isotopic molecule from the "ordinary" molecule, we may take the ratio of the classical vibrational frequencies to obtain

$$\frac{\nu^*}{\nu} = \frac{\mu}{\mu^*} = \rho \tag{3.26}$$

The heavier molecule has the lower frequency. For polyatomic molecules it is possible to give explicit formulae such as Eq. 3.26 when only one normal vibrational mode of a given species exists. For this case we have

$$\nu_i^* = \nu_i \rho'$$

where the factor ρ' depends on the masses of the nuclei and geometry of the molecule. If there are several vibrations, ν_1, ν_2, $\cdots$, ν_j, of a given species, then according to the Teller–Redlich product rule, it is only for the product of the vibrational frequencies that such a simple relation exists, that is,

$$\nu_1^* \nu_2^* \cdots \nu_j^* = \rho' \nu_1 \nu_2 \cdots \nu_j \tag{3.27}$$

Additional relations between the frequencies of isotopic molecules in a given electronic state have been discussed.[11]

An example of isotope effects in PE spectroscopy is provided by HF and DF (Fig. 3.7).[12] In this isotopic pair, the E_{I}s are nearly identical,

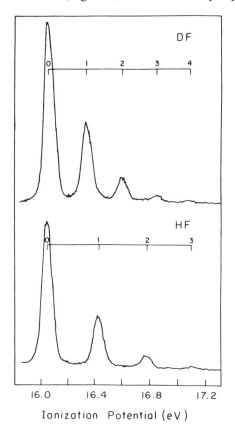

Fig. 3.7. He I photoelectron spectrum of the $^2\Pi_{3/2,1/2} \leftarrow {}^1\Sigma^+$ transition in HF and DF. Reproduced with permission from Ref. 12.

Ionization Potential (eV)

Table 3.2. Vibrational Intervals (cm^{-1}) in the $^2\Pi_{3/2, 1/2}$ States of HF and DF[12]

Vibrational Interval, $v'_i - v'_j$	HF (cm^{-1})	DF (cm^{-1})
0–1	2950 ± 50	2250 ± 50
1–2	2800 ± 50	2200 ± 50
2–3	2600 ± 50	2000 ± 50
3–4	—	1850 ± 50

$16.04(5) \pm 0.01$ for HF and $16.05(3) \pm 0.1$ for DF. The vibrational frequencies of the ions (Table 3.2) indicate a strong anharmonicity in the $^2\Pi_{3/2, 1/2}$ state. The observed frequency ratios are as expected from Eq. 3.26.

3.7. Electronic selection rules

Photoelectron transitions are allowed whenever the integrals of Eq. 3.13 are different from zero. The approximations used in expressing the wavefunctions and operators define the allowedness and forbiddenness of transitions. In Sections 3.7 to 3.10, a one-electron operator is used to discuss the most important type of photoelectron transition, that is, ejection of a single electron. Transitions involving ejection and/or excitation of more than one electron are possible through configuration interaction. These processes are considered in Section 3.11.

For photoelectron transitions the operator of Eq. 3.14 can be approximated as a one-electron dipole operator whose components transform in the molecular point group as the cartesian vectors x, y, and z. A transition is allowed if the product of the irreducible representations Γ of the species in Eq. 3.14, $\Gamma(\Psi_e^{*\prime\prime}) \times \Gamma(\mathbf{p}_e) \times \Gamma(\Psi'_e)$, is totally symmetric with respect to all symmetry elements of the molecular point group for at least one component of $\mathbf{p}_e$, or, in other words, the product $\Gamma(\Psi_e^{*\prime\prime}) \times \Gamma(\Psi'_e)$ belongs to the same symmetry species as one of the components of $\Gamma(\mathbf{p}_e)$. In photoelectron transitions the final state wavefunction Ψ'_e consists of antisymmetrized electronic component wavefunctions of the positive ion Ψ_e^+ and the unbound electron ϕ_e. The function ϕ_e is a one-electron continuum wave. The wavefunctions of the molecule and ion are constructed from the same set of molecular orbitals. We define a projection operator $\mathscr{P}_j(\lambda, m_s)$ which annihilates any electron occupying the orbital ϕ_j and having given orbital angular momentum λ and spin angular momentum m_s. The total electronic transition probability (Eq. 3.14) leading to a final state Ψ'_e of the ion may be factorized as

$$\mathscr{P} \propto \sum_{\Lambda, m_s} \langle |\Psi'_e(\Lambda, m_s| \sum_n \mathbf{p}_n |\Psi''_e(\Lambda'', m''_s)\rangle|^2 \tag{3.28}$$

$$\propto \sum_{n, j} |\langle \Psi^+_e(\Lambda', m'_s)| \mathscr{P}_j(\lambda'', m_s)|\Psi''_e(\Lambda'', m''_s)\rangle|^2 \cdot |\langle \phi_e(\lambda, m_s)| \mathbf{p}_n |\phi_j(\lambda'', m_s)\rangle|^2 \tag{3.29}$$

where the Λ and m_s represent orbital and spin angular momenta of electronic states and satisfy

$$\Lambda' = \Lambda'' - \lambda''; \quad \Lambda = \Lambda' + \lambda$$
$$M'_s = M''_s - m_s; \quad M_s = M'_s + m_s \tag{3.30}$$

with $m_s = \pm \frac{1}{2}$. Since the dipole operator in Eq. 3.29 can have components in the direction of $\lambda = -1$, 0, 1, the continuum wave ϕ_e can have $\lambda = \lambda_j - 1$, λ_j, $\lambda_j + 1$, where λ_j is the orbital angular momentum of the ϕ_j orbital. If Ψ^+_e and Ψ''_e are restricted to single electron configurational wavefunctions that can differ by only one molecular orbital ϕ_j, then the final one-electron transition probability is determined by the second integral of Eq. 3.29. This leads to the simple statistical intensity ratios for configurations that give rise to only one state of each term type, that is, if we assume that the value of the second integral in Eq. 3.29 is approximately constant for ionization of different orbitals,* the relative intensities of the spectral bands are proportional to the spin-orbital degeneracies of the ion states. If Eqs. 3.29 and 3.30 are applied to ionization of closed-shell molecules, we find that since $\Lambda'' = 0$ and $M''_s = 0$, $\Lambda' = \lambda''$ and $M'_s = m_s$. Therefore, ejection of an electron from a closed-shell molecule yields an ion with a doublet spin ($S = \frac{1}{2}$) and orbital angular momentum Λ' equal to that of the electron ejected λ''. From Eqs. 3.30 we see that *all one-electron photoionization transitions are allowed.*

Through use of a one-electron operator, transitions are limited to only those ionic states that are described by an electronic configuration differing from that of the molecule by a single-electron vacancy in a molecular orbital. We neglect, for the moment, the electronic states of a molecular ion that must be described by a configuration that differs from that of the ground state of the neutral molecule by two molecular orbitals or more, that is, ejection of one electron and simultaneous excitation of another electron into an unoccupied orbital.

If the equilibrium conformations of the molecule and ion have different symmetry, that is, if they belong to different point groups, then only

---

* This is a very crude approximation. The integral in Eq. 3.29 is evaluated in Chapter 6, where it is shown to vary considerably for different orbitals.

symmetry elements that are common to the two point groups can be considered in the integrals of Eq. 3.14. If one or both states involved in the transition are degenerate, the product $\Gamma(\Psi_e^{*\prime\prime}) \times \Gamma(\mathbf{p}_e) \times \Gamma(\Psi_e')$ will in general not contain the totally symmetric representation. However, linear combinations of the mutually degenerate eigenfunctions and of the dipole components can be found that make the product totally symmetric.

3.8. Spin selection rules

For systems in which spin-orbit interaction is small,* the electronic eigenfunctions including spin Ψ_{es} can be treated as a product of an orbital and a spin function $\Psi_{es} = \Psi_e \times \Psi_s$. Since the dipole operator does not operate on spin coordinates, the transition moment may be factorized as

$$\langle \Psi_e^{*\prime\prime}| \, \mathbf{p}_e \, |\Psi_e'\rangle \langle \Psi_s''| \, \Psi_s'\rangle \qquad (3.31)$$

The spin functions corresponding to different spin values are orthonormal to one another, therefore, the second integral in Eq. 3.31 vanishes for states of different spin. Photoelectron transitions are thus allowed only between initial and final states of the same total spin, that is, $\Delta S = 0$. In determining the spin of the final state, it is necessary to consider the spin of the positive ion and the photoelectron. This combined spin must be equal to that of the molecular state if the transition is to be allowed, therefore

$$M_{s, \text{molecule}} = M_{s, \text{ion}} + m_{s, \text{electron}} \qquad (3.32)$$

For molecules with closed-shell ground states, $M_{s, \text{mol.}} = 0$, corresponding to a singlet state. Since $m_{s, \text{electron}} = \frac{1}{2}$, the ionic state formed must be a doublet arising from $M_{s, \text{ion}} = \frac{1}{2}$. Thus, the *most common photoelectron spectra consist of transitions from the singlet molecular ground states to doublet states of the positive ion.*

3.9. Vibrational selection rules

For polyatomic molecules Eq. 3.22 must be expanded[11] to include different normal modes i and k of the molecular ion

$$E_{v'} = \sum_i h\nu_i'(v_i' + \tfrac{1}{2}) - \sum_i \sum_{k \geq i} h\nu_i' x_{ik}'(v_i' + \tfrac{1}{2})(v_k' + \tfrac{1}{2}) + \cdots \qquad (3.33)$$

Since $\mathbf{M}_e\,(R_0)$ of Eq. 3.19 is always finite for photoelectron transitions, whether or not a transition from a certain vibrational level of the initial

---

* Large spin-orbit interactions are discussed in Chapter 8.

state v_i'' to a certain vibrational level of the final state v_i' occurs depends entirely on the Franck–Condon overlap integral. In order for this integral to be nonzero, the integrand ($\Psi_v^{*''}\Psi_v'$) must be symmetric with respect to all symmetry operations of the point group, that is, $\Gamma(\Psi_v^{*''}) \times \Gamma(\Psi_v') = \Gamma$(totally symmetric). For degenerate vibrations the product of the appropriate linear combinations of the mutually degenerate vibrational eigenfunctions must contain a totally symmetric component in order for the transition to be allowed. Therefore, we reach the conclusion that only vibrational levels of the same vibrational species in the initial and final state can combine with each other. *Most molecules exist in the totally symmetric zero-point vibrational level of the ground state at room temperature, allowing photoelectron transitions only to totally symmetric vibrational levels of the ionic state.* In ordinary photoelectron transitions only progressions in single quanta of totally symmetric modes of the ionic state are observed. Since there may be several symmetric vibrational modes in a polyatomic molecule, interlocking progressions of these symmetric vibrations are often observed. This is illustrated in Fig. 3.8 for the $^2\Pi_g \leftarrow {}^1\Sigma_g^+$ transition of carbon suboxide.[10]

The higher vibrational levels of antisymmetric vibrations are symmetric for even v and antisymmetric for odd v. According to the selection rule, transitions from the symmetric zero-point vibrational level of the ground state of the molecule can only occur to those antisymmetric vibrational levels of the ionic state for which v is even. Therefore nontotally symmetric vibrations of the ionic state can be excited in double quanta,

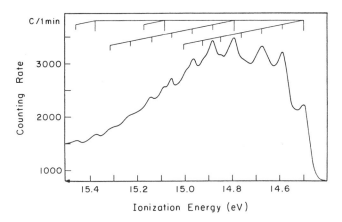

Fig. 3.8. The photoelectron band resulting from the $^2\Pi_g \leftarrow {}^1\Sigma_g^+$ transition of carbon suboxide C_3O_2. The lines above the spectrum indicate interlocking vibrational progressions. Reproduced with permission from Ref. 10.

that is, transitions are possible only to those levels for which $v' = 0, 2, 4,$ $\cdots$. The relative intensities of the bands in such a progression are determined from the Franck–Condon principle. If the symmetry of the molecule and ion are the same, the potential minima of both states involved in the transition occur at the same value of the antisymmetric coordinate regardless of any contraction or expansion that conserves symmetry. The situation is similar to that of a symmetric vibration for which the potential minima have the same position in the two states. The maximum intensity occurs for the $0 \leftarrow 0$ transition and appreciable intensity of the bands with $\Delta v \neq 0$ occurs only when the frequency of the antisymmetric vibration in the two states involved is very different. Progressions in double quanta of antisymmetric modes are usually extremely weak, for it has been shown[7] that even when the frequency changes by as much as a factor of two, the $\Delta v = 0$ transition contains 94.4% of the intensity.

In transitions to degenerate vibrational levels it is necessary to consider the quantum number $l_i = v_i, v_i - 2, \cdots, 1$ or 0, which characterizes the different possible sublevels that occur.[7, 13] For this quantum number we have the selection rule, $\Delta l = 0$. Therefore, since l is even or odd when v is even or odd, only the levels $v' = 0, 2, 4, \cdots$ are symmetric. This selection rule is rigorous for point groups such as D_{2d}, D_{4h}, D_{6h}, and $D_{\infty h}$, but is relaxed in point groups such as D_{3h}, C_{3v}, and T_d, where all overtones of degenerate vibrations contain at least one totally symmetric component. The rule remains rigorous for the $1 \leftarrow 0$ transition in all point groups because the $v = 1$ level of a degenerate vibration is never totally symmetric. If the interaction between vibrational and electronic motion is weak, the vibrational structure of electronic transitions involving degenerate electronic states is the same as that involving nondegenerate states; this is the situation for most degenerate states of diatomic molecules and ions. If the interaction between electronic and vibrational motion is not negligibly small, we must consider the effect of vibronic splittings on the vibrational structure of the band system. This gives rise to the dynamic Jahn–Teller and Renner splittings discussed in Chapter 9.

Until now we have considered only those transitions for which the molecule and ion have the same symmetry. In the event that there is a change of symmetry between the molecule and ion, we must apply the vibrational selection rules using only the common elements of symmetry. From the observed structure, conclusions can be drawn about the difference in symmetry of the molecule and ion in the two equilibrium positions. For example, the degenerate bending vibration of π symmetry in a linear molecule is split into two normal modes in the corresponding bent molecule. Since at least one of these split vibrations can be totally symmetric in the point group of the bent species, progressions in every quanta of this mode can be observed in the PE spectrum. Conversely, observation of a long progression in such a bending mode is good evidence that the ion is no longer linear.

The nature of the vibrational structure in a photoelectron band is often a clue to the assignment of the transition. For ejection of nonbonding electrons, one expects a strong $0 \leftarrow 0$ band accompanied by a very short progression of weak intensity. For ejection of bonding or antibonding electrons, the active vibrations are usually those which involve a change in the internuclear parameters in that region of the molecule where the ejected electron was most heavily localized. For example, ejection of an electron from a strongly C—C bonding orbital is expected to excite the C—C stretching mode, and ejection of a π electron in benzene excites predominantly the ring "breathing" mode.

3.10. Rotational selection rules

Sufficient resolution has been obtained in the PE spectra of H_2,[14] HF, and DF[15] in order to partially resolve the rotational structure of vibrational bands. For molecules composed of such light atoms, the electron spin is only very weakly coupled to the molecular axis, and both the initial molecular state, and final ionic state correspond to Hund's case (b).[7] Hund's case (b), in which the motions of the electrons are strongly coupled to the molecular axis, has been discussed by Dixon et al.[15] Their derivation of the rotational selection rules is as follows. Consider a symmetric or near symmetric top molecule with rotational quantum number K. For a neutral molecule with n electrons

$$\Psi'' = \Psi''_{ev}(1, \cdots, n)\Psi''_\tau = |\Psi''_{ev}(1, \cdots, n)> |N''K''> \qquad (3.34)$$

and for an ion with $n-1$ electrons

$$\Psi' = |\Psi'_{ev}(2, \cdots, n)> |N'K'> \qquad (3.35)$$

where Ψ_{ev} is the combined electronic and vibrational eigenfunction, Ψ_τ is the rotational eigenfunction, and N is the total angular momentum excluding spin. It is necessary to consider the transient intermediate state of the photoelectron transition that conforms to Hund's case (d), in which the orbital angular momentun l' of the ejected electron is weakly coupled to the total angular momentum N' of the ion. Dixon et al. introduce van Vleck's 'reversed' angular momentum $\bar{l}'$ for this electron so that standard vector coupling equations can be used as in Fig. 3.9. With the subscript i for the intermediate state, the antisymmetrized wavefunction is

$$\Psi_i = \mathscr{A}|\phi'_i(1)> |\Psi'_{ev}(2, \cdots, n)> |N'_i l' N' K'> \qquad (3.36)$$

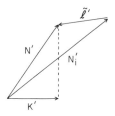

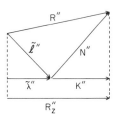

Fig. 3.9. The angular momentum of the intermediate state Ψ_i. Reproduced with permission from Ref. 15.

Fig. 3.10. The relationship between the angular momentum of the initial state Ψ'' and Hund's case (d). Reproduced with permission from Ref. 15.

where $\phi_l'(1)$ is the wavefunction for the ejected electron in the transient state. Rotational transitions are allowed for those cases in which the matrix element $\langle \Psi_i | \mathbf{p}_e | \Psi'' \rangle$ is nonzero. Since $\mathbf{p}_e$ is a one-electron operator, we define an "effective ionization orbital" $\phi''(1)$ for the emitted electron. For the case of single-configuration MO wavefunctions, $\phi''(1)$ is identical with the orbital to be vacated. This MO can be represented as a one-center expansion of atomic orbitals

$$\phi''(1) = \sum_{l''\gamma''} c_{l''\lambda''} \chi_{l''\lambda''}(1) \tag{3.37}$$

The transition moment can then be expressed as a sum of one-electron integrals

$$\langle \Psi_i | \mathbf{p}_e | \Psi'' \rangle = \sum_{l''\lambda''} c_{l''\lambda''} \langle N_i' l' N' K' | \langle \phi_i'(1) | \mathbf{p}_e(1) | \chi_{l''\lambda''}(1) \rangle | N'' K'' l'' \lambda'' \rangle \tag{3.38}$$

Each effective one-electron ground state function in Eq. 3.38 is then expressed in terms of Hund's case (d) in order to derive selection rules for N and K (Fig. 3.10). Then if R'' is the angular momentum vector of the ionic core in the initial molecular state:

$$\langle \Psi_i | \mathbf{p}_e | \Psi'' \rangle = \sum_{l''\lambda''} \sum_{R''=N''-l''}^{N''+l''} c_{l''\lambda''} \langle N_i' l' N' K' | \langle \phi_i'(1) | \mathbf{p}_e(1) | \chi_{l''\lambda''}(1) \rangle$$
$$\times | N'' l'' R'' R_z'' \rangle \cdot \langle N'' l'' R'' R_z'' | N'' K'' l'' \tilde{\lambda}'' \rangle \tag{3.39}$$

Each elementary integral in Eq. 3.39 is decoupled from the rotation of the ionic core so that allowed electronic transitions obey

$$N' = R'' = (N'' + l''), (N'' + l'' - 1), \cdots, (N'' - l'') \tag{3.40}$$
$$K' = R_z'' = (K'' + \tilde{\lambda}'') = (K'' - \lambda'')$$

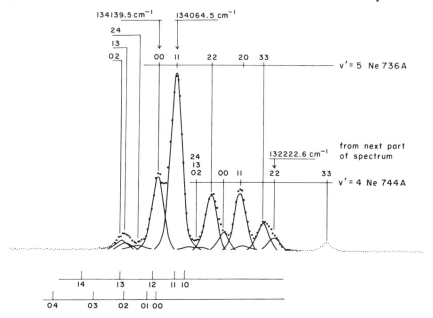

Fig. 3.11. Details of the photoelectron spectrum of H_2 showing the $v' = 5$ transitions excited with the Ne 736-Å line and the $v' = 4$ transitions excited with the Ne 744-Å line. The rotational lines are labeled $J''J'$. The lines that correspond to $\Delta J = 0, \pm 2$ have been indicated above the curve. Below the curve all lines with $J'' = 0$ or 1 have been indicated, provided that no selection rule is valid. The intensities of the rotational lines reflect the thermal distribution of the rotational levels at room temperature with about 70% orthohydrogen (odd J) and 30% parahydrogen (even J). Reproduced with permission from Ref. 14.

for each possible value of l'' and λ'', subject to the appropriate Wigner coefficient $\langle N''l''N'K' | N''K''l''\tilde{\lambda}''\rangle$ of Eq. 3.39 being nonzero.

For Hund's case (a) coupling in the molecule or ion, N is replaced by J and K by P (or Ω), and the effective ionization orbital is a spinorbital with quantum numbers j'', ω''.

A part of the high-resolution PE spectrum of hydrogen corresponding to the process $H_2^+, {}^2\Sigma_g^+ \leftarrow H_2, {}^1\Sigma_g^+$ is presented in Fig. 3.11. The effective ionization orbital is $1\sigma_g$, which may be expressed in a single-center expansion in terms of $1s\sigma_g$, $3d\sigma_g$, and so forth, with $1s\sigma_g$ as the largest contributor. Applications of Eqs. 3.40 leads to

$$\phi_{l''\lambda''} = 1s\sigma_g \qquad \Delta N = 0 \; (\Delta\Lambda = 0)$$

$$\phi_{l''\lambda''} = 3d\sigma_g \qquad \Delta N = 0, \; \pm 2 \; (\Delta\Lambda = 0)$$

$$(3.41)$$

where the exclusion of $\Delta N = \pm 1$ in the second line of Eq. 3.41 arises from the purity of the rotational levels as indicated by zero values of the Wigner coefficients. These equations are identical to those of Sichel,[16] who has considered the rotational selection rules for diatomic molecules in Hund's case (b) states. They are also in agreement with the high resolution spectrum obtained by Åsbrink (Fig. 3.11), who observed strong $\Delta N = 0$ transitions and weak $\Delta N = \pm 2$ transitions.

For bent AB_2 type molecules Eq. 3.41 leads to the following selection rules for various $\chi_{l''\lambda''}$ (with further restrictions as above for $K'' = 0$ and $\lambda'' = 0$):

$\chi_{l''\lambda''}$	ΔN	ΔK_a	ΔK_a
sa_1	0	0	0
pa_1	$0, \pm 1$	± 1	± 1
pb_1	$0, \pm 1$	± 1	0
pb_2	$0, \pm 1$	0	± 1
da_1	$0, \pm 1, \pm 2$	$0, \pm 1, \pm 2$	$0, \pm 1, \pm 2$
da_2	$0, \pm 1, \pm 2$	± 1	± 1
db_1	$0, \pm 1, \pm 2$	± 2	± 1
db_2	$0, \pm 1, \pm 2$	± 1	± 2

Since K_a and K_c cannot both be good quantum numbers for any given level, transitions may occur with further changes of 2, 4, $\cdots$ in ΔK_a or ΔK_c, as in optical spectra.

3.11. Configuration interaction

The selection rules discussed in the previous sections have been based on the assumption that photoionization is a single step one-electron process in which the interaction between a neutral molecule and a photon is described by a one-electron operator. If the wavefunctions for the neutral and ionized states are constructed from the same set of molecular orbitals and a one-electron operator is used, only states characterized by an electronic configuration differing from that of the neutral molecule by one molecular orbital will normally be predicted in PE spectroscopy. This situation has given rise to the usual language, according to which the measured E_is are equated to the different orbital energies of the neutral molecule on the basis of one band per orbital. Even in this simple description, some electronic states of a molecular ion have to be described by a configuration that differs from that of the ground state of the neutral molecule by two molecular orbitals or more, that is, ejection of one electron and simultaneous excitation of another electron into an unoccupied orbital.[17-24] We label these "doubly excited" states.

In order to describe such two-electron processes, replace the simple molecular orbital description by a more correct configuration interaction expansion of wavefunctions. The wavefunction Ψ_j of the jth electronic state can be expressed as a configuration interaction expansion over all electronic configurations Φ_i belonging to a certain spin and symmetry species.

$$\Psi_j = \sum_{i=1}^{\infty} c_{ij}\Phi_i \tag{3.42}$$

The configurations Φ_i may be singly and/or doubly excited. Substituting Eq. 3.42 into the transition moment integral, Eq. 3.14, we obtain

$$\mathbf{M_e} = \langle \Psi'' | \, \mathbf{p} \, | \sum_{i=1}^{\infty} c_{ij}\Phi_i \rangle = \sum_{i=1}^{\infty} c_{ij} \langle \Psi'' | \, \mathbf{p} \, | \Phi_i \rangle \tag{3.43}$$

Transitions to the state described by Ψ_j are allowed if any of the components in the summation of Eq. 3.43 are nonzero. Therefore, a *transition to a state described mainly by a doubly excited configuration can be allowed by the mixing of singly excited configurations into the expansion of Ψ_j.* The intensities of such transitions are determined by the square of Eq. 3.43, that is, the square of the configuration interaction coefficients and the transition moment integral, $\sum_k |c_{kj} \langle \Psi'' | \, \mathbf{p} \, | \Phi_k \rangle|^2$, for the allowed configurations Φ_k in the expansion over the i configurations.

Two-electron transitions have not been found to be common for ejection of an electron from the outer orbitals of the valence shell by means of He I radiation. Such transitions do become important when He II radiation is used, evidently because the increased photon energy makes possible transitions to the more highly excited states of molecular ions for which configuration interaction assumes a significant role. Their presence may even confuse the analysis of a simple spectrum in regions where band intensities for the two-electron processes become comparable with those involving only a single electron.

As an example of two-electron transitions we consider the He II PE spectra of N_2, CO, and HCN, an isoelectronic series that shows bands in excess of the number required to account for simple one-electron processes.[22] Since for this group of molecules the complete valence shell is expected to lie below 41 eV, autoionization cannot be the cause of the anomalous structure observed between 22 and 33 eV. Consider N_2 with the electron configuration

$$1\sigma_g^2 1\sigma_u^2 2\sigma_g^2 2\sigma_u^2 1\pi_u^4 3\sigma_g^2 1\pi_g^0 3\sigma_u^0, \; {}^1\Sigma_g^+ \tag{3.44}$$

The bands observed in the He II PE spectrum of N_2 and their assignments are listed in Table 3.3. We consider only those two-electron

Table 3.3.   Bands Observed in the He II PE Spectrum of N_2[a]

Ionization Energy (eV)	Relative Intensity of Band	Ionized Orbital or Ionization Process
15.58	1	$3\sigma_g$
16.98	1.53	$1\pi_u$
18.75	0.40	$2\sigma_u$
25.3	0.03	CI[b]
28.8	0.10	CI[b]
32.8	0.02	CI[b]
36.6	0.07	$2\sigma_g$

[a] From Ref. 22.
[b] Configuration interaction processes.

processes that involve ejection of one electron plus excitation of another. Even with these limitations, a large number of ionic states can be produced, for example, see Table 3.4. It is difficult to correlate these transitions with the observed bands because information on excited states of ions is very scarce. Generally, one makes the assumption of Siegbahn et al.[24] that the sum of the ionization energy and the excitation energy of a neutral molecule will not differ greatly from the energy of the excited ionized state. This means that the E_I associated with a filled orbital of an excited neutral molecule is taken to be the same as that of the ground state of the neutral molecule. Neglect of the change in nuclear shielding leads to low values for the energy of the excited ionized state. The energy

Table 3.4.   Ionic States Produced in Photoionization to Various Electronic Configurations of N_2^+.

Number of Electrons in MOs						
$2\sigma_g$	$2\sigma_u$	$1\pi_u$	$3\sigma_g$	$1\pi_g$	$3\sigma_u$	Final States
2	2	4	1	0	0	$^2\Sigma_g^+$
2	2	3	2	0	0	$^2\Pi_u^+$
2	1	4	2	0	0	$^2\Sigma_u^+$
1	2	4	2	0	0	$^2\Sigma_g^+$
2	2	4	0	0	1	$^2\Sigma_u^+$
2	2	3	1	1	0	$2\,^2\Sigma_u^+,\ ^4\Sigma_u^+,\ 2\,^2\Sigma_u^-,\ ^4\Sigma_u^-,\ 2\,^2\Delta_u,\ ^4\Delta_u$
2	1	3	2	1	0	$2\,^2\Sigma_g^+,\ ^4\Sigma_g^+,\ 2\,^2\Sigma_g^-,\ ^4\Sigma_g^-,\ 2\,^2\Delta_g,\ ^4\Delta_g$
2	2	4	0	1	0	$^2\Pi_g$
2	2	2	2	1	0	$4\,^2\Pi_g,\ 2\,^4\Pi_g,\ 2\,^2\Phi_g,\ ^4\Phi_g$

found may also correspond to a number of different states, presumably being a mean energy. These are clearly assumptions, and conclusions based upon them must be used with caution. It is clear that a large number of excited states become allowed, in fact, far more than are needed to explain the observed structure on the basis of one band per state. Two-electron processes have been observed and identifications attempted in N_2, CO, HCN, N_2O, O_2, C_2N_2, SO_2, CH_3I, CS_2, CO_2, and COS[18, 22, 23] despite this multitude of possible final states.

We conclude that normally forbidden transitions can become allowed through configuration interaction and can be observed in PE spectroscopy, particularly when the He II line is used for excitation. More accurate calculations are required, however, to identify the states with certainty.

3.12. Autoionization

The radiation used in PE spectroscopy is of sufficiently high energy to excite electrons, other than the most loosely bound one, to discrete neutral states which extend above the ionization threshold. If an atom or molecule is excited to such a highly excited state, radiationless transitions can take place from the discrete state to the ionization continuum. Such a process is called *autoionization* and is observed in PE spectroscopy as irregularities in the intensities of members of a vibrational progression and broadening of vibrational bands due to the relatively short lifetimes of the autoionizing states ($\sim 10^{-14}$ sec). Autoionization accompanying photoexcitation can be represented by

$$M + h\nu \rightarrow M^* \rightarrow M^+ + e^- \qquad (3.45)$$

where M^* is the atom or molecule in a highly excited state. Highly excited states with energies in this region are Rydberg states that converge to second, third, or higher E_1s. The energies of these discrete Rydberg states can be above the threshold of the ionization continumm. When a molecule is excited by photons whose energy coincides with that of an autoionized resonance, there is a high probability of exciting the molecule to this autoionizing state with a well defined vibrational quantum number. If the lifetime of this autoionizing state is greater than one vibrational period, then the photoelectron spectrum can be expected to be dependent on the Franck–Condon factors for radiationless transitions from the autoionizing state of the neutral molecule to the final state of the ion. In contrast with the direct ionization case, the value of v'' in this case is generally not zero. As a result, vibrational structure in autoionized photoelectron spectra may be dramatically extended in a manner determined by the autoionizing level.

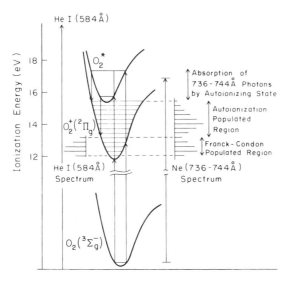

Fig. 3.12. Schematic diagram representing direct and autoionizing transitions of O_2.

In Fig. 3.12 we illustrate the way in which the autoionizing level of oxygen lying at the energy of the Ne I resonance line (right side) completely changes the shape of the first band observed in the PE spectrum compared to that observed using the He I line (left side) at which no autoionization occurs. In this figure, the discrete energy levels of the autoionizing state have the same energy as the upper vibrational levels of the $^2\Pi_g$ state of 0_2^+. The levels of the autoionizing state can assume some properties of the $^2\Pi_g$ state, that is, there is a mixing of the eigenfunctions of the two states. Because of this mixing of the eigenfunctions, the system will at some time find itself in the $^2\Pi_g$ state and will be ionized. The actual He 584-Å and Ne 736- to 744-Å spectra are shown in Fig. 3.13.

The selection rules for autoionization are derived from the expression for the radiationless transition probability γ from the autoionizing level Ψ_a to the final ionic state Ψ_f:

$$\gamma = \frac{4\pi^2}{h} |\langle\Psi_f| \mathbf{W} |\Psi_a\rangle|^2 \tag{3.46}$$

Here $\mathbf{W}$ is a perturbation function representing certain terms in the Hamiltonian. Since $\mathbf{W}$ is totally symmetric, it follows that the states Ψ_f

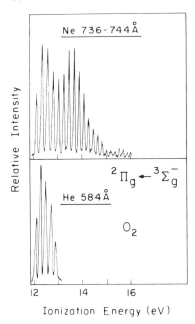

Fig. 3.13. He 584-Å and Ne 736–744-Å spectra of the $^2\Pi_g \leftarrow {}^3\Sigma_g^-$ transition in oxygen. Only bands from the more intense 736-Å line are visible in the Ne spectrum.

and Ψ_a must have common symmetry species in order for the radiationless transition probability to be nonzero, that is,

$$\Gamma(\Psi_f) = \Gamma(\Psi_a) \tag{3.47}$$

This selection rule holds rigorously for the overall wavefunction, but if we can separate electronic, vibrational and rotational motion as $\Psi = \Psi_e\Psi_v\Psi_\tau$, then the selection rule, Eq. 3.47, becomes

$$\Gamma(\Psi_e\Psi_v\Psi_\tau)_f = \Gamma(\Psi_e\Psi_v\Psi_\tau)_a \tag{3.48}$$

It is clear that information about the electronic states responsible for autoionization resonances can be obtained from a study of PE spectra exhibiting these effects and application of the selection rules, Eq. 3.48. Approximate values of the equilibrium internuclear distance R_e for autoionized states can be derived by comparing experimental intensity distributions for the resonance alone with calculated intensities and adjusting ΔR_e for the best fit.

The calculation of vibrational intensities in PE spectra perturbed by autoionization has received considerable attention. The configuration interaction method of Fano,[25-27] as applied by Smith,[28] has been reasonably successful in predicting these intensities. Two distinct cases are defined in this method: In case (1), the source has a narrow width and is tuned to the maximum of the autoionizing resonance. The intensity of the vibrational bands in the spectrum is given by

$$I \propto F_{if} + F_{ia} \cdot F_{af} \cdot q^2 \tag{3.49}$$

where F_{if} is the Franck–Condon factor for direct ionization from the initial ground molecular state i to the final ionic state f, F_{ia} is the Franck–Condon factor for the transition to the autoionizing level a, F_{af} is the Franck–Condon factor for the transition from the autoionizing level to the final ionic state, and q is the Fano line profile index. For case (2), the source linewidth $\delta(h\nu)$ is larger than the width of the autoionization line Γ_n. The intensities of the vibrational levels in this case are given by

$$I \propto F_{if} + F_{ia} \cdot F_{af} \cdot \frac{\pi q^2 \Gamma_n}{2 \cdot \delta(h\nu)} \tag{3.50}$$

In both cases the assumptions are made that the excited states have the same vibrational constants as the ionic state, only one electronic state is populated by the decay of the autoionizing level, and only one autoionizing state is involved in the ionization process. Case (1) is appropriate when using rare gas resonance lamps where the exciting linewidth is small. Case (2) is appropriate when using continuum radiation dispersed by a monochromator where the available intensity limits the resolution which may be available for a given experiment.

Most autoionization effects have been observed with Ne or Ar resonance lines or continuum sources in the region, ≈ 12 to $17\,\text{eV}$.[29-35] The density of excited and ionic states in this region gives a high probability for autoionization effects. Autoionization is usually not observed with the He I resonance line because of the low density of excited and ionic states in this energy region. One case of autoionization perturbations arising from He I photons has been reported[29] for CF_4. A comparison of the PE spectrum of CF_4 recorded with 21.22 and 40.81 eV photons shows that in the former there is anomalous structure in three bands between 16 and 19 eV. This structure is believed to result from interaction with an autoionizing state at 21.22 eV. This is supported by the photoabsorption spectrum[30] of CF_4, which shows a number of resonances superposed on the continuum between $h\nu = 20.3$ and $22.5\,\text{eV}$. It is very probable that one of these states is responsible for the observed anomaly in the PE spectrum of CF_4 excited with 21.22-eV photons. Autoionization effects in the molecules and atoms O_2, CO, NO, N_2, H_2, Xe, and Hg have been observed by several groups.[29-37] In all cases, the general effect of the autoionizing process is to change the vibrational intensity distributions in the PE spectrum.

3.13. Duration of photoelectron transitions

It is of interest to investigate the possibility of being able to specify the time duration of photoelectron transitions. In this investigation we apply the arguments of Schwartz[38] to photoelectron experiments. First, let us consider the common statement that electronic transitions take place in a time short compared to the period of molecular vibrations. We set the time scale by considering a vibrational frequency $\bar{\nu} = 1000 \text{ cm}^{-1}$ or $\nu = \bar{\nu}c = 3 \times 10^{13}$ Hz. This corresponds to a vibrational period of $\tau_{\text{vib}} = 3 \times 10^{-14}$ sec. Assume that we wish to measure the position of the nuclei on a time scale, $\tau \leq \tau_{\text{vib}}$. The energy uncertainty resulting from such a measurement is related to τ by the Heisenberg time-energy uncertainty expression, $\tau \cdot \Delta E \simeq h/2\pi$.[39] Since $E = h\nu$, then $\tau \cdot \Delta\nu \simeq 1/2\pi$. This is the "classical" uncertainty principle that relates the accuracy with which the frequency of a wave-train can be determined in a time τ. We wish to estimate the energy uncertainty (cm^{-1}) introduced by our measurement in a time τ such that $\tau \cdot \Delta\bar{\nu} \simeq 1/c = 3 \times 10^{-11} \text{ cm}^{-1}$ sec. For $\tau \leq 3 \times 10^{-14}$ sec, $\Delta\bar{\nu} \geq 1000 \text{ cm}^{-1}$, which is just the vibrational frequency that we have assumed. The proposed measurement will introduce an energy uncertainty that leads to complete indeterminateness of the phase of the vibration. *Thus the statement that an electron transition takes place in a time short compared to the period of a vibration cannot be tested by experiment for such a measurement violates the Heisenberg uncertainty principle.*

Although the uncertainty principle does not allow us to speak of the period of time in which a photoelectron transition "takes place," we can determine the *duration of the interaction of the molecules with the ionizing radiation.* Considering gas-phase molecules absorbing photons in the absence of perturbations, we must be careful not to disturb the system by any measurement during the time of interaction of the molecules with the radiation. In this case we may infer from the natural width of a photoelectron line in the spectrum (the term "natural" is used to rule out contributions to the width due to collisions, Doppler effects, etc.) that the duration of the average molecular interaction with the ionizing radiation is $\tau \simeq (\Delta\nu)^{-1}$, where $\Delta\nu$ is the width, in Hz, of the photoelectron line. But according to radiation theory the natural linewidth $\Delta\nu$ is equal to the inverse of the radiative lifetime, τ_{R}.[40] Thus we infer that, in this circumstance, the molecules have interacted with the radiation field for a time equal, on the average, to their radiative lifetime. A lower limit to radiative lifetimes may be set by the classical expression[41]

$$\tau_{\text{R}} \geq \frac{mc}{8\pi e^2}\bar{\nu}^{-2} = 1.51\bar{\nu}^{-2} \qquad (\tau_{\text{R}} \text{ in sec}, \bar{\nu} \text{ in cm}^{-1}) \qquad (3.51)$$

For a photoelectron transition at $12\,eV$, $\bar{\nu} \simeq 96{,}800\,cm^{-1}$ and $\tau_R \geq 1.6 \times 10^{-10}\,sec$, which is several orders of magnitude longer than the period of molecular vibrations, $\tau_{vib} \simeq 3 \times 10^{-14}\,sec$. For a molecule absorbing light in the presence of a disturbance (experimental intervention, collision, etc.) the absorption line may be broadened. Other exit channels for the excited state such as predissociation, dissociation, autoionization, and so forth shorten the lifetime and broaden the photoelectron lines. In such situations the molecule may interact with the exciting radiation for a shorter period of time. This period of time may be (1) long compared to a vibrational period—vibrational structure is evident, (2) of the order of a vibrational period—broad diffuse vibrational structure, or (3) shorter than a vibrational period—direct dissociation with no apparent vibrational structure.

References

1. See for example, H. Eyring, J. Walter, and G. E. Kimball, *Quantum Chemistry*, Chapter VIII, Wiley, New York, 1944.
2. G. Herzberg, *Molecular Spectra and Molecular Structure I. Spectra of Diatomic Molecules*, 2nd edit., Van Nostrand, Princeton, N.J. 1950.
3. J. Berkowitz, *J. Chem. Phys.*, **56,** 2766 (1972); J. Berkowitz and J. L. Dehmer, ibid., **57,** 3194 (1972).
a. T. D. Goodman, J. D. Allen, Jr., L. C. Cusachs, and C. K. Schweitzer, *J. Electron Spectrosc.*, **3,** 289 (1974).
4. J. B. Coon, R. E. Dewames, and C. M. Loyd, *J. Mol. Spectrosc.* **8,** 285 (1962); N. L. Shinkle and J. B. Coon, ibid., **40,** 217 (1971).
5. E. Heilbronner, K. A. Muszkat, and J. Schaublin, *Helv. Chim. Acta*, **53,** 331 (1970).
6. G. Herzberg, *Infrared and Raman Spectra*, Van Nostrand, Princeton, N.J., (1945).
7. G. Herzberg, *Molecular Spectra and Molecular Structure III. Electronic Spectra and Electronic Structure of Polyatomic Molecules*, Van Nostrand, Princeton, N.J. 1967.
8. J. W. Rabalais, L. O. Werme, T. Bergmark, L. Karlsson, and K. Siegbahn, *Int. J. Mass Spectrom. Ion Phys.*, **9,** 185 (1972).
9. B. R. Higginson, D. R. Lloyd, and P. J. Roberts, *Chem. Phys. Letters*, **19,** 480 (1973).
10. J. W. Rabalais, T. Bergmark, L. O. Werme, L. Karlsson, and K. Siegbahn, *Electron Spectroscopy*, D. A. Shirley (Ed.), p. 425, North–Holland, Amsterdam, 1972.
11. E. B. Wilson, J. C. Decius, P. C. Cross, *Molecular Vibrations*, McGraw-Hill, New York, 1955.
12. C. R. Brundle, *Chem. Phys. Letters*, **7,** 317 (1970).

13. H. C. Longuet–Higgins, M. H. L. Pryce, and R. A. Sack, *Proc. Roy. Soc.*, *Ser. A*, **244**, 1 (1958).
14. L. Åsbrink, *Chem. Phys. Letters*, **7**, 549 (1970).
15. R. N. Dixon, G. Duxbury, M. Horani, and J. Rostas, *Mol. Phys.*, **22**, 977 (1971).
16. J. M. Sichel, *Mol. Phys.*, **18**, 95 (1970).
17. J. C. Lorquet and C. Cadet, *Chem Phys. Letters*, **6**, 198 (1970).
18. J. C. Lorquet and C. Cadet, *Int. J. Mass. Spectrom. Ion Phys.* **7**, 245 (1971).
19. J. C. Lorquet and M. Desouter, *Chem. Phys. Letters*, **16**, 136 (1972).
20. R. N. Dixon and S. E. Hull, *Chem. Phys. Letters*, **3**, 367 (1969).
21. M. Okuda, *Discussions Faraday Soc.*, **54**, 140 (1972).
22. A. W. Potts and T. A. Williams, *J. Electron Spectrosc.*, **3**, 3 (1974).
23. M. Okuda and N. Jonathan, *J. Electron Spectrosc.*, **3**, 19 (1974).
24. C. A. Allan, U. Gelius, D. A. Allison, G. Johansson, H. Siegbahn, and K. Siegbahn, *J. Electron Spectrosc.*, **1**, 131 (1972/73).
25. U. Fano, *Phys. Rev.*, **124**, 1866 (1961).
26. U. Fano and J. W. Cooper, *Phys. Rev. A*, **137**, 1364 (1965); *Rev. Mod. Phys.*, **40**, 441 (1968).
27. F. H. Mies, *Phys. Rev.*, **175**, 164 (1968).
28. A. L. Smith, *J. Quant. Spectrosc. Radiat. Transfer*, **10**, 1129 (1970); *Phil. Trans. Roy. Soc. London*, **A268**, 169 (1970).
29. C. R. Brundle, M. B. Robin and H. Basch, *J. Chem. Phys.*, **53**, 2196 (1970).
30. J. L. Bahr, A. F. Blake, J. H. Carver, J. L. Gardner, and V. Kumar, *J. Quant. Spectrosc. Radiat. Transfer*, **11**, 1839 (1971); ibid., **11**, 1853 (1971); ibid., **12**, 59 (1972).
31. P. Natalis and J. E. Collin, *Chem. Phys. Letters*, **2**, 414 (1968); idem., *Int. J. Mass Spectrom. Ion. Phys.*, **2**, 231 (1969).
32. J. E. Collin, J. Delwicke, and P. Natalis, *Int. J. Mass Spectrom. Ion Phys.*, **7**, 19 (1971).
33. G. R. Branton, D. C. Frost, T. Makita, C. A. McDowell, and I. A. Stenhouse, *Phil. Trans. Roy. Soc. London*, **A268**, 77 (1970).
34. J. Berkowitz and W. P. Chupka, *J. Chem. Phys.*, **51**, 2341 (1969).
35. J. A. R. Samson and R. B. Cairns, *Phys. Rev.*, **173**, 80 (1968).
36. J. A. Kinsinger and J. W. Taylor, *Int. J. Mass Spectrom. Ion Phys.*, **2**, 231 (1969).
37. P. H. Doolittle and R. I. Schoen, *Phys. Rev. Letters*, **14**, 348 (1965).
38. S. E. Schwartz, *J. Chem. Educ.*, **50**, 608 (1973).
39. W. Heisenberg, *The Physical Principles of Quantum Theory*, Dover, New York, 1949.
40. W. Heitler, *The Quantum Theory of Radiation*, 3rd Edit., Chap. V, Oxford U.P., London, 1954; A. Messiah, *Quantum Mechanics*, Vol. I, Chap. 4, Sect. 10, Wiley, New York, 1961.
41. A. C. G. Mitchell and M. W. Zemansky, *Resonance Radiation and Excited Atoms*, Chap. III, Cambridge U.P., Cambridge, 1934.

4 Ionization of Closed-Shell Molecules

The single most important application of PE spectroscopy is the direct measurement of ionization energies of electrons from various energy levels of atoms and molecules. The technique is capable of providing ionization energies or potentials, E_{In}, $n = 1, 2, 3, \cdots$, corresponding to removal of single electrons from orbitals that are less energetic than $h\nu$, the excitation energy. By first, second, third, and so forth E_Is we refer to the energy required to remove a single electron, respectively, from the first, second, third, and so forth energy level of the species (labeling the first as the level that requires the least amount of energy for ionization). The positive ions formed have one less unit of negative charge than the species from which they were produced. The unique feature of PE spectroscopy is the clarity with which it reveals inner E_Is, that is, E_Is other than the first.

This chapter considers the interpretation of ionization energies and the theoretical models used to describe and calculate ionization energies and electron orbitals of molecules with closed-shell ground states. Considerable emphasis is placed on the type of orbitals (molecular or equivalent orbitals) used to describe ionization phenomena. This subject is discussed in detail because of the ambiguity and often misuse of terminology (particularly among nontheorists) in describing the ionization process, the type of electron ejected, and the particular orbital(s) from which the electron is ejected. Sections 1 and 2 define ionization energies and the basic theoretical concepts used for their interpretation. A qualitative discussion of the construction of simple molecular and equivalent orbitals is presented in Sections 3 and 4. Applications of these orbital theories to interpretation of general physical properties of molecules and ionization phenomena, in particular, are presented in Section 5. Koopmans' theorem, ubiquitous to PE spectroscopy, is developed in Section 6, and its limitations are presented. Section 7 introduces and contrasts some of the more common methods of calculating ionization energies.

4.1. Ionization energies

The ionization process may be interpreted as an assault on a molecule by an overwhelming quantity of energy, the result being disfiguration of the molecule by discarding one of its electrons and producing a positive ion. Such a qualitative view of ionization is sufficient to understand the importance and potential of PE spectroscopy. The ability to measure different ionization energies within an atom or a molecule is direct evidence for the shell nature of electronic structure. In fact, PE spectroscopy provides evidence for the existence of electrons in quantized atomic and molecular orbitals and in so doing gives reality to the concepts of orbitals, eigenvalues, and eigenfunctions.

The magnitudes of the E_Is are determined by the tenaciousness with which electrons are bound to the material, hence the "binding energy." The binding energy is related to the ability of an atom to attract electrons to itself, that is, its electronegativity. When atoms combine to form molecules or solids, those electrons that take part in the bonding, that is, electrons whose wavefunctions overlap with electronic wavefunctions from neighboring atoms, belong to the molecule or solid as a whole rather than to a single atom. They form molecular orbitals or valence bands that are unique to the nature of the material and about which PE spectroscopy provides valuable information. In the absence of two-electron processes, autoionization, and so forth, one spectral band arises for each occupied electron energy level in the molecule. The spectra, therefore, provide a mapping of the electron energy levels, or an energy level diagram.

4.2. Theoretical models for ionization energies

The ability to measure orbital energies has added new vigor to the field of quantum chemistry. PE spectroscopy and quantum chemistry are complementary techniques, each one benefiting from achievements of the other. Since quantum chemical concepts are an invaluable tool for PE spectroscopists, it is useful to digress into the interpretation and applications of certain types of orbital theories.

In the quantum theory of electronic structure it is convenient to express the electronic wavefunction of a molecule as a determinant of mutually orthogonal orbitals, each involving the coordinates of one electron only. There is a certain arbitrariness in the choice of orbitals used in the determinantal wavefunction. Two types of orbitals have special significance: (1) *Molecular Orbitals* are chosen so as to belong to the various irreducible representations of the molecular symmetry group. (2) *Equivalent Orbitals* are chosen such that all of the members of a set are identical in the sense that they are permuted among one another by a symmetry operation that permutes equivalent chemical bonds or equivalent spatial positions of a molecular system. The first type (MOs) is the basis of molecular orbital theory[1-3] in which electrons are delocalized over the whole molecule. The second type (EOs) is the basis for the electron pair method,[4-6] which is essentially a generalization of the Heitler–London calculation for the hydrogen molecule according to which the electrons belong to certain atoms or spatial regions of the molecule. Both theories have been invaluable in understanding the electronic structure of complex molecules. The fundamentals of these orbitals have been derived, and their appropriate usages have been discussed in a series of admirable papers.[7]

The chemist uses these orbital theories in order to assist in understanding, interpreting, and explaining molecular properties. However, there exists considerable ambiguity among nonspecialists in orbital theory concerning the best choice of orbitals to use for interpretation of molecular properties; this problem has been discussed in two recent papers.[8,9] For example, the chemical bond between two atoms, the electron density distribution, and the energies of the electron energy levels in a molecule are fundamental concepts in chemistry that can generally be interpreted best in terms of one or the other type of orbital. It is the object of these sections to present a simplified analysis of molecular and equivalent orbitals and to examine their effectiveness in describing various molecular properties. The analysis is applied to the isoelectronic series of first-row hydrides, HF, H_2O, NH_3, and CH_4. These molecules are constructed, hypothetically, by partitioning off protons from the relevant inert gas, Ne, to form polyatomic structures with symmetries $C_{\infty v}$, C_{2v}, C_{3v}, and T_d, respectively, as 1, 2, 3, or 4 protons are extracted. The construction of molecular orbitals and equivalent orbitals for these first-row hydrides is presented in Sections 4.3 and 4.4. The properties and applications of these MOs and EOs are discussed in Section 4.5; PE spectra of the first-row hydrides are used in this section to provide an example of the necessity for proper use of orbitals.

4.3. Construction of molecular orbitals

Molecular orbitals are conveniently formed as linear combinations of atomic orbitals. These combinations can be chosen such that they are symmetry orbitals, that is, they transform as an irreducible representation (IR) of the molecular point group. A method for constructing such orbitals is as follows: Observe the transformation properties of the various atomic orbitals at the atomic sites in the molecule. Atomic

Table 4.1.   Resolution of Atomic Orbital Species into Those of Point Groups of Lower Symmetry

$C_{\infty v}$		C_{2v}	C_{3v}	T_d
s	σ	a_1	a_1	a_1
p_x		b_2		
	π		e	
p_y		b_1		t_2
p_z	σ	a_1	a_1	

orbitals on different atoms of the same type that are at equivalent symmetry positions must be linearly combined (group orbitals) in order for them to transform according to the properties of the molecular point group. The molecular orbitals are then constructed as linear combinations of those atomic orbitals and/or group orbitals that transform according to the same IRs. The number of MOs of symmetry Γ that can be formed is equal to the number of AOs of symmetry Γ. The coefficients of these AOs in the MOs can be determined by solving the secular determinant.[6,10,11]

Consider the example of the first-row hydrides. Beginning with Ne, the 10 electrons fully occupy the $1s$, $2s$, and $2p$ AOs. When a proton is abstracted from the nucleus to form HF, the AOs transform according to the IRs of the new symmetry group $C_{\infty v}$ (Table 4.1). The σ AOs ($1s_F$, $2s_F$, $2p_{z_F}$, $1s_H$) combine to form four σ MOs. The π AOs ($2p_{x_F}$, $2p_{y_F}$) remain as degenerate atomic-like orbitals localized on the fluorine atom. In H_2O the oxygen AOs transform as a_1, b_1, and b_2 with positive and negative combinations of the two hydrogen AOs transforming as a_1 and b_2, respectively. These combine to form four a_1, two b_2, and one b_1 MO. The nitrogen AOs of NH_3 and appropriate combinations of the hydrogen AOs each transform as a_1 and e. These combine to form four a_1 and two e MOs. The carbon AOs of CH_4 and appropriate combinations of the hydrogen AOs each transform as a_1 and t_2. These combine to form three a_1 and two t_2 MOs.

The approximate energy ordering of these MOs, and consequently the occupied MOs in the molecular ground state, can be determined from a consideration of the AO composition of the MOs. The energy of an MO will be determined predominantly by the energy of the AO that constitutes its major character and by bonding and antibonding interactions with other orbitals. The molecular orbitals occupied in the ground states of the four molecules discussed above are presented in Table 4.2. The

Table 4.2. Occupied MOs in the Ground States^a of HF, H_2O, NH_3, and CH_4 and Relative Sizes of the AO Coefficients

Molecule	MOs as Linear Combination of AOs	Relative Size of AO Coefficients^b
HF	$\phi_{1\sigma} = C_{11}1s_F + C_{12}2s_F + C_{13}2p_{z_F} + C_{14}1s_H$	$C_{11} \gg C_{12} > C_{13} \sim C_{14}$
	$\phi_{2\sigma} = C_{21}1s_F + C_{22}2s_F + C_{23}2p_{z_F} + C_{24}1s_H$	$C_{22} > C_{21} \sim C_{23} \sim {}_{24}$
	$\phi_{3\sigma} = C_{31}1s_F + C_{32}2s_F + C_{33}2p_{z_F} + C_{34}1s_H$	$C_{33} \sim C_{34} \sim C_{32} > C_{31}$
	$\phi_{1\pi} = \begin{bmatrix} 2p_{x_F} \\ 2p_{y_F} \end{bmatrix}$	
H_2O	$\phi_{1a_1} = C_{11}1s_O + C_{12}2s_O + C_{13}2p_{z_O} + C_{14}(1s_{H_1} + 1s_{H_2})$	$C_{11} \gg C_{12} > C_{13} \sim C_{14}$
	$\phi_{2a_1} = C_{21}1s_O + C_{22}2s_O + C_{23}2p_{z_O} + C_{24}(1s_{H_1} + 1s_{H_2})$	$C_{22} > C_{21} \sim C_{24} \sim C_{23}$
	$\phi_{1b_2} = C_{31}2p_{y_O} + C_{32}(1s_{H_1} - 1s_{H_2})$	$C_{31} \sim C_{32}$
	$\phi_{3a_1} = C_{41}1s_O + C_{42}2s_O + C_{43}2p_{z_O} + C_{44}(1s_{H_1} + 1s_{H_2})$	$C_{43} \sim C_{42} > C_{44} > C_{41}$
	$\phi_{1b_1} = 2p_{x_O}$	
NH_3	$\phi_{1a_1} = C_{11}1s_N + C_{12}2s_N + C_{13}2p_{z_N} + C_{14}(1s_{H_1} + 1s_{H_2} + 1s_{H_3})$	$C_{11} \gg C_{12} > C_{14} \sim C_{13}$
	$\phi_{2a_1} = C_{21}1s_N + C_{22}2s_N + C_{23}2p_{z_N} + C_{24}(1s_{H_1} + 1s_{H_2} + 1s_{H_3})$	$C_{22} > C_{21} \sim C_{24} \sim C_{23}$
	$\phi_{1e} = \begin{bmatrix} C_{31}2p_{x_N} + C_{32}1s_{H_1} + C_{33}(1s_{H_2} + 1s_{H_3}) \\ C_{41}2p_{y_N} + C_{42}(1s_{H_2} + 1s_{H_3}) \end{bmatrix}$	$\begin{matrix} C_{31} \sim C_{32} > C_{33} \\ C_{41} \sim C_{42} \end{matrix}$
	$\phi_{3a_1} = C_{51}1s_N + C_{52}2s_N + C_{53}2p_{z_N} + C_{54}(1s_{H_1} + 1s_{H_2} + 1s_{H_3})$	$C_{53} > C_{52} > C_{54} \sim C_{51}$
CH_4	$\phi_{1a_1} = C_{11}1s_C + C_{12}2s_C + C_{13}(1s_{H_1} + 1s_{H_2} + 1s_{H_3} + 1s_{H_4})$	$C_{11} \gg C_{12} > C_{13}$
	$\phi_{2a_1} = C_{21}1s_C + C_{22}2s_C + C_{23}(1s_{H_1} + 1s_{H_2} + 1s_{H_3} + 1s_{H_4})$	$C_{22} > C_{21} \sim C_{23}$
	$\phi_{1t_2} = \begin{bmatrix} C_{31}2p_z + C_{32}(1s_{H_1} - 1s_{H_2} - 1s_{H_3} + 1s_{H_4}) \\ C_{41}2p_x + C_{42}(1s_{H_1} + 1s_{H_2} - 1s_{H_3} - 1s_{H_4}) \\ C_{51}2p_y + C_{52}(1s_{H_1} - 1s_{H_2} + 1s_{H_3} - 1s_{H_4}) \end{bmatrix}$	$\begin{matrix} C_{31} > C_{32} \\ C_{41} > C_{42} \\ C_{51} > C_{52} \end{matrix}$

^{a} The MOs are listed in order of decreasing binding energy.
^{b} The symbol > indicates approximately one order of magnitude difference between the size of the coefficients it separates. Coefficients separated by $\sim$ symbols are listed in order of decreasing size. The relative sizes of the coefficients were determined from the *ab initio* wavefunctions of Ref. 12.

relative sizes of the AO coefficients are listed in the right column of Table 4.2. The lowest energy MO is the symmetric $\phi_{1\sigma}$ or ϕ_{1a}. It consists of predominantly central-atom $1s$ AO character and is commonly called a core orbital because of its atomic-like character. The second orbital, considered a valence orbital, is the symmetric $\phi_{2\sigma}$ or ϕ_{2a_1}. This MO consists of mainly central atom $2s$ AO character with a small admixture of other orbitals arranged in a bonding manner. The next three MOs are the strongly bonding or nonbonding combinations. The bonding combinations consist of predominantly central atom $2p$ and hydrogen $1s$ character, for example 3σ of HF, $1b_2$ and $3a_1$ of H_2O, $1e$ of NH_3, and $1t_2$ of CH_4. The nonbonding orbitals consist of predominantly central atom $2p$ and $2s$ character, for example, 1π of HF, $1b_1$ of H_2O, and $3a_1$ of NH_3.

4.4. Construction of equivalent orbitals

Equivalent orbitals can be constructed as linear combinations of the MOs developed in the previous section. It is desired to determine EOs that are equivalent in the sense that they are interchangeable under the operations of the molecular symmetry group. One begins by finding the character system of the EOs under the operations of the group and then comparing the result to the character system of the MOs. The set of EOs will transform according to an irreducible representation or a reducible representation which can be expressed as a sum of MO representations. Usually one of the MOs in this sum is totally symmetric under all group operations, while the other is not totally symmetric and may be of the single or degenerate type. The set of EOs are chosen such as to localize the charge probability density of the bonding MOs in the spatial regions of molecular bonds and the nonbonding MOs in the spatial regions away from the molecular bonds.

Examples of the results of this procedure are provided in Table 4.3. In this table are listed the character system of the MOs and that of the EOs. The core MOs, 1σ and $1a_1$, are actually EOs χ^c, which are localized about the central atom; they are left out of the linear combination scheme. The character system of the "bonding" EOs χ^bs is found by observing the transformation properties of equivalent orbitals directed along the bonds. The character system of the "nonbonding" or "lone-pair" EOs χ^{lp} can be found by choosing linear combinations of the remaining unhybridized valence MOs. These combinations are chosen such that they are directed in symmetrical equivalent spatial positions away from the χ^bs. The choice of symmetrical MO to be used in the χ^bs or χ^{lp}s is determined by the form of the MO. For χ^bs one chooses that symmetrical MO that has most electron density in the spatial regions of the molecular bonds. For example, in CH_4 χ^b is clearly a superposition of ϕ_{2a_1} and ϕ_{1t_2}. In NH_3 χ^b is the sum of ϕ_{2a_1} and ϕ_{1e} and χ^{lp} is simply ϕ_{3a_1}. In H_2O χ^b is a superposition of ϕ_{3a_1} and ϕ_{1b_1} and χ^{lp} is a superposition of ϕ_{2a_1} and ϕ_{1b_1}. In HF, where there is only one bond, there can be only one χ^b, that is, $\phi_{3\sigma}$. The other valence MOs of HF can be combined to form three χ^{lp}s, which are a superposition of $\phi_{2\sigma}$ and $\phi_{1\pi}$.

We now know which MOs to combine to form EOs. It is desired to determine the coefficients in these linear combinations in order that the EOs be orthonormal with the bonding orbitals directed towards the hydrogen atoms and the "lone-pair" orbitals directed away from the hydrogen atoms. Since we require that the χ^bs point along the bond directions, we find the direction cosines of each bond direction l_{x_i}, l_{y_i}, and l_{z_i}. We then find the constants α and β for the expression

$$\chi_i = \alpha\phi_s + \beta(l_{x_i}\phi_x + l_{y_i}\phi_y + l_{z_i}\phi_z) \tag{4.1}$$

Table 4.3. Types of MOs[a] and EOs[b] and Their Character Systems for the First-Row 10-Electron Hydrides

HF($C_{\infty v}$)	E	$2C_\infty^\phi$	$\infty \sigma_v$
ϕ_σ	1	1	1
ϕ_π	2	$2\cos\phi$	0
χ^c, χ^b	1	1	1
χ^{lp}	3	$1 + 2\cos\phi$	1

H$_2$O(C_{2v})	E	C_2	$\sigma_v(xz)$	$\sigma_v(yz)$
ϕ_{a_1}	1	1	1	1
ϕ_{b_1}	1	-1	1	-1
ϕ_{b_2}	1	-1	-1	1
χ^b	2	0	0	2
χ^{lp}	2	0	2	0
χ^c	1	1	1	1

NH$_3$(C_{3v})	E	$2C_3$	$3\sigma_v$
ϕ_{a_1}	1	1	1
ϕ_e	2	-1	0
χ^b	3	0	1
χ^c, χ^{lp}	1	1	1

CH$_4$(T_d)	E	$8C_3$	$3C_2$	$6S_4$	$6\sigma_d$
ϕ_{a_1}	1	1	1	1	1
ϕ_{t_2}	3	0	-1	-1	1
χ^b	4	1	0	0	2
χ^c	1	1	1	1	1

[a] MOs are labeled as in Table 4.2.
[b] EOs are labeled as χ^c, core orbital; χ^{lp}, "lone-pair" orbital; χ^b, "bond-pair" orbital.

where ϕ_s is a symmetric MO (predominantly s-type AO character) and ϕ_x, ϕ_y, and ϕ_z are either symmetric or nonsymmetric MOs (predominantly p-type AO character of the indicated direction).* We assume that the AOs are orthonormal. Take two directions with direction cosines l_{x_i}, l_{y_i}, l_{z_i} and l_{x_j}, l_{y_j}, l_{z_j}. In order to normalize the χ^b function of Eq. 4.1, we multiply by its complex conjugate, integrate over space, and use the

---

* The EOs obtained in this manner may not be directed exactly along the bond directions. This is a result of some "directed" AOs in the predominantly spherical ϕ_s MO and some "spherical" AOs in the predominantly directed MOs.

normalization and orthogonality of the ϕ functions. Using the properties of directional consines, $l_{x_i}^2 + l_{y_i}^2 + l_{z_i}^2 = 1$, we obtain

$$\alpha^2 + \beta^2 = 1 \tag{4.2}$$

To orthogonalize the functions of Eq. 4.1 we take two such functions having the same α and β and two different sets of direction cosines as above. Multiply one function by the complex conjugate of the other and integrate over space. Use the property $l_{x_i}l_{x_j} + l_{y_i}l_{y_j} + l_{z_i}l_{z_j} = \cos\theta$, where θ is the angle between the two directions. Thus we find that the orthonormality of the two functions requires that

$$\alpha^2 + \beta^2 \cos\theta = 0$$

or

$$\frac{\alpha^2}{\beta^2} = -\cos\theta \tag{4.3}$$

From Eq. 4.3, α^2/β^2 can only be positive and hence α/β real. From this relation we see that the contribution of the symmetric MO ϕ_s gradually increases as θ increases from 90°, where it is zero, to 180°, where it equals the contribution of the directed MOs.

Values of direction cosines and the constants α and β determined in this manner for the χ^bs of the first-row hydrides are presented in Table 4.4. The coefficients of the χ^{lp}s are found in a similar manner. Finally, the orthonormal χ_is are presented in Table 4.5. The orbital χ^c in Table 4.5 is mainly a central atom 1s AO, that is, a core orbital. Therefore, excluding χ^c, we obtain one χ^b and three χ^{lp}s for HF, two χ^bs and two χ^{lp}s for

Table 4.4. Values of the Constants α and β and the Direction Cosines l_i Corresponding to the χ^b Orbitals of the 10-Electron First-Row Hydrides

	α	β		l_{x_i}	l_{y_i}	l_{z_i}
HF (180°)	$1/2^{1/2}$	$1/2^{1/2}$		0	0	1
H_2O (104.5°)	0.447	0.895	$i=1$	0	0.791	0.612
			$i=2$	0	-0.791	0.612
NH_3 (107.4°)	0.480	0.877	$i=1$	0.927	0	0.376
			$i=2$	-0.463	0.803	0.376
			$i=3$	-0.463	-0.803	0.376
CH_4 (109.5)	$1/2$	$3^{1/2}/2$	$i=1$	$1/3^{1/2}$	$1/3^{1/2}$	$1/3^{1/2}$
			$i=2$	$1/3^{1/2}$	$-1/3^{1/2}$	$-1/3^{1/2}$
			$i=3$	$-1/3^{1/2}$	$1/3^{1/2}$	$-1/3^{1/2}$
			$i=4$	$-1/3^{1/2}$	$-1/3^{1/2}$	$1/3^{1/2}$

Table 4.5.   EOs of the 10-Electron First-Row Hydrides

HF^a	$\chi^c = \phi_{1\sigma}$
	$\chi^b = \phi_{3\sigma}$
	$\chi_1^{lp} = 0.582\phi_{2\sigma} + 0.813\phi_{1\pi_x}$
	$\chi_2^{lp} = 0.582\phi_{2\sigma} - 0.406\phi_{1\pi_x} + 0.704\phi_{1\pi_y}$
	$\chi_3^{lp} = 0.582\phi_{2\sigma} - 0.406\phi_{1\pi_x} - 0.704\phi_{1\pi_y}$
H_2O	$\chi^c = \phi_{1a_1}$
	$\chi_1^b = (1/2^{1/2})(\phi_{3a_1} + \phi_{1b_2})$
	$\chi_2^b = (1/2^{1/2})(\phi_{3a_1} - \phi_{1b_2})$
	$\chi_1^{lp} = (1/2^{1/2})(\phi_{2a_1} + \phi_{1b_1})$
	$\chi_2^{lp} = (1/2^{1/2})(\phi_{2a_1} - \phi_{1b_1})$
NH_3	$\chi^c = \phi_{1a_1}$
	$\chi_1^b = 0.582\phi_{2a_1} + 0.813\phi_{1e_x}$
	$\chi_2^b = 0.582\phi_{2a_1} - 0.406\phi_{1e_x} + 0.704\phi_{1e_y}$
	$\chi_3^b = 0.582\phi_{2a_1} - 0.406\phi_{1e_x} - 0.704\phi_{1e_y}$
	$\chi^{lp} = \phi_{3a_1}$
CH_4	$\chi^c = \phi_{1a_1}$
	$\chi_1^b = 1/2(\phi_{2a_1} + \phi_{1t_{2x}} + \phi_{1t_{2y}} + \phi_{1t_{2z}})$
	$\chi_2^b = 1/2(\phi_{2a_1} + \phi_{1t_{2x}} - \phi_{1t_{2y}} - \phi_{1t_{2z}})$
	$\chi_3^b = 1/2(\phi_{2a_1} - \phi_{1t_{2x}} + \phi_{1t_{2y}} - \phi_{1t_{2z}})$
	$\chi_4^b = 1/2(\phi_{2a_1} - \phi_{1t_{2x}} - \phi_{1t_{2y}} + \phi_{1t_{2z}})$

^{a} The coefficients of χ_1^{lp}, χ_2^{lp} and χ_3^{lp} are chosen such that the "lonepairs" have angles of $\sim107°$ between them.

H_2O, three χ^bs and one χ^{lp} for NH_3, and four χ^bs for CH_4. An example of MOs and EOs constructed from them is presented in Fig. 4.1 for H_2O.

4.5.   Applications of molecular orbitals and equivalent orbitals

The MOs and EOs of the 10-electron first-row hydrides have now been constructed. In this section the possibilities of using these orbitals for an examination of molecular properties is considered. The molecular proper-ties are arranged into two groups: (1) *Invariant properties* are those that can be expressed equally well in terms of either type of orbital. (2) *Variant properties* are those that can be expressed best in terms of one or the other type orbital. These terms are chosen to represent the molecular properties because of the nature of the determinantal function represent-ing an approximate wavefunction of the molecule. MOs and EOs can be converted into one another by means of a unitary transformation. The value of the total determinantal function is unaltered by such a unitary

MOLECULAR   ORBITALS   OF H_2O

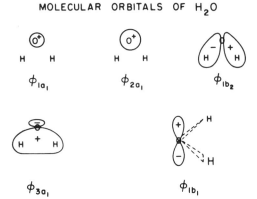

EQUIVALENT   ORBITALS   OF H_2O

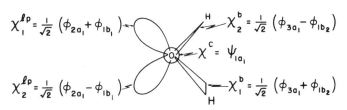

Fig. 4.1. Pictorial representation of the occupied MOs of H_2O and the EOs constructed from them.

transformation between the functions (orbitals); therefore, any properties of the molecule as a whole that can be deduced from the orbitals are expected to be *invariants* of the transformation.

INVARIANT PROPERTIES. Since the total determinantal wavefunction ψ is invariant, the total electron distribution function $\rho = \psi^*\psi$ and the various matrix elements of ρ are also invariant. Any quantities derived from these functions will also be invariants, for example, the joint probability densities of any number of electrons. A unitary transformation between MOs and EOs leaves the trace of the matrix unchanged. The physical significance of the invariance of the trace implies that the Coulomb potential at any point in space due to all electrons, the total electron kinetic energy, the total electronic potential energy, and the character of the set of occupied orbitals are invariant. These properties can be treated equally well using MOs or EOs.

VARIANT PROPERTIES. These are the properties pertaining to only selected portions of the total electrons. The terms in the molecular Hamiltonian are all invariant under an orthonormal transformation of orbitals. The problem of noninvariance arises from the physical interpretations of these terms. For example, the best computational scheme available at present is a self-consistent-field calculation using antisymmetrized spin-orbitals; this is called a *Hartree–Fock SCF* calculation.[12a] The total Hartree–Fock energy of a molecule with N electrons is

$$E = 2 \sum_{i=1}^{N} \varepsilon_i^0 + \sum_{i,j}^{N} (2J_{ij} - K_{ij}) \qquad (4.4)$$

and the Hartree–Fock eigenvalue for a particular eigenfunction is

$$\varepsilon_i = \varepsilon_i^0 + \sum_j (2J_{ij} - K_{ij}) \qquad (4.5)$$

In these equations, ε_i^0 represents a zeroth-order eigenvalue of the system, that is, the energy predicted by the independent-particle model, J_{ij} is a coulomb integral, and K_{ij} is an exchange integral. These integrals are defined as:

$$J_{ij} = \langle \phi_\mu(i)\phi_\nu(j) | \mathcal{H} | \phi_\mu(i)\phi_\nu(j) \rangle$$
$$K_{ij} = \langle \phi_\mu(i)\phi_\nu(j) | \mathcal{H} | \phi_\mu(j)\phi_\nu(i) \rangle \qquad (4.5a)$$

The J_{ij} represents electronic repulsion between pairs of electrons i and j on atoms μ and ν and the K_{ij} has no classical interpretation since it arises as a consequence of the antisymmetry principle. In deriving Eq. 4.4, one must consider the electronic interactions, or $1/r_{ij}$ terms. There are two types of electronic interaction or $1/r_{ij}$ terms, both of which are invariant under an orthonormal transformation of orbitals. Both of these terms contain a common part[7] that is cancelled to obtain J_{ij} and K_{ij} in the derivation of Eq. 4.4. The resulting J_{ij} and K_{ij} are not invariant when this common part has been taken away.

The noninvariance of the Hartree–Fock equations has important implications for electron energies calculated from them. In Eq. 4.5, ε_i^0 represents the average kinetic energy of the electron in spin-orbital ϕ_i and its average potential energy in the field of the nuclei, J_{ij} its average potential energy in the field of all electrons, and K_{ij} its exchange charge density. These energy terms are just those that would be missing if the electron with spin-orbital ϕ_i were removed from the system. Thus, the Hartree–Fock eigenvalue ε_i corresponds to the ionization potential of electron i. This is simply an extension of Koopmans' theorem. Koopmans proved[13] that if *ψ(2N) is a stationary state and ϕ_i is an eigenfunction of the*

Hartree–Fock operator, then ψ(2N − 1) is also a stationary state with respect to further variations in the orbital ϕ_i. Therefore, $E(2N − 1) − E(2N) = −\varepsilon_i$ is the ionization energy of orbital ϕ_i.* In order for ϕ_i to be an eigenfunction of the Hartree–Fock operator, it must have the transformation properties of the IRs of the symmetry group. Molecular orbitals always meet this restriction, but EOs in general do not. Since MOs are always eigenfunctions of the Hartree–Fock operator, the energy parameters obtained from the calculations represent stationary states. Equivalent orbitals are usually not eigenfunctions, hence they do not represent stationary states and consequently do not correspond to physical observables.

PE spectra of the 10-electron hydrides

An informative example of the use of MOs or EOs in determining electron energy levels is provided by the PE spectra of the 10-electron hydrides (Fig. 4.2). Each of the bands, or band envelopes, observed in these spectra is due to ejection of electrons from an electronic energy level. These bands can readily be interpreted in terms of the MOs of Table 4.2 as follows.

Solutions of the wave equation arising from progressive subdivision of the positive nuclear charge of Ne pass continuously into one another with a gradual splitting of degeneracy as the internal molecular fields are changed. It is, therefore, expected that the orbitals of the hydrides can be continuously developed from those of Ne. First, it is clear that the $1s$ orbital of Ne passes smoothly into the 1σ and $1a_1$ orbitals of the hydrides (Fig. 4.2). These are called core orbitals because they are highly localized about the heavy atomic nucleus. The progressive lowering of this E_I reflects the decreasing electronegativity along the series F, O, N, C. Similar behavior is exhibited by the $2s$ orbital of Ne as it passes smoothly to the 2σ and $2a_1$ orbitals of the hydrides. The 2σ and $2a_1$ bands and the remaining hydride bands at lower energy are considerably broadened from their atomic counterparts in Ne due to ionization into excited vibrational levels of the molecular ions. Next, the triple degeneracy of the p^6 shell of Ne is revealed by the $^2P_{3/2}$, $^2P_{1/2}$ lines of Ne$^+$. In forming HF, the $^2P_{1/2}$ and one of the degrees of freedom of the $^2P_{3/2}$ level are used in forming the two components of the 1π level, and the remaining degree of freedom is used in forming the 3σ level. In H_2O the degeneracy is completely split, and three band envelopes appear. In forming NH_3 from H_2O, the $1b_1$ (H_2O) orbital becomes the $3a_1$ (NH_3)

---

* Koopmans' theorem gives only approximate E_Is for it neglects electronic relaxation and correlation upon ionization (see Section 4.7).

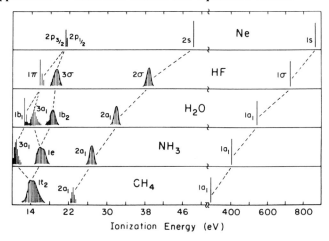

Fig. 4.2. Schematic photoelectron spectra of the 10-electron hydrides and neon.

orbital, and the $3a_1$ (H_2O) and $1b_2$(H_2O) orbitals combine to form the doubly degenerate $1e$ (NH_3) orbital. The $1e$ level assumes an E_I intermediate between those of the $3a_1$ and $1b_2$ levels of H_2O. Further mixing of the $3a_1$ and $1e$ orbitals of NH_3 leads to the triply degenerate $1t_2$ orbital of CH_4. It assumes an energy near the weighted mean of the energies of the species from which it has been derived. Vibrational frequencies of the molecular ions[14-16] and Jahn–Teller effects[15-16] in the degenerate bands are prominent in these spectra.

It is instructive to attempt an interpretation of these photoelectron spectra in terms of EOs. The equivalency of certain sets of EOs (Table 4.5) suggests that the photoelectron spectra of HF, H_2O, and NH_3 consist of only *three* bands, while that of CH_4 consists of only *two* bands, an obviously incorrect prediction. *The inconsistency arises in the assumption that EOs have eigenvalues corresponding to energy levels and that electrons in an equivalent set of orbitals all have the same energies.* The fact is that even though electrons occupy orbitals that are equivalent with respect to symmetry operations and equivalent in their spatial projections, the electrons do not necessarily have equivalent energies. From this example we see that the six electrons occupying the three χ^{lp}s of HF are not energetically equivalent; they consist of a degenerate pair at 16.0 eV (1π) and a nondegenerate pair at 39.0 eV (2σ). The two χ^{lp}s and two χ^{b}s of H_2O do not each contain two pairs of degenerate electrons; instead, the χ^{lp}s contain the $1b_1$ (12.6 eV) and $2a_1$ (32.2 eV) electrons, and the χ^{b}s contain the $3a_1$ (14.5 eV) and $1b_2$ (17.5 eV) electrons. The three χ^{b}s of NH_3 contain the degenerate $1e$ (14.7 eV) and nondegenerate $2a_1$

(27.0 eV) electrons. The four χ^bs of CH_4 do not contain eight energetically equivalent electrons; they contain the triply degenerate $1t_2$ (12.6 eV) and nondegenerate $2a_1$ (22.9 eV) electrons.

From the preceding example it is obvious that *EOs in general lead to totally incorrect conclusions regarding one-electron orbital energies*. It is necessary to use molecular or symmetry orbitals to obtain a realistic model for these orbital energies.

Equivalent orbitals do have useful qualities. For example, the respective values of coulomb J_{ij} and exchange K_{ij} contributions of Eq. 4.4 depend on the form of the orbitals used. The exchange part is small when EOs are used because the orbitals are localized in different spatial regions. Since the coulomb part gives the major contribution to the energy, the interactions of electrons may be interpreted as due largely to the repulsion of charges between EOs. It can be illuminating to use such a model to understand localized electron distributions and molecular geometries. This is the origin of Valence-Shell-Electron-Pair-Repulsion Theory[17] in which "lone-pair" and "bond-pair" interactions are used to estimate molecular geometries.

A useful feature of localized EOs is their approximate transferability from molecule to molecule. For example, having calculated the localized EO for the C—H bond in methane, we expect to find this localized CH orbital to be about the same in any hydrocarbon. Transferable EOs can be determined for various types of localized bonds. In this manner, EOs reconcile MO theory with the chemist's intuitive picture of chemical bonding.

The concept of equivalent orbitals should be used with caution for it has shortcomings: (1) The choice of EOs used to describe a system is not unique, for there are generally more than one equally valid set of EOs that can be formed. (2) Equivalent orbitals cannot be used to describe states with open-shell configurations (unpaired electrons) because the orbital with the vacancy does not satisfy the orthonormality properties described in Section 4.4.

Conclusions

From the above analysis we conclude that both MOs and EOs are useful in studying electronic structures; however, they should be used appropriately. Either type of orbital can be used to describe molecular properties that are invariant to unitary transformations between the orbitals. Such properties involve only the ground state wavefunction of a closed-shell molecule, for example, total electron probability density, dipole moment, geometry, heat of formation, and so forth. In describing properties that are noninvariant to such transformations, it is important to use MOs and EOs selectively. *Molecular orbitals* can be used to describe those properties that are either invariant or variant to unitary transformations between the orbitals. They are usually employed to describe those properties that involve the molecule as a whole entity and are described by both ground- and excited-state wavefunctions. Examples are spectroscopic properties involving stationary states and electronic transitions involving excitation or ejection of electrons. *Equivalent orbitals* should be used to describe localized properties of molecules, for example, the properties pertaining to selected bonds or localized spatial regions such as charge distributions in particular bonds, the spatial arrangement of one localized bond relative to another, or electronic repulsions between two moieties.

A manifestation of the uncertainty principle arises from a consideration of these two types of orbitals. Whereas MOs provide eigenvalues that are useful approximations for electron energy levels, they allow the electrons to be delocalized over the whole molecule. Equivalent orbitals can be interpreted in terms of the spatial position of electrons (chemical bonds and "lone-pairs"), but at the loss of any meaningful energy values.

4.6. Koopmans' theorem

A molecular orbital calculation that provides accurate reliable values for orbital energies and the correct orbital ordering is invaluable as a guide for interpretation of PE spectra. Conversely, experimental E_Is provide a unique test for theoretical procedures and the reliability of calculations.

The most rigorous approach for calculating ionization energies within the Hartree–Fock approximation is the *direct method* by which separate calculations are performed on the ground state of the neutral molecule and the various states of the molecular ion. Since the ionization energies correspond to transitions from the molecular ground state to the ionic states, differences in the calculated total energies of the states should correspond to the experimental E_Is. This procedure requires inclusion of correlation effects, for these effects are usually smaller in the ion than in the molecule; calculated E_Is are usually too small when correlation energies are neglected. Such calculations normally require an open-shell treatment[18] for the ion radical. Open-shell iterative calculations have not been very popular, for they often diverge or converge very slowly.[19] The adiabatic E_I can be calculated by varying the geometry of the ion to achieve an energy minimum.[20–22]

Most of the photoelectron investigations to date have employed calculations on only the ground state of the neutral molecule. Such calculations are used by applying *Koopmans' theorem,*[13] *which states that the molecular E_is are equal to the negatives of the MO eigenvalues.* This approximation has been widely employed by PE spectroscopists and deserves detailed consideration in order that its limitations and virtues be made explicit.

The applications of Koopmans' theorem[13] have been discussed by Richards.[23] It is useful to follow his development of Koopmans' theorem for the general case. Consider a closed-shell molecule with n doubly occupied molecular orbitals ϕ_i. The determinantial wavefunction for the molecular ground state is

$$\psi = |\phi_1^2 \cdots \phi_{n-1}^2 \phi_n^2| \tag{4.6}$$

The total energy of the system E, as determined by *ab initio* SCF calculations,[10,11] is expressed in Eq. 4.4. The E is therefore a sum of three different terms: (1) ε_i^0, the energy each electron would have alone in the nuclear framework, its kinetic energy plus potential energy of nuclear attraction; (2) J_{ij}, a coulomb repulsion integral between every pair of electrons; (3) K_{ij} an exchange interaction between every pair of electrons of the same spin.

The expression for the energy of the ion formed by removing an electron from the orbital ϕ_n is similar to Eq. 4.4, with the exception of missing terms in the summations due to the absence of electron n. If we assume that the quantities ε_i^0, J_{ij}, and K_{ij} are the same in the molecule and ion (frozen core approximation), then the energy difference between the molecule and ion, hence the ionization energy, is

$$\varepsilon_n^0 + \sum_{i=1}^{n-1} (2J_{ni} - K_{ni}) + J_{nn} \tag{4.7}$$

The orbital energy $\varepsilon_n^{\text{SCF}}$ is determined by solution of the SCF equations, that is

$$\left[\mathcal{H}^{\text{N}} + \sum_{j=1}^{n} (2\mathcal{J}_j - \mathcal{K}_j) \right] \phi_n = \varepsilon_n^{\text{SCF}} \phi_n \tag{4.8}$$

From this equation, $\varepsilon_n^{\text{SCF}}$ can be expressed as a sum of integrals,

$$\varepsilon_n^{\text{SCF}} = \varepsilon_n^0 + \sum_{j=1}^{n-1} (2J_{nj} - K_{nj}) + J_{nn} \tag{4.9}$$

this sum being identical to those for the ionization energy expressed in Eq. 4.7. This is a general proof of Koopmans' theorem, which states that *the orbital energy $\varepsilon_n^{\text{SCF}}$ for a closed-shell molecule computed in an ab initio SCF calculation is exactly equal to the ionization energy of an electron from that orbital in the frozen core approximation.*

As an example of the application of the above equations, consider the LiH molecule, for which the ground-state Hartree–Fock wave function is of the form

$$\psi_{HF} = 1\sigma\alpha \; 1\sigma\beta \; 2\sigma\alpha \; 2\sigma\beta \tag{4.10}$$

According to Eq. 4.9, the 1σ orbital energy is

$$\varepsilon_{1\sigma} = \langle 1\sigma| \mathcal{H} |1\sigma\rangle + \langle 1\sigma(1)1\sigma(2)| \mathcal{H} |1\sigma(1)1\sigma(2)\rangle$$
$$+ 2\langle 1\sigma(1)2\sigma(2)| \mathcal{H} |1\sigma(1)2\sigma(2)\rangle - \langle 1\sigma(1)2\sigma(2)| \mathcal{H} |1\sigma(2)2\sigma(1)\rangle \tag{4.11}$$

From Eq. 4.4, the total HF energy for LiH is

$$E_{LiH} = 2\langle 1\sigma(1)| \mathcal{H} |1\sigma(1)\rangle + 2\langle 2\sigma(2)| \mathcal{H} |2\sigma(2)\rangle$$
$$+ \langle 1\sigma(1)1\sigma(2)| \mathcal{H} |1\sigma(1)1\sigma(2)\rangle + 4\langle 1\sigma(1)2\sigma(2)| \mathcal{H} |1\sigma(1)2\sigma(2)\rangle$$
$$- 2\langle 1\sigma(1)2\sigma(2)| \mathcal{H} |1\sigma(2)2\sigma(1)\rangle$$
$$+ \langle 2\sigma(1)2\sigma(2)| \mathcal{H} |2\sigma(1)2\sigma(2)\rangle + \frac{3}{R_{LiH}} \tag{4.12}$$

For the positive ion LiH^+ in which one electron has been removed, the HF wave function takes the form

$$E_{LiH^+} = \langle 1\sigma(1)| \mathcal{H} |1\sigma(1)\rangle + 2\langle 2\sigma(2)| \mathcal{H} |2\sigma(2)\rangle$$
$$+ 2\langle 1\sigma(1)2\sigma(2)| \mathcal{H} |1\sigma(1)2\sigma(2)\rangle - \langle 1\sigma(1)2\sigma(2)| \mathcal{H} |1\sigma(2)2\sigma(1)\rangle$$
$$+ \langle 2\sigma(1)2\sigma(2)| \mathcal{H} |2\sigma(1)2\sigma(2)\rangle + \frac{3}{R_{LiH}} \tag{4.13}$$

We can see by inspection that the energy difference $E_{LiH} - E_{LiH^+}$, the ionization potential, is just the 1σ orbital energy given in Eq. 4.11. One of the most important features of Koopmans' theorem is that the ionization energy it refers to is that obtained from a calculation on the positive ion using orbitals variationally determined for the neutral molecule.

When applying Koopmans' theorem to general molecular photoionization phenomena, there are three additional approximations that are necessary. These assumptions are:[23]

1. The Reorientation or Frozen Core Approximation—The general proof of Koopmans' theorem uses the assumption that all the orbitals, $\phi_1 \cdots \phi_{n-1}$, are unaltered when going from molecule to ion. In order for the integrals of Eqs. 4.7 and 4.9 to be identical, it is essential that the orbitals of the ion be identical to those of the molecule, that is, there is *no reorientation* of orbitals upon ionization.

2. The Relativistic Energy Approximation—Since Hartree–Fock theory does not consider relativistic effects, the use of Koopmans' theorem is

equivalent to assuming that the relativistic energy is the same in both the molecule and ion, and the E_I is given by a difference in electrostatic energy expressions. According to the virial theorem, inner electrons have large kinetic energies. The magnitude of the relativistic effects increases as does the electron kinetic energies. The magnitude of the relativistic contributions of various subshells to the total energy of argon has been given by Hartman and Clementi:[24] *1s*, 1.2256 a.u.; *2s*, 0.2353 a.u.; *2p*, 0.2574 a.u.; *3s*, 0.0255 a.u.; *3p*, 0.0224 a.u. (1 a.u. = 27.21 eV).

3. The Correlation Energy Approximation—Electrons tend to keep apart, that is, their motions are correlated. Koopmans' theorem takes the difference between two energy expressions, neither of which include correlation effects. The assumption therefore, is, made that the correlation energy is the same in both the molecule and ion. Correlation effects arise largely from pair interactions between electrons, and since the ion has less electrons than the parent molecule, this correlation energy will certainly be different and generally less in the ion than its parent molecule. Generally, separate Hartree–Fock calculations on the molecule and ion yield total energy differences that are smaller than the E_Is due to this neglect of correlation effects (Fig. 4.3). An interesting example is that of molecular nitrogen, where an extensive *ab initio* LCAO-MO SCF study[25] of N_2 and N_2^+ failed to predict the correct ionic ground state. The failure is presumably due to the neglect of correlation terms.

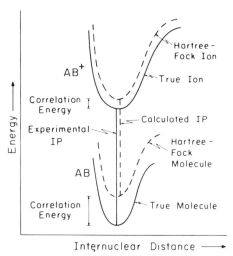

Fig. 4.3. Molecular and ionic potential energy and Hartree-Fock potential energy curves. The correlation energies and experimental and calculated E_Is are indicated. Adapted with permission from Ref. 23.

Koopmans' theorem used in its simple form, that is, taking the MO eigenvalue as the E_I, often overestimates the E_Is. This is obvious if one considers the relevant structures of the molecule $(\cdots \phi_n^2)$ and ion $(\cdots \phi_n^1)$; the orbital energy ε_n^{SCF} of ϕ_n in the molecular calculation where there are two ϕ_n electrons is invariably less than in the ionic calculation where there is only one electron.[25] For a simple example consider H_2 where reorientation and relativistic effects should be minimal. The total energy of the ion and molecule is

$$E(H_2^+) = \varepsilon_{1\sigma_g}^0 \quad \text{and} \quad E(H_2) = 2\varepsilon_{1\sigma_g}^0 + J_{1\sigma_g, 1\sigma_g} \quad (4.14)$$

Using Koopmans' theorem the orbital energy is

$$\varepsilon_{1\sigma_g}^{SCF} = \varepsilon_{1\sigma_g}^0 + J_{1\sigma_g, 1\sigma_g} = -E_I \quad (4.15)$$

The calculated vertical E_I using Eq. 4.15 is 16.11 eV, while the observed value is 15.88 eV. Orbital eigenvalues obtained from *ab initio* LCAO-MO SCF calculations using minimal basis set Slater-type orbitals are compared to experimental E_Is of some selected small molecules in Table 4.6.

4.7. Methods of calculating ionization energies

It is of considerable importance to have a theoretical method of calculating E_Is that will provide reliable guidance for interpreting photoelectron spectra. The method should ideally satisfy four criteria:

1. It should predict the correct order of E_Is and approximate spacings of these E_Is for comparison to PE spectral bands.
2. It must include all valence electrons for all atoms in the molecule. Preferably, it should include both valence and core electrons.
3. The calculations must be easy to perform and not require too much computer time.
4. The calculation should be versatile; it should be usable for molecules containing several atoms from different rows of the periodic table.

In this section we discuss the various methods of calculation that have been employed for interpretation of PE spectra and provide some typical examples.

Ab Initio SCF Methods

The general theory of *ab initio* calculations for closed-shell molecules was derived by Roothaan[26] and by Hall[27] in 1951. With the advances in

Table 4.6. Comparison of Experimental E_Is and Koopmans' Theorem E_Is Obtained from *ab initio*, CNDO, INDO, and Extended Huckel Calculations

Molecule	Orbital	Experimental[a] E_Is (eV)	Eigenvalues $(-\epsilon_i)$ ab initio	CNDO[b]	INDO[b]	Extended Huckel[b]
H_2	$1\sigma_g$	15.4 (A) 15.88 (V)	16.11[c]	20.9	20.9	15.1
HF	1π	16.06	12.65[d]	21.3	20.1	14.7
	3σ	18.6 (A) 19.5 (V)	15.39[d]	23.1	22.3	18.8
	2σ	39.0 (V)	40.19[d]	45.5	45.3	38.7
	1σ	686	711.52[d]	—	—	—
N_2	$3\sigma_g$	15.5	14.82[d]	18.3	16.0	13.5
	$1\pi_u$	16.7 (A) 16.9 (V)	15.77[d]	20.4	20.2	17.0
	$2\sigma_u$	18.8	19.89[d]	25.2	22.8	20.3
	$2\sigma_g$	37.3 (V)	39.53[d]	43.3	44.7	31.1
	$1\sigma_u, 1\sigma_g$	410	427.95[d]	—	—	—
CO	5σ	14.0	13.09[d]	17.3	15.2	13.4
	1π	16.5 (A) 16.9 (V)	15.88[d]	21.1	20.8	17.1
	4σ	19.7	19.93[d]	24.7	22.8	19.3
	3σ	38.3 (V)	40.80[d]	45.3	46.4	33.2
	2σ	296	309.03[d]	—	—	—
	1σ	542	563.62[d]	—	—	—
CH_4	$1t_2$	12.6 (A) 14.0 (V)	14.75[e]	19.7	19.5	13.5
	$2a_1$	23.0 (V)	25.37[e]	34.5.	35.1	21.9
	$1a_1$	290	206.80[e]	—	—	—
NH_3	$3a_1$	10.95 (A) 10.88 (V)	9.97[e]	16.3	14.2	11.1
	$1e$	14.98 (A) 16.00 (V)	15.85[e]	20.2	19.9	14.8
	$2a_1$	27.0 (V)	29.97[e]	37.0	37.7	25.3
	$1a_1$	406	422.54[e]	—	—	—
H_2O	$1b_1$	12.6	10.96[f]	17.8	16.3	11.2
	$3a_1$	13.8 (A) 14.7 (V)	12.69[f]	19.4	17.6	12.1

Table 4.6. (*continued*)

Molecule	Orbital	Experimental[a] E_is (eV)	Eigenvalues $(-\epsilon_i)$			
			ab initio	CNDO[b]	INDO[b]	Extended Huckel[b]
H_2O (*continued*)	$1b_2$	17.2 (A) 18.5 (V)	16.99^f	21.4	21.2	15.3
	$2a_1$	32.2 (V)	34.98^f	40.4	40.9	30.2
	$1a_1$	540	559.53^f	—	—	—
H_2CCH_2	$1b_{2u}$	10.51	10.10^e	15.9	15.4	12.5
	$1b_{2g}$	12.5 (A) 12.9 (V)	13.78^e	16.0	15.8	12.0
	$3a_g$	14.4 (A) 14.7 (V)	15.29^e	19.0	18.5	12.9
	$1b_{3u}$	15.7 (A) 15.9 (V)	17.52^e	25.3	25.0	15.3
	$2a_u$	18.8 (A) 19.1 (V)	21.29^e	27.5	27.7	17.8
	$2a_g$	23.4 (V)	27.61^e	39.2	39.9	24.2
	$1a_u, 1a_g$	291	307.25^e	—	—	—
H_2S	$2b_1$	10.5	9.42^g	13.4	—	10.8
	$5a_1$	12.8 (A) 13.3 (V)	12.68^g	16.0	—	13.0
	$2b_2$	14.8 (A) 15.5 (V)	15.27^g	17.4	—	14.4
	$4a_1$	22.2 (V)	25.45^g	27.2	—	23.1
	$1b_1, 3a_1, 1b_2$	170.2; 171.5 171.5	175.49^g	—	—	—
	$2a_1$	235	238.86^g	—	—	—
	$1a_1$	—	2501.57^g	—	—	—

[a] A = adiabatic IP; V = vertical IP; no letter indicates that A = V.
[b] J. W. Rabalais, unpublished results.
[c] L. C. Snyder and H. Basch, *Molecular Wave Functions and Properties*, Wiley, New York, 1972.
[d] B. J. Ransil, *Rev. Mod. Phys.*, **32**, 245 (1960).
[e] W. E. Palk and W. N. Lipscomb, *J. Amer. Chem. Soc.*, **88**, 2384 (1966).
[f] S. Aung, R. M. Pitzer, and S. I. Chan. *J. Chem. Phys.*, **49**, 2071 (1968).
[g] F. P. Boer and W. N. Lipscomb, *J. Chem. Phys.*, **50**, 989 (1969).

computer technology these *ab initio* calculations for polyatomic molecules are becoming increasingly more common.[28,29] The term *ab initio is usually reserved for those calculations in which the wavefunction is determined by minimizing the total energy of the system and all integrals are evaluated rather than approximated or neglected.* Most *ab initio* calculations used in conjunction with PE spectra are symmetry- and spin-restricted in that the molecular orbitals are constrained to transform as irreducible representations of the molecular point group, and the molecular orbitals occupied by α and β electrons are constrained to have the same functional form. Orbital- and spin-unrestricted calculations[30] are much less common than the restricted versions, although substantial improvements in calculated E_Is are obtainable by means of these unrestricted calculations.

Using restricted *ab initio* calculations, the accuracy is usually determined by the proximity to the Hartree–Fock limit. This accuracy is normally determined by the choice of basis functions. The most common type of basis functions are the Slater-type orbitals (STOs),[31] which have the general form

$$\chi = Nr^{n-1} \exp\left(-\zeta r\right) \qquad (4.16)$$

These functions are nodeless; in order to reproduce the nodal properties of hydrogenic orbitals of the type *2s, 3p, 5d,* and so forth, it is necessary to take linear combinations of these STOs. Combinations involving one, two, or three STOs are commonly called single-, double-, or triple-zeta functions.[32,33] Clementi has determined the best STO combinations which approximate exact Hartree–Fock functions for atoms.[34] The major difficulty in using STOs is that electron repulsion and exchange integrals involving different atomic centers in a molecule are difficult to evaluate and can be time consuming.

Gaussian-type functions (GTF) of the form

$$\chi = N'r^{n-1} \exp\left(-\zeta r^2\right) \qquad (4.17)$$

can also be used as basis functions. The advantage of GTFs is that evaluation of repulsion and exchange integrals is rather simple and requires much less time than STOs.[35] The disadvantage of GTFs is that they do not provide as satisfactory a physical representation of the basis wavefunctions as do STOs. The physical representation can be improved by expanding the best available Slater functions as a linear combination of gaussians using a least-squares fit[36] or to use a linear combination of gaussians whose exponents have been optimized.[37–39] The complete wavefunction is generated by multiplying the radial function by a spherical harmonic. An alternative approach is to simulate the spherical harmonic by using gaussion lobe functions[40-42] in which the origins of the gaussions are not at the nucleus but slightly removed from it.

The total size of the basis functions can be important in predicting molecular properties. *A minimum basis set is one which contains only functions representing the classical core and valence orbitals.* For example, such functions for CH_4 would include representations for the hydrogen *1s* and carbon *1s*, *2s*, and *2p* AOs. *An extended basis set includes representations for orbitals outside the valence shell.* For CH_4, such orbitals could be hydrogen *2s* and *2p* and carbon *3s*, *3p*, *3d*, and *4f*. Such extensions are mathematical devices used to improve the representation of the MOs; they can be of major importance in predicting electronic properties.

Ab initio calculations have been used rather heavily for the assignment of PE spectra.[43-47] Generally they are considered only to provide a rough ordering of ion state energies rather than detailed quantitative predictions of E_Is. *Ab initio* and experimental E_Is are compared in Table 4.6.

Semiempirical SCF methods

Several SCF-MO methods have been used to aid in interpretation of PE spectra. The most popular has been the CNDO (complete neglect of differential overlap) SCF-MO method developed by Pople and coworkers.[48-51] This method includes all valence electrons and incorporates the *1s* electrons of the first-row atoms in their cores. All integrals involving differential overlap including those between orbitals on the same center are neglected; however, the few remaining electron-electron repulsion integrals are evaluated theoretically. A later version of CNDO called CNDO/2 employs empirical atomic matrix elements and neglects certain penetration terms.[50] This method is capable of handling open-shell systems such as ion radicals. An outgrowth of the CNDO method is the INDO (intermediate neglect of differential overlap) SCF-MO method.[52] The primary difference between the INDO and CNDO methods is that INDO takes some account of exchange terms by retaining monatomic differential overlap in one-center integrals.

E_Is calculated by CNDO and INDO methods using Koopmans' theorem usually average about 3 eV higher than experimental E_Is.[53] Even allowing for this error, the scatter between calculated and experimental E_Is is quite large, rendering the methods unreliable for determining the orderings of orbitals that are energetically close. Cox et al.[53] have investigated these problems extensively and find that the main difficulty is in the calculation of the diagonal matrix elements. The manner in which these vary with the charge distribution in the molecule is extremely sensitive to the choice of electron repulsion integrals. Pople and Segal[50] evaluated these integrals theoretically using single STO functions. Whitehead[54] obtained the one-center integrals empirically from atomic data and used the empirical formula of Mataga[55] and Ohno[56] for the two-center integrals. Eigenvalues from CNDO and INDO calculations are listed in Table 4.6 for a variety of molecules. CNDO and INDO calculations have been used by most PE spectroscopists as a model for obtaining an approximate MO ordering.

Dewar and coworkers[57] have modified the INDO method, hence MINDO. The MINDO method differs from INDO in the following way: (1) The Slater–Condon parameters used in the one-center electron-interaction integrals are all evaluated empirically from atomic spectral data. (2) The two-center integrals are treated empirically rather than by direct quadrature. (3) The two-center resonance integrals H_{ij} are evaluated empirically and parameterized to give correct ground state energies for a number of reference molecules rather than to reproduce *ab initio* methods as should INDO. These parameterizations yield unexpectedly favorable agreement between MINDO E_Is and experimental E_Is. The method may be used for open-shell systems by employing the "half-electron" approach.[58]

Lindholm and coworkers[59] have parameterized the INDO method to obtain agreement with the ionization energies for some hydrocarbons. This method, called SPINDO (spectroscopic-potential-adjusted INDO), provides a distribution of orbital energies which usually corresponds closely to that of the photoelectron bands.

Hückel and extended Hückel methods

The Hückel molecular orbital (HMO) method[60] is patterned on quantum mechanics, but is no more than a useful empirical scheme that rationalizes chemical experience concerning π-electron molecules. This method has not been popular for assigning photoelectron bands because it allows estimates of only the E_Is of π orbitals and not those of σ orbitals. The *extended Hückel molecular orbital (EHMO) methods* as developed by Hoffmann[61] include all valence electrons and are of considerable use in assigning spectra. EHMO methods use the LCAO form for MOs

$$\phi_i = \sum_\mu c_{i\mu}\chi_\mu \tag{4.18}$$

with the MO energy given as

$$\varepsilon_i = \langle \phi_i | \mathcal{H} | \phi_i \rangle = \sum_\mu \sum_\nu c_{i\mu}^* c_{i\nu} \langle \chi_\mu | \mathcal{H} | \chi_\nu \rangle \tag{4.19}$$

The MOs are orthonormal, that is

$$\langle \phi_i \mid \phi_j \rangle = \sum_\mu \sum_\nu c_{i\mu}^* c_{j\nu} \langle \chi_\mu \mid \chi_\nu \rangle = \delta_{i,j} \qquad (4.20)$$

The secular determinant is obtained by minimizing the energy with respect to the AO coefficients,

$$\left| H_{\mu\nu} - \varepsilon_i S_{\mu\nu} \right| = 0 \qquad (4.21)$$

where

$$\begin{aligned} H_{\mu\nu} &= \langle \chi_\mu \mid H \mid \chi_\nu \rangle \\ S_{\mu\nu} &= \langle \chi_\mu \mid \chi_\nu \rangle \end{aligned} \qquad (4.22)$$

EHMO methods vary in the manner in which they evaluate the diagonal matrix elements $H_{\mu\mu}$, the off-diagonal matrix elements $H_{\mu\nu}$, and the overlap integrals $S_{\mu\nu}$. One of the popular EHMO methods is the Mulliken–Wolfsberg–Helmholtz (MWH) parameterization.[62] In this method the coulomb integral $H_{\mu\mu}$ is equated to the negative value of the valence-orbital-ionization-energy (VOIE) of the χ_μ AO. The resonance integral $H_{\mu\nu}$ can be evaluated by a number of different approximations,[62] some of which are the Wolfsberg–Helmholtz formula[63]

$$H_{\mu\nu} = \frac{k S_{\mu\nu} (H_{\mu\mu} + H_{\nu\nu})}{2} \qquad (4.23)$$

and the Cusachs formula[64]

$$H_{\mu\nu} = \frac{(2 - |S_{\mu\nu}|) S_{\mu\nu} (H_{\mu\mu} + H_{\nu\nu})}{2} \qquad (4.24)$$

The overlap integrals S_{ij} are evaluated analytically. The inner-shell or non-valence electrons are considered to be a nonpolarizable core; however, they need not be neglected if desired.

The empirically parameterized EHMO methods are simple, flexible, and can provide helpful models for assigning photoelectron bands. Orbital energies obtained from EHMO are generally too high by 1 to 3 eV. Some representative EHMO eigenvalues are listed in Table 4.6.

The X_α method

The recently developed SCF-X_α-SW (self-consistent-field-X_α-scattered wave) method[65] is based upon the arbitrary division of matter into component clusters of atoms. The cluster is further partitioned into three basic types of regions: (1) atomic—regions within nonoverlapping spheres surrounding each constituent atom; (2) interatomic—region between atomic spheres and an outer, usually spherical, boundary defining the entire cluster; and (3) extramolecular—the region outside of the cluster. The one-electron Schrödinger equation is numerically integrated within each region using the method of partial waves and using spherically averaged and volume averaged potentials. The difficult exchange integral problem is replaced by Slater's statistical X_α approximation. Wavefunctions and their first derivatives are joined continuously throughout via the multiple-scattered-wave formalism.[66]

The statistical total energy for a molecule or cluster goes to the proper separated atom limit as the system dissociates; this is not generally true for the Hartree–Fock energy. The SCF-X_α-SW method provides for nonintegral occupation numbers and satisfies Fermi statistics for systems with partly filled shells; it is referred to as the Hyper-Hartree–Fock Method in contrast to the ordinary Hartree–Fock approach, which is limited to single determinant closed-shell systems. The X_α calculation gives results that are in close agreement with experimental ionization energies and is becoming increasingly important[67] as an interpretive and predictive approach for PE spectroscopy.

Conclusions

From the above discussions it is obvious that different types of MO calculations provide different orbital energies and can even change the orderings of orbitals that are energetically close together. The calculations can resolve difficulties in the interpretations of PE spectra if they are performed carefully.

The best results are obtained by using *ab initio* SCF-MO calculations for the energy of the molecule and each separate state of the ion; these energies must be computed as close to the Hartree–Fock limit as possible, for the ε_n^{SCF}s have a definite value at this limit. The computed ionization energies should be accurate if calculations are performed in this rigorous manner and estimates of the differences in correlation energy and relativistic energy between the various states are made. The calculated potential curves and surfaces of the ionic states are frequently more realistic than the absolute energies, allowing interpretation of the fine structure in photoionization bands. Koopmans' theorem considerations are helpful because they give hints to interpretation; however, it must be kept in mind that the theorem may provide an incorrect ordering for levels that are energetically close together.

From the preceding discussion it is obvious that computed ionization energies for any one molecule may be neither particularly credible nor useful. However, if calculations are applied to a series of molecules and if used to predict trends in electronic structure, they can be of great assistance in correlating and assigning the observed photoelectron bands. It is in this facet of empirical quantum-chemical schemes that their importance lies. Examples of the application of such calculations to predicting trends in several series of molecules are presented in Chapter 10.

References

1. J. E. Lennard-Jones, *Trans. Faraday Soc.*, **25,** 668 (1929).
2. E. Hückel, *Z. Phys.*, **60,** 423 (1930); *Z. Electrochem.* **43,** 752, 827 (1937).
3. R. S. Mulliken, *Phys. Rev.*, **40,** 55 (1932); ibid., **41,** 751 (1932); ibid., **43,** 279 (1933); R. S. Mulliken, *Rep. Progr. Phys.*, **8,** 231 (1941).
4. L. Pauling, *J. Amer. Chem. Soc.*, **53,** 1367 (1931); *The Nature of the Chemical Bond*, Cornell U. P, Ithaca, N.Y., 1946.
5. J. C. Slater, *Phys. Rev.*, **37,** 481 (1931).
6. J. C. Slater, *Quantum Theory of Molecules and Solids. Vol. I: Electronic Structure of Molecules*, McGraw-Hill, New York, 1963.
7. J. E. Lennard-Jones, *Proc. Roy. Soc., Ser. A*, **198,** 1 (1949a); ibid., **198,** 14 (1949b); J. E. Lennard-Jones and J. A. Pople, ibid., **202,** 166 (1950); ibid., **210,** 190 (1951); G. G. Hall and J. E. Lennard-Jones, ibid., **202,** 155 (1950); ibid., **205,** 357 (1951); G. G. Hall, ibid., **202,** 336 (1950); ibid., **205,** 541 (1951); ibid., **213,** 102 (1952); ibid., **213,** 113 (1952).
8. D. A. Sweigart, *J. Chem. Educ.*, **50,** 322 (1973).
9. I. Cohen and J. DelBene, *J. Chem. Educ.*, **46,** 487 (1969).
10. F. L. Pilar, *Elementary Quantum Chemistry*, McGraw-Hill, New York, 1968.
11. I. N. Levine, *Quantum Chemistry. Vol. I: Quantum Mechanics and Molecular Electronic Structure*, Allyn and Bacon, Boston, 1970.
12. W. E. Palk and W. N. Lipscomb, *J. Amer. Chem. Soc.*, **88,** 2384 (1966); B. J. Ransil, *Rev. Mod. Phys.*, **32,** 245 (1960); S. Aung, R. M. Pitzer, and S. I. Chan, *J. Chem. Phys.*, **49,** 2071 (1968).
   a. V. Fock, *Z. Phys.*, **61,** 126 (1930); J. C. Slater, *Phys. Rev.*, **35,** 210 (1930).
13. T. A. Koopmans, *Physica*, **1,** 104 (1933).
14. L. Åsbrink and J. W. Rabalais, *Chem. Phys. Letters*, **12,** 1821 (1971).
15. J. W. Rabalais, T. Bergmark, L. O. Werme, L. Karlsson, and K. Siegbahn, *Phys. Scripta*, **3,** 13 (1971).
16. J. W. Rabalais, L. Karlsson, L. O. Werme, T. Bergmark, and K. Siegbahn, *J. Chem. Phys.*, **58,** 3370 (1973).
17. R. J. Gillespie and R. S. Nyholm, *Quart. Rev.*, **11,** 339 (1957); R. J. Gillespie, *J. Chem. Educ.*, **40,** 295 (1953); R. J. Gillespie, *Angew. Chem. (Int. Edit.)*, **6,** 819 (1967); R. J. Gillespie, *J. Chem. Educ.*, **47,** 18 (1970).
18. A. Brickstock and J. A. Pople, *Trans. Faraday Soc.*, **50,** 901 (1954).
19. T. H. Brown and W. M. Myszkowski, *J. Chem. Phys.*, **52,** 4918 (1970).

20. M. J. S. Dewar, J. A. Hashmall, and C. G. Venier, *J. Amer. Chem. Soc.*, **90,** 1953 (1968).

21. W. A. Lathan, L. A. Curtiss, and J. A. Pople, *Mol. Phys.*, **22,** 1081 (1971).

22. H. F. Schaefer, *The Electronic Structure of Atoms and Molecules*, Addison-Wesley, Reading, Mass., 1972.

23. W. G. Richards, *Int. J. Mass. Spectrom. Ion Phys.*, **2,** 419 (1969).

24. H. Hartman and E. Clementi, *Phys. Rev.*, **133,** A1295 (1964).

25. P. E. Cade, K. S. Sales, and A. C. Wahl, *J. Chem. Phys.*, **44,** 1973 (1966).

26. C. C. J. Roothaan, *Rev. Mod. Phys.*, **23,** 69 (1951).

27. G. G. Hall, *Proc. Roy. Soc., Ser. A*, **205,** 541 (1951); ibid., **213,** 102, 113 (1952).

28. A. Golebiewski and H. S. Taylor, *Annu. Rev. Phys. Chem.*, **18,** 353 (1952).

29. E. Clementi, *Chem. Rev.*, **68,** 341 (1968).

30. A. T. Amos and G. G. Hall, *Proc. Roy. Soc., Ser. A*, **263,** 483 (1961).

31. J. C. Slater, *Phys. Rev.*, **35,** 210 (1930).

32. E. Clementi and D. L. Radimondi, *J. Chem. Phys.*, **38,** 2686 (1963).

33. E. Clementi, *J. Chem. Phys.*, **40,** 1944 (1964).

34. E. Clementi, *I.B.M. J. Res. Develop.*, **9** (Suppl.), 1 (1965).

35. S. F. Boys, *Proc. Roy. Soc., Ser. A*, **200,** 542 (1950).

36. W. H. Hehre, R. Ditchfield, R. F. Stewart, and J. A. Pople, *J. Chem. Phys.*, **52,** 2769 (1970); ibid., **51,** 2657 (1969).

37. H. Bash, C. J. Hornback, and J. W. Moskowitz, *J. Chem. Phys.*, **51,** 1311 (1969).

38. R. F. Stewart, *J. Chem. Phys.*, **52,** 431 (1970).

39. A. J. H. Wachtess, *J. Chem. Phys.*, **52,** 1033 (1970).

40. J. L. Whitten, *J. Chem. Phys.*, **44,** 359 (1966).

41. S. Shih, R. J. Buenker, S. D. Peyerimoff, and B. Wirsam, *Theor. Chim. Acta*, **18,** 277 (1970).

42. A. A. Frost, *J. Chem. Phys.*, **47,** 3707 (1967).

43. C. R. Brundle, D. Neumann, W. C. Price, D. Evans, A. W. Potts, and D. G. Streets, *J. Chem. Phys.*, **53,** 705 (1970); C. R. Brundle, M. B. Robin, and H. Bash, *J. Chem. Phys.*, **53,** 2196 (1970); C. R. Brundle, M. B. Robin, and G. R. Jones, *J. Chem. Phys.* **52,** 3383 (1970).

44. P. J. Bassett and D. R. Lloyd, *Chem. Phys. Letters*, **6,** 166 (1970); D. R. Lloyd and N. Lynaugh, *Phil. Trans. Roy. Soc. London*, **A268,** 97 (1970).

45. W. E. Bull, B. P. Pullen, F. A. Grimm, W. E. Moddeman, G. K. Schweitzer, and T. A. Carlson, *Inorg. Chem.* **9,** 2474 (1970).

46. G. R. Branton, D. C. Frost, T. Makita, C. A. McDowell, and I. A. Stenhouse, *Phil. Trans. Roy. Soc. London*, **A268,** 77 (1970); ibid., *J. Chem. Phys.*, 52, 802 (1970).

47. H. Bash, M. B. Robin, N. A. Kuebler, C. Baker, and D. W. Turner, *J. Chem. Phys.*, **51,** 52 (1969).

48. J. A. Pople and D. L. Beveridge, *Approximate Molecular Orbital Theory*, McGraw-Hill, New York, 1970.

49. J. A. Pople, D. P. Santry, and G. A. Segal, *J. Chem. Phys.*, **43,** S129 (1965).

50. J. A. Pople and G. A. Segal, *J. Chem. Phys.*, **43,** S136 (1965); ibid., **44,** 3289 (1966).

51. D. P. Santry and G. A. Segal, *J. Chem. Phys.*, **47,** 158 (1967).

52. J. A. Pople, D. L. Beveridge and P. A. Dobosh, *J. Chem. Phys.*, **47,** 2026 (1967).

53. P. A. Cox, S. Evans, A. F. Orchard, N. V. Richardson, and P. J. Roberts, *Faraday Discussions Chem. Soc.*, **54,** 26 (1972).

54. M. A. Whitehead, *Sigma Molecular Orbital Theory*, Yale U.P., New Haven, 1970.

55. N. Mataga, *Bull. Chem. Soc. Japan*, **31,** 453 (1958).

56. K. Ohno, *Theor. Chim. Acta*, **2,** 219 (1964).

57. N. C. Baird and M. J. S. Dewar, *J. Chem. Phys.*, **50,** 1262 (1969); N. C. Baird, M. J. S. Dewar, and R. Sustmann, *J. Chem. Phys.*, **50,** 1275 (1969); M. J. S. Dewar and E. Haselbach, *J. Amer. Chem. Soc.*, **92,** 590 (1970); N. Bodor, M. J. S. Dewar, A. Harget, and E. Haselbach, *J. Amer. Chem. Soc.*, **92,** 3854 (1970); M. J. S. Dewar and G. Klopman, *J. Amer. Chem. Soc.*, **89,** 3089 (1967).

58. M. J. S. Dewar, J. A. Hashmall and C. G. Venier, *J. Amer. Chem. Soc.*, **90,** 1953 (1968).

59. C. Fridh, L. Åsbrink, and E. Lindholm, *Chem. Phys. Letters*, **15,** 408 (1972).

60. A. Streitwiesser, Jr., *Molecular Orbital Theory for Organic Chemists*, Wiley, New York, 1961.

61. R. Hoffman, *J. Chem. Phys.*, **39,** 1397 (1963).

62. S. P. McGlynn, L. G. Vanquickenborne, M. Kinoshita, and D. G. Carroll, *Introduction to Applied Quantum Chemistry*, Holt Rinehart, New York, 1972.

63. M. Wolfsberg and L. Helmholz, *J. Chem. Phys.*, **20,** 837 (1952).

64. L. C. Cusachs, *J. Chem. Phys. Suppl.*, **43,** S157 (1965).

65. J. C. Slater, *J. Chem. Phys.*, **43,** 5228 (1965); J. C. Slater, *Advan. Quantum Chem.*, **6,** 1 (1972).

66. K. H. Johnson, *J. Chem. Phys.*, **45,** 3085 (1966); K. H. Johnson, *Wave Mechanics—The First Fifty Years, A Tribute to Louis DeBroglie*, pp. 332–356, Wiley, New York, (1973).

67. J. C. Slater and K. H. Johnson, *Phys. Today*, 34 (*Oct.* 1974).

5 Ionization of Open-Shell Molecules

Molecules with unpaired electrons in their ground states (open shells) yield PE spectra that are significantly more complicated than that of their closed-shell counterparts. In some cases ejection of a single electron from an open-shell molecule can produce a molecular ion in any of several ionic states; all or only a selected few of these transitions may be observed in the PE spectrum depending upon the selection rules for the process. The first three sections of this chapter provide some background on electronic configurations and electronic states of open-shell molecules. Section 1 describes the basic terminology for classifying electronic states. Sections 2 and 3 discuss the procedures for determining the electronic states of linear and nonlinear molecules. Section 4 provides a treatment of the ionization processes of simple and complex open-shell systems. The ionization processes of the open-shell species in this section are treated in the order of their complexity: First, we consider ionization of a closed subshell of a molecule with a single open shell. Second, the open shell itself is ionized in a molecule with a single open shell. Third, ionization of molecules with two or more open shells is treated according to vector coupling methods. Finally, Section 5 provides examples of the ionization patterns of some open-shell molecules.

5.1. Electrons in molecules

In order to understand the occurrence of different electronic states upon ionization, it is instructive to obtain the discrete energy levels of a molecule from those of the united atom. Electrons in a united atom are characterized by their principal n and azimuthal l quantum numbers. The l indicates the orbital angular momentum in units of $h/2\pi$. If the united atom is hypothetically partitioned into smaller atomic units, that is, if portions of the nucleus are pulled apart to form atoms within a molecule, the electrons revolving about these nuclei generate an electric field. The orbital angular momentum vector is quantized in this electric field and can only assume those orientations with regard to the field direction for which its component in that direction is $m_l h/2\pi$, where

$$m_l = l, l-1, l-2, \cdots, -l \tag{5.1}$$

The electronic states i resulting from this quantization, to a first approximation, have energies $E_i = Km_{li}^2$ where K is a constant. States of the electron differing only in the sign of m_l have the same energy, that is, they are degenerate. In order to distinguish electronic states of given n and l and different $|m_l|$ values, one introduces the quantum number

$$\lambda = |m_l| = l, l-1, \cdots, 0 \tag{5.2}$$

Values of λ greater than $\lambda = 0$ occur only in molecules with a rotational symmetry axis greater than twofold. For linear molecules there is an infinite-fold axis, hence an infinite number of λ values. In this case, the orbital wavefunctions of one-electron states with $\lambda = 0, 1, 2, \cdots$ are called $\sigma, \pi, \delta, \cdots$ orbitals, and the electrons in these orbitals are $\sigma, \pi, \delta, \cdots$ electrons. This is analogous to the designation in atoms where $s, p, d, \cdots$ are used to denote orbitals or electrons with $l = 0, 1, 2, \cdots$. Molecular orbitals with $\lambda \neq 0$ have two component functions corresponding to $m_l = \pm \lambda$, hence they are doubly degenerate. Orbitals with $\lambda = 0$ have only one component function; they are nondegenerate. For homonuclear diatomic molecules the orbital wavefunctions can be symmetric (even) or antisymmetric (odd) with respect to the center of symmetry that exists in the molecule. This symmetry property is denoted by the subscript g or u, respectively (from the German "gerade" or "ungerade"), and the molecular orbitals are described as $\delta_g, \sigma_u, \pi_g, \pi_u$, and so forth. In the united atom the corresponding orbital is even (g) for even l and odd (u) for odd l.

For nonlinear molecules, nondegenerate states have zero electronic angular momentum and are called a or b, respectively, if they are symmetric or antisymmetric with respect to rotation about the principal axis. For nonlinear molecules in degenerate electronic states, the angular momentum is generally less than that of linear molecules due to the presence of the off-axis nuclei which impede the orbital motion of the electrons. As a result, the angular momentum is in general nonintegral and may be positive or negative; hence it is no longer a good quantum number. In nonlinear molecules, doubly degenerate orbitals are called e orbitals and triply degenerate orbitals are called t orbitals.

Molecules are many-electron systems where, in a first approximation, each electron may be considered separately as moving in the field of the nuclei and the other electrons. Each electron i can be described by the quantum number λ_i, n_i and l_i, where n_i and l_i refer to either the united atom or separated atoms. The total electronic wavefunction Ψ is made up of the antisymmetrized product of the individual orbital wavefunctions $\phi_j(q_j)$ as

$$\Psi = \mathcal{A}[\phi_1(q_1)\phi_2(q_2)\phi_3(q_3)\cdots] \tag{5.4}$$

where q_j denotes the coordinates of electron j. *A product of orbital wavefunctions with the electron occupancy of each orbital denoted as a right superscript describes the electron configuration.* A given electron configuration corresponds to at least one (and maybe several) electronic states.

5.2.   Electronic configurations and electronic states of linear molecules

The electronic states of linear molecules are characterized by a total orbital angular momentum Λ (in units of $h/2\pi$), which is determined from the one-electron orbital angular momenta λ_i by

$$\Lambda = \sum_i \lambda_i \tag{5.5}$$

Since all angular momenta lie in the internuclear axis, the sum over the λ_i is an ordinary algebraic one. The direction of λ_i in the internuclear axis must be considered by using the signed quantity $m_{li} = \pm\lambda_i$ in the summation. The possibility of different signs for the λ_is in the summation can generate several different states from a single electron configuration. The resultant of this summation can have $\Lambda = 0, 1, 2, \cdots$, which is denoted by Σ, Π, Δ, $\cdots$. All states with $\Lambda \neq 0$ are doubly degenerate. These electronic states in homonuclear diatomic molecules are either even (g) or odd (u) depending on whether there is an even or odd number of odd orbitals (σ_u, π_u, $\cdots$); the resulting states are Σ_g, Σ_u, Π_g, Π_u, $\cdots$. The total spin S of the system is obtained by adding the spins of the individual electrons vectorially as

$$S = \sum_i s_i \tag{5.6}$$

Since $s_i = \pm\frac{1}{2}$, for two electrons $S = 1, 0$, for three electrons $S = \frac{3}{2}, \frac{1}{2}$, and so on. *The resulting spin degeneracy of a particular state is called its multiplicity and is expressed as $2S + 1$.*

Inequivalent electrons

First, we consider systems of inequivalent electrons, for example, one δ_u and one π_g electron. According to Eq. 5.5, the resulting state must be a Π_u state, that is, $\Lambda = 1$. There are two electrons, hence $S = 1, 0$, and the resulting states have triplet or singlet multiplicity. The configuration $\sigma_g^1 \pi_g^1$, therefore, yields $^1\Pi_u$ and $^3\Pi_u$ states, the u arising from the product of u and g functions.

For two inequivalent π electrons $\Lambda = 2, 0$ corresponding to Δ and Σ states. The Δ state arises from coupling of the one electron orbital angular momenta as $\lambda_1 + \lambda_2$ and $-\lambda_1 - \lambda_2$; this state is doubly degenerate. The Σ states arise from $\lambda_1 - \lambda_2$ and $-\lambda_1 + \lambda_2$ couplings; these levels are not degenerate. They split into two states whose wavefunctions are the sum and difference of the functions corresponding to $\lambda_1 - \lambda_2$ and $-\lambda_1 + \lambda_2$. These functions are denoted by Σ^+ and Σ^- according to whether they are symmetric or antisymmetric, respectively, with respect to a reflection at any plane through the internuclear axis. Since the two electrons are in different π orbitals, there are no restrictions on the spin quantum numbers and $S = 1, 0$. Therefore, the $\pi^1 \pi^1$ configuration yields $^1\Sigma^+$, $^3\Sigma^+$, $^1\Sigma^-$, $^3\Sigma^-$, $^1\Delta$, and $^3\Delta$ states.

Equivalent electrons

If the electrons are equivalent, that is, if they have the same n, l, and λ, we must take account of the Pauli exclusion principle, which requires that no two electrons have the same set of four quantum numbers n, l, m_l, and m_s. Orbitals of σ type can have a maximum of two electrons differing by $m_s = \pm\frac{1}{2}$, and orbitals of π, δ, $\cdots$ type can have a maximum of four electrons differing by $m_l = \pm\lambda$ and $m_s = \pm\frac{1}{2}$. Thus the configurations σ^2 and π^4 always have $S = 0$ and $\lambda = 0$ yielding only one $^1\Sigma^+$ state each. The configuration π_u^2 gives three states $^3\Sigma_g^-$, $^1\Delta_g$, $^1\Sigma_g^+$. This is just half the number of states given above for the configuration $\pi^1 \pi^1$. The states of the π_u^2 configuration can be derived as follows: The total electronic wavefunction (orbital × spin) must be antisymmetric with respect to exchange of the coordinates of any two electrons. The spin functions of singlet states are antisymmetric, while those of triplet states are symmetric with respect to such an exchange. In order for the total wavefunction to be antisymmetric, obviously the orbital parts of the functions must have opposite symmetry from the spin parts. Taking the symmetric direct product* of two π representations we obtain $\Pi \times \Pi \overset{s}{=} \Sigma^+$, Δ. Since these

---

* The *direct product* of two functions is obtained by multiplying the characters of the representations of the two functions. The resulting representation is in general reducible and can be reduced to the sum of characters of several irreducible representations. These irreducible representations are unique functions resulting from the product of the original functions. The direct product can be separated into symmetric and antisymmetric parts. This is achieved by forming symmetric and antisymmetric linear combinations of the original functions. One then obtains the *symmetric direct product* by taking the product of the symmetrical species with itself and the *antisymmetric direct product* by taking the product of the antisymmetric species with itself. This procedure is described in detail by Hochstrasser[1a] and by Herzberg.[1a]

orbital functions are symmetric, their corresponding spin parts must be antisymmetric, hence we have $^1\Sigma^+$ and $^1\Delta$ states. Taking the antisymmetric direct product of two π representations we obtain $\Pi \times \Pi \overset{as}{=} \Sigma^-$. This antisymmetric orbital function must be combined with a symmetrical spin function, hence we have $^3\Sigma^-$.

States of the molecular ion

In order to determine the electronic states of a given linear molecule or ion, it is necessary to consider the various possible electron configurations. The molecular ground state is obtained by distributing all electrons into the lowest energy orbitals to the extent allowed by the Pauli principle. The ground electronic state is determined from this N electron configuration as discussed above. The possible electronic states of the molecular ion, with $N-1$ electrons, can be determined by removing an electron from an occupied orbital and determining the states of the $N-1$ electron configuration. All of the $N-1$ states of the singly positive ion are generated by removing a single electron at a time from each of the occupied orbitals. States of the doubly positive ion are formed by removing two electrons from either the same or different orbitals resulting in an $N-2$ electron configuration. These states are usually of considerably higher energy than those of the singly positive ion.

Ejection of an electron from a molecule with a closed-shell ground state configuration yields an ionic configuration with a single unpaired electron. Thus ejection of a single electron from each occupied orbital yields a doublet ionic state for each ionization process. This simple one-orbital–one-state relationship makes spectra of closed-shell molecules easier to interpret than those of open-shell molecules. An example of the complexity of open-shell spectra is provided by the open-shell species NO and O_2. The electron configurations and electronic states observed in the PE spectra of these two molecules are listed in Table 5.1.[1–3]

Consider the NO molecule with one unpaired electron. The ground state is $^2\Pi$. Ejection of the single unpaired electron leaves the ion with a closed-shell $^1\Sigma^+$ ground state. Ionization of the 1π orbital produces an ion with the configuration $\cdots 1\pi^3 2\pi^1$. The two inequivalent unpaired π electrons couple to produce $^3\Sigma^+$, $^3\Delta$, $^3\Sigma^-$, $^1\Sigma^-$, $^1\Delta$, $^1\Sigma^+$ ionic states, in that energetic order. Transitions to all six of these ionic states are observed in the PE spectrum of NO. Ionization of each of the remaining σ orbitals produces the configuration $\cdots \sigma^1 \cdots 2\pi^1$, hence $^1\Pi$ and $^3\Pi$ ionic states. The PE spectrum of NO is shown in Fig. 5.1.

The oxygen molecule has two unpaired electrons in the $1\pi_g$ orbital. These two equivalent π_g electrons couple to produce $^3\Sigma_g^-$, $^1\Sigma_g^+$, and $^1\Delta_g$

Table 5.1. Electron Configurations and Corresponding Electronic States Observed[1-3] in PE Spectroscopy for NO and O_2

Molecule or Ion	Electron Configuration	Electronic States and Energies (eV) Above Molecular Ground State
NO	$1\sigma^2 2\sigma^2 3\sigma^2 4\sigma^2 5\sigma^2 1\pi^4 2\pi^1$	$^2\Pi$
NO^+	$1\sigma^2 2\sigma^2 3\sigma^2 4\sigma^2 5\sigma^2 1\pi^4$	$^1\Sigma^+ (9.27)$
	$1\sigma^2 2\sigma^2 3\sigma^2 4\sigma^2 5\sigma^2 1\pi^3 2\pi^1$	$^3\Sigma^+ (15.65)$
		$^3\Delta\ (16.86)$
		$^3\Sigma^- (17.59)$
		$^1\Sigma^- (17.82)$
		$^1\Delta\ (18.07)$
		$^1\Sigma^+ (23.3)$
	$1\sigma^2 2\sigma^2 3\sigma^2 4\sigma^2 5\sigma^1 1\pi^4 2\pi^1$	$^3\Pi\ (16.56)$
		$^1\Pi\ (18.32)$
	$1\sigma^2 2\sigma^2 3\sigma^2 4\sigma^1 5\sigma^2 1\pi^4 2\pi^1$	$^3\Pi\ (21.72)$
		$^1\Pi\ (21.72)$
	$1\sigma^2 2\sigma^2 3\sigma^1 4\sigma^2 5\sigma^2 1\pi^4 2\pi^1$	$^3\Pi\ (40.6)$
		$^1\Pi\ (43.8)$
	$1\sigma^2 2\sigma^1 3\sigma^2 4\sigma^2 5\sigma^2 1\pi^4 2\pi^1$	$^3\Pi\ (410.3)$
		$^1\Pi\ (411.8)$
	$1\sigma^1 2\sigma^2 3\sigma^2 4\sigma^2 5\sigma^2 1\pi^4 2\pi^1$	$^3\Pi\ (543.3)$
		$^1\Pi\ (544.0)$
O_2	$1\sigma_g^2 1\sigma_u^2 2\sigma_g^2 2\sigma_u^2 3\sigma_g^2 1\pi_u^4 1\pi_g^2$	$^3\Sigma_g^-$
O_2^+	$1\sigma_g^2 1\sigma_u^2 2\sigma_g^2 2\sigma_u^2 3\sigma_g^2 1\pi_u^4 1\pi_g^1$	$X\ ^2\Pi_g\ (12.07)$
	$1\sigma_g^2 1\sigma_u^2 2\sigma_g^2 2\sigma_u^2 3\sigma_g^2 1\pi_u^3 1\pi_g^2$	$a\ ^4\Pi_u\ (16.10)$
		$A\ ^2\Pi_u\ (17.05)$
		$^2\Pi_u\ (24.0)$
	$1\sigma_g^2 1\sigma_u^2 2\sigma_g^2 2\sigma_u^2 3\sigma_g^1 1\pi_u^4 1\pi_g^2$	$b\ ^4\Sigma_g^-\ (18.17)$
		$B\ ^2\Sigma_g^-\ (20.30)$
	$1\sigma_g^2 2\sigma_u^2 2\sigma_g^2 2\sigma_u^1 3\sigma_g^2 1\pi_u^4 1\pi_g^2$	$c\ ^4\Sigma_u^-\ (24.58)$

electronic states. The molecular ground state is $^3\Sigma_g^-$ in accordance with Hund's rule, which specifies that the state of highest multiplicity is of lowest energy. Ejection of a $1\pi_g$ electron yields a $\cdots 1\pi_g^1$ configuration, hence a $^2\Pi_g$ state. Ejection of a $1\pi_u$ electron produces a $\cdots 1\pi_u^3 1\pi_g^2$ configuration, which can yield several ionic states. These are determined as follows: The $1\pi_g^2$ configuration yields $^3\Sigma_g^-$, $^1\Sigma_g^+$, and $^1\Delta_g$ terms. Coupling the unpaired $1\pi_u$ electron to these terms yields $^4\Pi_u$, (3) $^2\Pi_u$, and $^2\Phi_u$ states. This process is illustrated in Fig. 5.2. From this figure it is obvious that ionization of a $1\pi_u$ electron of O_2 in its $^3\Sigma_g^-$ ground state produces

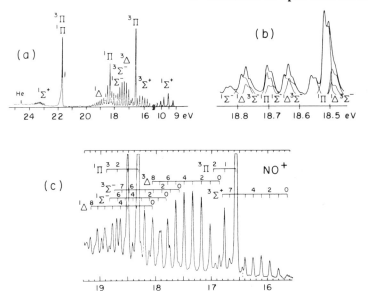

Fig. 5.1. Photoelectron spectrum of NO. (*a*) He II spectrum. (*b*) Deconvolution of overlapping peaks in the region of 1π ionization (resolution, 10 meV). (*c*) Expanded portion of the spectrum between 15.6 eV and 19.3 eV showing complexity due to various ionic states produced by 1π ionization. Reproduced with permission from Ref. 1b.

only two ionic states, $^4\Pi_u$ and $^2\Pi_u$. The $^2\Phi_u$ state cannot be reached in a one-electron transition from the ground state and, therefore, is absent in the spectrum. The other two $^2\Pi_u$ ionic states produced by ejection of a $1\pi_u$ electron of O_2 in its $^1\Sigma_g^+$ and $^1\Delta_g$ states can be reached through configuration interaction with the $^2\Pi_u$ state produced from the $^3\Sigma_g^-$ state ionization. Dixon and Hull[4] have calculated the wavefunctions for all of

$$\pi_g \times \pi_g = {}^3\Sigma_g^- , {}^1\Sigma_g^+ , {}^1\Delta_g \left.\right\} \begin{array}{l} \text{States resulting} \\ \text{from two equivalent} \\ \pi_g \text{ electrons} \end{array}$$

$$^3\Sigma_g^- \times \pi_u = {}^4\Pi_u , {}^2\Pi_u$$

$$^1\Sigma_g^+ \times \pi_u = {}^2\Pi_u$$

$$^1\Delta_g \times \pi_u = {}^2\Pi_u , {}^2\Phi_u$$

$$\left.\right\} \begin{array}{l} \text{States resulting} \\ \text{from coupling a} \\ \pi_u \text{ electron to the} \\ \text{terms of the } \pi_g^2 \\ \text{configuration} \end{array}$$

Fig. 5.2. Derivation of the electronic states resulting from the $\cdots 1\pi_u^3 1\pi_g^2$ electron configuration.

the $^2\Pi_u$ states produced by the $\cdots 1\pi_u^3 1\pi_g^2$ configuration using configurational mixing. They found that the three $^2\Pi_u$ states are heavily mixed. Using the coefficients of the configurational wavefunctions to approximate the relative transition intensities, their calculations lead to a ratio of the $^4\Pi_u$ and the three $^2\Pi_u$ states of 2:0.34:0.001:0.64. The spectrum of O_2 is shown in Fig. 5.3. Two $^2\Pi_u$ states and one $^4\Pi_u$ state are observed in

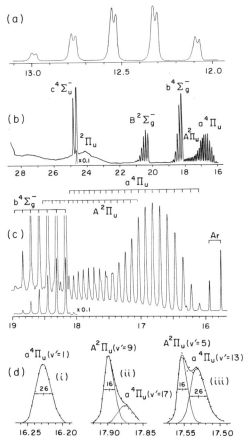

Fig. 5.3. Photoelectron spectrum of O_2. (a) He I spectrum showing the transition to the ground state of O_2^+, $^2\Pi_g \leftarrow {}^3\Sigma_g^-$. The spin–orbit splitting of the vibrational bands is clearly visible. (b) He II spectrum of O_2 from 16 to 28 eV. (c) High resolution He I spectrum of O_2 in the 15.7- to 19.0-eV region. (d) Parts of the He I spectrum illustrating deconvolution of overlapping $^4\Pi_u$ and $^2\Pi_u$ vibrational bands. (i) A single $^4\Pi_u$ peak with a halfwidth of 26 meV. (ii) An unresolved band. When the $^4\Pi_u$ peak is subtracted, one obtains a $^2\Pi_u$ peak with 16-meV halfwidth. (iii) An unresolved band that can be deconvoluted when the halfwidths of the two peaks are known. Reproduced with permission from Ref. 2.

this spectrum with intensity ratios of $^4\Pi_u : {}^2\Pi_u : (17.05 \text{ eV}) : {}^2\Pi_u$ (24.0 eV) of 2:0.3:0.75 in reasonable agreement with the calculated intensities. The other $^2\Pi_u$ state with the low calculated intensity is not observed. Ionization of the remaining σ orbitals produces the configuration $\cdots \sigma^1 \cdots 1\pi_g^2$. The ground-state term $^3\Sigma_g^-$ from the $1\pi_g^2$ configuration couples with the σ electron to form $^4\Sigma^-$ and $^2\Sigma^-$ states. Some of these higher energy states can be observed in Fig. 5.3b.

5.3. Electronic configurations and electronic states of nonlinear molecules

For nonlinear molecules the electronic angular momentum Λ is, in general, not a good quantum number. In this case, the wavefunctions are classified according to their symmetry species, or irreducible representations (IRs) of the molecular point group. The electronic states are determined from the symmetric and antisymmetric direct products of the IRs of the coupled electrons.

Consider the example of NO_2, a nonlinear open-shell molecule with the electron configuration

$$1a_1^2 1b_2^2 2a_1^2 3a_1^2 2b_2^2 4a_1^2 3b_2^2 5a_1^2 1b_1^2 1a_2^2 4b_2^2 6a_1^1$$

giving a $^2A_1$ ground state. Ejection of the $6a_1$ electron produces the ground state of NO_2^+, $\cdots 4b_2^2$, $^1A_1$. Ionization of any of the filled orbitals produces an ionic configuration with two unpaired inequivalent electrons, hence singlet and triplet states. The symmetry species are determined from the direct products of the IRs of the coupled electrons. For example, the configuration, $\cdots 1b_1^1 1a_2^2 4b_2^2 6a_1^1$, yields $^1B_1$ and $^3B_1$ ionic states. The spectrum of NO_2 is presented in Fig. 5.4, and the ionic state assignments[5,6] are indicated above the bands.

For nonlinear open-shell molecules with degenerate electronic states, one can proceed in the manner described for linear molecules. This procedure is cumbersome, and it becomes expedient to use a vector-coupling approach as developed in the Section 5.4.

5.4. Vector-coupling method for open-shells

In order to determine the accessible ionic states and the relative intensities of photoelectron transitions in complex open-shell molecules, it is desirable to use the vector-coupling approach. The ionic states produced and the relative transition intensities to these states are determined by the occupancy of the orbital ionized, the coupling between the unpaired electrons, the degeneracies of the final ionic states, and the one-electron

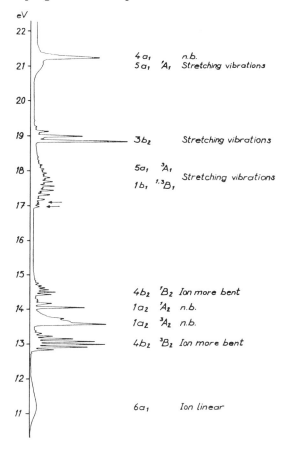

Fig. 5.4 Photoelectron spectrum of NO_2 using the He I line below 20 eV at moderate resolution and the He II line between 20 and 22 eV. The two arrows indicate peaks produced by the 537-Å line. Reproduced with permission from Ref. 5.

photoionization cross-section matrix elements. In this section we consider the first three items; cross sections are treated in Chapter 6.

Intensity ratios for configurations

In discussing intensities of photoelectron transitions, the photoionization transition probability, as given by the square of Eq. 3.14, must be defined in more specific terms. Following the treatment of Cox and Orchard,[7] we consider the pth photoionization of a molecule exposed to radiation of frequency ν. The kinetic energy of the photoelectrons is given by $E_k^p = h\nu - E_1^p$, where E_1^p is the pth vertical ionization energy corresponding to the $(p-1)$th excited state of the molecular ion. The photoionization process involves transfer of an electron to the continuum state $|\chi_{p\sigma}\rangle$, characterized by kinetic energy E_k^p and spin $\sigma = \pm\frac{1}{2}$. Rewriting Eq. 3.28 in these terms, we obtain

$$\mathscr{P} \propto \sum_{j,\,k} \sum_{m,\,m'} |\langle \Psi_{km}'' | \sum_n \mathbf{p}_n |\Psi_{p-1,\,jm'}^+ \chi_{p\sigma}\rangle|^2 \tag{5.7}$$

where $m = m' + \sigma$ is the magnetic spin quantum number, $|\Psi_{km}''\rangle$ is the wavefunction for the kth orbital component of the ground state of the neutral molecule, and $|\Psi_{p-1,\,jm'}^+\rangle$ is the wavefunction of the excited state of the molecular ion. The Dirac brackets imply antisymmetrization of all electron pairs. Consider a molecule with a single open shell r^n, and assume that there is no orbital rescaling or reorientation in the molecular ion. Let the ground state Russell–Saunders term transform as the degenerate irreducible representation Γ_c of the molecular point group, that is, $|\Psi_k''\rangle = |S\Gamma_{ck}(r^n)\rangle$.

We first consider ionization of a *closed subshell a* of degeneracy ω_a, the orbitals of which transform together as $\Gamma_a(\Gamma_1)$. According to Eq. 5.7 the probability for ionization of the *configuration* is proportional to

$$\mathscr{P} \propto \sum_{i,\,\sigma} |\langle O\Gamma_1(a^{2\omega_a})| \sum_n \mathbf{p}_n |(\tfrac{1}{2})\Gamma_{ai}(a^{2\omega_a-1})\chi_{p\sigma}\rangle|^2 \tag{5.8}$$

where the index i runs over the degenerate orbitals ϕ_{ai} of the subshell a. The above expression, after reduction of the many-electron matrix elements becomes

$$\mathscr{P} \propto 2\sum_i |\langle\phi_a| \mathbf{p}_n |\chi_{p\sigma}\rangle|^2 = 2\omega_a|\langle\phi_{ai}| \mathbf{p}_n |\chi_{p\sigma}\rangle|^2 \tag{5.9}$$

where $\langle\phi_{ai}| \mathbf{p}_n |\chi_{p\sigma}\rangle$ is the representative one-electron transition matrix element for the transition from the subshell a. Thus, the total configurational cross section is proportional to the degeneracy ω_a of the subshell a.

In an open-shell molecule, the states of the molecular ion, $|\Psi_{p-1,\,j}^+\rangle = |S'\Gamma_{bj}\rangle$, with $S' = S \pm \frac{1}{2}$, arise from the coupling of the states $|(\tfrac{1}{2})\Gamma_{ai}\rangle$ with the ground state $|S\Gamma_{ck}\rangle$ of the open-shell r^n.[8–10] The states $|S'\Gamma_{bj}\rangle$ are related to the states $|(\tfrac{1}{2})\Gamma_{ai}\rangle$ by a unitary transformation. It follows[11] that Eq. 5.8, summed over the states $|S'\Gamma_{bj}\rangle$, still yields $2\omega_a|\langle\phi_{ai}| \mathbf{p}_n |\chi_{p\sigma}\rangle|^2$. The total ionization cross section for production of a particular ion configuration, therefore, is just the same as in the case of a closed-shell molecule.

When the open-shell itself is ionized, by similar arguments we expect the probability for ionization to be proportional to $n|\langle \phi_r| \mathbf{p}_n |\chi_{p\sigma}\rangle|^2$, thus reflecting the *occupancy* n of the subshell r. Therefore, if primitive matrix elements such as $\langle \phi_a| \mathbf{p}_n |\chi_{p\sigma}\rangle$ and $\langle \phi_r| \mathbf{p}_n |\chi_{p\sigma}\rangle$ are approximately equal, the probability of realizing different ion configurations should be roughly proportional to the occupancies of the subshells ionized, irrespective of whether these subshells are fully or partially occupied. Applying this to the ionization of NO, the relative probabilities of producing the ion configurations in which one electron is removed from the 2π orbital, one electron from the 1π orbital, or one electron from the 5σ orbital will be in the ratio 1:4:2, respectively. For O_2 the relative probabilities of producing ion configurations in which one electron is removed from the $1\pi_g$ orbital, one electron from the $1\pi_u$ orbital, or one electron from the $3\sigma_g$ orbital will be in the ratio 2:4:2, respectively.

Terms of a single configuration and their relative probabilities

Each configuration of a molecular ion can give rise to one or more Russell–Saunders terms. It remains to consider the relative probability for production of these different ion states. First consider ionization of a closed subshell a, which produces ion states $|S'\Gamma_b\rangle$. We initially examine only the *orbital* aspect of the problem. The formation of the Γ_{bj} states by coupling of the orbital states Γ_{ai} and Γ_{ck} may be represented by[9,10]

$$|bj\rangle = \sum_{i,k} \langle ac, ik| bj\rangle\, | ai\rangle\, |ck\rangle \qquad (5.10)$$

where the summation extends over the components of Γ_a and Γ_c. The $\langle ac, ik| bj\rangle$ are the *Clebsch–Gordan* or *vector-coupling* coefficients for the appropriate molecular point group. These coefficients are the elements of the transformation matrix obtained by adding the angular momenta of the states Γ_{ai} and Γ_{ck}. They have the normalization property[9]

$$\sum_{i,k} |\langle ac, ik| bj\rangle|^2 = 1 \qquad (5.11)$$

Introducing the expansion (Eq. 5.10) into a purely orbital expression as Eq. 5.7 and neglecting orbital rescaling so that the many-electron matrix elements may be reduced in the usual way, we obtain

$$\mathscr{P}_{\text{orbit}} \propto \sum_{jk} \left| \sum_{ik} \langle \phi_{ai}| \mathbf{p}_n |\chi_{p\sigma}\rangle \langle ac, ik| bj\rangle \right|^2 \qquad (5.12)$$

Using the normalization condition of Eq. 5.11, this equation simplifies to

$$\mathcal{P}_{\text{orbit}} \propto \sum_{jk} |\langle \phi_a | \, \mathbf{p}_n \, | \chi_{p\sigma} \rangle|^2 \tag{5.13}$$

Therefore, considering the orbital aspect of the problem, the probability of photoionization is simply proportional to the orbital degeneracy of the particular term $|S'\Gamma_b\rangle$ in which the molecular ion is produced.

Next we consider the spin aspect of the problem. The space and spin coupling problems may be treated separately since the subshell ionized Γ_a and the ground state of the open-shell Γ_c relate to different orbitals. This eliminates all equivalence restrictions on the ion states Γ_b that may be generated according to Eq. 5.10. If the arguments used above are applied to the spin aspect of the problem,[8] we find that $\mathcal{P}_{\text{spin}}$ contains the factor $2S'+1$, the spin degeneracy of the ion state $|S'\Gamma_b\rangle$.

Combining the results from orbital and spin considerations, the relative probabilities for photoionization to distinct terms arising from the ion configuration $a^{(2\omega_a-1)}r^n$ should reflect the total spin-orbital degeneracies of the ion terms in question, that is

$$\mathcal{P}_{\text{tot}} = \mathcal{P}_{\text{orbit}} \cdot \mathcal{P}_{\text{spin}} \tag{5.14}$$

Consider, for example, the NO molecule. Ejection of a 1π electron to give the ion configuration $\cdots 1\pi^3 2\pi^1$ produces $^3\Sigma^+$, $^3\Delta$, $^3\Sigma^-$, $^1\Sigma^-$, $^1\Delta$, $^1\Sigma^+$ ion states. Using the spin-orbital degeneracies of these states, the relative intensities of the six photoelectron bands should be $3:6:3:1:2:1$. As another example, we use the ionization process $t_1^6 t_2^4 \rightarrow t_1^5 t_2^4$ in the cubic molecule PtF_6. The ion states are formed by coupling the ground state of the $t_1^6 t_2^4$ configuration, $^3T_1$, with the term corresponding to the electron ejected, t_1. This yields the following eight distinct ion states with relative intensities of photoelectron transitions as indicated.

$^4T_1$	$^4T_2$	^4E	$^4A_1$	$^2T_1$	$^2T_2$	^2E	$^2A_1$
6	6	4	2	3	3	2	1

Finally, we consider the case in which the open shell itself is ionized. We cannot use the previous coupling argument because of the antisymmetry restriction on the wavefunctions $|S_q\Gamma_q(r^{n-1})\rangle$ of the molecular ion states formed. Instead we expand the wavefunction of the neutral molecule ground state in terms of the possible states of the ion as

$$|S\Gamma_c(r^n)\rangle = \sum_q [\langle S_q\Gamma_q(r^{n-1})| \,) S\Gamma_c(r^n)\rangle] \, |S_q\Gamma_q(r^{n-1})\rangle \, |(\tfrac{1}{2})\Gamma_r(r)\rangle \tag{5.15}$$

where the first term in brackets is the *coefficient of fractional parentage.*[*8-10,12,13] These are the coefficients of an antisymmetrical combination of parent wave functions used to represent a configuration of equivalent electrons. Using Eq. 5.7 and the orthogonality of the different ion states $|S_q\Gamma_q\rangle$, it is easily shown that the relative probability of photoionization to a particular ion state is proportional to the square of the fractional parentage coefficient $\langle S_q\Gamma_q(r^{n-1})|\,)S\Gamma_c(r^n)\rangle$.

For simple cases where the molecular configuration r^n generates only one Russell–Saunders term $|S\Gamma_c\rangle$, the squares of the fractional parentage coefficients simply reflect the overall spin-orbital degeneracies of the ion states $|S_q\Gamma_q\rangle$. The probabilities for production of these states are the same as in the case of ionization of a closed subshell. An example[7] of this is the photoionization of V(CO)$_6$ whose ground state configuration, $\cdots t_2^5$, yields only one Russell–Saunders term, $^2T_2$. The photoionization transition $\cdots t_2^5 \rightarrow \cdots t_2^4$ produces ion states $^1A_1$, ^1E, $^1T_2$, and $^3T_1$ with expected intensities $1:2:3:9$, respectively.

In cases where the ground state $|S\Gamma_c\rangle$ is but one of a family of terms arising from the configuration r^n, the intensity pattern of the photoelectron bands may be quite different. For such cases, the squares of the fractional parentage coefficients do not, in general, correspond to the spin-orbital degeneracies of the possible ion states. As an example, consider the case of ionization of the open shell of PtF$_6$. The ground state $^3T_1$ of the molecular configuration $\cdots t_1^6 t_2^4$ is one of a family of four terms. In the transition, $\cdots t_1^6 t_2^4 (^3T_1) \rightarrow \cdots t_1^6 t_2^3$, we expect photoelectron bands corresponding to each of the ion states $^4A_2$, ^2E, $^2T_1$, and $^2T_2$ with relative intensities $4:2:3:3$, respectively. (The fractional parentage coefficients are available from several sources.[8-10,12,13]) Another interesting case is ionization of the configuration $\cdots t_2^3(^4A_2)$ for which the $^4A_2$ state is one of a family of four terms. The only ionic state that should be observed in the transition $\cdots t_2^3(^4A_2) \rightarrow \cdots t_2^2$ is the $^3T_1$ state, for the fractional parentage coefficients of all other ion states are zero. There are many examples of these effects.

From the above discussion it can be seen that for simple cases, the number of ion states produced and their expected relative intensities are easy to determine. However, for complex cases, for example when the open-shell itself is ionized, the vector coupling approach is particularly useful. The states of the molecular ion and their relative intensities (given as squares of the coefficients of fractional parentage) produced by one-electron ionization processes from the ground state of a unique open-shell of a cubic molecule are listed in Table 5.2.

---

* The symbol $|\,)$ represents $(\frac{1}{2})R(r)$, where R is the irreducible representation to which the subshell r belongs.

Table 5.2. States of the Molecular Ion and Their Relative Intensity[a] (Squares of the Coefficients of Fractional Parentage) Produced by One-Electron Ionization Processes from the Ground State of a Unique Open Shell of a Cubic Molecule

Initial Molecular State	Final Ionic State	Relative Intensity[a]
$\cdots t_2^1, {}^2T_2$	${}^1A_1$	1(1)
$\cdots t_2^2, {}^3T_1$	$\cdots t_2^1, {}^4A_2, {}^2E, {}^2T_1, {}^2T_2$	2(4, 2, 3, 3)
$\cdots t_2^3, {}^4A_2$	$\cdots t_2^2, {}^3T_1$	3(1)
$\cdots t_2^4, {}^3T_1$	$\cdots t_2^3, {}^4A_2, {}^2E, {}^2T_1, {}^2T_2$	4(4, 2, 3, 3)
$\cdots t_2^5, {}^2T_2$	$\cdots t_2^4, {}^1A_1, {}^1E, {}^3T_1, {}^1T_2$	5(1, 2, 9, 3)
$\cdots e^1, {}^2E$	${}^1A_1$	1(1)
$\cdots e^2, {}^3A_2$	$\cdots e^1, {}^2E$	2(1)
$\cdots e^3, {}^2E$	$\cdots e^2, {}^1A_1, {}^1E, {}^3A_2$	3(1, 2, 3)

[a] The first number represents the relative probability for ionization of the configuration, whereas the numbers in parenthesis represent the distribution of this intensity among the ion states produced.

Terms of systems with two open shells and their relative probabilities

The spectra of molecules with two or more open shells can be considerably more complicated than those of molecules with only one open shell. Configurations with two or more open shells have been treated by Cox, Evans, and Orchard;[14] this discussion follows their treatment. An essential point in handling configurations with more than one open shell is to consider the coupling that already exists between the different open shells in the molecule. This is amply illustrated as follows: Consider a high-spin octahedral d^5 complex with the ground state $\cdots t_2^3 e_g^2 ({}^6A_{1g})$. Ejection of a single electron produces a configuration with four unpaired electrons, hence a quintet state. However, if we consider ionization from uncoupled t_{2g} and e_g subshells separately, we would predict the concomitant production of singlet and triplet states of d^4. Clearly, the fact that the two subshells are coupled together with parallel spins has a crucially important effect on the pattern of ionization. This coupling must be considered in order to predict the resulting ionic states that are formed.

To analyze the ionization processes in a molecule having more than one open shell, Cox et al.[14] employ the *annihilation operator* method. We introduce this method by rewriting Eqs. 3.29 and 5.7 as

$$\mathscr{P} \propto \sum_{jk} \sum_{mm'} \sum_{\sigma} \left| \langle \Psi''_{km} | \sum_n \mathbf{p}_n | \Psi^+_{jm'} \chi_{p\sigma} \rangle \right|^2 \tag{5.16}$$

$$\mathscr{P} \propto \sum_{jk} \sum_{mm'} \sum_{\rho\sigma} \left| \langle \phi^r_{p\sigma} | \mathbf{p}_n | \chi_{p\sigma} \rangle \langle \Psi^+_{jm'} | \mathbf{a}^r_{\rho\sigma} | \Psi''_{km} \rangle \right|^2 \tag{5.17}$$

where the summation runs over the degenerate spin (m, m') and orbital (k, j) components of the ground state of the molecule $|\Psi''_{km}\rangle$ and the ion state $|\Psi^+_{jm'}\rangle$. The annihilation operator $\mathbf{a}^r_{\rho\sigma}$ annihilates an electron with spin component σ from the orbital component ρ of the subshell r. We make the assumption that the various matrix elements for the r subshell are equal and can be represented by $\langle\phi^r|\,\mathbf{p}_n\,|\chi_p\rangle$. This is equivalent to a spherical averaging over the continuum angular functions. Using this, Eq. 5.17 can be factorized as

$$\mathcal{P} \propto |\langle\phi^r|\,\mathbf{p}_n\,|\chi_p\rangle|^2 \sum_{jk} \sum_{mm'} \sum_{\rho\sigma} |\langle\Psi^+_{jm'}|\,\mathbf{a}^r_{\rho\sigma}\,|\Psi''_{km}\rangle|^2 \tag{5.18}$$

Treating the annihilation operator $\mathbf{a}^r_{\rho\sigma}$ as an *irreducible tensor operator*,* we apply the *Wigner–Eckart Theorem*[8,9,15] to express the second term in the brackets of Eq. 5.18 as

$$\langle\Psi''_{jm'}|\,\mathbf{a}^r_{\rho\sigma}\,|\Psi''_{km}\rangle = \langle rK\rho k|\,Jj\rangle\langle(\tfrac{1}{2})S\sigma m\,|S'm'\rangle\,\langle\Psi^+(S'J)\|\mathbf{a}^r\|\Psi''(SK)\rangle$$
$$\tag{5.19}$$

where SK and $S'J$ label the total spin and orbital symmetries of $|\Psi''\rangle$ and $|\Psi^+|$, respectively. The Wigner–Eckart theorem expresses the physical properties of a matrix element in terms of a reduced matrix element ($\langle\Psi^+(S'J)\|\mathbf{a}^r\|\Psi''(SK)\rangle$ of Eq. 5.19) of the irreducible tensor operator $\mathbf{a}^r$. Substituting this relationship in Eq. 5.18 and using the normalization properties of the coupling coefficients yields

$$\mathcal{P} \propto |\langle\phi^r|\,\mathbf{p}_n\,|\chi_p\rangle|^2 |\langle\Psi^+(S'J)\|\mathbf{a}^r\|\Psi''(SK)\rangle|^2 \omega_{S'J} \tag{5.20}$$

where $\omega_{S'J}$ is the total spin-orbital degeneracy of the ion state $|\Psi^+\rangle$. If $|\Psi''\rangle$ and $|\Psi^+\rangle$ are states arising from the single open-shell configurations, r^n and r^{n-1}, respectively, then $|\Psi''(SK)\rangle$ may be expanded in terms of the *coefficients of fractional parentage*, $\langle S'J(r^{n-1})|\,)SK(r^n)\rangle$, so that

$$\langle\Psi^+(S'J)\|\mathbf{a}^r\|\Psi''(SK)\rangle = \pm(n\omega_{SK}/\omega_{S'J})^{1/2}\langle S'J(r^{n-1})|\,)SK(r^n)\rangle \tag{5.21}$$

The phase factor is not important here because the square of the reduced matrix element of $\mathbf{a}^r$ will be used in what follows. Introducing Eq. 5.21 into Eq. 5.20 yields

$$\mathcal{P} \propto n|\langle\phi^r|\,\mathbf{p}_n\,|\chi_p\rangle|^2 |\langle S'J(r^{n-1})|\,)SK(r^n)\rangle|^2 \omega_{SK} \tag{5.22}$$

for ionization of a unique open-shell. The degeneracy factor ω_{SK} appears because of the summation over all components of the molecular state $|\Psi''\rangle$. The ionization probability from just one of these components would

---

* An irreducible tensor operator is a set of n operators $T_1, T_2, \cdots, T_n$, which have the property that their rotational transforms are linear functions of the n operators; the tensor itself transforms according to an irreducible representation of its respective point group.

be $\mathscr{P}/\omega_{SK}$. Eq. 5.22 is equivalent to the result obtained in the previous section, that is, the relative probabilities of producing different ion states will reflect the squares of the fractional parentage coefficients.

Now consider the case in which the states $|\Psi''(SK)\rangle$ and $|\Psi^+(S'J)\rangle$ arise from the coupling of *two* open shells. One of the open shells remains unchanged after ionization, its particular state being of symmetry S_1L, while the other undergoes a transition from its ground state $|S_2P(r^n)\rangle$ to a particular ionized state $|S_3Q(r^{n-1})\rangle$. The specific coupling relations are

$$|\Psi''(SKmk)\rangle = \sum_{lp}\sum_{m_1m_2} \langle LPlp|\,Kk\rangle\langle S_1S_2m_1m_2|\,Sm\rangle|S_1Lm_1l\rangle|S_2Pm_2p\rangle \tag{5.23}$$

and

$$|\Psi^+(S'Jm'j)\rangle = \sum_{lq}\sum_{m_1m_3} \langle LQlq|\,Jj\rangle\langle S_1S_3m_1m_3|\,S'm'\rangle|\,S_1Lm_1l\rangle|\,S_3Qm_3q\rangle \tag{5.24}$$

The annihilation operators $\mathbf{a}^r_{\rho\sigma}$ have matrix elements only between components of the states $|S_2P\rangle$ and $|S_3Q\rangle$. The reduced matrix elements of $\mathbf{a}^r$ between the total states $|\Psi''\rangle$ and $|\Psi^+\rangle$ may be expressed in terms of those for the subshell r by

$$\langle\Psi^+(S'J)\|\,\mathbf{a}^r\,\|\Psi''(SK)\rangle = \pm(\omega_{SK}\omega_{S_3Q})^{1/2}$$
$$\times\langle S_3Q\|\,a^r\,\|S_2P\rangle W(S_3S_2S'S;(\tfrac{1}{2})S_1)\cdot W(QPJK;rL) \tag{5.25}$$

where $W(ABCD;EF)$ stands for the *Racah W-coefficients*.*[8,9,13] Inserting Eq. 5.25 into Eqs. 5.20 and 5.21 produces the following expression for the probability of ionization to the state Ψ^+:

$$\mathscr{P} \propto n|\langle\phi^r|\,\mathbf{p}_n\,|\chi_p\rangle|^2\,|\langle S_3Q(r^{n-1})|\,)S_2P(r^n)\rangle|^2\,\omega_{SK}\omega_{S'j}\omega_{S_2P}$$
$$\times W^2(S_3S_2S'S;(\tfrac{1}{2})S_1)\cdot W^2(QPJK;rL) \tag{5.26}$$

In the special case of closed shell molecules, r is fully occupied and ω_{SK}, ω_{S_2P}, the coefficient of fractional parentage, and the Racah W-coefficients reduce to unity. Therefore, this complex expression for the probability of ionization, Eq. 5.26, correctly reduces to a simple form such as Eq. 5.9, which is proportional to $\omega_{S'J}$, the total spin-orbital degeneracy of the ion state produced. Another special case is the situation in which $|S'J\rangle$ is the *only* ionic state formed by coupling of two open shells $|S_1L\rangle$ and $|S_2P\rangle$. In this case the Racah coefficients cancel with the degeneracy factors in Eq. 5.26 because $W^2(QPJK;rL)W^2(S_3S_2S'S;(\tfrac{1}{2})S_1) = (\omega_{S'J}\omega_{S_2P})^{-1}$.

* The coefficient $W(ABCD;EF)$ is equivalent, apart from a possible phase factor, to the coefficients $W(^{ABE}_{DCF})$ of Griffith[9] for the finite point groups. Racah coefficients result from taking the scalar product of tensors. They are discussed in detail by J. C. Slater.[13]

Now consider application of Eq. 5.26 to a typical example. The coefficients of fractional parentage determine the states arising from the open shell that is ionized, and the Racah coefficients take account of the initial coupling between the two different open-shells. To illustrate the significance of these terms, we return to the example of the high-spin octahedral d^5 complex with the ground state $\cdots t_2^3 e_g^2(^6A_{1g})$. The initial coupling conditions are defined by $S = \frac{5}{2}$, $S_1 = \frac{3}{2}$, $S_2 = 1$, and $S_3 = \frac{1}{2}$. Considering just the spin aspect of the problem, the probability of an e_g ionization, for example, realizing an ion state with spin S' will depend on the Racah coefficient $W[(\frac{1}{2})(1)(S')(\frac{5}{2}); (\frac{1}{2})(\frac{3}{2})]$. This coefficient is zero unless $S' = 2$, so that only quintet states can be produced.

Another interesting example[14] is the molecule chromocene, $(C_5H_5)_2Cr$. Chromocene probably has D_{5d} symmetry and a ground state electronic configuration $\cdots a_1^1 e_{2g}^3, {}^3E_{2g}$ resulting from a perturbed d^4 configuration. The probabilities of ionization of the a_{1g} and e_{2g} subshells are expected to be approximately proportional to their occupancies, that is $1:3$. Ionization of the e_{2g} subshell can produce several distinct states of the molecular ion. First consider ionization of the $\cdots e_{2g}^3, {}^2E_{2g}$ configuration, on its own, to yield $\cdots e_{2g}^2, {}^1A_{1g}, {}^1E_{1g}, {}^3A_{2g}$ states of the ion. In this simple case, the squares of the coefficients of fractional parentage happen to be proportional to the spin-orbital degeneracies of the ion states, and the relative probabilities for producing the ${}^1A_{1g}$, ${}^1E_{1g}$, and ${}^3A_{2g}$ states are $1:2:3$. We must now couple these states with the undisturbed open shell $a_{1g}^1, {}^2A_{1g}$. The ${}^1A_{1g}$ and ${}^1E_{1g}$ states can form only one state each, namely, ${}^2A_{1g}$ and ${}^2E_{1g}$, upon coupling with the undisturbed open shell. In this case the Racah coefficients cancel with the degeneracy factors of Eq. 5.26. The ${}^3A_{2g}$ state can form two states, ${}^4A_{2g}$ and ${}^2A_{2g}$, upon coupling with the undisturbed open shell. In this case,[14] the Racah orbital coefficients are unity and the spin coefficients are ${}^4A_{2g}: W^2[(\frac{1}{2})(1)(\frac{3}{2})(1); (\frac{1}{2})(\frac{1}{2})] = 1/9$ and ${}^2A_2: W^2[(\frac{1}{2})(1)(\frac{1}{2})(1); (\frac{1}{2})(\frac{1}{2})] = 1/36$. The coupling scheme used to produce the final ionic states is illustrated in Fig. 5.5. By incorporating the degeneracy factors, it is possible to predict the pattern of photoelectron band intensities as:

configuration	$\cdots e_{2g}^3$	$\cdots a_{1g}^1 e_{2g}^2$			
resulting states	${}^2E_{2g}$	${}^2A_{1g},$	${}^2A_{2g},$	${}^2E_{1g},$	${}^4A_{2g}$
predicted relative intensities	$\simeq 1$	$1/2$	$1/6$	1	$4/3$

Summary of vector coupling rules for open shells

The general rules for ionization of open-shell molecules, for which several states of the molecular ion may be produced on ionization of any given

Ionization of Chromocene

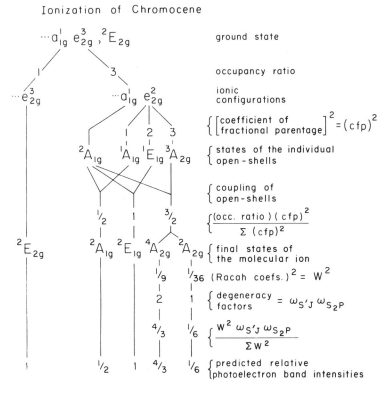

Fig. 5.5. Coupling scheme for production of ionic states by one-electron ionization processes from the ground state, $\cdots a_{1g}^1 e_{2g}^3$, $^2E_{2g}$, of $(C_5H_5)_2Cr$.

electronic subshell, may be summarized as follows:[7,14]

1. If orbitals belonging to different subshells are assumed to have the same one-electron cross sections, the total probability for ionization of a particular subshell is simply proportional to the occupancy of that subshell in the molecule. Experimentally, the total probability for ionization of a particular subshell is the sum of the areas of all photoelectron bands arising from the ionization of that subshell. This rule is the essence of Eq. 5.9.

2. If a closed subshell is ionized, all states arising from the coupling of the positive hole with the open-shell ground state will be realized, the relative probabilities for production of these states being in proportion to their spin-orbital degeneracies. This rule is determined from Eqs. 5.13 and 5.14.

3. For molecules with a single open shell, ionization of this unique open shell may produce several ion states with relative probabilities in proportion to the squares of the fractional parentage coefficients of these respective states. The states of the ion are determined by coupling the ground-state term representation with the representation of the electron ejected. The squares of the fractional parentage coefficients may, but in general will not, be proportional to the spin-orbital degeneracies of the ion states. This rule is determined from the coupling expression, Eq. 5.15.

4. For molecules with more than one open shell, it is necessary to take into account the predetermined coupling that exists between the different open shells. This involves Racah coefficients as additional proportionality factors in the expressions for the probability of ionization. The probability for production of particular ion states, therefore, is dependent on the squares of the fractional parentage coefficients and the Racah parameters, as expressed in Eq. 5.26.

There are several assumptions made in deriving these rules that should be kept in mind. (1) It is assumed that both molecule and ion are well described by restricted Hartree–Fock wavefunctions. (2) All molecular orbitals are assumed to have the same photoionization cross sections, that is, the matrix elements $\langle \phi_j | \mathbf{p}_n | \chi_p \rangle$ are assumed to be identical for all one-electron ionizations from the various MOs ϕ_j. (3) Configuration interaction is completely neglected in the derivation. It is known that CI can be of major significance in open-shell molecules and ions (for example, see O_2, Section 5.2). All three of these approximations can lead to inaccuracies in predicted intensities; therefore, intensities determined from the above rules should be considered as crude approximations and should be used with caution when interpreting spectra. The spectra themselves are not easily interpreted, for most of these open-shell molecules are large and their photoelectron bands are broad, overlapping, and structureless.

5.5. Molecules with two or more open shells

Some of the most common types of molecules with two or more open shells in the ground state are the complexes of transition and rare earth elements. The bis(π-cyclopentadienyl) transition metal complexes, $(C_5H_5)_2M$, constitute an interesting series with both closed- and open-shell species. Most of these metallocenes have D_{5d} symmetry, that is, the sandwich structure in which the cyclopentadiene rings are staggered. The outer molecular orbitals of these complexes, as predicted from recent

Table 5.3. Outer Molecular Orbitals of Metallocenes as Predicted from Recent Calculations[17]

	Orbital Symmetry	Orbital Type	Approximate Description
	.	.	
	.	.	
	.	.	
	e_{1g}	$3d$	Metal-ring π
Binding energy	a_{1g}	$3d$	Metal
	e_{2g}	$3d$	Metal-ring π
	e_{1g}	$\pi, \sigma, 3d$	Metal-ring
	e_{1u}	π	Ring π
	.	.	σ orbitals and one
	.	.	π orbital localized
	.	.	on the rings

Table 5.4. Molecular Ground-State Electron Configurations and Ionic States Accessible as a Result of One-Electron Ionization Processes for the Series $(C_5H_5)_2M$, Where $M = V$, Cr, Mn, Fe, Co, and Ni (Alternative Ground-State Configurations are Listed for Chromocene and Manganocene)

Molecule	Ground State	Accessible Ionic States
$(C_5H_5)_2V$	$(a_{1g})^1(e_{2g})^2, {}^4A_{2g}$	$\rightarrow \cdots (a_{1g})^1(e_{2g})^1, {}^3E_{2g}$
		$\rightarrow \cdots (e_{2g})^2, {}^3A_{2g}$
$(C_5H_5)_2Cr$	$\cdots (a_{1g})^2(e_{2g})^2, {}^3A_{2g}$	$\rightarrow \cdots (a_{1g})^2(e_{2g})^1, {}^2E_{2g}$
		$\rightarrow \cdots (a_{1g})^1(e_{2g})^2, {}^2A_{2g}, {}^4A_{2g}$
	$\cdots (e_{2g})^3(a_{1g})^1, {}^3E_{2g}$	$\rightarrow \cdots (e_{2g})^3, {}^2E_{2g}$
		$\rightarrow \cdots (e_{2g})^2(a_{1g})^1, {}^2A_{1g}, {}^2A_{2g}, {}^2E_{1g}, {}^4A_{2g}$
	$\cdots (e_{2g})^4, {}^1A_{1g}$	$\rightarrow \cdots (e_{2g})^3, {}^2E_{2g}$
$(C_5H_5)_2Mn$	$\cdots (a_{1g})^2(e_{2g})^3, {}^2E_{2g}$	$\rightarrow \cdots (a_{1g})^2(e_{2g})^2, {}^1A_{1g}, {}^3A_{2g}, {}^1E_{1g}$
		$\rightarrow \cdots (a_{1g})^1(e_{2g})^3, {}^1E_{2g}, {}^3E_{2g}$
	$\cdots (e_{2g})^4(a_{1g})^1, {}^2A_{1g}$	$\rightarrow \cdots (e_{2g})^4, {}^1A_{1g}$
		$\rightarrow \cdots (e_{2g})^3(a_{1g})^1, {}^1E_{2g}, {}^3E_{2g}$
	$\cdots (a_{1g})^1(e_{2g})^2(e_{1g})^2, {}^6A_{1g}$	$\rightarrow \cdots (e_{2g})^2(e_{1g})^2, {}^5A_{1g}$
		$\rightarrow \cdots (a_{1g})^1(e_{2g})^1(e_{1g})^2, {}^5E_{2g}$
		$\rightarrow \cdots (a_{1g})^1(e_{2g})^2(e_{1g})^1, {}^5E_{1g}$
$(C_5H_5)_2Fe$	$\cdots (a_{1g})^2(e_{2g})^4, {}^1A_{1g}$	$\rightarrow \cdots (a_{1g})^2(e_{2g})^3, {}^2E_{2g}$
		$\rightarrow \cdots (a_{1g})^1(e_{2g})^4, {}^2A_{1g}$
$(C_5H_5)_2Co$	$\cdots (a_{1g})^2(e_{2g})^4(e_{1g})^1, {}^2E_{1g}$	$\rightarrow \cdots (a_{1g})^2(e_{2g})^4, {}^1A_{1g}$
		$\rightarrow \cdots (a_{1g})^2(e_{2g})^3(e_{1g})^1, {}^1E_{1g}, {}^1E_{2g}, {}^3E_{1g}, {}^3E_{2g}$
		$\rightarrow \cdots (a_{1g})^1(e_{2g})^4(e_{1g})^1, {}^1E_{1g}, {}^3E_{1g},$
$(C_5H_5)_2Ni$	$\cdots (a_{1g})^2(e_{2g})^4(e_{1g})^2, {}^3A_{2g}$	$\rightarrow \cdots (a_{1g})^2(e_{2g})^4(e_{1g})^1, {}^2E_{1g}$
		$\rightarrow \cdots (a_{1g})^2(e_{2g})^3(e_{1g})^2, {}^2E_{2g}, {}^4E_{2g}$
		$\rightarrow \cdots (a_{1g})^1(e_{2g})^4(e_{1g})^2, {}^2A_{2g}, {}^4A_{2g}$

calculations,[17] are listed in Table 5.3. The group of orbitals, $\cdots e_{2g}a_{1g}e_{1g}$, are predominantly metal d-atomic orbitals localized on the central metal atom and are the lowest energy occupied orbitals in the ground state. The orbitals of the cyclopentadiene moieties are considerably more tightly bound than these metal d-orbitals. The most likely molecular ground states have been determined from magnetic susceptibility measurements.[16] The molecular ground state electron configurations and ionic states accessible as a result of one-electron ionization processes from the ground state are listed in Table 5.4 for the series $(C_5H_5)_2M$, where $M = V$, Cr, Mn, Fe, Co, and Ni. Alternative ground-state configurations are listed for chromocene and manganocene. The spectra of these six metallocenes are presented in Fig. 5.6. The simplest spectrum to analyze is that of the closed-shell species ferrocene, whose electron configuration is expected to be[17]

$$\cdots (e_{1u})^4(e_{1g})^4(e_{2g})^4(a_{1g})^2, \, {}^1A_{1g} \tag{5.27}$$

The low energy bands of ferrocene are expanded in Fig. 5.7. The two lowest energy bands centered at 6.9 eV and 7.25 eV have intensity ratios

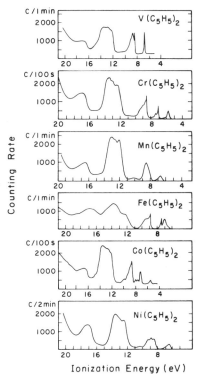

Fig. 5.6. He I photoelectron spectra of the metallocenes $(C_5H_5)_2M$ where $M = V$, Cr, Mn, Fe, Co, and Ni.

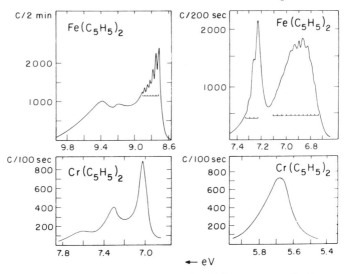

Fig. 5.7. Low-energy photoelectron bands of ferrocene and chromocene on expanded energy scales.

of $\sim 2.2:1.0$, respectively. The band shapes indicate that the 6.9-eV band results from ejection of a bonding electron and the 7.25-eV band results from ejection of a primarily nonbonding electron. Vibrational progressions of $\sim 282\ cm^{-1}$ are resolved corresponding to the symmetric ring-metal stretching frequency, which is $303\ cm^{-1}$ in the molecular ground state. These experimental observations lead to an assignment of the ionic ground state of $(C_5H_5)_2Fe^+$ as $^2E_{2g}$ and the first excited ionic state as $^2A_{1g}$, resulting from ionization of the bonding e_{2g} orbital and the non-bonding a_{1g} orbital, respectively. This experimentally predicted ordering for these two lowest energy ionic states, $^2E_{2g} < ^2A_{1g}$, is inconsistent with the calculated ordering of the ground-state orbitals, Eq. 5.27. The inconsistency must be explained either as a failure of Koopmans' theorem or inaccuracies in the computed orbital energies. Considering the remaining bands of the ferrocene spectrum, the structure centered at ~ 9 eV is very complex, although vibrational structure can be resolved. The energy of this complex band is not sensitive to the central metal atom, for its position remains relatively constant in all of the metallocene spectra. It is most likely due to ionization of the e_{1g} and e_{1u} orbitals, which are metal-ring bonding and ring-π bonding (Table 5.3). The complex structure between ~ 11.7 eV and ~ 15 eV is common to all the metallocene spectra and is undoubtedly due to ejection of electrons from the π and σ orbitals of the cyclopentadiene rings.

The remaining metallocenes all have from one to three electrons more or less than ferrocene. Their open-shell ground states provide multiple ionic states upon ionization. These multiple states are overlapped, making the analysis exceedingly complex. Identifications of the ionization bands have been attempted,[17,18] although the assignments must still be considered tentative. For molecules such as $(C_5H_5)_2Cr$ and $(C_5H_5)_2Mn$, the spectral interpretations are complicated by uncertainties in the ground state electron configurations. Attempts have been made to determine these ground-state configurations from the photoelectron spectra.[17,18] The assignment of the low energy bands of chromocene, Fig. 5.7, has not been deciphered. Cobaltocene has the lowest first E_I of the series, 5.5 eV, which is a result of its alkali metal-like properties conferred from the single electron outside of a closed-shell configuration (Table 5.4).

The relative energies of the molecular ion states resulting from ejection of these metal d electrons can be estimated from ligand-field theory. Such energy considerations using the ligand-field parameters can be of considerable assistance in assigning the multiplicity of d ionizations.

The lanthanide compounds form an interesting series in view of the possibility of several unpaired electrons in the $4f$ orbitals. Spectra of the closed $4f$ shells of lutetium(III) and hafnium(IV) show only two bands[19] corresponding to the spin-orbit components, $4f_{7/2}$ and $4f_{5/2}$. When the $4f$ orbitals are only partially filled, ionization of a $4f^n$ configuration to produce the ionic configuration $4f^{n-1}$ can yield many different ionic states corresponding to the large number of possible J values. Spectra of the lanthanide fluorides, oxides, and other nonconducting compounds[20,21] and of the metallic elements[22] show complicated structure in the region of $4f$ ionization. Most of these spectra have been studied with X-ray excitation where the resolution is insufficient for observation of the individual Russell–Saunders terms of the ion.

References

1. a. R. M. Hochstrasser, *Molecular Aspects of Symmetry*, p. 169 Benjamin, New York, 1966; G. Herzberg, *Electronic Spectra and Electronic Structure of Polyatomic Molecules*, p. 332, Van Nostrand, New York, 1967.
   b. O. Edqvist, L. Åsbrink, and E. Lindholm, *A. Naturforsch. A*, **26,** 1407 (1971); O. Edqvist, E. Lindholm, L. E. Selin, H. Sjögren, and L. Åsbrink, *Ark. Fys.*, **40,** 439 (1970).
2. O. Edqvist, E. Lindholm, L. E. Selin, and L. Åsbrink, *Phys. Scripta*, **1,** 25 (1970).
3. C. R. Brundle, *Chem. Phys. Letters*, **5,** 410 (1970).
4. R. N. Dixon and S. E. Hull, *Chem. Phys Letters*, **3,** 367 (1969).

5. O. Edqvist, E. Lindholm, L. E. Selin, L. Åsbrink, C. E. Kuyatt, S. R. Mielczarek, J. A. Simpson, and I. Fischer–Hjalmars, *Phys. Scripta*, **1,** 172 (1970).

6. C. R. Brundle, D. Neumann, W. C. Price, D. Evans, A. W. Potts, and D. G. Streets, *J. Chem. Phys.*, **53,** 705 1970).

7. P. A. Cox and A. F. Orchard, *Chem. Phys. Letters*, **7,** 273 (1970).

8. D. M. Brink and G. R. Satcher, *Angular Momentum*, Oxford U.P., London, 1968.

9. J. S. Griffith, *The Irreducible Tensor Method for Molecular Symmetry Groups*, Prentice-Hall, London 1962.

10. J. S. Griffith, *The Theory of Transition Metal Ions*, Cambridge U. P. London, 1961.

11. M. Born, W. Heisenberg, and P. Jordan, *Z. Phys.* **35,** 557 1926).

12. G. Racah, *Phys. Rev.*, **63,** 367 (1943).

13. J. C. Slater, *Quantum Theory of Atomic Structure*, Vol. II, McGraw-Hill, New York, 1960.

14. P. A. Cox, S. Evans, and A. F. Orchard, *Chem. Phys. Letters*, **13,** 386 1972).

15. J. B. French, E. C. Halbert, J. B. McGory, and S. S. M. Wong, *Advan. Nucl. Phys.*, **3,** 193 (1969).

16. M. L. H. Green, *Organometallic Compounds*, Vol. II, Methuen, London, 1967, and references therein.

17. J. W. Rabalais, L. O. Werme, T. Bergmark, L. Karlsson. M. Hussain, and K. Siegbahn, *J. Chem. Phys.*, **57,** 1185 (1972).

18. S. Evans, M. L. H. Green, B. Jewitt, G. H. King, and A. F. Orchard, *J. C. S. Faraday II*, **70,** 356 (1974).

19. P. A. Cox, Y. Baer, and C. K. Jörgensen, *Chem. Phys. Letters*, **22,** 433 (1973).

20. G. K. Wertheim, A. Rosencwaig, R. L. Cohen, and H. J. Guggenheim, *Phys. Rev. Letters*, **27,** 505 (1971).

21. C. Bonnell, R. C. Karanatak, and C. K. Jörgensen, *Chem. Phys. Letters*, **14,** 145 (1972).

22. P. O. Hedén, H. Löfgren, and S. B. M. Hagström, *Phys. Rev. Letters*, **26,** 432 (1971).

6 Photoionization Cross Sections

The intensities of photoelectron bands are most conveniently described in terms of specific differential cross sections $d\sigma/d\Omega$, which are a measure of the probability of photoionization from a specific molecular orbital. These cross sections have units of area and are a function of the incident photon energy. In Sections 1 to 4, we derive the equations necessary to describe photoionization of a general polyatomic molecule in a closed-shell electronic state (with nuclei fixed according to the Born–Oppenheimer approximation) to form an ionic state in which one electron is excited into the continuum. All bound orbitals are assumed identical in the initial and final states and the unbound orbital is approximated either by a plane wave or by a plane wave orthogonalized to all occupied bound orbitals. General equations are derived in Section 5 for the photoionization cross section of randomly oriented polyatomic molecules as a function of angle between the polarization vector of the photon beam and the propagation vector of photoejected electrons. Also derived is the cross section of the same sample as a function of angle between an unpolarized photon beam and the propagation vector of photoejected electrons. The application of these equations is discussed in Section 6. Simplified equations describing the cross sections of linear molecules are derived in Section 7, and the use of orthogonalized atomic orbitals is discussed in Section 8. Interpretation of the results of these computations is considered in Chapter 7.

6.1. Theoretical method

Considerable theoretical work on atomic photoionization intensities has been accomplished.[1] For molecules, calculations on H_2^+ have been reported by Bates, Öpik, and Poots[2a] and by Cohen and Fano,[2b] on H_2 by Flannery and Öpik[3a] and by Kelly,[3b] on CH_4 by Dalgarno,[4] on N_2 by Schneider and Berry[5a] and by Tuckwell,[5b] on benzene by Johnson and Rice,[6] and on benzene, H_2, ethylene, and butadiene by Kaplan and Markin.[7] Extensive calculations on π-electron systems have been done by Lohr and Robin[8] and on various molecules by Thiel and Schweig.[9] A most general formulation of the photoionization cross section for diatomic molecules has been derived by Tully, Berry, and Dalton.[10] Detailed consideration of the angular dependence of the photoionization cross section has been given by Cooper and Zare[11a] and by Buckingham, Orr, and Sichel.[11b]

Primary ingredients in a cross-section calculation are initial-state and final-state wavefunctions. One component of the latter is the one-electron function representing the ejected electron. A plane wave is the easiest and simplest approximation to this one-electron continuum function. Lohr has called attention to the neglect of orthogonality inherent in usual plane-wave calculations[12] and proposed a correction involving Schmidt orthogonalization of the plane wave to the occupied ground-state bound molecular orbitals. Applications become much more complicated with this correction, but since its effect may be quite significant in certain cases, general formulas for this extension are included. The derivation presented here follows the detailed treatment of Ellison.[13] Begin by defining the laboratory coordinate system (x, y, z), and take the exciting radiation to be directed along the y axis with its unit polarization vector $\mathbf{u}$ directed along the z axis. Let the propagation vector for the ejected electron be $\mathbf{k}$. Define a molecular coordinate system (x', y', z'); the orientation of this molecular coordinate system (x', y', z') relative to the laboratory system (x, y, z) is specified by the Euler angles (α, β, γ), which we abbreviate by the vector argument $(\boldsymbol{\beta})$. Spherical harmonics $Y_{l,m'}(\theta', \phi')$ written in terms of spherical polar coordinates in the molecular system can be expressed as linear combinations of spherical harmonics involving corresponding angles in the laboratory system:

$$Y_{l,m'}(\theta', \phi') = \sum_{m} \mathbf{D}^{l}_{m,m'}(\boldsymbol{\beta}) Y_{l,m}(\theta, \phi) \qquad (6.1)$$

The $\mathbf{D}^{l}_{m,m'}(\boldsymbol{\beta})$ are elements of the usual rotation matrix. We follow conventions given by Rose,[14] and list in Appendix I for easy reference several relations involving $\mathbf{D}^{l}_{m,m'}$, the spherical harmonics, and the Clebsch–Gordon coefficients.

6.2. General equations for cross sections in plane-wave and orthogonalized plane-wave approximations

The quantum mechanical expression for the photoionization cross section σ is related to the Golden Rule rate

$$W = \frac{2\pi}{\hbar} |\mathscr{H}'_{ab}|^2 \rho(E) \qquad (6.2)$$

as obtained from time-dependent perturbation theory.[15] Here $\mathcal{H}'_{ab}$ is the interaction energy of the initial state $|\Psi_0\rangle$ and final state $|\Psi_j\rangle$ due to the presence of the radiation field, and $\rho(E)$ is the density of final states with energy in the vicinity of E_j. Treating the interaction in the electric-dipole approximation, which is equivalent to neglecting photon momentum and valid for low photon energies,

$$W = \frac{2\pi e^2 A_0^2}{\hbar m^2 c^2} |\mathbf{u} \cdot \langle \Psi_0| \sum_n \mathbf{p}_n |\Psi_j\rangle|^2 \rho(E) \tag{6.3}$$

where A_0 is the magnitude of the vector potential, $\mathbf{u}$ is a unit vector in the polarization direction, and $\mathbf{p}_n$ is the linear momentum operator for the nth electron. The number rate is converted to a cross section by multiplying by the photon energy $\hbar\omega$ and equating the result to the product of the cross section σ and the magnitude of the Poynting vector $\omega^2 A_0^2/2\pi c$. Dividing the expression for σ by 4π results in the differential cross section for producing photoelectrons in the solid angle $d\Omega$

$$\frac{d\sigma}{d\Omega} = \frac{\pi e^2}{m^2 \omega c} |\mathbf{u} \cdot \langle \Psi_0| \sum_n \mathbf{p}_n |\Psi_j\rangle|^2 \rho(E) \tag{6.4}$$

A useful first step is to consider the ionization process as a pure electronic transition and to evaluate the cross section for fixed nuclear positions; thus we initially ignore vibrational and rotational factors.

Consider a molecule initially in a closed-shell ground state Ψ_0, which we shall approximate by a Slater determinant of doubly occupied orthonormal MOs ϕ_l. Photoionization with quanta $\hbar\omega$ leads to a final state Ψ_j, which is approximated by a linear combination of two Slater determinants giving a spin singlet in which one electron has been excited from the MO ϕ_j to an unbound plane wave orbital

$$|PW(\mathbf{k})\rangle = L^{-3/2} e^{i\mathbf{k} \cdot \mathbf{r}} \tag{6.5}$$

which has been normalized in a large cubic box of edge length L. The accompanying density of states is

$$\rho(E) = \frac{mkL^3}{2\pi^2 \hbar^2} \tag{6.6}$$

where

$$k = \frac{[2m(\hbar\omega - E_I^j)]^{1/2}}{\hbar} \tag{6.7}$$

is the magnitude of $\mathbf{k}$ and E_I^j is the ionization energy for the MO ϕ_j. With the preceding restrictions on Ψ_0 and Ψ_j, substitution of Eqs. 6.5 and 6.6

into Eq. 6.4 yields

$$\frac{d\sigma}{d\Omega} = \frac{e^2 k L^3}{2\pi m c \hbar^2 \omega} |\mathbf{u} \cdot \mathbf{P}_{0j}|^2 \tag{6.8}$$

$$\mathbf{P}_{0j} = 2^{1/2}\hbar[\mathbf{k}\langle \phi_j | PW(\mathbf{k})\rangle + i\sum_l \langle \phi_j | \mathbf{\nabla} | \phi_l\rangle \langle \phi_l | PW(\mathbf{k})\rangle] \tag{6.9}$$

in which we have used the fact that the plane-wave function, Eq. 6.5, is an eigenfunction of the momentum operator $\mathbf{p} = (\hbar/i)\mathbf{\nabla}$. Since the mean value of the momentum in a bound state is zero, the term $l = j$ may be excluded from the sum over all occupied MOs ϕ_l. The important point concerning Eq. 6.9 is that it properly accounts for the fact that a plane wave is not necessarily orthogonal to any of the occupied MOs;[12,13] Equation 6.9 is equivalent to representing the unbound electron, not by Eq. 6.5, but rather by an *orthogonalized plane wave*.

We assume that the MOs are defined so as to be real. The gradient matrix elements

$$\mathbf{G}_{jl} = \langle \phi_j | \mathbf{\nabla} | \phi_l \rangle \tag{6.10}$$

appearing in Eq. 6.9 will thus be real vectors. We employ the LCAO (linear combination of atomic orbitals) approximation

$$\phi_l = \sum \mathbf{b}_{l\mu} \chi_\mu^{(r)} \tag{6.11}$$

Define the initial basis $\chi^{(r)}$ to be AOs, either Slater-type orbitals (STOs), orthogonalized STOs, or Gaussian-type orbitals (GTOs), written in their most commonly used *real* forms. The subsequent analysis is heavily dependent upon rotations and thus intimately connected with the properties of angular momentum. It thus becomes expeditious to transform to a new AO basis χ defined with angular factors as spherical harmonics written in the usual complex form. If $\chi^{(r)} = \mathbf{a}\chi$, then

$$\phi_l = \sum c_{lp}\chi_p \tag{6.12}$$

where $\mathbf{c} = \mathbf{ba}$. Details are provided in Appendix II.

Equations 6.9, 6.10, and 6.12 combine to give

$$\mathbf{u} \cdot \mathbf{P}_{0j} = 2^{1/2}\hbar \sum_p F_{pk}[c_{jp}^* k \cos\theta_k + i\gamma_{jp}] \tag{6.13}$$

in which

$$F_{pk} = \langle \chi_p | PW(\mathbf{k})\rangle \tag{6.14}$$

$$\gamma_{jp} = \sum_l (\mathbf{u} \cdot \mathbf{G}_{jl}) c_{lp}^* \tag{6.15}$$

and θ_k is the angle between the ejected electron and the electric vector $\mathbf{u}$ of the incoming photon. Then

$$|\mathbf{u} \cdot \mathbf{P}_{0j}|^2 = 2\hbar^2 \sum_p \sum_q [F_{pk}^* F_{qk}(c_{jp}c_{jq}^* k^2 \cos^2 \theta_k$$
$$+ \gamma_{jp}^* \gamma_{jq}) - 2k \cos \theta_k \mathscr{I}(c_{jp}F_{pk}^* F_{qk}\gamma_{jq})] \qquad (6.16)$$

in which we have used the identity $f - f^* = 2i\mathscr{I}(f)$ where $\mathscr{I}(f)$ is the imaginary component of f. Furthermore, we note that

$$\sum_p \sum_q f_p^* f_q = \sum_p f_p^* f_p + \sum_p \sum_{q>p} (f_p^* f_q + f_q^* f_p)$$
$$= \sum_p f_p^* f_p + 2 \sum_p \sum_{q>p} \mathscr{R}(f_p^* f_q) \qquad (6.17)$$

where $\mathscr{R}(f)$ is the real component of f. Thus we write

$$|\mathbf{u} \cdot \mathbf{P}_{0j}|^2 = 2\hbar^2(S_1 + S_2 + S_3 + S_4 + S_5) \qquad (6.18)$$

Where

$$S_1 = k^2 \cos^2 \theta_k \sum_p F_{pk}^* F_{pk} c_{jp} c_{jp}^*$$

$$S_2 = 2k^2 \cos^2 \theta_k \sum_p \sum_{q>p} \mathscr{R}(F_{pk}^* F_{qk} c_{jp} c_{jq}^*)$$

$$S_3 = \sum_p F_{pk}^* F_{pk} \gamma_{jp}^* \gamma_{jp} \qquad (6.19)$$

$$S_4 = 2 \sum_p \sum_{q>p} \mathscr{R}(F_{pk}^* F_{qk} \gamma_{jp}^* \gamma_{jq})$$

$$S_5 = -2k \cos \theta_k \sum_p \sum_q \mathscr{I}(c_{jp} F_{pk}^* F_{qk} \gamma_{jq})$$

The first two sums $S_1 + S_2$ yield the plane-wave approximation; adding in $S_3 + S_4 + S_5$ gives the orthogonalized plane-wave approximation.

At this stage, if Eqs. 6.8, 6.18, and 6.19 are combined, we will have the cross section for the following conditions: exciting radiation directed along the y axis with polarization vector along the z axis, molecule fixed with its coordinate system specified by the Euler angles $(\boldsymbol{\beta}) = (\alpha, \beta, \gamma)$ relative to the laboratory system, and photo-ejected electrons propagated in the direction $\mathbf{k}$. The functions $F_p^* F_q$ and $\gamma_p^* \gamma_q$ that appear in Eqs. 6.19 depend (as we see in Section 6.4) explicitly upon the Euler angles $(\boldsymbol{\beta})$, which specify the molecular orientation; the $F_p^* F_q$ also depend upon the direction of $\mathbf{k}$ in the laboratory system. If we are dealing with randomly oriented molecules, as is usually the case, we must average Eq. 6.18 over all space defined by the Euler angles ($\boldsymbol{\beta}$). We signify the quantities so averaged as follows:

$$|\mathbf{u} \cdot \mathbf{P}_{0j}|^2_{av} = 2\hbar^2(\bar{S}_1 + \bar{S}_2 + \bar{S}_3 + \bar{S}_4 + \bar{S}_5) \tag{6.20}$$

Mathematically, one might expect that the cross section at this stage, hence Eq. 6.20, would depend upon the two remaining laboratory variables θ_k and ϕ_k. It turns out that explicit dependence is obtained only upon θ_k, the angle between the polarization vector $\mathbf{u}$ (that is, the laboratory z axis) and the propagation vector $\mathbf{k}$. If one is dealing with unpolarized light, a further average over all polarization angles is appropriate.

6.3. Basic components of cross-section equations

In Eqs. 6.19, the indices p and q refer to AOs χ_p, defined in Appendix II. Each AO is specified by the set of parameters $n(p)$, $l(p)$, $m(p)$, $\zeta(p)$ and $\mathbf{R}_p$: the principal quantum number, azimuthal quantum number, magnetic quantum number, orbital exponent, and the vector directed from the laboratory origin to the orbital origin.

The two basic factors appearing in Eqs. 6.19 are the overlap integrals F_{pk}, connecting the AO χ_p with the plane-wave $PW(\mathbf{k})$, and the γ_{jp}, which are related to a linear combination of components (along the polarization axis $\mathbf{u}$) of transition moments connecting the MO ϕ_j with all other occupied MOs ϕ_l. It follows from Eq. 6.15 that

$$\begin{aligned}
\gamma_{jp} &= \sum_l G_{jl} \cos \theta_l c^*_{lp} \\
&= \sum_l G_{jl} c^*_{lp} P_1(\cos \theta_l) \\
&= \left(\frac{4\pi}{3}\right)^{1/2} \sum_l G_{jl} c^*_{lp} Y_{1,0}(\theta_l, \phi_l)
\end{aligned} \tag{6.21}$$

If Eq. I.9 (see Appendix I) is introduced, we obtain

$$\gamma_{jp} = \left(\frac{4\pi}{3}\right)^{1/2} \sum_{m=-1}^{1} \mathbf{D}^{1*}_{0,m}(\boldsymbol{\beta}) \sum_l G_{jl} c^*_{lp} Y_{1,m}(\theta'_l, \phi'_l) \tag{6.22}$$

The arguments (θ'_l, ϕ'_l) refer to spherical polar coordinates of the vector $\mathbf{G}_{jl}$ in the molecular coordinate system; that is, (θ'_l, ϕ'_l) specify the direction of the transition moment $\mathbf{G}_{jl}$. Equation 6.22 will be useful directly in analyzing $\bar{S}_5$.

The product $\gamma^*_{jp}\gamma_{jq}$ is easily formulated using Eq. 6.22. In this expression there will appear factors of the form $\mathbf{D}^1_{0,m}(\boldsymbol{\beta})\mathbf{D}^{1*}_{0,m}(\boldsymbol{\beta})$; we impose Eq.

I.11 upon the second factor so that Eq. I.10 can be applied. After this, we use Eq. I.11 again, in reverse, to finally yield

$$\gamma_{jp}^*\gamma_{jq} = \left(\frac{4\pi}{3}\right)\sum_L C(11L;00) \sum_{m=-1}^{1} \sum_{m'=-1}^{1} (-1)^m C(11L; m, -m')$$

$$\times \mathbf{D}_{0,m'-m}^{L*}(\boldsymbol{\beta})\sum_l\sum_{l'} G_{jl}G_{jl'}c_{lp}c_{l'q}^* Y_{1,m}^*(\theta_l',\phi_l')Y_{1,m'}(\theta_{l'}',\phi_{l'}') \quad (6.23)$$

This equation will be useful in analyzing $\bar{S}_3$ and $\bar{S}_4$.

In Appendix III, we show that

$$F_{pk} = T_p i^{l(p)} \exp(i\mathbf{k}\cdot\mathbf{R}_p) Y_{l(p),m(p)}^*(\theta_k',\phi_k') \quad (6.24)$$

An explicit expression is derived there for T_p, a real function that depends upon k as well as the specific AO parameters $n(p)$, $l(p)$, and $\zeta(p)$. The arguments (θ_k',ϕ_k') refer to spherical polar coordinates of the propagation vector $\mathbf{k}$ in the molecular coordinate system. We now formulate the product $F_{pk}^*F_{qk}$; Eq. I.4 is used to give

$$F_{pk}^*F_{qk} = T_pT_q i^{l(p)+l(q)}(-1)^{l(p)+m(q)}\exp(i\mathbf{k}\cdot\mathbf{R}_{pq})$$

$$\times\sum_\lambda Q_{l(p),l(q),\lambda}C(l_pl_q\lambda; m_p, -m_q)C(l_pl_q\lambda; 0,0)Y_{\lambda,m(p)-m(q)}(\theta_k',\phi_k') \quad (6.25)$$

Here, $\mathbf{R}_{pq} = \mathbf{R}_q - \mathbf{R}_p$ is the vector directed from the atom to which χ_p belongs to the atom to which χ_q belongs.

For the special case $R_{pq}=0$, the exponential factor in Eq. 6.25 is unity, and we introduce Eq. 6.1 to give

$$F_{pk}^*F_{qk} = T_pT_q(-1)^{[l(q)-l(p)]/2}(-1)^{m(q)}\sum_\lambda Q_{l(p),l(q),\lambda}C(l_pl_q\lambda; m_p, -m_q)$$

$$\times C(l_pl_q\lambda; 0,0)\sum_\mu \mathbf{D}_{\mu,m(p)-m(q)}^\lambda(\boldsymbol{\beta})Y_{\lambda,\mu}(\theta_k,\phi_k) \qquad R_{pq}=0 \quad (6.26)$$

If $R_{pq}\neq 0$, we first introduce Eq. I.8 into Eq. 6.25, followed by the coupling Eq. I.4, with $(\theta,\phi)=(\theta_k',\phi_k')$, then Eq. 6.1, and finally the symmetry Eq. I.16. The final expression is complicated:

$$F_{pk}^*F_{qk} = 4\pi T_pT_q(-1)^{l(p)+m(p)}\sum_\lambda Q_{l(p),l(q),\lambda}C(l_pl_q\lambda; m_p, -m_q)C(l_pl_q\lambda; 0,0)$$

$$\times\sum_{\lambda'} i^{\lambda'+l(p)+l(q)}j_{\lambda'}(kR_{pq})(2\lambda'+1)^{-1}\sum_{\mu'=-\lambda'}^{\lambda'}(-1)^{\mu'}Y_{\lambda',\mu'}(\theta_{pq}',\phi_{pq}')\sum_{\lambda''}Q_{\lambda,\lambda',\lambda''}$$

$$\times(2\lambda''+1)C(\lambda\lambda''\lambda'; m_p-m_q, \mu'-m_p$$

$$+m_q)C(\lambda\lambda''\lambda'; 0,0)\sum_{\mu''}\mathbf{D}_{\mu'',m(p)-m(q)-\mu'}^{\lambda''}(\boldsymbol{\beta})Y_{\lambda'',\mu''}(\theta_k,\phi_k) \qquad R_{pq}\neq 0$$

$$(6.27)$$

This expression will be useful in analyzing each one of the sums appearing in Eqs. 6.18 to 6.19.

6.4. Derivation of the sums $\bar{S}_n$

The Sum $\bar{S}_1$

We see from Eq. 6.19 that $\bar{S}_1$ contains the factor $F_{pk}^*F_{pk}$. Equation 6.26 with $p=q$ is clearly applicable; we multiply first by $\mathbf{D}_{0,0}^{0*}(\boldsymbol{\beta})=1$, and then use Eqs. I.12 and I.13 to average over all angles $\boldsymbol{\beta}$. The condition $\lambda=0$ and $\mu=0$ result. Since $Y_{0,0}(\theta_k,\phi_k)=(4\pi)^{-1/2}$,

$$F_{pk}^*F_{pk}=(-1)^{m(p)}T_p^2 Q_{l(p),l(p),0}C(l_pl_p0;m_p,-m_p)C(l_pl_p0;0,0)(4\pi)^{-1/2} \tag{6.28}$$

Making use of Eqs. I.5 and I.18, we finally derive

$$\bar{S}_1=(4\pi)^{-1}k^2\cos^2\theta_k\sum_p T_p^2|c_{jp}|^2 \tag{6.29}$$

This term contains a sum of contributions T_p^2 from each AO χ_p, each weighted by its net atomic population arising from the ionized MO ϕ_j.

The Sum $\bar{S}_2$

We first consider contributions to $\bar{S}_2$ arising from orbitals χ_p and χ_q belonging to the same atom, so that $R_{pq}=0$. We multiply Eq. 6.26 by $\mathbf{D}_{0,0}^{0*}(\boldsymbol{\beta})=1$, and use Eqs. I.12 and I.13 to average over all $\boldsymbol{\beta}$. Again, we obtain $\lambda=0$, $\mu=0$, and $m_p-m_q=0$. But by Eq. I.18, $C(l_pl_q0;m_p,-m_p)=0$ unless $l_p=l_q$. Thus, using Eq. I.5, we find that the one-center contributions to $\bar{S}_2$ are

$$\bar{S}_2=(2\pi)^{-1}k^2\cos^2\theta_k\sum_p\sum_{q>p}T_pT_q\mathfrak{R}(c_{jp}c_{jp}^*)\delta_{l(p),l(q)}\delta_{m(p),m(q)}\qquad R_{pq}=0 \tag{6.30}$$

We now proceed to analyze contributions from orbitals χ_p and χ_q not belonging to the same center.

We multiply Eq. 6.27 by $\mathbf{D}_{0,0}^{0*}(\boldsymbol{\beta})=1$, and use Eqs. I.12 and I.13 to average over all $\boldsymbol{\beta}$. We find that $\lambda''=0$, $\mu''=0$, and $\mu'=m_p-m_q$. Making use of Eqs. I.5 and I.17, it can be shown that

$$\bar{S}_2=2k^2\cos^2\theta_k\sum_p\sum_{q>p}T_pT_q(-1)^{l(p)+m(q)}\sum_\lambda(-1)^{[\lambda+l(p)+l(q)]/2}$$
$$\times Q_{l(p),l(q),\lambda}C(l_pl_q\lambda;m_p-m_q)C(l_pl_q\lambda;0,0)j_\lambda(kR_{pq})$$
$$\times\mathfrak{R}[c_{jp}c_{jq}^*Y_{\lambda,m(p)-m(q)}(\theta_{pq}',\phi_{pq}')]\qquad R_{pq}\neq0 \tag{6.31}$$

Now, the Clebsch–Gordon coefficient

$$C(l_p l_q \lambda; 0, 0) = 0$$

unless

$$\lambda = l_p + l_q, \; l_p + l_q - 2, \cdots, |l_p - l_q| \tag{6.32}$$

Also, it follows that $\lambda + l_p + l_q$ is necessarily even. Thus the sum over λ in Eq. 6.31 is limited, and the exponent of -1 is integral.

The Sum $\bar{S}_3$

We first rewrite Eq. 6.17 using different symbols:

$$\sum_l \sum_{l'} f_l^* f_{l'} = \sum_l f_l^* f_l + 2 \sum_l \sum_{l'>l} \Re(f_l^* f_{l'}) \tag{6.33}$$

Then we set $p = q$ in Eq. 6.26 and note that the double sum over l and l' in Eq. 6.23 is the same form as Eq. 6.33. Keeping this in mind, we now multiply Eq. 6.23 by Eq. 6.26, each with $p = q$, and then average over all β using Eqs. I.12 and I.13; $\lambda = L$, $\mu = 0$, and $m' = m$ result. We thus obtain

$$\bar{S}_3 = \left(\frac{4\pi}{3}\right) \sum_\lambda P_\lambda(\cos \theta_k)(2\lambda + 1)^{-1} N_{\lambda,0} \sum_p (-1)^{m(p)} T_p^2 Q_{l(p),l(p),\lambda}$$

$$\times C(l_p l_p \lambda; m_p, -m_p) C(l_p l_p \lambda; 0, 0) C(11\lambda; 0, 0) \sum_{m=-1}^{1} (-1)^m C(11\lambda; m, -m).$$

$$\times \left\{ \sum_l G_{jl}^2 |c_{lp}|^2 |Y_{1,m}(\theta_l', \phi_l')|^2 + 2 \sum_l \right.$$

$$\left. \times \sum_{l'>l} G_{jl} G_{jl'} \Re[c_{lp} c_{l'p}^* Y_{1,m}^*(\theta_l', \phi_l') Y_{1,m}(\theta_{l'}', \phi_{l'}')] \right\} \tag{6.34}$$

Applying Eq. I.15 to $C(11\lambda; 0, 0)$ and to $C(l_p l_p \lambda; 0, 0)$, we see that the sums over λ and p reduce to terms for which

$$\lambda = 0, 2$$
$$\lambda \leq 2l_p \tag{6.35}$$

We note that the plane-wave approximation, which included the sums $\bar{S}_1$ and $\bar{S}_2$, gave a strictly $\cos^2 \theta_k$ angular dependence in the cross section. The orthogonalized plane-wave approximation, one term of which is $\bar{S}_3$, contains spherical terms as well as $\cos^2 \theta_k$ terms on account of the factor $Y_{\lambda,0}(\theta_k, \phi_k)$ with $\lambda = 0, 2$; $\lambda \leq 2l_p$.

The Sum $\bar{S}_4$

We first consider contributions to $\bar{S}_4$ rising from AOs χ_p and χ_q that belong to the same atom, hence $R_{pq} = 0$. We multiply Eq. 6.23 by Eq. 6.26 and then average over all $\boldsymbol{\beta}$ using Eq. I.12 and I.13; $\lambda = L$, $\mu = 0$, $m' = m + m_p - m_q$. Using Eq. 6.19, we thus derive

$$
\begin{aligned}
\bar{S}_4 = \left(\frac{8\pi}{3}\right) &\sum_\lambda P_\lambda(\cos\theta_k)(2\lambda+1)^{-1}N_{\lambda,0} \sum_p \sum_{q>p} (-1)^{m(q)} \\
&\times (-1)^{[l(p)-l(q)]/2} T_p T_q Q_{l(p),l(q),\lambda} C(l_p l_q \lambda; m_p, -m_q) \\
&\times C(l_p l_q \lambda; 0, 0) C(11\lambda; 0, 0) \\
&\times \sum_{m=-1}^{1} (-1)^m C(11\lambda; m, m_q - m_p - m) \\
&\times \sum_l \sum_{l'} G_{jl} G_{jl'} \Re[c_{lp} c_{l'q}^* Y_{1,m}^*(\theta_l', \phi_l') \\
&\times Y_{1,m+m(p)-m(q)}(\theta_{l'}', \phi_{l'}')] \qquad R_{pq} = 0
\end{aligned}
\qquad (6.36)
$$

Applying Eqs. I.14 and I.15 to $C(11\lambda; 0, 0)$, $C(l_p l_q \lambda; 0, 0)$, and $C(l_p l_q \lambda; m_p, -m_q)$, we find that sums over λ and p are limited to terms for which all of the following conditions apply:

$$
\begin{aligned}
&\lambda = 0, 2 \\
&\lambda = l_p + l_q, l_p + l_q - 2, \cdots, |l_p - l_q| \\
&\lambda \geq |m_p - m_q|
\end{aligned}
\qquad (6.37)
$$

The latter two conditions may be replaced by the following conditions upon terms entering the sum over p:

$$
\begin{aligned}
&\text{For } \lambda = 0, l_p = l_q \quad \text{and} \quad 0 = |m_p - m_q| \\
&\text{For } \lambda = 2, l_p = l_q \pm 2 \quad \text{and} \quad 2 \geq |m_p - m_q|
\end{aligned}
\qquad (6.38)
$$

Clearly, $l_p + l_q$ must be even; thus $l_p - l_q$ must be even, $(l_p - l_q)/2$ is integral, and the factor (-1) raised to that power in Eq. 6.36 is real.

We now consider contributions to $\bar{S}_4$ from AOs χ_p and χ_q for which $R_{pq} \neq 0$. We multiply Eq. 6.23 by Eq. 6.27 and then average over all $\boldsymbol{\beta}$ using Eqs. I.12 and I.13; $\lambda'' = L$, $\mu'' = 0$, and $\mu' = m_p - m_q + m - m'$.

$$
\begin{aligned}
\bar{S}_4 = \left(\frac{32\pi^2}{3}\right) &\sum_L P_L(\cos\theta_k) C(11L; 0, 0) N_{L,0} \sum_p \sum_{q>p} T_p T_q (-1)^{l(p)+m(q)} \\
&\times \sum_\lambda Q_{l(p),l(q),\lambda} C(l_p l_q \lambda; 0, 0) C(l_p l_q \lambda; m_p, -m_q)
\end{aligned}
$$

$$\times \sum_{m=-1}^{1} \sum_{m'=-1}^{1} (-1)^{m'} C(11L; m, -m')$$

$$\times \sum_{\lambda'} (-1)^{[\lambda'+l(p)+l(q)]/2} j_{\lambda'}(kR_{pq})(2\lambda'+1)^{-1} Q_{\lambda,\lambda',L} C(\lambda L\lambda'; 0, 0)$$

$$\times C(\lambda L\lambda'; m_p - m_q, m - m') \sum_{l} \sum_{l'} G_{jl} G_{jl'} \mathscr{R}[Y_{\lambda',m(p)-m(q)+m-m'}$$

$$\times (\theta'_{pq}, \phi'_{pq}) c_{lp} c^*_{l'q} Y^*_{1,m}(\theta'_l, \phi'_l) Y_{1,m'}(\theta'_{l'}, \phi'_{l'})] \qquad R_{pq} \neq 0 \qquad (6.39)$$

Application of Eqs. I.15 to the C coefficients requires that

$$L = 0, 2$$
$$\lambda = l_p + l_q, l_p + l_q - 2, \cdots, |l_p - l_q| \qquad (6.40)$$
$$\lambda' = \lambda + L, \lambda + L - 2, \cdots, |\lambda - L|$$

Also,

$$L \geq |m - m'|$$
$$\lambda \geq |m_p - m_q| \qquad (6.41)$$
$$\lambda' \geq |m_p - m_q + m - m'|$$

by virtue of Eqs. I.14.

The Sum $\bar{S}_5$

Once again, we must initially consider the special case concerning AOs χ_p and χ_q belonging to a common center. We multiply Eq. 6.22 by Eq. 6.24 and then average over all $\boldsymbol{\beta}$; $\lambda = 1$, $\mu = 0$, and $m = m_p - m_q$ result. We note that application of Eq. I.14 to the factor $C(l_p l_q \lambda; m_p, -m_q)$ with $\lambda = 1$ limits terms appearing in $\bar{S}_5$ to those for which

$$l_q = l_p \pm 1$$
$$|m_p - m_q| \leq 1 \qquad (6.42)$$

We finally use the last of Eqs. 6.19, as well as Eq. I.1 with $P_1^0(\cos \theta_k) = \cos \theta_k$, to obtain

$$\bar{S}_5 = -\tfrac{2}{3} k \cos^2 \theta_k \sum_{p} \sum_{q} (-1)^{m(q)} T_p T_q Q_{l(p),l(q),1} C(l_p l_q 1; m_p, -m_q)$$

$$\times C(l_p l_q 1; 0, 0) \sum_{l} G_{jl} \mathscr{I}[c_{jp} c^*_{lq} i^{l(q)-l(p)} Y_{1,m(p)-m(q)}(\theta'_l, \phi'_l)] \qquad R_{pq} = 0$$

$$(6.43)$$

We now turn to the case, $R_{pq} \neq 0$. We multiply Eq. 6.22 by Eq. 6.27 and average over all $\boldsymbol{\beta}$; $\lambda'' = 1$, $\mu'' = 0$, and $\mu' = m_p - m_q - m$. We note that application of Eq. I.15 to the factor $C(\lambda\lambda''\lambda'; 0, 0)$ with $\lambda'' = 1$ limits terms to those for which

$$\lambda' = \lambda \pm 1 \tag{6.44}$$

Application of Eq. I.15 to the factor $C(l_p l_q \lambda; 0, 0)$ limits

$$\lambda = l_p + l_q, l_p + l_q - 2, \cdots, |l_p - l_q| \tag{6.45}$$

and $\lambda + l_p + l_q$ must consequently be even. Finally, we obtain

$$\bar{S}_5 = -8\pi k \cos^2 \theta_k \sum_p \sum_q (-1)^{l(p)+m(q)} T_p T_q \sum_\lambda Q_{l(p),l(q),\lambda} C(l_p l_q \lambda; 0, 0)$$

$$\times C(l_p l_q \lambda; m_p, -m_q) \sum_{m=-1}^{1} (-1)^m \sum_{\lambda'} (2\lambda'+1)^{-1} Q_{\lambda,\lambda',1} j_{\lambda'}(kR_{pq})$$

$$\times C(\lambda 1\lambda'; m_p - m_q, -m) C(\lambda 1\lambda'; 0, 0) \sum_l G_{jl} \mathcal{I}[Y_{\lambda',m(p)-m(q)-m}$$

$$\times (\theta'_{pq}, \phi'_{pq}) c_{jp} c^*_{lq} i^{\lambda'+l(p)+l(q)} Y_{1,m}(\theta'_l, \phi'_l)] R_{pq} \neq 0 \tag{6.46}$$

In addition to the limitations on sums imposed by Eqs. 6.44 and 6.45, we also have

$$\lambda \geq |m_p - m_q|$$
$$\lambda' \geq |m_p - m_q - m| \tag{6.47}$$

as required by Eq. I14.

6.5. Angular distributions

It is well known that the differential cross-section angular distribution for polarized light and randomly oriented molecules is of the form

$$\frac{d\bar{\sigma}}{d\Omega} = \left(\frac{\sigma}{4\pi}\right)[1 + \beta P_2(\cos \theta_k)] \tag{6.48}$$

where θ_k is the angle between the ejected electron and the electric vector **u** of the exciting radiation and $P_2(\cos \theta)$ is the second-order Legendre polynomial $\frac{1}{2}(3\cos^2 \theta - 1)$. If one integrates $d\sigma$ over all $d\Omega = \sin \theta_k\, d\theta_k\, d\phi_k$, the total cross section σ is obtained; β is defined as the asymmetry parameter.

The theoretical sums $\bar{S}_n$ can all be expressed in the general form

$$\bar{S}_n = t_n \cos^2 \theta_k + u_n P_0(\cos \theta_k) + v_n P_2(\cos \theta_k) \tag{6.49}$$

in which $u_n = v_n = 0$ for $n = 1, 2,$ and 5, and $t_n = 0$ for $n = 3$ and 4. Thus spherical contributions to the distribution of photoelectrons are obtained only from the sums S_3 and S_4, that is, only after the plane-wave approximation is corrected for orthogonality.

One could carry through the numerical calculation first with $\theta_k = 0$, then with $\theta_k = \pi/2$:

$$\bar\sigma_\| = \left(\frac{d\bar\sigma}{d\Omega}\right)_\| = \left(\frac{\sigma}{4\pi}\right)(1+\beta), \qquad \theta_k = 0$$

$$\bar\sigma_\perp = \left(\frac{d\bar\sigma}{d\Omega}\right)_\perp = \left(\frac{\sigma}{4\pi}\right)(1-\tfrac{1}{2}\beta), \qquad \theta_k = \frac{\pi}{2} \tag{6.50}$$

From these, the asymmetry parameter and total cross section are easily obtained:

$$\beta = \frac{2(\bar\sigma_\| - \bar\sigma_\perp)}{(\bar\sigma_\| + 2\bar\sigma_\perp)} \tag{6.51}$$

$$\sigma = \frac{4\pi(\bar\sigma_\| + 2\bar\sigma_\perp)}{3}$$

In practice, one would normally carry through the calculation once only, retaining the factors t_1, t_2, t_5, u_3, u_4, v_3, and v_4 defined in Eq. 6.49. Then, if we define

$$a = u_3 + u_4 - \frac{(v_3 + v_4)}{2} \tag{6.52}$$

$$b = t_1 + t_2 + t_5 + \frac{3(v_3 + v_4)}{2}$$

Equations 6.8, 6.18, 6.19, 6.48, and 6.49 combine to give

$$\beta = \frac{2b}{(3a+b)} \tag{6.53}$$

$$\sigma = \frac{4e^2 kL^3(3a+b)}{(3mc\omega)} \tag{6.54}$$

The expression for σ in atomic units is given in Appendix IV.

In most experiments, it is not possible to use plane-polarized light, and so the angle θ_k cannot be defined. Consider new spherical coordinates (θ'', ϕ'') with the laboratory y axis (the photon-beam axis) as the polar axis and the yz plane defining the zero meridian. If in Eq. I.6 we let $(\theta_1, \phi_1) = (\theta_z'', \phi_z'') = (90°, 0°)$ [that is, (θ_z'', ϕ_z'') specify the direction of the laboratory z axis in the new (θ'', ϕ'') coordinate system], $(\theta_2, \phi_2) = (\theta_k'', \phi_k'')$, $\theta = \theta_k$ and $l = 1$, we find that

$$\cos\theta_k = -\sin\theta_k'' \cos\phi_k'' \tag{6.55}$$

thus

$$\frac{d\bar{\sigma}}{d\Omega} = \left(\frac{\sigma}{4\pi}\right)[1 + \tfrac{1}{2}\beta(3\sin^2\theta_k''\cos^2\phi_k'' - 1)] \qquad (6.56)$$

For unpolarized light, we must average over all $0 \le \phi_k'' \le 2\pi$. Thus

$$\langle\bar{\sigma}\rangle_{Av} = \left\langle\frac{d\bar{\sigma}}{d\Omega}\right\rangle_{Av} = \left(\frac{\sigma}{4\pi}\right)[1 + \tfrac{1}{2}\beta(\tfrac{3}{2}\sin^2\theta_k'' - 1)]$$

$$= \left(\frac{\sigma}{4\pi}\right)[1 - \tfrac{1}{4}\beta(3\cos^2\theta_k'' - 1)] \qquad (6.57)$$

Using unpolarized light

$$\sigma_\parallel = \langle\bar{\sigma}\rangle_{Av,\parallel} = \left(\frac{\sigma}{4\pi}\right)(1 - \tfrac{1}{2}\beta) \qquad \theta_k'' = 0 \qquad (6.58)$$

$$\sigma_\perp = \langle\bar{\sigma}\rangle_{Av,\perp} = \left(\frac{\sigma}{4\pi}\right)(1 + \tfrac{1}{4}\beta) \qquad \theta_k'' = \frac{\pi}{2} \qquad (6.59)$$

We define here the simpler notation $\sigma_\parallel$ and $\sigma_\perp$ in which all average signs are suppressed; $\sigma_\perp$ corresponds to the most frequently used observation geometry.

6.6. Application of cross-section equations

The general equations for calculating photoionization cross sections of polyatomic molecules in the plane-wave and orthogonalized plane-wave approximations have been programmed for the electronic computer. The following equations in Section 6.2 are employed for this calculation: Eq. 6.59 for $\sigma_\perp$, Eq. 6.54 or Eq. IV.7 for σ, Eq. 6.53 for β, and Eq. 6.52 for a and b. The t_n, u_n, and v_n are defined by Eq. 6.49, and the computations are done finally according to Eqs. 6.29 to 6.32 and Eqs. 6.34 to 6.47. The integrals T_p appearing in these expressions are obtained from Eqs. III.3 to III.10. Also appearing in the orthogonalized plane-wave method are the gradient matrix elements $\mathbf{G}_{jl}$ defined by Eq. 6.10; our approach for computing these is discussed below.

A computer program is formulated in terms of subroutines for various standard mathematical functions: spherical harmonics $Y_{l,m}(\theta, \phi)$, spherical Bessel Functions $j_\lambda(kR)$, Clebsch–Gordon coefficients $C(l'l''l; m', m'')$, and gamma functions $\Gamma(x)$. The ultimate input for these equations consists of specifying the real AO basis $\chi_\mu^{(r)}$ and the LCAO MO coefficients $b_{l,\mu}$ (Eq. 6.11), the molecular coordinates (X_a', Y_a', Z_a') for each atom in the polyatomic, the energy $\hbar\omega$ of the exciting radiation and the ionization energy E_I^j of a designated MO ϕ_j (Eq. 6.7). Molecular coordinates and ionization energies can be obtained from various sources. The computational results described in Chapter 7 employ *experimental* ionization energies and ground-state molecular coordinates.

Molecular wave functions used in the calculations are of the *ab initio* SCF LCAO MO type employing *minimal* bases of nonorthogonal real Slater-type orbitals (STOs) $\chi_\mu^{(r)}$;[17-21] orbital exponents ζ are, in all cases, chosen according to the usual Slater's rules. Input consists first of identifying for each STO its principal quantum number n_μ, azimuthal quantum number l_μ, "pseudomagnetic" quantum number m_μ (for example, $m_\mu = +1$ and -1 for p_x and p_y, respectively), orbital exponent ζ_μ, and the atom to which the STO belongs. As detailed in Section 6.1 we transform to a basis of complex STOs χ_p; the LCAO coefficients c_{lp} that actually enter the cross-section calculations are complex, and are related to the original LCAO coefficients $b_{l\mu}$ by the transformation $\mathbf{c} = \mathbf{ba}$, where $\mathbf{a}$ is given by Eq. II.3.

The molecular geometry enters the working formulae in terms of R_{pq}, θ'_{pq} and ϕ'_{pq}: $\mathcal{R}_{pq} = \mathcal{R}_q - \mathcal{R}_p$ is the vector directed from the atom to which χ_p belongs to the atom to which χ_q belongs; $(\theta'_{pq}, \phi'_{pq})$ are spherical polar coordinates (in the molecular coordinate system) that specify the direction of $\mathcal{R}_{pq}$. Or,

$$R_{pq} = (X'^2_{pq} + Y'^2_{pq} + Z'^2_{pq})^{1/2}$$

$$\theta'_{pq} = \cos^{-1}\left(\frac{Z'_{pq}}{R_{pq}}\right) \tag{6.60}$$

$$\phi'_{pq} = \tan^{-1}\left(\frac{Y'_{pq}}{X'_{pq}}\right)$$

in which $X'_{pq} = X'_q - X'_p$, and so forth.

Finally, as mentioned above, in the orthogonalized plane-wave method we need the gradient integrals, related to the dipole transition probabilities $\phi_j \rightarrow \phi_l$:

$$\mathbf{G}_{jl} = \langle \phi_j | \boldsymbol{\nabla} | \phi_l \rangle \tag{6.61}$$

Actually, the magnitude G_{jl} and direction (θ'_l, ϕ'_l) enter our formulations, and these are given as follows:

$$G_{jl} = (G^2_{jlx'} + G^2_{jly'} + G^2_{jlz'})^{1/2}$$

$$\theta'_l = \cos^{-1}\left(\frac{G_{jlz'}}{G_{jl}}\right) \tag{6.62}$$

$$\phi'_l = \tan^{-1}\left(\frac{G_{jly'}}{G_{jlx'}}\right)$$

in which

$$\mathbf{G}_{jlx'} = \langle \phi_j | \frac{\partial}{\partial x'} | \phi_l \rangle \tag{6.63}$$

and so forth. We first expanded these in terms of the original real STOs $\chi_\mu^{(r)}$; for example

$$\mathbf{G}_{jlx'} = \sum_\mu \sum_\nu b_{l\mu} b_{l\nu} \gamma_{\mu\nu x'} \tag{6.64}$$

$$\gamma_{\mu\nu x'} = \langle \chi_\mu^{(r)} | \frac{\partial}{\partial x'} | \chi_\nu^{(r)} \rangle \tag{6.65}$$

For the general nonlinear polyatomic molecule, the STOs $\chi_\mu^{(r)}$ belonging to different atoms will not necessarily bear a σ, π, δ, and so forth, relationship to each other. One would need to rotate these STOs $\chi_\mu^{(r)}$ to new sets that are so coherently oriented; the resulting "standard" gradient integrals over STOs could then, in fact, be reduced to ordinary overlap integrals over STOs (however, including *0s*, *1p*, etc.).

We chose, however, to utilize expansions of the real STOs in terms of *1s*–Gaussian-type orbitals (GTO). This so-called Gaussian-lobe expansion method[26] eliminates the need for performing complicated rotations to obtain values of gradient elements over STOs; all two-center integrals between *1s*–GTOs depend only upon the distance between the centers. Detailed formulas for calculating integrals $\gamma_{\mu\nu x'}$ are developed in Appendix V.

6.7. Cross sections for linear molecules in terms of atomic subshell cross sections and diffraction effects

For linear molecules it is simple to express the photoionization cross section in terms of atomic subshell cross sections and diffraction effects.[22] In the plane-wave approximation and in the LCAO-MO scheme, the *normal* differential cross section for photoionization of an electron from a molecular orbital has the general form

$$\sigma^{MO} = \sum_i c_i^2 AF_i^2 + 2 \sum_{j>i} c_i c_j AF_i F_j f_{ij} \tag{6.66}$$

where $AF_i^2 = \sigma_i^{AO}$ is the *normal* differential cross section for the i-th atomic orbital.

When Slater-type AOs are used, we obtain[23] for *normal* differential cross sections of atomic subshells

$$\sigma_{n1} = AF_{n1}^2 \tag{6.67}$$

where

$$A = \frac{32e^2k}{mc\omega} \tag{6.68}$$

$$F_{1s} = \frac{\alpha_{1s}^{5/2}k}{(\alpha_{1s}^2 + k^2)^2} \tag{6.69}$$

$$F_{2s} = \frac{\alpha_{2s}^{5/2}k(3\alpha_{2s}^2 - k^2)}{(\alpha_{2s}^2 + k^2)^3}\left(\frac{1}{3}\right)^{1/2} \tag{6.70}$$

$$F_{2p} = \frac{\alpha_{2p}^{7/2}k^2}{(\alpha_{2p}^2 + k^2)^3}\left(\frac{16}{3}\right)^{1/2} \tag{6.71}$$

For *orthogonalized* 2s Slater-type orbitals of the form

$$\chi_{2s}^{\text{ortho}} = \frac{\chi_{2s} - S_{12}\chi_{1s}}{(1 - S_{12}^2)^{1/2}} \tag{6.72}$$

with $S_{12} = \langle \chi_{1s} | \chi_{2s} \rangle$, we have

$$\sigma_{2s}(\text{ortho}) = \frac{e^2 L^3 k^3}{2\pi mc\omega}|\langle \chi_{2s}^{\text{ortho}} | PW \rangle|^2$$

$$= \frac{A(F_{2s}^2 + S_{12}^2 F_{1s}^2 - 2S_{12}F_{1s}F_{2s})}{1 - S_{12}^2} \tag{6.73}$$

But from Eqs. 6.67 to 6.71, it is obvious that $F_{1s} = (\sigma_{1s}/A)^{1/2}$ and $F_{2s} = \pm(\sigma_{2s}/A)^{1/2}$, depending on the condition $3\alpha_{2s}^2 > k^2$ or $3\alpha_{2s}^2 < k^2$. Therefore

$$\sigma_{2s}(\text{ortho}) = \frac{\sigma_{2s} + S_{12}^2 \sigma_{1s} \mp 2S_{12}(\sigma_{1s}\sigma_{2s})^{1/2}}{1 - S_{12}^2} \cdot \tag{6.74}$$

Again the $\mp$ signs depend upon the relative magnitudes of $3\alpha_{2s}^2$ and k^2. The corresponding F_{2s}^{ortho} to be used in Eq. 6.66 is then

$$-F_{2s}^{\text{ortho}} = \left(\frac{F_{2s}^2 + S_{12}^2 F_{1s}^2 - 2S_{12}F_{1s}F_{2s}}{1 - S_{12}^2}\right)^{1/2} \tag{6.75}$$

The F_i terms are already given in Eqs. 6.69 and 6.70.

The factor f_{ij} in Eq. 6.66 contains diffraction effects in the form of spherical Bessel functions specific to the type of AOs involved. Geometric factors are also contained in f_{ij}, but in a somewhat cumbersome way. For linear molecules, however, they can be reduced to simple terms;

for example,

$$
f_{ij} =
\begin{cases}
j_0(kR_{ij}) & \text{for} \quad (i, j) = (s, s) \\
3^{1/2} j_1(kR_{ij}) & \text{for} \quad (2p\sigma, s) \\
-3^{1/2} j_1(kR_{ij}) & \text{for} \quad (s, 2p\sigma) \\
0 & \text{for} \quad (s, p\pi) \quad \text{and} \quad (p\pi, s) \\
0 & \text{for} \quad (2p, 2p) \quad \text{if} \quad 2p_i \perp 2p_j \\
j_0(kR_{ij}) - 2j_2(kR_{ij}) & \text{for} \quad (2p\sigma, 2p\sigma) \\
j_0(kR_{ij}) + j_2(kR_{ij}) & \text{for} \quad (2p\pi, 2p\pi)
\end{cases}
\tag{6.76}
$$

where we use right-handed Cartesian coordinate systems on each atom, with similar axes directed parallel to each other with identical positive directions; atomic z axes are aligned with the molecular σ axis and, in the molecular coordinates, $R_{ij} = z_j - z_i$ with $z_j > z_i$. The spherical Bessel functions involved in Eq. 6.75 are

$$
j_0(x) = \frac{\sin x}{x}
$$

$$
j_1(x) = \frac{\sin x}{x^2} - \frac{\cos x}{x}
\tag{6.77}
$$

$$
j_0(x) - 2j_2(x) = \frac{3(\sin x)}{x} - \frac{6(\sin x)}{x^3} + \frac{6(\cos x)}{x^2}
$$

$$
j_0(x) + j_2(x) = \frac{3(\sin x)}{x^3} - \frac{3(\cos x)}{x^2}
$$

These $j_\lambda(x)$ functions are oscillating functions of x. Note that $j_0(0) = 1$ and $j_1(0) = j_2(0) = 0$, whereas all $j_\lambda(x) \to 0$ as $x \to \infty$.

In the application of Eq. 6.66, the LCAO-MO coefficients can be obtained from *ab initio* SCF calculations using, for example, a minimal or extended basis of nonorthogonal Slater AOs. In this case, the F_i appearing in Eq. 6.66 would be taken directly from Eqs. 6.67 to 6.71. Alternatively, for simplicity and economy, one might want to use results of a less sophisticated valence-electrons-only calculation, such as the CNDO/2 method developed by Pople and coworkers.[24] But then, how are the AOs χ_i to be interpreted? We explore[22] this question in Section 6.8.

6.8. Orthogonal slater atomic orbitals

It is important to use *orthogonalized 2s* atomic orbital cross sections, especially for high photon energies, when employing approximate MO wavefunctions in a cross-section calculation. Let us consider a two-step transformation from the nonorthogonal Slater AO basis to an orthogonalized CNDO AO basis. We first separate the Slater AO basis χ_i into two parts: *valence* AOs η_i and *inner-shell* AOs ξ_i. We transform the η_i to a new basis $\mathbf{w}_i$, which is orthogonal to *all* inner-shell AOs ξ_i:

$$\mathbf{w}_i = N(\eta_i - \sum_j \langle \eta_i \mid \xi_j \rangle \xi_j) \tag{6.78}$$

The overlap integrals $\langle \eta_i \mid \xi_j \rangle$ should all be very small except for the case in which ξ_j and η_i belong to the same atom. If we neglect all but one-center overlap integrals, which are large, $\mathbf{w}_i$ reduces to the usual *orthogonalized* Slater AO basis, that is, all AOs on a *given* atom are mutually orthogonal. Next, we apply the symmetric orthogonalization procedure proposed by Löwdin:[25]

$$\mathbf{w}' = \mathbf{w} \mathbf{S}^{-1/2} \tag{6.79}$$

where $\mathbf{S}$ is the overlap matrix using the $\mathbf{w}_i$ basis.

Since the CNDO AO basis is an orthogonalized basis, it is appropriate to associate the LCAO coefficients from CNDO theory with the orthogonalized AOs $\mathbf{w}'$. This new basis $\mathbf{w}'$ is the *closest possible* set of orthogonalized orbitals, in the least square sense, to the orthogonalized Slater AO basis $\mathbf{w}$. Therefore, when we use CNDO coefficients for the wavefunctions, we use σ_i^{AO} or F_i^{ortho} computed with orthogonalized Slater AOs.

References

1. G. V. Marr, *Photoionization Processes in Gases*, Academic, New York, 1967; H. A. Bethe and E. E. Salpeter, *Quantum Mechanics of One- and Two-Electron Atoms*, pp. 295–322, Academic, New York, 1957; D. R. Bates, *Monthly Notices Roy. Astron. Soc.*, **106**, 432 (1946); A. Burgess and M. J. Seaton, ibid., **120**, 121 (1960); A. L. Stewart, *Advances in Atomic and Molecular Physics*, Vol. 3, pp. 1–51, D. R. Bates and I. Estermann (Eds.), Academic, 1967; G. McGinn, *J. Chem. Phys.*, **53**, 3635 (1970).

2. a. D. R. Bates, U. Öpik, and G. Poots, *Proc. Phys. Soc., Ser. A*, **66**, 1113 (1953).
   b. H. D. Cohen and V. Fano, *Phys. Rev.*, **150,** 30 (1966).

3. a. M. R. Flannery and U. Öpik, *Proc. Roy. Soc.*, **86**, 491 (1965).
   b. H. P. Kelley, *Chem. Phys. Letters*, **20**, 547 (1973).

4. A. Dalgarno, *Proc. Phys. Soc., Ser. A*, **65**, 663 (1952).

5. a. B. Schneider and R. S. Berry, *Phys. Rev.*, **182**, 141 (1969).
   b. H. C. Tuckwell, *J. Phys. B*, **3**, 293 (1970).

6. P. M. Johnson and S. A. Rice, *J. Chem. Phys.*, **49**, 2734 (1968).

7. I. G. Kaplan and A. P. Markin, *Opt. Spectrosc.*, **24,** 475 (1968); idem., **25,** 275 (1968).

8. L. L. Lohr and M. B. Robin, *J. Amer. Chem. Soc.*, **92,** 7241 (1970).

9. W. Thiel and A. Schweig, *Chem. Phys. Letters*, **12,** 49 (1971); idem., **16,** 409 (1972); ibid., **21,** 541 (1973).

10. J. C. Tully, R. S. Berry, and B. J. Dalton, *Phys. Rev.*, **176,** 95 (1968).

11. a. J. Cooper and R. N. Zare, *J. Chem. Phys.*, **48,** 942, 4252 (1968).
    b. A. D. Buckingham, B. J. Orr, and J. M. Sickel, *Phil. Trans. Roy. Soc. London*, **A268,** 147 (1970); J. M. Sickel, *Mol. Phys.* **18,** 95 (1970).

12. L. L. Lohr, Jr., *Electron Spectroscopy*, D. A. Shirley (Ed.), pp. 245–258, North-Holland, Amsterdam, 1972.

13. F. O. Ellison, *J. Chem. Phys.*, **61,** 507 (1974).

14. M. E. Rose, *Elementary Theory of Angular Momentum*, Wiley, New York, 1957.

15. E. Merzbacher, *Quantum Mechanics*, pp. 439–481, Wiley, New York, 1961.

16. J. W. McGowan, D. A. Vroom, and A. R. Comeaux, *J. Chem. Phys.*, **51,** 5626 (1969).

17. F. P. Boer and W. N. Lipscomb, *J. Chem. Phys.*, **50,** 989 (1969).

18. W. E. Palk and W. N. Lipscomb, *J. Amer. Chem. Soc.*, **88,** 2384 (1966).

19. S. Aung, R. M. Pitzer, and S. I. Chan, *J. Chem. Phys.*, **49,** 2071 (1968).

20. B. J. Ransil, *Rev. Mod. Phys.*, **32,** 245 (1960).

21. H. Preuss, *Z. Naturforsch.*, **11,** 823, (1956); J. L. Whitten, *J. Chem. Phys.*, **39,** 349 (1963).

22. J. T. J. Huang, F. O. Ellison, and J. W. Rabalais, *J. Electron Spectrosc.*, **3,** 339 (1974).

23. J. T. J. Huang and F. O. Ellison, *Chem. Phys. Letters*, **25,** 43 (1974).

24. J. A. Pople and D. L. Beveridge, *Approximate Molecular Orbital Theory*, McGraw-Hill, New York, 1970.

25. P. O. Löwdin, *J. Chem. Phys.*, **18,** 365 (1950); idem., *Advan. Quantum Chem.*, **5,** 185 (1970).

7 Interpretation of Cross Sections and Angular Distributions

Correlation of observed intensities of photoelectron bands with theoretical intensities has either been done very qualitatively or else not at all in most of the early PE spectroscopic research; the principal emphasis has been on interpreting band energies. Measurements of the angular distributions of photoelectrons have been made for only a few molecules, and theoretical values for these distributions have only recently become available. In this chapter we consider the results and interpretation of the cross-section and angular-distribution calculations described in Chapter 6. Examples of the variation of photoelectron band intensities as a function of excitation energy are presented in Section 1. The behavior of the cross-section and the angular-distribution parameter β as a function of incident photon energy is discussed in Sections 2 and 4, respectively, for several small molecules. The results of these calculations are compared to relative experimental photoionization band intensities and experimental β determinations in Sections 3 and 5, respectively. Models for interpretation of the results are proposed and discussed. Vibrational transition probabilities in photoionization processes are considered in Section 6. The variation of the photoionization cross section over the vibrational envelope of a photoelectron transition is described, and its consequences are considered.

7.1. Cross section versus excitation energy

Variations in the intensities of photoelectron bands as a function of excitation energy have been observed by several groups.[1–7] The three readily available uv sources, Ne I (16.8 eV), He I (21.2 eV), and He II (40.8 eV), can be used for this purpose along with the conventional X-ray sources such as MgKα (1254 eV) and AlKα (1487 eV) radiation. In order to illustrate the variation of photoelectron band intensities with incident photon energies, the Ne I, He I, and He II spectra of several molecules

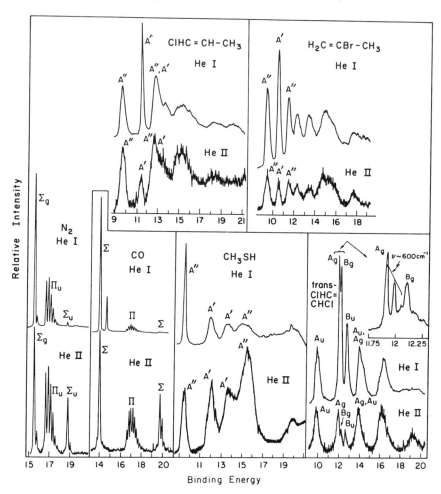

Fig. 7.1. He I and He II photoelectron spectra of N_2, CO_2, CH_3SH, $ClCH{=}CH{-}CH_3$, $H_2C{=}CBr{-}CH_3$, and *trans*—$ClHC{=}CHCl$. Reproduced with permission from Ref. 1.

are presented in Figs. 7.1 and 7.2. Even larger intensity variations can be observed when uv and X-ray excited spectra are compared. It is evident from the spectra in Figs. 7.1 and 7.2 that the variation of the differential photoionization cross section as a function of the energy of the incident photons can be very significant and thus can constitute a useful criterion for identification and assignment of photoelectron bands. The usefulness of this criterion depends on the degree of reliability and accuracy of

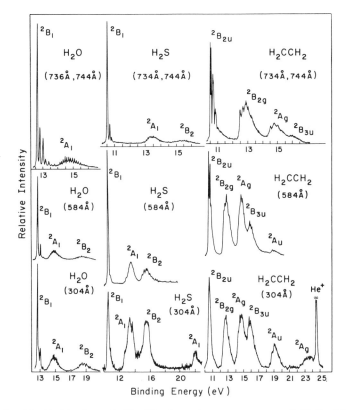

Fig. 7.2. Ne I, He I, and He II photoelectron spectra of H_2O, H_2S, and H_2CCH_2. Reproduced with permission from Ref. 2.

theoretical cross sections. It should be kept in mind that such theoretical quantities can be useful even though their absolute magnitudes may be considerably different from the experimental quantities, for it is only necessary to predict the variation in the ratio of the intensities for various ionization bands as a function of photon energy. In the following section we analyze these variations in intensity for several small molecules.

7.2. Results of cross-section calculations

The most frequently employed experimental arrangement in PE spectroscopy involves a gaseous sample (randomly oriented molecules) and collection of photoelectrons through a slit centered normal to an incident unpolarized photon beam. The corresponding averaged differential cross section $\sigma_\perp$ is the quantity of major interest to the photoelectron spectroscopist. This cross section is a measure of the number of electrons emitted per unit time per unit solid angle Ω, with Ω normal to the incident unpolarized photon beam. In order to observe the dependence of $\sigma_\perp$ on photon energy, calculations are presented for various photon energies between threshold and 1500 eV. Plots of $\sigma_\perp$ in the plane-wave (PW) approximation (terms S_1 and S_2, Eq. 6.18) versus $\hbar\omega$ are shown in Figs. 7.3 to 7.12 for ionization of electrons from the filled molecular orbitals of H_2, Ne, HF, H_2O, NH_3, CH_4, N_2, CO, H_2S, and H_2CCH_2. *Due to the longer computer times required, the orthogonalized plane wave (OPW) calculations (terms $S_3 + S_4 + S_5$; Eq. 6.18) were carried out at only those photon energies for which conventional excitation sources are available: 16.8, 21.2, 40.8, 1254, and 1487 eV. The corrections that the OPW places on the PW method are indicated as arrows at the appropriate energies in Figs. 7.3 to 7.12.* For some curves, no arrows are indicated at these energies because the OPW correction is negligible.

In the following discussion, the general features of the cross-section curves are discussed with particular reference to the agreement between theory and experiment.

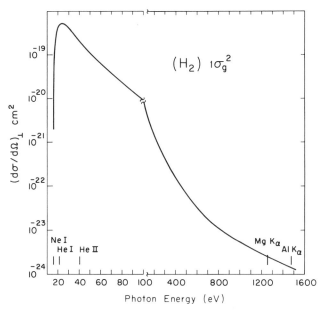

Fig. 7.3. Differential photoionization cross section of H_2 as a function of incident photon energy. Reproduced with permission from Ref. 2.

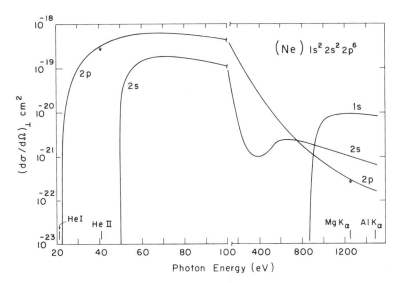

Fig. 7.4. Specific differential photoionization cross sections of Ne as a function of incident photon energy. Arrows indicate orthogonalized plane wave corrections. Reproduced with permission from Ref. 2.

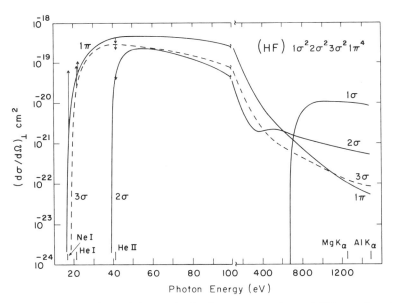

Fig. 7.5. Specific differential photoionization cross sections of HF as a function of incident photon energy. Arrows indicate orthogonalized plane wave corrections. Reproduced with permission from Ref. 2.

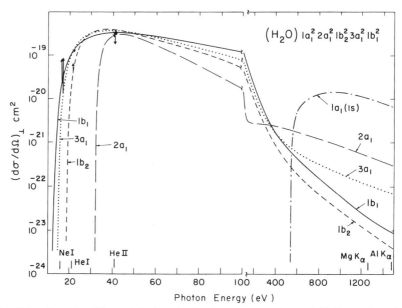

Fig. 7.6. Specific differential photoionization cross sections of H_2O as a function of incident photon energy. Arrows indicate orthogonalized plane wave corrections. Reproduced with permission from Ref. 2.

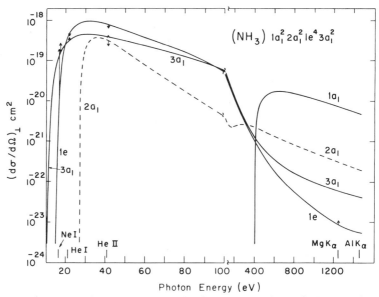

Fig. 7.7. Specific differential photoionization cross sections of NH_3 as a function of incident photon energy. Arrows indicate orthogonalized plane wave corrections. Reproduced with permission from Ref. 2.

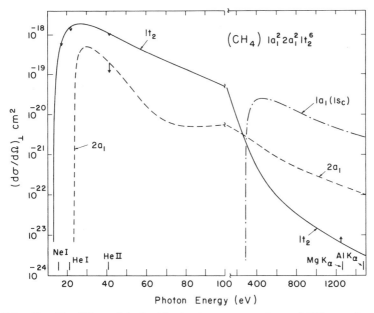

Fig. 7.8. Specific differential photoionization cross sections of CH_4 as a function of incident photon energy. Arrows indicate orthogonalized plane wave corrections. Reproduced with permission from Ref. 2.

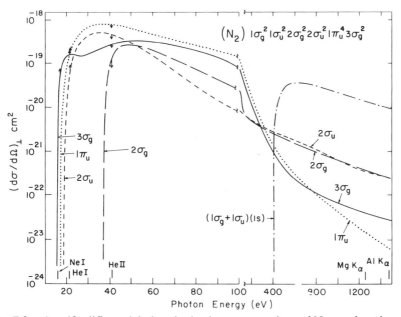

Fig. 7.9. Specific differential photoionization cross sections of N_2 as a function of incident photon energy. Arrows indicate orthogonalized plane wave corrections. Reproduced with permission from Ref. 2.

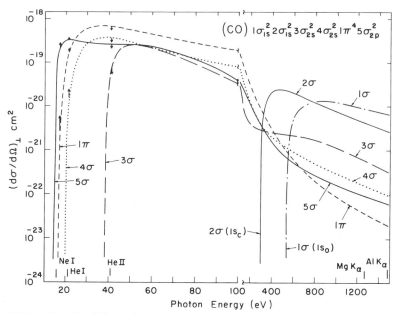

Fig. 7.10. Specific differential photoionization cross sections of CO as a function of incident photon energy. Arrows indicate orthogonalized plane wave corrections. Reproduced with permission from Ref. 2.

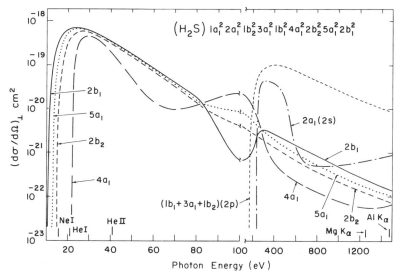

Fig. 7.11. Specific differential photoionization cross sections of H_2S as a function of incident photon energy. Reproduced with permission from Ref. 2.

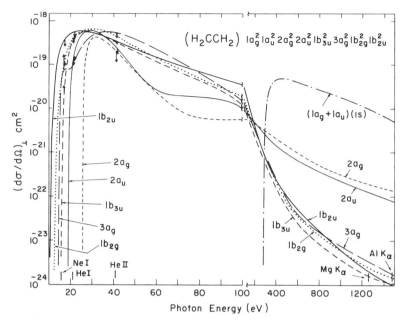

Fig. 7.12. Specific differential photoionization cross sections of H_2CCH_2 as a function of incident photon energy. Arrows indicate orthogonalized plane wave corrections. Reproduced with permission from Ref. 2.

General shapes of cross-section curves

The cross sections rise sharply near the ionization threshold, go through a maximum (or in some cases maxima), and then decrease (usually monotonically) to small values for excitation energies much greater than the ionization threshold. The initial maximum in the cross-section curves, sometimes called the "spectral head," is located at $\hbar\omega \simeq (2.0 \pm 0.5)\, E_I$, where E_I is the ionization threshold. Thus as the binding energy of the orbitals increases this peak maximum is shifted further from the ionization threshold. The rise to the maximum at the ionization threshold is much too slow in these plane-wave calculations, for it is known[8] that experimental cross sections behave more like step functions at the threshold.

Interpretation of the plane-wave cross-section curves

Calculated cross sections do not all decrease monotonically after going through an initial maximum; in many cases, there are additional maxima and minima or changes in curvature that occur at increasing photon energies. It is tempting to interpret these structures in terms of AO compositions of the MO being ionized. In doing so, we must keep in mind that such theoretical explanations may be partially artifactitious resulting from approximate LCAO formulation of bound MOs and/or approximate plane-wave formulation of the ejected electron. The OPW correction is often very significant near threshold, and can thus modify our calculated multiple maxima curves.

There appear in the general equations two possible sources for these maxima and minima. First, spherical Bessel functions $j_\lambda(kR_{pq})$ appearing in two-center contributions to the sum $\bar{S}_2$ in Eq. 6.31 could impose a damped sine-wave modulation upon the cross section. For example, the internuclear distance is $1.4b$ (see Appendix IV for atomic unit definitions) in H_2. Since $j_0(x)$ displays its initial minimum at $x \sim 4.5$, one might expect a corresponding minimum in the cross section for H_2 at $k = (4.5/1.4b)$, which corresponds to an electron energy $E_e = \frac{1}{2}k^2 \sim 5.1\,H = 142\,eV$. Such a minimum is not observed in the calculated cross section. The fact that $j_0(x)$ becomes strongly damped, and small compared with unity, after passing the initial minimum reduces the likelihood of the minimum being seen. In fact, at such higher values of k_2, we find that two-center contributions to $\bar{S}_2$ become negligible compared to one-center contributions; this is the basis of atomic intensity models for predicting and interpreting photoionization intensities in the X-ray region.[9]

The second possible source of multiple maxima in cross-section curves is probably best understood by considering first contributions from $\bar{S}_1$ alone (Eq. 6.29). It is essentially a sum of AO terms $k^3T_p^2$ (see Eqs. 6.8 and 6.18 for factor k^3), each multiplied by the net population of AO χ_p in MO ϕ_j. These atomic cross sections $k^3T_p^2$ display characteristic maxima. In fact, using best atom STOs,[10] one can show that $k^3T_{2s}^2$ display maxima at $E_e = \frac{1}{2}k^2 \sim 10$, 15, 20, 25, and 30 eV for C, N, O, F, and Ne, respectively; $k^3T_{2p}^2$ display maxima at $E_e \sim 24$, 36, 48, 60, and 72 eV for C, N, O, F, and Ne, respectively. Extrema in the cross sections for STOs of H, C, N, O, F, Ne, and S are listed in Table 7.1. Consider the following example of the use of Table 7.1 in interpreting the cross-section curves. The MOs $3\sigma_g$ in N_2 and 5σ in CO are mixtures of predominantly $2s$ and $2p$ STOs; for CO, the net populations on C exceed those on O. We speculate that the low-energy cross-section maxima in Figs. 7.9 and 7.10 at $E_e = \hbar\omega - E_I^j \sim 6\,eV$ in $3\sigma_g$ N_2 and 5σ CO are due primarily to $2s$ C and N STO contributions and that the high-energy maxima at $E_e \sim 50\,eV$ are due to $2p$ STO contributions.

As noted, contributions of AO cross sections to an MO cross section are regulated by the net population $|c_{jp}|^2$ of the AO χ_p in the MO ϕ_j. We

Table 7.1. Extrema in $k^3 T_p^2$ as a Function of Electron Energy for Best-Atom STOs

Atom	STO	Principal Maxima (eV)	Minima (eV)	Minor Maxima (eV)
H	1s	8	—	—
C	1s	265	—	—
	2s	10	106	229
	2p	24	—	—
N	1s	365	—	—
	2s	15	151	328
	2p	36	—	—
O	1s	485	—	—
	2s	20	—	—
	2p	48	206	448
F	1s	585	—	—
	2s	25	—	—
	2p	60	251	567
Ne	1s	685	—	—
	2s	30	—	—
	2p	72	296	695
S	1s	1970	—	—
	2s	108	1153	2505
	2p	350	—	—
	3s	10	61	126
	3p	19	228	397

would expect to see distinct maxima only from AOs that have comparable populations and have "characteristic maxima" that are sufficiently separated.

Interpretation of cross sections in uv region

Restricting consideration to low photon energies for the moment, we observe no double maxima in the N_2 curves other than $3\sigma_g$ discussed above. The $2\sigma_u$ and $2\sigma_g$ MOs both contain $2s$ and $2p$ components, but in both cases the net population in $2p$ is less than 0.1. There appears a slight hint of a second maximum in 4σ CO; the populations of carbon $2s$ and oxygen $2s$ and $2p$ all exceed 0.25 in this MO. It does not seem surprising that these *three* contributions add up so that distinct maxima do not appear.

For compounds containing hydrogen, we note that the calculated characteristic maximum in $k^3 T_p^2$ for $1s$–H is at $E_e \sim 8$ eV. This is nearly identical with the carbon $2s$-STO maximum (10 eV) but rather separated from C, N, O, and F $2p$-STO maxima (24–60 eV). If the above interpretative analysis is valid, one might expect to see low-energy double maxima in cross-section curves for hydrides if the MO contains significant net populations in $1s$-H and $2p$-C, -N, -O, or -F. We see such a structure only in $3a_g$ for C_2H_4. We assume that other likely cases, such as $2t_2$ in CH_4 and $1b_2$ in H_2O, are absent because relative populations of heavy atom $2p$-STO and $1s$-H are not properly equalized.

Quite dramatic variations are observed in some of the cross sections for H_2S (Fig. 7.11). The nonbonding $2b_1$ MO is primarily sulfur $2p$ and $3p$ STOs, the mixture being essentially a $3p$ *orthogonalized* STO. The $2b_1$ cross section displays maxima at $E_e \sim 10$ eV and 300 eV separated by a minimum at $E_e \sim 100$ eV. According to Table 7.1, the principal maxima in $3p$ and $2p$ STO cross sections are at 20 and 350 eV, respectively. To give assurance that the predicted variations are not necessarily simple artifacts brought about by the limited STO basis set, calculations were performed on the sulfur atom using Clementi's extended basis set wavefunctions.[11] The fluctuations observed with this extended basis set were very similar to those obtained with the minimal basis set.

The $4a_1$ H_2S MO is also nonbonding, containing principally sulfur $3s$ and $2s$ STOs. According to Table 7.1, we would expect maxima at 10 and 108 eV; these are also evident in Fig. 7.11. The $5a_1$ MO is bonding and contains mainly hydrogen $1s$, sulfur $3p$ and $3s$, with minor amounts of sulfur $2p$ and $2s$ AOs. Here, the minor high energy maximum in the $3s$-STO at 126 eV plus the $2s$-STO maximum at 108 eV (see Table 7.1) are probably responsible for the slight maximum in the $5a_1$ cross section at $E_e \sim 100$ eV. Finally, the $2a_1$ nonbonding MO is mainly sulfur $2s$ and $1s$. The corresponding atomic STO peaks are at $E_e \sim 108$ and 1970 eV, respectively; peaks near these values are also observed in the $2a_1$ cross section.

The interpretive approach outlined above is based on contributions to the cross section from one-center terms S_1 alone. The same integrals T_p appear in the sum S_2 (Eqs. 6.30 and 6.31), but they also appear as cross terms $T_p T_q$. We speculate that multiple-maxima structure may be modulated by these cross terms.

Interpretation of cross sections in soft X-ray region

At high electron energies, the spherical Bessel functions $j_\lambda(kR)$ all tend to zero, rendering two-center contributions (Eq. 6.31) largely unimportant in determining relative intensities.

Inner *1s* electrons display intense characteristic maxima at $E_e \sim 265$, 365, 485, 585, and 685 eV for C, N, O, F, and Ne, respectively. It has been pointed out[12] that since $\sigma_{1s} \gg \sigma_{2s}$ or σ_{2p} in the X-ray region, relative intensities of MOs are determined mainly by their *1s* character. In the 3σ curve for CO we actually detect a slight high energy maximum due to the carbon and oxygen *1s* (net populations ~ 0.01 and ~ 0.04, respectively) AO character in 3σ; a similar maximum is observed in the 2σ curve of HF due to the small *1s* population of the 2σ orbital.

The $3\sigma_g$ and $1\pi_u$ MOs in N_2, 5σ and 1π MOs in CO, and 3σ and 1π MOs in HF are of *2p* character with the σ MOs having a small admixture of *2s* and *1s* character. For large *k*, the terms $k^3 T_p^2$ are much larger for *1s* than for *2s* or *2p* electrons. Therefore, even though the intensities of σ and π valence MOs are comparable in the low energy region, the small amount of *1s* character is sufficient to make σ MOs approximately twice as intense as π MOs in the soft X-ray region, notwithstanding the twofold degeneracy of π MOs. We assert that this distinction between σ and π MOs displayed in the soft X-ray region is not an artifact of the LCAO or plane wave approximations. Similarly, various MOs having *a* symmetry in H_2O, CH_4, NH_3, and C_2H_4 contain small amounts of *1s*-core forced hybridization, which sometimes yields high-energy maxima and always yields higher cross sections than *b*, *e*, and *t* MOs in which *1s* core is absent.

In these calculations we have used "all-electron" MOs. If one were to use "valence-electrons-only" MOs, such as CNDO wave functions, one should at least use orthogonalized STOs for calculating cross sections in the soft X-ray region.[12] Only then will σ MOs contain a forced quantity of *1s*-core-electron character responsible for the intensity distinction that σ MOs display over π MOs at high energies.

Orthogonalized plane-wave contributions

Inspection of Eqs. 6.8, 6.18 and 6.19 shows that the PW contribution contains the factor k^3 and the remaining OPW terms contain k^2 and k factors. Thus, for high *k*, the OPW corrections should become small compared to the regular PW term. On the other hand, for very small *k*, the OPW may strongly correct the calculated curves so that they more closely approximate the step function that is observed experimentally. The OPW terms S_3, S_4, and S_5 all contain gradient elements $\mathbf{G}_{jl}$, related to bound-bound dipole transition probabilities connecting the ionized MO ϕ_j and a different occupied MO ϕ_l. Thus, we have a *"selection rule" for OPW corrections; the cross section for photoejecting an electron from ϕ_j will be modified by orthogonalization of the PW to all occupied MO's ϕ_l that combine with ϕ_j under a dipole operator.*

Inspection of numerical results for S_3, S_4, and S_5 near threshold reveals that often, but not always, S_3 is predominant. We note from Eq. 6.34 that S_3, like S_1, is a sum of terms, each one containing an atomic factor T_p^2. We might expect significant contributions to the OPW corrections from factors kT_p^2, just as we found contributions to the PW results from factors $k^3 T_p^2$; these contributions should be similarly located with respect to threshold. In the PW results, the weighting factor of $k^3 T_p^2$ was the net population of AO χ_p in the photoionized MO ϕ_j. In the OPW correction, the main part of the weighting factor of kT_p^2 is the gradient element $\mathbf{G}_{jl}^2$ multiplied by the net population of AO χ_p in the occupied MO ϕ_l.

For low electron energies, the terms kT_p^2 are larger for valence orbitals than for core orbitals. Therefore, near threshold, the largest OPW corrections arise from valence orbitals. In this region, the OPW correction is sometimes comparable to or even larger than the PW contribution. In particular, near the threshold of the core orbitals ($1s$-C, N, O, F, or Ne), the OPW correction is sometimes 100 times larger than the PW terms; this is ultimately due to the large gradient element $\mathbf{G}_{1s,2p}$. For high electron energies, the terms kT_p^2 are larger for core orbitals than for valence orbitals. Therefore, far above threshold the largest OPW corrections arise from core orbitals. *If there is an intense bound-bound dipole transition connecting a given valence MO ϕ_j with a core MO, there should be a strong OPW correction to the valence orbital cross section in the soft X-ray region.* For example, the OPW correction to $1\pi_u$ N_2 is almost 15% of the PW results; this is still not sufficient to show on the logarithmic scale of Fig. 7.9.

Absolute values of cross sections

The unit of 10^{-24} cm^2 is called a "barn." Photoionization cross sections are normally much larger than this when low photon energies are used for ionization. Consequently the unit of 10^{-18} cm^2, known as a "megabarn" or "Mbn," has been widely used in cross-section studies. We use the unit "cm^2" throughout these discussions.

The value of $\sigma_\perp$ is in the range of 10^{-19}–10^{-18} cm^2 in the low-energy region. In the high-energy region, $\sigma_\perp$ is considerably lower, being 10^{-21}–10^{-20} cm^2 for core orbitals and 10^{-24}–10^{-23} cm^2 for valence orbitals. Cross sections in the low-energy region for the assorted molecules studied here are relatively large; furthermore, cross sections for individual levels differ from one another by at most one order of magnitude. *Thus it is highly unlikely that any specific cross sections in other molecules will be so low as to escape measurement in uv PE spectroscopy.* Variations in specific cross sections by several orders of magnitude *are* apparent in the high-energy region. Hence, the possibility exists that certain bands in X-ray PE spectroscopy may actually escape detection; however, long accumulating times should render all bands measureable.

It is observed that the cross-section curves cross one another at various places so that the orbital with highest cross section at a given incident photon energy is not necessarily the most intense at some other energy. This observation is particularly useful as a means of orbital identification and assignment.[1]

Total differential $\sigma_\perp^{tot}$ and total σ cross sections

Different types of cross sections can be obtained from the equations of Chapter 6. Here we discuss two types that are of particular interest. The *total differential cross section $\sigma_\perp^{tot}$ is the sum of the specific differential cross sections, that is, $\sigma_\perp^{tot} = \sum_i \sigma_{\perp i}$ where i represents a summation over all occupied orbitals.* Plots of the total normal differential cross sections $\sigma_\perp^{tot}$ versus $\hbar\omega$ are shown in Fig. 7.13 for ejection of electrons from the filled orbitals of Ne and the isoelectronic first-row hydrides. The cross sections in these curves are a sum of the specific differential cross sections $\sigma_\perp$. The other type of cross section of interest is the *total specific cross section σ, which is a measure of the number of electrons emitted from an orbital per unit time in all directions.* We have the relative magnitudes $\sigma > \sigma_\perp^{tot} > \sigma_\perp$. The low-energy $\sigma_\perp^{tot}$ maximum ($\hbar\omega < 60$ eV) in Fig. 7.13 is due to the maximum in the sum of the valence MO $\sigma_\perp$s. As noted earlier, the maximum in this region is due to the overlap of $2s$ and $2p$ and hydrogen

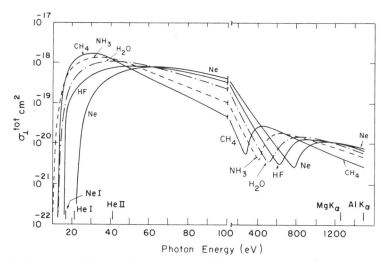

Fig. 7.13. Total differential cross section $\sigma_\perp^{tot}$ of Ne, HF, H_2O, NH_3, and CH_4 as a function of incident photon energy. Reproduced with permission from Ref. 2.

$1s$ AOs with PW. As expected from Table 7.1, this maximum shifts to higher energies with increasing atomic number of the central atom. The value of the maximum decreases progressively across the series $CH_4 >$ $NH_3 > H_2O > HF > Ne$. This trend is a result of two factors: (1) The number of hydrogen AOs in the molecule is significant in this region because of their low-energy principal maximum (Table 7.1). (2) The radial extent of the $2s$ and $2p$ AOs in the valence MOs increase as $Ne < F < O < N < C$, providing a larger cross section.

In the region, 50 to 80 eV, the curves cross each other resulting in an inversion of the lower-energy ordering. This ordering remains inverted until the region of the $1s$ ionization energies is reached. In this high-energy region, the $\sigma_\perp^{tot}$ are determined mainly by the $1s$, 1σ, or $1a_1$ MO, which is essentially a central atom $1s$ core AO. The position of the maximum in this region moves to higher energy as the atomic number of the central atom increases, in accord with expectations from Table 7.1. As in the low-energy region, the intensity of this maximum increases with decreasing atomic number of the central atom. However, at the excitation energies used in X-ray electron spectroscopy (1254 and 1487 eV), the $\sigma_\perp^{tot}$ increase with increasing atomic number. The total cross sections σ exhibit similar trends as the $\sigma_\perp^{tot}$.

It is obvious from the above discussion that the molecule with the highest cross section in one spectral region may have the lowest cross section in some other region and that these relative intensities change in a complex manner.

7.3. Comparison of experimental intensities and theoretical cross sections

Relative experimental band intensities and calculated differential photo-ionization cross sections $\sigma_\perp$ are compared in Table 7.2. Agreement between calculated and relative experimental intensities is good in the soft X-ray region, but only satisfactory in the uv region. It is well known that the PW approximation is rather crude when the electron kinetic energy is low; the OPW correction helps to reflect the influence of the attractive molecular potential. Nevertheless, the steep slopes in $\sigma_\perp$ versus $\hbar\omega$ at low electron energy render the possibility of large errors in comparing relative intensities. In general, agreement between experiment and theory improves as $\hbar\omega$ increases above threshold. The data in Table 7.2 are now discussed in detail.

H_2. The spectral head for ionization of H_2 is computed to be at ~ 23 eV with $\sigma_\perp = 5.26 \times 10^{-19}\,cm^2$ and $\sigma = 4.41 \times 10^{-18}\,cm^2$. The cross section in the high-energy region is the lowest of all molecules studied with $\sigma_{\perp} = 2.53 \times 10^{-24}$ cm^2 and $\sigma = 2.12 \times 10^{-23}$ cm^2 at 1254 eV; consequently the X-ray PE spectrum of H$_2$ has never been measured. The photoelectron cross section of H$_2$ has been calculated by Flannery and Öpik[14] and more recently by Kelly,[15] and experimental measurements were made by Cook and Metzger.[16] The calculations presented here are in fair agreement with these theoretical and experimental cross sections for photon energies beyond ~ 21 eV. For lower photon energies the plane-wave approximation[17] does not rise fast enough to reproduce the spectral head as observed at ~ 17.5 eV.[16] More refined approximations are necessary for calculating accurate cross sections very near the ionization threshold.

CH$_4$. Calculated relative cross sections for methane are in good agreement with relative experimental intensities. Dalgarno[18] has calculated the photoionization cross section of methane and found the spectral head at ~ 13.1 eV with $\sigma = 9.4 \times 10^{-17}$ cm^2. Our Calculation yields a spectral head for the $1t_2$ orbital at ~ 25 eV with $\sigma_{\perp} = 1.9 \times 10^{-18}$ cm^2 and $\sigma = 1.6 \times 10^{-17}$ cm^2.

H$_2$O AND H$_2$S. H$_2$O and H$_2$S exhibit a similar orbital structure and provide an excellent example of cross-section dependence on principal quantum number. For H$_2$O, agreement between experiment and theory is good except for the $1b_2$ band with He II excitation. For H$_2$S, agreement between experiment and theory is not as good as H$_2$O; however, only the PW contribution for H$_2$S is listed in Table 7.2.

N$_2$ AND CO. The cross-section curves for nitrogen and carbon monoxide are very similar, the major difference being in the splitting of the N$_{1s}$ core orbital of N$_2$ into C$_{1s}$ and O$_{1s}$ core orbitals in CO. The experimental and calculated intensities are in good agreement with the exception of the $2\sigma_u$ at 40.8 eV and the 5σ at 21.2 eV. These discrepancies cannot be immediately rationalized. Theoretical specific cross sections for N$_2$ have been presented by Tuckwell[19] and experimental total cross sections by Samson and Cairns[20] and Wainfain et al.[21] The calculations of Tuckwell are in reasonable agreement with these, both predicting the intensities of the outer bands at 21.2 eV excitation to be in the order $1\pi > 3\sigma_g > 2\sigma_u$.

H$_2$CCH$_2$. Agreement between theory and experiment in ethylene is rather good for He II excitation, but poor for Ne I and He I excitation. As noted in Fig. 7.12 there are several E_Is that are very close, resulting in several cross-section curves of near-equal intensity in the low-energy region. Considering the complexity of the situation, the poor agreement near threshold is not unexpected. Unfortunately, intensities in the X-ray region are not available.

Table 7.2. Relative Experimental Intensities[a] and Calculated Differential Photoionization Cross Sections of H_2, Ne, HF, H_2O, NH_3, CH_4, N_2, CO, H_2S, and H_2CCH_2 as Produced by Various Modes of Excitation

		Differential Photoionization Cross Sections $\sigma_\perp$							
Assignment	$E_I(eV)^b$	NeI (16.87 eV)		HeI (21.22 eV)		He II (40.81 eV)		MgK_α X-ray (1254 eV)	
		exp	theo^c	exp	theo^c	exp	theo^c	exp	theo^d
H_2									
$(1\sigma_g)\,^2\Sigma_g$	15.4 A								
	15.9 V	—	1.54	—	5.12	—	1.97	—	0.025
Ne									
$(2p)^2P$	22.0	—	—	—	—	—	1.67	0.09	1.70
$(2s)^2S$	49.0	—	—	—	—	—	—	0.34	9.44
$(1s)^2S$	870	—	—	—	—	—	—	1.00	49.7
HF									
$(1\pi)^2\pi$	16.1	—	1.37	—	1.69	1.00	2.20	—	0.69
$(3\sigma)^2\Sigma$	18.6 A	—	—	—	1.26	0.41	1.60	—	1.23
	19.5 V								
$(2\sigma)^2\Sigma$	39.0	—	—	—	—	—	0.370	—	7.43
$(1\sigma)^2\Sigma$	686	—	—	—	—	—	—	—	87.7
H_2O									
$(1b_1)^2B_1$	12.6	0.81	1.31	0.85	0.90	0.96	1.96	0.095	0.189
$(3a_1)^2A_1$	13.8 A	1.0	1.54	1.0	1.01	1.0	2.29	0.26	1.05
	14.7 V								
$(1b_2)^2B_2$	17.2 A	—	—	0.85	0.89	0.80	2.53	0.081	0.089
	18.5 V								
$(2a_1)^2A_1$	32.2 V	—	—	—	—	—	0.60	1.0	4.72
$(1a_1)^2A_1$	540	—	—	—	—	—	—	19	89.4
NH_3									
$(3a_1)^2A_1$	10.2 A	—	3.74	1.00	4.34	1.00	4.00	—	0.63
	10.9 V								
$(1e)^2E$	15.0 A	—	4.01	2.20	3.01	1.90	5.78	—	0.11
	16.0 V								
$(2a_1)^2A_1$	27.0 V	—	—	—	—	—	1.23	—	2.92
$(1a_1)^2A_1$	406	—	—	—	—	—	—	—	70.4
CH_4									
$(1t_2)^2T_2$	12.6 A	—	3.84	—	8.96	10	7.81	0.27	0.112
	14.0 V								
$(2a_1)^2A_1$	23.0 V	—	—	—	—	1.4	0.36	1.0	1.67
$(1a_1)^2A_1$	290	—	—	—	—	—	—	20	44.1
N_2									
$(3\sigma_g)^2\Sigma_g$	15.5	—	0.398	1.0	1.29	1.0	2.78	0.20	0.698
$(1\pi_u)^2\Pi_u$	16.7 A	—	—	1.4	3.72	2.1	6.41	0.10	0.191
	16.9 V								
$(2\sigma_u)^2\Sigma_u$	18.8	—	—	0.32	0.292	0.47	2.49	1.0	3.63

Table 7.2 (continued)

Assignment	$E_1(eV)$[b]	NeI (16.87 eV) exp	NeI (16.87 eV) theo[c]	HeI (21.22 eV) exp	HeI (21.22 eV) theo[c]	He II (40.81 eV) exp	He II (40.81 eV) theo[c]	MgK$_\alpha$ X-ray (1254 eV) exp	MgK$_\alpha$ X-ray (1254 eV) theo[d]
N$_2$ (continued)									
$(2\sigma_g)^2\Sigma_g$	37.3 V	—	—	—	—	—	0.148	0.70	3.83
$(1\sigma_g, 1\sigma_u)^2\Sigma_u, {}^2\Sigma_g$	410	—	—	—	—	—	—	29	140.6
CO									
$(5\sigma)^2\Sigma_{2p}$	14.0	—	3.00	0.80	3.89	0.26	1.44	0.40	0.981
$(1\pi)^2\Pi_{2p}$	16.5 A 16.9 V	—	1.93	1.0	2.38	1.0	4.14	0.20	0.312
$(4\sigma)^2\Sigma_{2s}$	19.7	—	—	0.21	0.394	0.52	1.80	1.0	1.57
$(3\sigma)^2\Sigma_{2s}$	38.3 V	—	—	—	—	—	0.082	2.0	5.07
$(2\sigma)^2\Sigma_{1s_c}$	296	—	—	—	—	—	—	22	44.62
$(1\sigma)^2\Sigma_{1s_0}$	542	—	—	—	—	—	—	48	89.61
H$_2$S									
$2b_1)^2B_1$	10.5	1.2	4.53	0.88	6.78	0.58	3.30	1.4	2.47
$(4a_1)^2A_1$	12.8 A 13.3 V	1.0	2.66	1.0	6.07	1.0	2.99	1.8	1.69
$(2b_2)^2B_2$	14.8 A 15.5 V	0.11	0.608	1.1	4.31	0.97	2.93	1.0	1.22
$(3a_1)^2A_1$	22.3 V	—	—	—	—	0.23	1.71	1.2	0.58
$(1b_1, 1b_2, 2a_1)^2B_1,$ ${}^2A_1, {}^2B_2,$	171	—	—	—	—	—	—	65.4	146.0
$(1a_1)^2A_1$	235	—	—	—	—	—	—	7.4	8.61
H$_2$CCH$_2$									
$(1b_{2u})^2B_{2u}$	10.51	0.68	1.69	0.71	2.79	0.98	2.43	—	0.039
$(1b_{2g})^2B_{2g}$	12.5 A 12.9 V	1.0	0.724	0.93	2.58	1.1	3.65	—	0.042
$(3a_g)^2A_g$	14.4 A 14.7 V	0.94	2.54	1.00	1.23	1.3	3.80	—	0.059
$(1b_{3u})^2B_{3u}$	15.7 A 15.9 V	0.18	0.429	0.63	2.22	1.0	3.41	—	0.117
$(2a_u)^2A_u$	18.8 A 19.1 V	—	—	0.14	0.51	0.21	1.10	—	1.36
$(2a_g)^2A_g$	23.4 V	—	—	—	—	0.24	0.28	—	2.00
$(1a_g, 1a_u)^2A_g, {}^2A_u$	291	—	—	—	—	—	—	—	84.6

[a] The relative intensities of the bands from Ne I, He I, and He II excitation sources were determined as A/E, where A is the total area of the band obtained by integrating over all its vibrational components and E is the electron kinetic energy at the band maximum. The MgK$_\alpha$ intensities were obtained by integrating the spectra in Ref. 13.
[b] A = adiabatic E_1; V = vertical E_1; no letter indicates A = V.
[c] $\sigma_\perp$ in units of 10^{-19} cm^2.
[d] $\sigma_\perp$ in units of 10^{-22} cm^2.

Ne, HF, NH_3. The calculations on Ne using MgK_α excitation energy predict the correct experimental intensity ordering. Experimental data on HF and NH_3 are incomplete, however, in the few cases where comparisons can be made, the agreement is encouraging.

The results presented show that the PW approximation is useful for understanding variations in photoionization cross sections as a function of photon energy. The OPW corrections to the PW approximation are largest near the ionization threshold where the electron kinetic energy is low. It is generally adequate (within at most about 15% to use the simple PW method for calculating cross sections with photon energies much greater than threshold.

7.4. Angular-distribution calculations

The angular distribution of photoelectrons emitted from an ensemble of randomly oriented molecules has important implications for understanding the nature of the states involved in photoelectron transitions.[2] The characterization of orbitals according to their asymmetry parameters β and an understanding of the variation of β with photon energy can yield useful information about ionic states. The differential cross section $d\bar{\sigma}/d\Omega$ for linearly polarized light and randomly oriented molecules is often written in terms of the total cross section σ and the asymmetry parameter β as expressed in Eqs. 6.48 or 6.56. In these equations, θ_k is the angle between the propagation vector $\mathbf{k}$ of the ejected electron and the electric vector $\mathbf{u}$ of the exciting radiation, and (θ_k'', ϕ_k'') are polar coordinates of $\mathbf{k}$ using the photon beam as a polar axis. Since $d\bar{\sigma}/d\Omega \geq 0$, it follows that $2 \geq \beta \geq 1$. For unpolarized light, Eq. 6.56 is averaged over all orientations of $\mathbf{u}$, that is, averaged over all $0 \leq \theta_k'' \leq 2\pi$, to give Eq. 6.57

$$\langle \bar{\sigma} \rangle_{Av} = \left\langle \frac{d\bar{\sigma}}{d\Omega} \right\rangle_{Av} = \left(\frac{\sigma}{4\pi} \right) \left[1 - \frac{1}{4} \beta (3 \cos^2 \theta_k'' - 1) \right] \qquad (7.1)$$

Here, $\beta = 2$ corresponds to a pure $\sin^2 \theta_k''$ distribution, $\beta = 0$ corresponds to an isotropic distribution, and $\beta = -1$ corresponds to a $(1 + \cos^2 \theta_k'')$ distribution.

We now apply the equations of Chapter 6 to the determination of angular distributions. In the PW method, only S_1 and S_2 of Eq. 6.18 are retained, giving a pure $\cos^2 \theta_k$ or a pure $\sin^2 \theta_k''$ dependence. In the OPW method all terms in Eq. 6.18 are retained, and these provide spherical as well as angular dependent terms. The PW method yields $\beta = 2$ with lower values of β appearing only when the OPW extended calculation is accomplished. In Section 7.2 it has been shown that the OPW correction to the *normal* differential cross section $\sigma_\perp$ may be very important near threshold, but it becomes negligible far above threshold. It is interesting to note that this fact does not carry over to β. The deviation of β from 2 is often important near threshold; however, it may also be extremely significant far above threshold. It may seem strange that for high electron energy the OPW correction may only affect $\sigma_\perp$ by a few percent and simultaneously reduce β by over 100%. This paradox is examined in Appendix VI. Plots of β versus photon energy ($0 < \hbar\nu < 1250$ eV) for CH_4, H_2O, and N_2 are presented in Figs. 7.14 to 7.16.

Nearly all experimental measurements of β have been obtained with He I excitation. From Figs. 7.14 to 7.16, it is just in this region that most of the computed β are changing rapidly with photon energy. Also, of course, the OPW approximation does not properly reflect the attractive molecular potential, the influence of which is greatest near threshold. Certainly, β can be an important parameter for identifying transitions if we can reproduce it theoretically.

Two different approaches are followed for interpreting the calculated $\beta(E)$ curves.[2] The first extends a model originally proposed by Carlson for linear molecules: β values are correlated with most probable values of the angular momentum quantum number l associated with the MO being ionized.[22] The second approach directly relates deflections of β from 2 (that is, the β obtained from the PW approximation), to OPW corrections of the PW approximation.

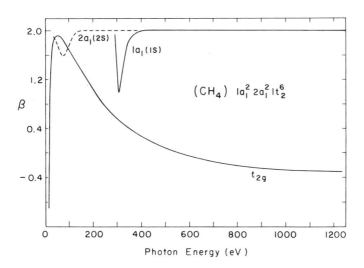

Fig. 7.14. The asymmetry parameter β as a function of photon energy for the occupied orbitals of CH_4. Reproduced with permission from Ref. 2.

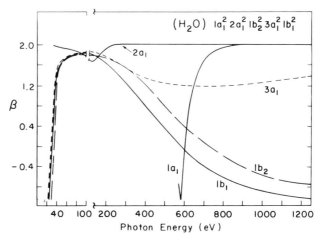

Fig. 7.15. The asymmetry parameter β as a function of photon energy for the occupied orbitals of H_2O. Reproduced with permission from Ref. 2.

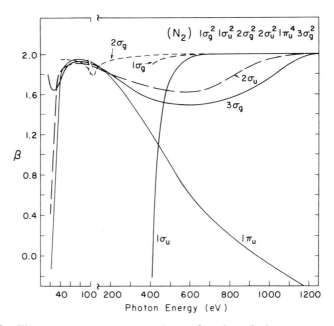

Fig. 7.16. The asymmetry parameter β as a function of photon energy for the occupied orbitals of N_2. Reproduced with permission from Ref. 2.

Interpretation of β in terms of most probable angular momentum l_{mp}

For atoms, it is known that $\beta = 2$ for ionization of an s electron; in the limit of high photon energy, it appears that β approaches 0 for $l = 1$ (that is, p electrons).[23] Krauss has noted that the larger the quantum number l, the lower the value of β.[24] Carlson has used this fact as a possible probe for understanding β values of linear molecules.[22] His approach is extended here to include the general nonlinear molecule.

Consider the expansion of an MO ϕ_j in terms of spherical harmonics $Y_{lm}(\theta, \phi)$. We shall be interested in an approximate measure of the orbital angular momentum associated with the MO ϕ_j. Angular momentum requires specification of an orgin; it seems reasonable to select this origin at or near the center of electron density. Such a point will necessarily be in common with all symmetry elements of the molecule (for example, for H_2O, the origin may lie somewhere along the C_2 axis; for CH_4, at the C atom; for N_2, at the inversion center). We thus write the expansion

$$\phi_j(r, \theta, \phi) = \sum_{l=0}^{\infty} \sum_{m=-l}^{+l} C_{lm}(r) Y_{l,m}(\theta, \phi) \tag{7.2}$$

$$C_{lm}(r) = \int\int Y_{l,m}^{*}(\theta, \phi) \phi_j(r, \theta, \phi) \sin\theta \, d\theta \, d\phi \tag{7.3}$$

We shall now see how symmetry may require some of the leading coefficients in this expansion to vanish, that is,

$$C_{lm}(r) = 0 \quad \text{if} \quad l \le l_{min} \tag{7.4}$$

If, for example, $l_{min} = 1$ for a given MO ϕ_j, we may expect $\beta \sim 0$ at least in the high-energy limit. If, on the other hand, $l_{min} = 0$, it is conceivable that $\beta = 2$. Given an *ab initio* MO ϕ_j, one could actually compute the coefficients C_{lm} using Eq. 7.3, and thus determine the most probable value l_{mp}.

Any specific MO ϕ_j can be characterized by the number of nodes n_j required by symmetry: we shall call these "symmetry nodes." Any given symmetry node necessarily is in common with at least one symmetry element. Molecular orbitals also may contain additional nodes, which we may call "energy" or "orthogonality" nodes; such nodes normally will not be in common with any symmetry element, except accidentally at various crossing points. For the group C_{2v}, MOs belonging to the symmetry species a_1, b_1, b_2, and a_2 necessarily contain 0, 1, 1, and 2 symmetry nodes, respectively. The $1a_1$, $2a_1$, and $3a_1$ MOs in H_2O will necessarily contain 0, 1, and 2 energy nodes, but no symmetry nodes.

Now, spherical harmonics also contain nodes; in fact, the number of nodes is equal to l. If we expand the $1b_1$ MO of H_2O in terms of spherical harmonics, it is clear that the coefficient $C_{00}(r) = 0$; the function $Y_{00}(\theta, \phi)$ is nodeless, $1b_1$ contains one symmetry node, and the integral of their product vanishes. Thus, $l_{min} = 1$ for $1b_1$, and we expect $\beta \sim 0$ at high energy; see Fig. 7.15. If we expand the $2a_1$ MO of H_2O, it is very unlikely that $C_{00}(r) = 0$; $2a_1$ possesses one energy node, approximately the radial node of the $2s$ oxygen AO, but this should not cause the integration over $\sin \theta \, d\theta \, d\phi$ to vanish.

The corresponding values of l_{min} obtained by these symmetry arguments are listed in Table 7.3 for each MO. The most probable value of angular momentum l_{mp} is that one corresponding to the maximum value of $|C_{lm}(r)|^2$. It is certainly not necessary that the lead term be maximum, that is, $l_{min} = l_{mp}$. On the other hand, necessarily $l_{min} \leq l_{mp}$. Therefore, it is clear that high values of l_{min} predicate low β.

The correct *high-energy* behavior of β is predicted by correlation with l_{min} in all cases except for $1\sigma_u$ and $2\sigma_u$ N_2. If the curves in Fig. 7.16 were actually experimental rather than theoretical, we could propose an explanation for $\beta \to 2.0$ for these two MOs for which $l_{mp} \geq 1$. Suppose that the final states in the photoionizations were localized hole states (that is, approximately $1s_A^2 1s_B$ or $2s_A^2 2s_B$).[25] Then, instead of expanding the initial state in terms of spherical harmonics centered at the molecular midpoint, one should expand about one atom or the other, from where the photoejected electron departed. We would then find $C_{00}(r) \neq 0$, $l_{min} = 0$, and correct β behavior is predicted. Unfortunately, the curves displayed are for calculated delocalized final states and not experimental. Therefore, this nice rationalization for the model's incorrect predictions for $1\sigma_u$ and $2\sigma_u$ is not pertinent.

Interpretation of β in terms of OPW corrections

The asymmetry parameter β is, of course, not a directly observed physical quantity, but rather a parameter that may be determined by fitting Eqs. 7.1, 6.48, or 6.56 to at least two experimental points: $\langle \bar{\sigma} \rangle_{Av}$ measured at two or more different angles θ_k''. The understanding of β as a function of photon energy $\hbar\omega$, or photoelectron energy $E_e = \frac{1}{2}k^2$, ultimately rests upon our understanding of $\langle \bar{\sigma} \rangle_{Av}$ as a function of energy. For the latter, our arguments will depend upon Eqs. 6.8 and 6.9; we rewrite the transition moment function given by Eq. 6.9:

$$\mathbf{P}_{0j} = 2^{1/2}\hbar[\mathbf{k}\langle \phi_j | \text{PW}(\mathbf{k})\rangle + i\sum_l \langle \phi_j | \boldsymbol{\nabla} | \phi_l \rangle \langle \phi_l | \text{PW}(\mathbf{k})\rangle] \qquad (7.5)$$

The first term comes from the pure PW formulation; the sum over all occupied MOs ϕ_l is a consequence of the OPW correction. Including only the first term in $\mathbf{P}_{0j}$ gives $\beta = 2$; all deviations of β from 2 obtained in the OPW approximation arise from the sum appearing in Eq. 7.5 above, which we rewrite in a condensed form

$$\alpha_j = \sum_l \mathbf{G}_{jl}\langle\phi_l \mid \mathrm{PW}(\mathbf{k})\rangle \tag{7.6}$$

Now, the gradient elements $\mathbf{G}_{jl}$ are related to bound-bound transition probabilities connecting the ionized MO ϕ_j and a different occupied MO ϕ_l. Thus *there will be contributions to α_j from all occupied MOs ϕ_l that combine with ϕ_j under a dipole operator.*

In Table 7.3, we list for each specific photoionization all of the occupied ϕ_l for which $\mathbf{G}_{jl} \neq 0$ by symmetry. In addition, for all MOs ϕ_j and ϕ_l, we indicate the most prominent AO in the LCAO. This enables us to roughly assess the strength of the bound-bound transition, and thus the weighting factor $\mathbf{G}_{jl}$ of the overlap integral $\langle\phi_l|\mathrm{PW}\rangle$, which ultimately contributes to α_j. If $\mathbf{G}_{jl}$ mainly reduces to two-center transition moments over AOs, then the $\mathbf{G}_{jl}$ are weak (w) or very weak (vw). If $\mathbf{G}_{jl}$ contains one-center transition moments connecting $1s$ and $2p$ or $2s$ and $2p$, then the $\mathbf{G}_{jl}$ are very strong (vs) or strong (s), respectively.

The asymmetry parameter β is a function of electron energy $E_e = \frac{1}{2}k^2$, this dependence being reflected by the fact that α_j is a function of electron energy. The overlap integrals $\langle\phi_l \mid \mathrm{PW}\rangle$ will display maxima at electron energies characteristic of the most prominent AOs χ_p contained in the LCAO MO ϕ_l. These prominent AOs have already been listed in Table 7.3. As discussed in Section 7.2, *valence* AOs yield large OPW corrections in the low-electron-energy region, and *core* AOs yield large OPW corrections in the high-electron-energy region. These energy regions are listed in Table 7.3 for every occupied MO ϕ_l contributing to the OPW correction α_j for photoionization of a given ϕ_j.

If $\mathbf{G}_{jl}$ for a given orbital ϕ_l is strong or very strong, we expect to see a contribution to α_j from $\langle\phi_l \mid \mathrm{PW}\rangle$ in the energy region listed; in the same energy region, we expect β to swerve from its PW value of 2. We indicate in the last column of Table 7.3 all deflections predicted by this argument that we actually observe in the calculated curves displayed in Figs. 7.14 to 7.16.

As discussed above, in the high-energy region, β for valence MOs ϕ_j is affected exclusively by core MOs. For $3\sigma_g$ and $2\sigma_g$ N_2 and for $3a_1$ H_2O, we observe high-energy saucer-like depressions, with $\beta \to 2$ at very high energy. For t_{2g} CH_4, $1\pi_u$ N_2, $1b_1$ and $1b_2$ H_2O, we observe β steadily decreasing to a low positive or to a negative value as E_e increases. The

Table 7.3. Terms Involved in Interpretation of β Values for Ionization from Specific Orbitals ϕ_j

Molecule	$\phi_j{}^a$	$l_{min}{}^b$	$\phi_l{}^c$	$G_{jl}{}^d$	$\langle \phi_l | PW(k) \rangle^e$	Observed^f
CH_4	$1a_1 \sim 1s_c$	0	$1t_2 \sim 2p$	vs	low	Yes
	$2a_1 \sim 2s$	0	$1t_2 \sim 2p$	s	low	Yes
	$1t_2 \sim 2p$	1	$1a_1 \sim 1s_c$	vs	high	Yes
			$2a_1 \sim 2s$	s	low	Yes
H_2O	$1a_1 \sim 1s_0$	0	$2a_1 \sim 2s$	vw	low	—
			$3a_1 \sim 2p$	vs	low ⎫	
			$1b_1 \sim 2p$	vs	low ⎬	Yes
			$1b_2 \sim 2p$	vs	low ⎭	
	$2a_1 \sim 2s$	0	$1a_1 \sim 1s_0$	vw	high	—
			$3a_1 \sim 2p$	s	low ⎫	
			$1b_1 \sim 2p$	s	low ⎬	Yes
			$1b_2 \sim 2p$	s	low ⎭	
	$1b_2 \sim 2p$	1	$1a_1 \sim 1s_0$	vs	high	Yes
			$2a_1 \sim 2s$	s	low	Yes
			$3s_1 \sim 2p$	w	low	—
	$3a_1 \sim 2p$	0	$1a_1 \sim 1s_0$	vs	high	Yes
			$2a_1 \sim 2s$	s	low	Yes
			$1b_1 \sim 2p$	w	low	—
			$1b_2 \sim 2p$	w	low	—
	$1b_1 \sim 2p$	1	$1a_1 \sim 1s_0$	vs	high	Yes
			$2a_1 \sim 2s$	s	low	Yes
			$3a_1 \sim 2p$	w	low	—
N_2	$1\sigma_g \sim 1s$	0	$1\sigma_u \sim 1s$	vw	high	—
			$2\sigma_u \sim 2s$	w	low	—
			$1\pi_u \sim 2p$	vs	low	Yes
	$1\sigma_u \sim 1s$	1	$1\sigma_g \sim 1s$	vw	low	—
			$2\sigma_g \sim 2s$	w	low	—
			$3\sigma_g \sim 2p$	vs	low	Yes
	$2\sigma_g \sim 2s$	0	$1\sigma_u \sim 1s$	w	high	—
			$2\sigma_u \sim 2s$	w	low	—
			$1\pi_u \sim 2p$	s	low	Yes
	$2\sigma_u \sim 2s$	1	$1\sigma_g \sim 1s$	w	high	Yes
			$2\sigma_g \sim 2s$	w	low	—
			$3\sigma_g \sim 2p$	s	low	Yes
	$1\pi_u \sim 2p$	1	$1\sigma_g \sim 1s$	vs	high	Yes
			$2\sigma_g \sim 2s$	s	low	Yes
			$3\sigma_g \sim 2p$	w	low	—
	$3\sigma_g \sim 2p$	0	$1\sigma_u \sim 1s$	vs	high	Yes
			$2\sigma_u \sim 2s$	s	low	Yes
			$1\pi_u \sim 2p$	w	low	—

method of interpretation suggested here is not sufficient to distinguish these two distinct behaviors. With this reservation, the model does provide an agent for every deflection observed in β from its PW value of 2.

7.5. Comparison of experimental and calculated β values

Experimental measurements of the angular distribution of photoelectrons have been made by several groups,[22,26-29] and theoretical calculations (particularly for atoms) of the asymmetry parameter β have recently appeared.[30-33] In Table 7.4, the calculated β and σ values for Ne, HF, H_2O, NH_3, CH_4, and N_2 at Ne I, He I, He II, and MgK_α excitation energies are presented along with some experimental β values. Agreement between calculated and experimental β values for He I excitation is marginal; this is not unexpected considering the steep slopes of β versus $\hbar\omega$ in this region. We expect better agreement in the soft X-ray region; however, experimental molecular β values are not yet available in this region.

Theoretical remarks concerning experimental determination of the asymmetry parameter β

It is clear from Eq. 7.1 that the measured $d\sigma/d\Omega$ (abbreviated as σ^d) will be independent of β, and equal to $(\sigma/4\pi)$ at the "magic angle" $\theta_k'' = \cos^{-1}(1/3)^{1/2} = 54°44'$ or $125°16'$. Samson[34] has previously pointed out this fact. Figure 7.17 illustrates further how relative intensities of photo-electron bands of different typical β are determined by the collection angles θ. The most intense band measured at one angle may be the least intense at some other angle. However, if observation is made in the region $90° \pm 45°$, no bands will escape detection due to the β factor.

The principal concern here is to relate the probable error in β to the

---

footnotes to **Table 7.3.**

[a] MO ionized; $\sim$ most prominent AO in ϕ_j.

[b] Minimum value allowed by symmetry for most probable angular momentum quantum number l associated with MO ϕ_j.

[c] Occupied MO ϕ_l that may combine with ϕ_j under dipole operator $\sim$ most prominent AO in ϕ_l.

[d] Approximate values for gradient matrix elements in Eq. 7.6: vw = very weak, w = weak, s = strong, vs = very strong.

[e] Photoelectron energy region in which overlap integral $\langle\phi_j|PW(k)\rangle$ is maximum: low means $\hbar\omega - E_1^j < 50$ eV; high means $\hbar\omega - E_1^j > 50$ eV.

[f] Predicted deflections that are observed in calculated curves displayed in Figs. 7.14 to 7.16.

Table 7.4. Total Specific Photoionization Cross Section σ and Asymmetry Parameters β for Ne, HF, H_2O, NH_3, CH_4, and N_2

| Assignment | $E_I(eV)^b$ | Theoretical σ^a and β | | | | | | | | |
| | | Ne I(16.8 eV) | | He I(21.2 eV) | | | He II(40.8 eV) | | MgK_α(1254 eV) | |
		σ^c	β	σ^c	β	$\beta(exp)^e$	σ^c	β	σ^d	β
Ne										
$(2p)^2P$	22.0	—	—	—	—	—	2.42	−0.54	2.57	0.61
$(2s)^2S$	49.0	—	—	—	—	—	—	—	7.91	2.00
$(1s)^2S$	870.0	—	—	—	—	—	—	—	41.6	2.00
HF										
$(1\pi)^2\Pi$	16.1	1.76	−0.097	2.62	−0.762	—	2.42	0.58	1.11	−0.85
$(3\sigma)^2\Sigma$	18.6 A, 19.5 V	—	—	1.63	−0.134	—	1.59	1.05	1.13	1.35
$(2\sigma)^2\Sigma$	39.0 V	—	—	—	—	—	0.394	0.75	6.22	2.00
$(1\sigma)^2\Sigma$	686.0	—	—	—	—	—	—	—	73.5	2.00
H_2O										
$(1b_1)^2B_1$	12.6	2.02	−0.73	1.47	−0.92	1.0±0.1	1.84	1.36	2.99	−1.00
$(3a_1)^2A_1$	13.8 A, 14.7 V	1.14	−0.40	1.61	−0.84	0.3±0.1	2.10	1.49	9.78	1.42
$(1b_2)^2B_2$	17.2 A, 18.5 V	—	—	1.43	−0.88	−0.1±0.2	2.32	1.47	1.32	−0.66
$(2a_1)^2A_1$	32.2 V	—	—	—	—	—	0.53	1.77	40.1	2.00
$(1a_1)^2A_1$	540	—	—	—	—	—	—	—	89.4	2.00

NH₃, CH₄, and N₂ — ionization data[a]

Orbital	IP (eV)[b]									
NH₃										
$(3a_1)^2\,A_1$	10.2 A / 10.9 V	4.37	0.30	3.99	1.47	—	3.48	1.95	0.342	1.81
$(1e)^2\,E$	15.0 A / 16.0 V	5.64	−0.43	4.54	−0.66	—	5.14	1.65	0.013	0.23
$(2a_1)^2\,A_1$	27.0 V	—	—	—	—	—	1.31	1.07	2.50	2.00
$(1a_1)^2\,A_1$	406	—	—	—	—	—	—	—	59.0	2.00
CH₄										
$(1t_2)^2\,T_2$	12.6 A / 14.0 V	5.50	−0.81	8.92	1.05	0.6±0.1	6.69	1.87	0.145	−0.121
$(2a_1)^2\,A_1$	23.0 V	—	—	—	—	—	0.323	1.59	1.39	2.00
$(1a_1)^2\,A_1$	290	—	—	—	—	—	—	—	37.0	2.00
N₂										
$(3\sigma_g)^2\,^2\Sigma_g$	15.5	0.338	1.93	1.10	1.91	0.5±0.1	2.34	1.97	0.585	2.00
$(1\pi_u)^2\,^2\Pi_u$	16.7 A / 16.9 V	—	—	3.96	0.72	0.3±0.1	7.23	0.46	0.317	−0.98
$(2\sigma_u)^2\,^2\Sigma_u$	18.8	—	—	0.467	−0.86	1.25±0.1	2.24	1.59	3.05	2.00
$(2\sigma_g)^2\,^2\Sigma_g$	37.3	—	—	—	—	—	0.142	1.23	3.21	2.00
$(1\sigma_u)^2\,^2\Sigma_u$	410	—	—	—	—	—	—	—	61.3	2.00
$(1\sigma_g)^2\,^2\Sigma_g$	410	—	—	—	—	—	—	—	56.7	2.00

[a] σ values include degeneracy factors for the orbital ionized.

[b] A = adiabatic IP; V = vertical IP; no letter indicates A = V.

[c] in units of 10^{-18} cm².

[d] σ in units of 10^{-21} cm².

[e] T. A. Carlson, G. E. McGuire, A. E. Jonas, K. L. Cheng, C. P. Anderson, C. C. Lu, and B. P. Pullen, *Electron Spectroscopy*, D. A. Shirley (Ed.), p. 207, North-Holland, Amsterdam, 1972.

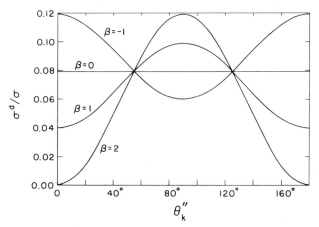

Fig. 7.17.   The ratio $\sigma^{\mathrm{d}}/\sigma$ as a function of θ''_k for $\beta = 2, 1, 0$, and -1. Reproduced with permission from Ref. 35.

probable error in the measured cross section σ^{d}, and to indicate how to minimize the probable error $\Delta\beta$ by judicious choice of measurement angles θ''. The treatment follows that of Huang et al.[35] It is true that statistical factors undoubtedly are not primary causes of uncertainty in β; in such experiments, systematic errors may be especially serious and can be avoided only by careful calibration and consideration of all possible corrections. Nevertheless, once systematic errors are accounted for, one seeks to reduce random errors in the most economical manner.

If the average of n independent measurements σ^{d}_i are accumulated at the magic angle, the probable error[36] in $(\sigma/4\pi)$ is

$$\left(\frac{\Delta\sigma}{4\pi}\right) = \frac{\Delta\sigma^{\mathrm{d}}}{n^{1/2}} \tag{7.7}$$

where $\Delta\sigma^{\mathrm{d}}$ is the probable error in the measured differential cross sections. Alternatively, one might fit Eq. 7.1 to independently measured intensities at two or more angles not necessarily equal to $54°44'$ or $125°16'$. With $f_i = -\frac{1}{4}(3\cos^2\theta''_k - 1)$, there exists a linear relationship between σ^{d}_i and f_i; a plot of σ^{d}_i versus f_i should give a straight line with intercept $(\sigma/4\pi)$ and slope $(\sigma\beta/4\pi)$. From standard formulas[37] relating to least-squares fits, we find the intercept

$$\left(\frac{\sigma}{4\pi}\right) = \sum_i \sum_j \frac{\sigma_i f_j (f_j - f_i)}{D} \tag{7.8}$$

$$D = \frac{1}{2}\sum_i \sum_j (f_i - f_j)^2 \tag{7.9}$$

If we assume that no errors are contained in the σ_i^d and f_i and that the probable errors $\Delta\sigma_i^d$ in the differential cross sections measured at different angles θ_i'' are all equal to $\Delta\sigma^d$, it can be shown that the consequent probable error[36] in $(\sigma/4\pi)$ is

$$\left(\frac{\Delta\sigma}{4\pi}\right) = \frac{\Delta\sigma^d\left(\sum_i g_i^2\right)^{1/2}}{D} \tag{7.10}$$

$$g_i = \sum_j f_j(f_j - f_i) \tag{7.11}$$

Because of the periodicity and symmetry of $f_i(\theta_i'')$, we may limit consideration to the interval $0 \leq \theta_k'' \leq \pi/2$, for which $-\frac{1}{2} \leq f_i \leq \frac{1}{4}$. In the initial columns of Table 7.5, we list $(\Delta\sigma/4\pi)/\Delta\sigma^d$ for several typical sets θ_k''. It appears that, for a given number of points n, $(\Delta\sigma/4\pi)$ is lowered the more closely all θ_k'' approach the magic angle. Nevertheless, combining measurements at $\theta'' = 20°$ and $\theta'' = 90°$ yields acceptable errors.

Again, from a least-squares fitting of Eq. 7.1, one finds

$$\beta = \frac{\text{slope}}{\text{intercept}}$$

$$= \frac{\sum_i \sum_j \sigma_i^d(f_i - f_j)}{\sum_i \sum_j \sigma_i^d f_j(f_j - f_i)} \tag{7.12}$$

If we define the relative cross section

$$r_i = \frac{\sigma_i^d}{\sigma_\perp} \tag{7.13}$$

$$= \frac{1 + \beta f_i}{1 + \frac{1}{4}\beta} \tag{7.14}$$

where $\sigma_\perp$ is the differential cross section measured at $\theta'' = \pi/2$, and if we again assume all $\Delta\sigma_i^d = \Delta\sigma^d$, then the probable error[36] in β can be shown to be

$$\Delta\beta = \left(\frac{\Delta\sigma^d}{\sigma_\perp}\right)\frac{\left(\sum_i h_i^2 + \beta^2\sum_i g_i^2\right)^{1/2}}{\sum_i r_i g_i} \tag{7.15}$$

Table 7.5.   Ratios $(\Delta\sigma/4\pi)\Delta\sigma^{\mathrm{d}}$ and $\Delta\beta/(\Delta\sigma^{\mathrm{d}}/\sigma_\perp)$ Calculated for σ_i^{d} Measured at Angles θ_k'' for Various β

Angles (deg)[a]	$(\Delta\sigma/4\pi)/\Delta\sigma^{\mathrm{d}}$ All β	$\Delta\beta/(\Delta\sigma^{\mathrm{d}}/\sigma_\perp)$ $\beta=2$	$\beta=1$	$\beta=0$	$\beta=-1$
55	1^b	—	—	—	—
55,55	0.707^b	—	—	—	—
0,90	0.745	3.61	2.52	1.88	1.49
20,90	0.728	3.87	2.83	2.14	1.69
55,55,55	0.577^b	—	—	—	—
0,0,90	0.707	3.24	2.22	1.63	1.33
0,90,90	0.577	3.00	2.17	1.63	1.30
20,20,90	0.677	3.44	2.46	1.85	1.48
20,90,90	0.580	3.27	2.42	1.85	1.45
20,55,90	0.589	3.63	2.74	2.11	1.65
55,55,90	0.706	7.64	6.18	4.89	3.71
55,90,90	0.998	7.92	6.24	4.89	3.74
55,55,55,55	0.500^b	—	—	—	—
20,90,90,90	0.521	3.05	2.27	1.74	1.36
20,20,90,90	0.515	2.74	1.99	1.51	1.20
20,20,20,90	0.660	3.28	2.33	1.74	1.40
55,90,90,90	0.998	7.54	5.90	4.61	3.54
55,55,90,90	0.706	6.35	5.07	3.99	3.04
55,55,55,90	0.576	7.13	5.81	4.61	3.48
20,40,55,90	0.532	3.47	2.65	2.05	1.59
20,45,60,90	0.513	3.43	2.63	2.04	1.58

[a] 55° angle listed in table was actually 54°44′.
[b] Computed by Eq. 7.7.

in which

$$h_i = \sum_j (f_i - f_j) \tag{7.16}$$

and g_i has been defined in Eq. 7.11 and r_i in Eq. 7.13. By Eqs. 7.14 and 7.15, $\Delta\beta$ is seen to be proportional to the fractional error $\Delta\sigma^{\mathrm{d}}/\sigma_\perp$ and to be dependent upon β. The calculated $\Delta\beta/(\Delta\sigma^{\mathrm{d}}/\sigma_\perp)$ as a function of β for the various sets of angles θ_k'' are listed in Table 7.5; for $n=2$ and $\Delta\sigma^{\mathrm{d}}/\sigma_\perp = \pm 0.1$, see Fig. 7.18.

To visualize the meaning of Eq. 7.15, we consider one special example. For the case $\beta=0$ and an *even* number of points n, half of which are measured at θ_1'' and half at θ_2'', for which $\Delta_f = f_2 - f_1$, Eq. 7.15 reduces to

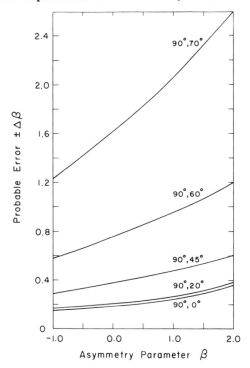

Fig. 7.18. Probable error $\pm \Delta\beta$ as a function of β for several selections of two collection angles θ_1, θ_2, and $\Delta\sigma/\sigma = \pm 0.1$. Reproduced with permission from Ref. 35.

the simple form

$$\frac{\Delta\beta}{(\Delta\sigma^d/\sigma_\perp)} = \frac{2n^{-1/2}}{|\Delta f|} \tag{7.17}$$

Numerically, if $\Delta\sigma^d/\sigma_\perp \sim \pm 0.1$, which is not unusual, a computed $\beta = 0$ contains an associated probable error $0.2 n^{-1/2}/|\Delta f|$.

Generally, the following conclusions can be made:

1. It is more direct to determine β from a linear least-squares fit of σ_i^d versus f_i, rather than from a nonlinear fit of σ_i^d versus θ_i.

2. Since β depends intrinsically upon a slope, $\Delta\beta$ is increased as the measuring angles θ_i'' are brought closer together. From the trend indicated in Table 7.5, using *many* data points σ_i^d at just two angles $0°$ (or $20°$) and $90°$ gives lower $\Delta\beta$ than using the same total number of σ_i^d at many different angles θ_i''. Measuring σ_i^d at two angles only

may even be more convenient in the operational sense. As an example, if $\Delta\sigma^d/\sigma_\perp = \pm 0.1$ and if one calculates $\beta = 0$ using just one measurement at $\theta'' = 20°$ and one at $\theta'' = 90°$, Eq. 7.17 gives $\Delta\beta \pm 0.19$; for ten measurements at $\theta'' = 20°$ and ten at $\theta'' = 90°$, then $\Delta\beta = \pm 0.06$. This may explain why scattered results for β for the same systems have been reported by different experimental groups.[38]

All of the above remarks are conditioned by the assumption of electric dipole absorption only. Contribution from electric quadrupole, magnetic dipole, and two-photon processes results in a more complicated angular dependence. These terms would result in a deviation form the assumed Eq. 7.1 angular dependence.

7.6. Variation of cross section over the vibrational envelope of a photoelectron transition

Vibrational transition probabilities accompanying photoelectron transitions have been measured ever since the resolving power of spectrometers has been sufficient to distinguish vibrational structure. This has prompted many researchers to match experimental vibrational intensities with calculated Franck-Condon factors in order to deduce the nuclear coordinates of the ionic species. While this procedure will yield approximate changes in nuclear coordinates, its accuracy is dependent upon the constancy of the electronic transition moment across the vibrational progression. Although variations in the electronic transition moment across vibrational progressions of discrete electronic transitions are well known,[39-41] corresponding variations in photoelectron spectra have been studied for only a few molecules.[42-45] The first example of this is H_2, for which Frost et al.[42] noted a discrepancy between calculated Franck–Condon factors and vibrational band intensities. Later, Itikawa[43] and Berkowitz and Spohr[44] showed that the electronic transition moment is not constant over the vibrational bands observed in the He I PE spectrum of H_2, HD, and D_2.

In this section we outline a method for calculation of vibrational band intensities in photoelectron transitions.[45] Variations of the electronic transition moment as a function of kinetic energy of ejected electrons and changes in internuclear distance (approximated according to the r centroid) over vibrational envelopes are considered. Franck–Condon factors coupled with varying transition moments are used to calculate the relative vibrational band intensities in the He I spectrum of N_2 and the He I and He II spectra of H_2.

We begin by extracting the transition moment integral from Eq. 6.4, rewriting it as

$$P_{0j} = \langle \Psi_0(q, Q) | \sum \mathbf{p}_n | \Psi_j(q, Q) \rangle \qquad (7.18)$$

The q and Q refer to electronic and nuclear coordinates, respectively. The operator $\mathbf{p}_n$ is the momentum operator $(\hbar/i)\nabla$ summed over all electrons n. Summation of the operator over nuclei is not included because these terms vanish due to orthogonality of the electronic components of $\Psi_0(q, Q)$ and $\Psi_j(q, Q)$ (see Chapter 3). Using the Born–Oppenheimer approximation, the eigenfunctions can be written as a product of electronic $\Psi_e(q, Q)$, vibrational $\Psi_v(Q)$, and rotational $\Psi_\tau(Q)$ wavefunctions. We assume that the rotational wavefunctions are uncoupled, that is, we can integrate over rotational coordinates explicitly. With the above assumptions we rewrite Eq. 7.18 as

$$P_{0j} = \langle \Psi_{0e}(q, Q)\Psi_{0v}(Q) | \sum \mathbf{p}_n | \Psi_{je}(q, Q)\Psi_{jv}(Q) \rangle \qquad (7.19)$$

In the previous sections the ionization process was considered as a pure electronic transition at fixed nuclear positions, Q_0, where Q_0 corresponds to the ground-state equilibrium nuclear configuration. The transition energy was chosen to correspond to the adiabatic process, thus fixing the kinetic energy of the photoejected electrons as E_0. Using these approximations and recognizing that p_n operates only on electronic coordinates, the cross section for a vibrational band $(d\sigma_v/d\Omega)$ in an electronic transition can be expanded from Eq. 6.4 and expressed as

$$\frac{d\sigma_v}{d\Omega} = \frac{\pi e^2}{m^2 \omega c} |\mathbf{u} \cdot \langle \Psi_{0e}(q, Q_0) | \sum \mathbf{p}_n | \Psi_{je}(q, Q_0) \rangle|^2$$

$$|\langle \Psi_{0v}(Q) | \Psi_{jv}(Q) \rangle|^2 \rho(E) = \frac{d\sigma}{d\Omega}(Q_0, E_0) \cdot F_v . \qquad (7.20)$$

where F_v is the square of the vibrational overlap integral. This type of treatment assumes that the electronic transition moment P_{0j} is constant across an envelope of vibrational bands. The results of Section 7.2 have shown that $d\sigma(Q_0, E_0)/d\Omega$ provides an approximation to the total integrated intensity of a photoelectron transition. This integrated intensity is simply the electronic intensity subdivided between the various vibrational bands according to the relative magnitude of the Franck–Condon factors for the individual bands.

We now consider the dependence of the electronic transition moment on the nuclear coordinates Q and the kinetic energy of the ejected electrons E. The dependence on Q can be treated by two different methods. First, the Q dependence can be retained in the electronic wavefunctions of Eq. 7.20, and integration can be performed over q and

Q. This method is difficult to apply to polyatomic or even most diatomic molecules whose wavefunctions can hardly be described by sophisticated mathematical forms allowing easy variation of Q. Second, the dependence of the transition moment on Q can be considered in terms of the r centroid ($\bar{r}_{v_0 v_j}$) method.[46,47] The r centroid

$$\bar{r}_{v_0 v_j} = \frac{\langle \Psi_{0v} | r | \Psi_{jv} \rangle}{\langle \Psi_{0v} | \Psi_{jv} \rangle} \tag{7.21}$$

is *the expectation value or weighed average with respect to* $\langle \Psi_{0v} | \Psi_{jv} \rangle$ *of the range of* r *values experienced by a molecule in both states of a transition* $\Psi_{jv} \leftarrow \Psi_{0v}$. In this method the transition probabilities are calculated at the r-centroid coordinates, and Eq. 7.20 becomes

$$\frac{d\sigma_v}{d\Omega} = \frac{d\sigma}{d\Omega} (\bar{r}_{v_0 v_j}, E_0) \cdot F_v \tag{7.22}$$

The method of r centroids is used in this treatment.

Next we consider the cross-section dependence on kinetic energy of the emitted electrons. In typical photoionization experiments the frequency of the incident radiation is fixed, and vibrational bands of photoelectron transitions are detected by measuring the kinetic energies of the emitted electrons. In such an experiment the kinetic energy of the photoelectrons varies with vibrational states and may introduce an enhancement (or decrement) of its own, for it is well known that photoelectron cross sections rise to a maximum near the ionization threshold where the kinetic energies are low and then decrease as the kinetic energies increase. This dependence can be determined by calculating the cross sections at the kinetic energies E_k corresponding to each vibrational transition; thus Eq. 7.20 becomes

$$\frac{d\sigma_v}{d\Omega} = \frac{d\sigma}{d\Omega} (Q_0, E_k) \cdot F_v \tag{7.23}$$

Finally, the combined effect of variations in nuclear coordinates and electron kinetic energies is determined by evaluation of Eq. 7.20 in the form

$$\frac{d\sigma_v}{d\Omega} = \frac{d\sigma}{d\Omega} (\bar{r}_{v_0 v_j}, E_k) \cdot F_v \tag{7.24}$$

The necessity of obtaining wavefunctions at many $\bar{r}_{v_0 v_j}$ values makes it much more economical to employ INDO wavefunctions rather than the *ab initio* wavefunctions used in Sections 7.2 to 7.4.

Calculation of Franck–Condon factors and r centroids

Franck–Condon overlaps and r centroids require vibrational wavefunctions; these can be expressed in varying degrees of sophistication. In the data to be presented, we use the approximate but general Morse oscillator wavefunctions[48] of the form

$$\Psi_v = \left(\frac{\beta}{N_v}\right)^{1/2} \exp\left(\frac{-X}{2}\right)(X)^{\alpha/2} L_v^\alpha(X) \tag{7.25}$$

in which

$$N_v = \sum_{s=0}^{v} \frac{\Gamma(s+\alpha)}{s!} \tag{7.26}$$

$$\alpha = \frac{\omega_e}{\omega_e x_e} - 2\overset{\text{.}}{v} - 1 \tag{7.27}$$

$$\beta = 1.2177 \times 10^{-1}(4\mu\omega_e x_e)^{1/2} \tag{7.28}$$

and

$$X = \frac{\omega_e}{\omega_e x_e} \exp\left[-\beta(r - r_e)\right]. \tag{7.29}$$

In Eqs. 7.25–7.29, r_e is the equilibrium internuclear distance, ω_e is the harmonic vibrational frequency, $\omega_e x_e$ is the anharmonicity constant of a given electronic state, and μ is the reduced mass of the species. Generalized Laguerre polynomials $L_v^\alpha(X)$ are of the form[48]

$$L_v^\alpha(X) = \sum_{s=0}^{v} (-1)^{v-s} \frac{\Gamma(\alpha+1+v)(X)^{v-s}}{s!(v-s)!\Gamma(\alpha+v+1-s)} \tag{7.30}$$

The Franck–Condon overlap and r-centroid integrals can be evaluated numerically according to Simpson's Rule and the calculation programmed closely to the manner suggested by Halmann and Laulicht.[49] Required spectroscopic constants for H_2 and N_2 are available and are reproduced in Table 7.6 together with the corresponding units being used in this formulation.

Cross-section dependence on kinetic energy E_k and nuclear coordinates Q

A variation in electronic transition moment over the envelope of vibrational bands of N_2 and H_2 is indicated from the following treatment. Experimental vibrational band intensities divided by calculated Franck–Condon factors, normalized to unity for the most intense band, are plotted against vibrational quantum number in Fig. 7.19. Since the intensities depend on the transition moment and Franck–Condon factors,

Table 7.6.   Spectroscopic Constants[a,b] for N_2, N_2^+, H_2, and H_2^+

	μ(a.u.)	ω_e(cm^{-1})	$\omega_e x_e$(cm^{-1})	r_e(Å)
N_2 $^1\Sigma_g^+$	7.00377	2358.07	14.19	1.0976
N_2^+ $^2\Sigma_g^+$	7.00363	2207.19	16.14	1.118
N_2^+ $^2\Pi_u$	7.00363	1902.84	14.91	1.177
N_2^+ $^2\Sigma_u^+$	7.00363	2419.84	23.19	1.075
H_2 $^1\Sigma_g^+$	0.50407	4395.24	117.995	0.74166
H_2^+ $^2\Sigma_g^+$	0.50407	2297	62	1.060

[a] R. W. Nicholls, *J. Res. Nat. Bur. Stand.*, *A*, **65,** 451 (1961).
[b] R. W. Nicholls and W. R. Jarmain, *Proc. Roy. Soc.*, *Ser. A*, **69,** 253 (1956).

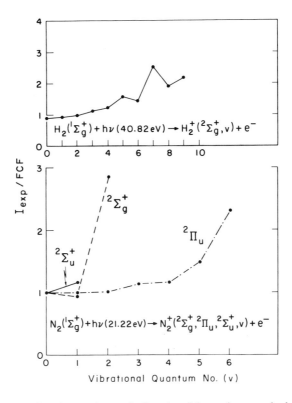

Fig. 7.19.   The ratio of experimental vibrational intensity to calculated Franck–Condon factor for N_2 and H_2 plotted as a function of vibrational quantum number. Reproduced with permission from Ref. 45.

a horizontal line at 1.0 is expected if the transition moment is constant throughout the progression. All of the plots show positive slopes indicating that the transition moment is increasing across the progression.

Calculations are presented within three different frameworks in order to test the dependence of cross section on Q and E_k and to compare calculated and observed vibrational band intensities.

1. Q is maintained constant and $(d\sigma/d\Omega)(Q_0, E_k)$ is calculated for each vibrational band with its corresponding E_k value, Eq. 7.23.
2. E_k is maintained constant and $(d\sigma/d\Omega)(\bar{r}_{v_0 v_j}, E_0)$ is calculated over a wide range of r values, Eq. 7.22.
3. Equation 7.24 is evaluated for each vibrational transition $v_j \leftarrow v_0$ with $\bar{r}_{v_0 v_j}$ and E_k corresponding to that vibrational transition and INDO wavefunctions determined at $r = \bar{r}_{v_0 v_j}$.

METHOD 1. Table 7.7 gives $(d\sigma/d\Omega)(Q_0, E_k)$ for each vibrational transition of N_2 and H_2. Considering N_2, the cross section increases across the envelope of the $^2\Sigma_g^+ \leftarrow {}^1\Sigma_g^+$ transition but decreases across the envelope of the $^2\Pi_u \leftarrow {}^1\Sigma_g^+$ and $^2\Sigma_u^+ \leftarrow {}^1\Sigma_g^+$ transitions. This behavior is comprehensible from the cross section versus $\hbar\omega$ curves where it is observed that cross sections rise sharply near the ionization threshold, go through a maximum, and then decrease to small values for excitation energies much greater than the ionization threshold. These curves can also be used as cross section versus E_k by using the conversion

$$\hbar\omega = E_I + E_k \tag{7.31}$$

The cross-section dependence on E_K can vary drastically depending upon the magnitude of E_k. That the cross section is increasing across the $^2\Sigma_g^+$ band but decreasing across the $^2\Pi_u$ and $^2\Sigma_u^+$ bands (Table 7.7) indicates that the E_ks of these bands are on opposite sides of the maximum in the cross section versus E_k curves.

A similar behavior is observed in the H_2 cross sections. Using He II excitation, the kinetic energies are high and the cross section is increasing towards lower E_K values. With He I excitation the kinetic energies are on the opposite side of the cross section maximum where the cross sections decrease with decreasing E_k. This is, in part, an artifact due to the approximate nature of the orthogonalized plane-wave (OPW) calculations. It is known that in OPW calculations, the rise to the maximum at the ionization threshold is much slower than the stepfunction that is observed experimentally. More sophisticated cross section calculations using, for example, coulomb waves may shift this maximum to lower E_k. Despite the approximate nature of the OPW calculations, the general trends of $(d\sigma_v/d\Omega)(Q_0, E_k)$ versus E_k are generated: For $E_k > \sim 5$ eV,

Table 7.7. Calculated Differential Photoionization Cross Sections $(d\sigma/d\Omega)(Q_0, E_k)$ (OPW Approximation) Evaluated at the Kinetic Energy Corresponding to each Vibrational Transition in the PE Spectra of H_2 and N_2

Transition	v	Ionization Potential	$(d\sigma/d\Omega)(Q_0, E_k)^a$	
			($h\nu = 21.22$ eV)	
$N_2\ {}^2\Sigma_g^+$	0	15.58	1.921	
	1	15.85	2.041	
	2	16.11	2.011	
			($h\nu = 21.22$ eV)	
$N_2\ {}^2\Pi_u$	0	16.70	2.010	
	1	16.93	1.939	
	2	17.16	1.873	
	3	17.38	1.817	
	4	17.61	1.767	
	5	17.83	1.723	
	6	18.04	1.683	
			($h\nu = 21.22$ eV)	
$N_2\ {}^2\Sigma_u^+$	0	18.75	0.392	
	1	19.05	0.318	
			($h\nu = 21.22$ eV)	($h\nu = 40.82$ eV)
$H_2\ {}^2\Sigma_g^+$	0	15.43	10.54	2.464
	1	15.70	10.41	2.501
	2	15.96	10.26	2.536
	3	16.20	10.10	2.570
	4	16.42	9.924	2.601
	5	16.63	9.740	2.631
	6	16.83	9.545	2.660
	7	17.01	9.352	2.687
	8	17.18	9.154	2.712
	9	17.34	8.952	2.736

^{a} Cross sections in units of 10^{-19} cm^2; *ab initio* wavefunctions used for N_2.

$(d\sigma/d\Omega)(Q_0, E_k)$ increases across a band envelope as E_k decreases. For $E_k < \sim 5$ eV, $(d\sigma_v/d\Omega)(Q_0, E_k)$ decreases across a band envelope as E_k decreases. From the curves of Section 7.2 it is obvious the the maximum in the curves [or turning point for $(d\sigma_v/d\Omega)(Q_0, E_k)$] will vary for each transition.

METHOD 2. In order to test the cross section dependence on Q, $(d\sigma/d\Omega)(\bar{r}_{v_0 v}, E_0)$ was calculated for N_2 and H_2O using INDO wavefunctions generated at various r values while keeping E_k constant. Plots of

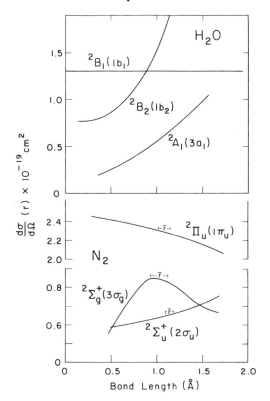

Fig. 7.20. Calculated differential cross section $(d\sigma/d\Omega)(r)$ (PW approx) versus bond length r for N_2 and H_2O using the adiabatic E_is and 21.22-eV excitation energy. The orbital ionized in producing the final ionic state is indicated in brackets. The region of r covered by the r centroid is indicated for the transitions of N_2. Wavefunctions used in the calculation were INDO functions. Reproduced with permission from Ref. 45.

$(d\sigma/d\Omega)(r, E_0)$ versus r are presented in Figs. 7.20 and 7.21. For N_2 all three curves exhibit a different behavior that is largely determined by changes in the INDO wavefunctions with r. We find that $(d\sigma/d\Omega)(r, E_0)$ follows the coefficient of the AO with the highest cross section. In this particular kinetic energy region, the $2s$ AO cross section is higher than that of the $2p$ AO; thus $(d\delta/d\Omega)(r, E_0)$ increases or decreases as does the $2s$ AO coefficient. In the $2\sigma_u$ MO, the $2s$ coefficients increase consistently at the expense of the $2p$ coefficients as r increases. In the $3\sigma_g$ MO, the $2s$ coefficients increase initially and then begin to decrease again as r becomes large, producing a maximum in the curve (Fig. 7.20). In the $1\pi_u$

Table 7.8. Calculated Differential Photoionization Cross Sections $(d\sigma/d\Omega)(\bar{r}_{v0vj}, E_k)$ (OPW Approximation) and Total Transition Probabilities

$$\frac{d\sigma}{d\Omega}(\bar{r}_{v0vj}, E_k) \cdot F_v, \quad \frac{d\sigma}{d\Omega}(Q_0, E_k) \cdot F_v, \quad \text{and} \quad \frac{d\sigma}{d\Omega}(\bar{r}_{v0vj}, E_0) \cdot F_v$$

for N_2 and H_2. (The r centroid, $\bar{r}_{v0vj}$, Franck–Condon factor, F_v, and Experimental Vibrational Intensities Are Also Listed.)

Transition	v	$\bar{r}_{v0vj}$ (Å)	$(d\sigma/d\Omega)(\bar{r}_{v0vj}, E_k)$[a]	Franck–Condon Factor[b] (F_v)	Experimental Vibrational Intensity[c]	$(d\sigma/d\Omega)F_v$[d]		
						$(\bar{r}_{v0vj}, E_k)$	(Q_0, E_k)	$(\bar{r}_{v0vj}, E_0)$
$N_2\ ^2\Sigma_g^+$ $(\hbar\nu = 21.22$ eV$)$	0	1.112	1.474	0.9023(100)	100.0	100	100	100
	1	1.013	1.554	0.0906(10.0)	6.94±0.66	10.6	10.6	10.3
	2	0.925	1.620	0.0065(0.72)	0.32±0.21	0.85	0.75	0.73
$N_2\ ^2\Pi_u$ $(\hbar\nu = 21.22$ eV$)$	0	1.139	8.492	0.2447(78.8)	87.1±3.1	76.1	81.6	76.1
	1	1.114	8.784	0.3107(100)	100.0±2.8	100	100	100
	2	1.090	9.046	0.2253(72.5)	76.3±1.5	74.5	70.0	72.7
	3	1.068	9.302	0.1236(39.8)	43.5±2.9	42.2	37.3	40.0
	4	1.047	9.546	0.0574(18.5)	19.1±2.4	19.9	16.8	18.7
	5	1.027	9.784	0.0238(7.7)	7.09±0.77	7.6	6.8	7.8
	6	1.010	9.978	0.0095(3.1)	2.51±0.40	2.3	2.7	2.9
$N_2\ ^2\Sigma_u^+$ $(\hbar\nu = 21.22$ eV$)$	0	1.091	1.682	0.8912(100)	100.0	100	100	100
	1	1.186	1.384	0.1070(12.0)	9.84±2.51	9.9	9.6	9.5
$H_2\ ^2\Sigma_g^+$ $(\hbar\nu = 40.82$ eV$)$	0	0.895	2.181	0.0921(47.0)	40±3	42.4	45.7	42.4
	1	0.847	2.305	0.1750(89.3)	83±3	84.9	88.1	86.3

2	0.803	2.422	0.1959(100)	100	100	100	100
3	0.763	2.530	0.1670(85.2)	96±3	89.0	83.4	87.8
4	0.726	2.631	0.1274(65.0)	81±4	70.6	66.7	69.2
5	0.691	2.727	0.0873(44.6)	65±4	50.2	46.2	48.5
6	0.659	2.815	0.0566(28.9)	47±4	33.6	30.3	32.2
7	0.629	2.896	0.0354(18.1)	32±4	21.6	19.1	20.5
8	0.600	2.973	0.0218(11.1)	21±4	13.7	11.9	12.9
9	0.573	3.045	0.0133(6.8)	15±4	8.5	7.3	7.7
$H_2\ ^2\Sigma_g^+$ ($h\nu = 21.22$ eV)							
0	0.895	10.21	0.0921(47.0)	46.3	47.3	48.3	47.3
1	0.847	10.20	0.1750(89.3)	86.3	89.8	90.6	88.5
2	0.803	10.15	0.1959(100)	100	100	100	100
3	0.763	10.06	0.1670(85.2)	88.4	84.5	83.9	85.9
4	0.726	9.948	0.1274(65.0)	70.0	63.7	62.9	65.9
5	0.691	9.811	0.0873(44.6)	57.9	43.1	42.3	45.5
6	0.659	9.651	0.0566(28.9)	47.2	27.5	26.9	29.6
7	0.629	9.485	0.0354(18.1)	31.3	16.9	16.5	18.6
8	0.600	9.308	0.0218(11.1)	24.1	10.2	9.9	11.5
9	0.573	8.915	0.0133(6.8)	13.7	6.3	5.9	7.1

[a] Cross section in units of 10^{-19} cm².

[b] Franck–Condon factors obtained from Morse wavefunctions, Eq. 7.25. Values in parenthesis are normalized with the highest intensity equal to 100.

[c] Vibrational band intensities were obtained as follows: N_2 ($\hbar\omega = 21.22$ eV), Ref. 50; H_2($\hbar\omega = 40.82$ eV), Ref. 45; H_2($\hbar\omega = 21.22$ eV), Ref. 44.

[d] Total transition probabilities $d\sigma/d\Omega$ from Eqs. 7.22 to 7.24. Values are normalized with the highest intensity equal to 100.

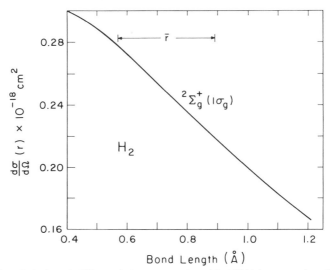

Fig. 7.21. Calculated differential cross section $(d\sigma/d\Omega)(r)$ versus bond length r for H_2 using the adiabatic E_I and 40.82-eV excitation energy. The region of r covered by the r centroid is indicated. The corresponding curve using 21.22-eV excitation energy is parallel to the one in this figure. Reproduced with permission from Ref. 45.

MO the $2p_x$ and $2p_y$ character of the MO begins to segregate as r increases, eventually separating to pure $2p_x$ and $2p_y$ AOs; consequently $(d\sigma/d\Omega)(r, E_0)$ approaches that of a nitrogen $2p$ AO.

The cross section versus r_{H-O} curves for H_2O reflect the hydrogen $1s$ AO character of the MOs. The $1b_1$ MO is simply a pure oxygen $2p$ AO and is unaffected by changes in r_{H-O}. The $3a_1$ and $1b_2$ MOs are mixtures of oxygen $2s$ and $2p$ and hydrogen $1s$ character. The H $1s$ character increases at the expense of O $2s$ and $2p$ character as r_{H-O} increases. Since the H $1s$ cross section is higher than that of O $2s$ and $2p$ in this kinetic energy region, the corresponding MO cross sections increase drastically.

In the case of H_2, the character of the MO is independent of the bond length. The computed variation in $(d\sigma/d\Omega)(r, E_0)$ is thus due entirely to the variation in r. The r dependence enters the cross section calculation in the argument of the spherical Bessel functions used to expand the plane waves. From Fig. 7.21, it is obvious that the r dependence of the cross section may be substantial even though the MO wavefunctions remain unchanged.

METHOD 3. The calculation is carried out on the three transitions of N_2 and the transition of H_2 using Eq. 7.24. The $(d\sigma/d\Omega)(\bar{r}_{v_0v_j}, E_k)$ are evaluated at $\bar{r}_{v_0 v_j}$ and E_k corresponding to the $v_j \leftarrow v_0$ transition using INDO wavefunctions determined at $\bar{r}_{v_0 v_j}$. These results are listed in Table 7.8 along with the total transition probabilities determined from Eqs. 7.22, 7.23, and 7.24, the r centroids, the calculated Franck–Condon factors from Eq. 7.25, and the experimental vibrational band intensities. The value of the electronic cross section $(d\sigma/d\Omega)(\bar{r}_{v_0 v_j}, E_k)$ increases across all of the vibrational envelopes except that of the N_2 $^2\Sigma_u^+$ band and the H_2 $^2\Sigma_g^+$ band with $\hbar\omega = 21.22$ eV. These exceptions are probably a result of the inadequacy of the OPW approximation for low kinetic energy electrons. The variation in the cross section is largest for H_2 ($\hbar\omega = 40.82$ eV), where the value increases by nearly 40% across the band envelope. With the exception of the two cases mentioned above, the data in Table 7.8 indicate that the Franck–Condon factors alone predict intensities that are lower than the vibrational band intensities for high vibrational quantum numbers. The total transition probability $(d\delta/d\Omega)(\bar{r}_{v_0 v_j}, E_k) \cdot F_v$ usually provides a better representation of the vibrational intensities than the Franck–Condon factor F_v alone. Despite this improvement, the calculated intensities remain low for higher members of the vibrational progressions. We feel that the major source of error is in the cross-section calculation, which is not able to reproduce the full increase in the transition moment as observed in Fig. 7.19.

The calculations presented here indicate that electronic cross sections are generally not constant over the vibrational envelope of a photoelectron transition. These electronic cross sections, or transition moments, are a function of the electron kinetic energy and average nuclear coordinates corresponding to each vibrational band. Simple Franck–Condon factor calculations alone can be misleading in determining nuclear coordinates of molecular ions when the assumption of constant cross section is used. The cross-section calculations coupled with Franck–Condon factors generally lead to improved agreement with experiment. Since the cross-section calculation is capable of handling general polyatomic molecules, the method proposed here for calculating vibrational band intensities is extendable to polyatomics providing that vibrational wavefunctions and the average normal coordinates associated with a vibrational-photoelectron transition are available.

References

1. A. Katrib, T. P. Debies, R. J. Colton, T. H. Lee, and J. W. Rabalais, *Chem. Phys. Letters*, **22**, 196 (1973).
2. J. W. Rabalais, T. P. Debies, J. L. Berkosky, J. T. J. Huang, and F. O. Ellison, *J. Chem. Phys.*, **61**, 516 (1974); ibid., **61**, 529 (1974); T. P. Debies and J. W. Rabalais, *J. Amer. Chem. Soc.*, **97**, 487 (1975).

3. M. B. Robin, N. A. Kuebler, and C. R. Brundle, *Electron Spectroscopy*, D. A. Shirley (Ed.), p. 351, North-Holland, Amsterdam, 1972.

4. W. C. Price, A. W. Potts, and D. G. Streets, *Electron Spectroscopy*, D. A. Shirley (Ed.), p. 187, North-Holland, Amsterdam, 1972.

5. L. L. Lohr, Jr. and M. B. Robin, *J. Amer. Chem. Soc.*, **92**, 7241, (1970).

6. A. Schweig and W. Thiel, *Chem. Phys. Lett.*, **16**, 409 (1972); ibid., **21**, 541 (1973); ibid., **12**, 49 (1971); A. Schweig and W. Thiel, *J. Chem. Phys.*, **60**, 951 (1974); idem., *J. Electron Spectrosc.*, **2**, 199 (1973); ibid., **3**, 27 1974).

7. I. G. Kaplan and A. P. Markin, *Opt. Spectrosc.*, **24**, 475 (1968); idem., **25**, 275 (1968).

8. G. V. Marr, *Photoionization Processes in Gases*, Chap. 6, Academic, New York, 1967.

9. J. T. J. Huang and F. O. Ellison, *Chem. Phys. Letters*, **25**, 43 (1974).

10. E. Clementi and D. L. Raimondi, *J. Chem. Phys.*, **38**, 2686 (1963).

11. E. Clementi, *IBM J. Res. Develop.*, **9**, (Suppl.), 2 (1965).

12. J. T. J. Huang, F. O. Ellison, and J. W. Rabalais, *J. Electron Spectrosc.*, **3**, 339 (1974).

13. K. Siegbahn, C. Nordling, G. Johansson, J. Hedman, P. F. Heden, K. Hamrin, U. Gelius, T. Bergmark, L. O. Werme, R. Manne, and Y. Baer, *ESCA Applied to Free Molecules*, North-Holland, Amsterdam, 1971.

14. M. R. Flannery and U. Öpik, *Proc. Phys. Soc. London*, **86**, 491 (1965).

15. H. P. Kelly, *Chem. Phys. Letters*, **20**, 547 (1973).

16. G. R. Cook and P. H. Metzger, *J. Opt. Soc. Amer.* **54**, 968 (1964).

17. For H_2 it should be noted that the OPW method reduces to the PW method because there is only one MO.

18. A. Dalgarno, *Proc. Phys. Soc.* **A65**, 663 (1952).

19. H. C. Tuckwell, *J. Phys. B*, **3**, 293 (1970).

20. J. A. R. Samson and R. B. Cairns, *J. Geophys. Res.*, **69**, 4583 (1964).

21. N. Wainfain, W. C. Walker, and G. L. Weissler, *Phys. Rev.*, **99**, 542 (1955).

22. T. A. Carlson, G. E. McGuire, A. E. Jonas, K. L. Cheng, C. P. Anderson, C. C. Lu, and B. P. Pullen, *Electron Spectroscopy*, D. A. Shirley (Ed.), p. 207, North-Holland, Amsterdam, 1972.

23. H. A. Bethe and E. E. Salpeter, *Quantum Mechanics of One- and Two-Electron Atoms*, Academic, New York, 1957.

24. M. O. Krauss, *Phys. Rev.*, **177**, 151 (1968).

25. P. S. Bagus and H. F. Schaefer, *J. Chem. Phys.*, **56**, 224 (1972); L. C. Snyder, *J. Chem. Phys.*, **55**, 95 (1971); M. E. Schwartz, J. D. Switalski, and R. S. Strouski, *Electron Spectroscopy*, D. A. Shirley (Ed.) p. 605, North-Holland, Amsterdam, 1972.

26. T. A. Carlson and G. E. McGuire, *J. Electron Spectrosc.*, **1**, 209 (1972/1973); T. A. Carlson and C. P. Anderson, *Chem. Phys. Letters*, 10, **561** (1971); T. A. Carlson and A. E. Jonas, *J. Chem. Phys.*, **55**, 4913 (1971); T. A. Carlson, *Chem. Phys. Letters*, **9**, 23 (1971).

27. J. W. McGowan, D. A. Vroom, and A. P. Comeaux, *J. Chem. Phys.*, **51**, 5626 (1969); D. A. Vroom, A. P. Comeaux, and J. W. McGowan, *Chem. Phys. Letters*, **3**, 476 (1969).

28. J. A. R. Samson, *Phil. Trans. Roy. Soc. London*, **A268**, 141 (1970).
29. J. Berkowitz and H. Ehrhardt, *Phys. Letters*, **21**, 531 (1966); J. Berkowitz, H. Ehrhardt, and T. Tekaat, *Z. Phys.* **200**, 69 (1967).
30. J. W. Cooper and R. N. Zare, *J. Chem. Phys.*, **48**, 942 (1968), ibid., **49**, 4252 (1968); J. W. Cooper and S. T. Manson, *Phys. Rev.*, **177**, 157 (1969); S. T. Manson and J. W. Cooper, *Phys. Rev. A*, **2**, 2170 (1970).
31. J. C. Tully, R. S. Berry, and B. J. Dalton, *Phys. Rev.*, **176**, 95 (1968).
32. A. O. Buckingham, B. J. Orr, and J. M. Sickel, *Phil. Trans. Roy. Soc. London*, **A268**, 147 (1970); J. M. Sickel, *Mol. Phys.*, **18**, 95 (1970).
33. B. Ritchie, *J. Chem. Phys.*, **60**, 898 (1974).
34. J. A. R. Samson, *J. Opt. Soc. Amer.*, **59**, 356 (1959).
35. J. T. J. Huang, J. W. Rabalais, and F. O. Ellison, *J. Electron Spectrosc.*, **6**, 85 (1975).
36. H. Margenau and G. M. Murphy, *The Mathematics of Physics and Chemistry*, p. 498, Van Nostrand, New York, 1943.
37. See for example, E. Bright Wilson, Jr., *An Introduction to Scientific Research*, pp. 217f, McGraw-Hill, New York, 1952.
38. For example, see the comparisons displayed by T. A. Carlson, G. E. McGuire, A. E. Jonas, K. L. Cheng, C. P. Anderson, C. C. Lu, and B. P. Pullen in Ref. 22.
39. M. J. Mumma, E. J. Stone, and E. C. Zipf, *J. Chem. Phys.*, **54**, 2627 (1971).
40. S. Durmaz, J. N. Murrell, J. M. Taylor, and R. R. Suffolk, *Mol. Phys.*, **19**, 533 (1970).
41. R. W. Nicholls, *Proc. Roy. Soc. London*, **A69**, 741 (1956).
42. D. C. Frost, C. A. McDowell, and D. A. Vroom, *Proc. Roy. Soc. London*, **A296**, 566 (1967).
43. Y. Itikawa, *J. Electron Spectrosc.*, **2**, 125 (1973).
44. J. Berkowitz and R. Spohr, *J. Electron Spectrosc.*, **2**, 143 (1973).
45. T. H. Lee and J. W. Rabalais, *J. Chem. Phys.*, **61**, 2747 (1974).
46. P. A. Fraser, *Can. J. Phys.*, **32**, 515 (1954).
47. R. W. Nicholls and W. R. Jarmain, *Proc. Roy. Soc. London*, **A69**, 253 (1956).
48. H. A. Ory, A. P. Gittleman, and J. P. Maddox, *Astrophys. J.*, **139**, 346 (1964); H. A. Ory, *J. Chem. Phys.*, **40**, 562 (1964).
49. H. Halmann and I. Laulicht, *J. Chem. Phys.*, **43**, 438 (1965).
50. J. L. Gardner and J. A. R. Samson, *J. Chem. Phys.*, **60**, 3711 (1974).

8 Spin-Orbit Coupling in Molecular Ions

Ejection of an electron from a degenerate orbital of a closed-shell atom or molecule produces an ion in an electronic state characterized by orbital angular momentum greater than zero and electron spin angular momentum $+\frac{1}{2}\hbar$. The degeneracy of this ionic state can be lifted by interaction between spin and orbital angular momenta; that is, spin-orbit coupling. This chapter is devoted to the phenomenon of spin-orbit coupling in molecular ions. A model for spin-orbit interaction is derived in Section 1, and spin-orbit operators are introduced in Section 2. The model is applied to molecular ions in Section 3. Examples of spin-orbit coupling in the halogen acids, the halogens, and the PX_3, AsX_3, and SbX_3 series are presented in Sections 4 to 7. Spin-orbit interaction in low-symmetry molecules is discussed in Section 8, and finally, the conclusions are presented in Section 9.

The Russell-Saunders states of atoms are designated by the term symbol $^{2S+1}\Gamma_J$ where S is the total electron spin, Γ the orbital angular momentum $L(S=0, P=1, D=2, \cdots)$, and J the total angular momentum, which may assume the range of values $|L+S|, |L+S-1|, \cdots, |L-S|$. For example, ejection of a p electron from a closed-shell rare gas atom produces an ion with $S=\frac{1}{2}$ and $L=1$, hence $^2P_{3/2}$ and $^2P_{1/2}$ states; the energy difference between these two states is called the spin-orbit splitting. The spin-orbit splittings resulting from removal of a valence p electron of Ne, Ar, Kr, and Xe are 0.097, 0.178, 0.665, and 1.306 eV, respectively. Statistically, the relative intensities of the photoelectron bands are given by $2J+1$, that is, a ratio of $2:1$ in this particular case.

For molecules Γ must be replaced by the corresponding symmetry representation of the molecular point group for which the symbols are $^2\Sigma_{1/2}$, $^2\Pi_{3/2}, \cdots$ for linear molecules and ^2A, $^2E, \cdots$ for nonlinear molecules. If an electron is ejected from a nondegenerate orbital (for example, a_1 or a_2) two possible degenerate ion states result, depending upon whether the remaining odd electron has α or β spin. Since the electron spin angular momentum operators do not commute with the spin-orbit interaction operator $\mathbf{H}_{so}$, a proper group-theoretical description of these molecular ion states requires the use of *extended point groups* (or *double groups*).[1] States with $^2A_1$, $^2A_2$, or $^2\Sigma$ symmetry become $E_{1/2}$ with spin-orbit interaction; the twofold degeneracy persists. If an electron is ejected from a doubly degenerate e or π orbital, four ion states naturally result. Spin-orbit interaction causes partial splitting of the fourfold degeneracy: ^2E becomes $E_{3/2} + E_{1/2}$, each of which is doubly degenerate. For example, spin-orbit coupling can split the quadruply degenerate ^2E state of an AB_3^+ molecular ion into two components $E_{1/2} + E_{3/2}$, each of which is doubly degenerate. The magnitude of splitting is a function of the spin-orbit coupling parameters of the constituent atoms in the molecular ion and the localization properties of the odd electron(s).

The assignment of PE spectra can be a very difficult problem. If heavy atoms are present, the spin-orbit splitting of a band may be very useful in confirming an assignment. To make certain that the splitting is, in fact, due to spin-orbit interaction, it is desirable to have an independent theoretical estimate of the effect. Since rigorous calculation of spin-orbit interactions is quite complicated,[2] several semiempirical calculations have appeared.[3-5] In this chapter, we present, in general but practicable form, an approximate semiempirical model[6,7] for estimating spin-orbit coupling in molecular ion states that are derived by ionization of closed-shell neutral-molecule states. Spin-orbit splittings in the photoelectron spectra of several molecules are illustrated, and the model is applied to their interpretation.

8.1. Model for spin-orbit interaction energies

The theoretical interpretation of molecular ion states is ordinarily given in terms of molecular orbital theory. The spin-orbit interaction operator is usually not included in the effective Hamiltonian or Fock operator in ordinary SCF MO models. The magnitude of spin-orbit coupling effects in molecules increases as the atomic number of the atomic constituents increases. For molecules composed of light atoms, the splitting of a configuration into different terms through electron repulsion is usually very much larger than the further splitting of these terms into levels through spin-orbit interaction. For this reason, even when spin-orbit coupling is allowed for, L and S (for atoms) are *almost* good quantum numbers. Therefore, we can treat spin-orbit interactions as small perturbations and, to first order, we find (for atoms) that a given L-S term is split so that there is one level for each J value. This perturbation approach for molecules will now be developed.

According to MO theory, a reasonably good approximation to the exact wave function for a closed shell molecular ground state with $2N$ electrons is obtained by allocating one electron of each spin to each of the space orbitals $\phi_1, \phi_2, \cdots, \phi_N$ and combining the products into a single Slater determinantal wave function:

$$D_0 = \det\{\phi_1(1)\alpha(1)\phi_2(2)\alpha(2) \cdots \phi_N(N)\alpha(N)\phi_1(N+1)$$
$$\times \beta(N+1) \cdots \phi_N(2N)\beta(2N)\} \quad (8.1)$$

α-spin has been assigned to electrons 1 through N appearing in the initial product of diagonal elements of the determinant; β spin has been reserved for electrons $N+1$ through $2N$. The optimum SCF MOs satisfy the usual Hartree–Fock equations

$$\mathbf{F}\phi_i = \varepsilon_i\phi_i \quad (8.2)$$

where $\mathbf{F}$ is the Hartree–Fock Hamiltonian operator and ε_i is the orbital energy.

The molecular spin orbitals (MSO) are denoted as follows:

$$\psi_i = \phi_i\alpha, \quad i = 1, 2, \cdots, N$$
$$\psi_{i+N} = \phi_i\beta, \quad i = 1, 2, \cdots, N \quad (8.3)$$

Equation 8.1 can then be written

$$D_0 = \det\{\psi_1(1)\psi_2(2) \cdots \psi_N(N)\psi_{N+1}(N+1) \cdots \psi_{2N}(2N)\} \quad (8.4)$$

Ion states may be derived from the closed-shell ground state by removing one electron from any one of the occupied MSOs ψ_j; we may write the resulting determinant

$$D_j = \det\{\psi_1(1)\psi_2(2) \cdots \psi_{j-1}(j-1)\psi_{j+1}(j+1) \cdots \psi_{2N}(2N)\} \quad (8.5)$$

We define D_j so that the product of diagonal elements is ordered such that electron labels (which are identical to the MSO labels) are placed in increasing order.

The Hamiltonian for the ion is separated in the usual way $\mathbf{H} = \mathbf{H}° + \mathbf{H}_{so}$, where $H°$ is the spin-independent electronic Hamiltonian and $\mathbf{H}_{so}$ is an effective spin-orbit interaction operator.[8] Matrix elements of the Hamiltonian $\mathbf{H}°$ between any two determinants D_j and D_k, as defined above, will vanish if the MOs ϕ_i for the cation are chosen to be identical to the optimum Hartree–Fock MOs ϕ_i for the parent neutral molecule; that is, satisfying Eq. 8.2.[9]

The eigenfunctions of H are approximated by the linear variation functions

$$\Omega_n = \sum_j a_{nj}D_j \quad (8.6)$$

Let M equal the total number of determinants included in the sum. Minimizing the expectation energy E with respect to simultaneous variation of the coefficients a_{nj} leads to the $M \times M$ secular determinant[10]

$$|\mathscr{H}_{jk} - \mathscr{S}_{jk} E| = 0 \tag{8.7}$$

in which

$$\mathscr{S}_{jk} = \langle D_j \mid D_k \rangle = \delta_{jk} \tag{8.8}$$

$$\mathscr{H}_{jk} = \langle D_j \mid \mathbf{H}^\circ + \mathbf{H}_{so} \mid D_k \rangle$$
$$= E_j \mathscr{S}_{jk} + \langle D_j \mid \mathbf{H}_{so} \mid D_k \rangle$$
$$= E_j \delta_{jk} + \mathscr{H}'_{jk} \tag{8.9}$$

Since we are primarily interested in the spin-orbit perturbations, we choose to *equate E_j with the weighted mean empirical energy for each given multiplet.* The roots $E_1, E_2, \cdots, E_M$ represent theoretical predictions of the ion state levels including spin-orbit interaction.

One may want to perform calculations in two stages. First, consider only mixing of ion states D_j having a given electron configuration: M will equal the total degeneracy of that multiplet neglecting spin-orbit interaction. Second, consider the complete mixing of all possible ion states D_j: M will equal $2N$ in this case. Intermediate cases may also be considered.

In the above analysis, we have *not* included configurations of the ion in which electrons have been promoted to virtual MOs ϕ_i with $i > N$. Some of these ion configurations will clearly be energetically similar to certain states D_j, $j \le N$, as defined above in accordance with Koopmans' theorem. The resulting configuration interaction could, in fact, have a marked effect upon predicted spin-orbital energies for ion excited states.

8.2. Spin-orbit operators

We follow Misetich and Bush[8] in formulating the spin-orbit interaction operator $\mathbf{H}_{so}$ as a sum of effective operators $h_{so}(t)$ for each atom t; the $h_{so}(t)$, in turn, are expressed as sums of one-electron operators:

$$\mathbf{H}_{so} = \sum_t \mathbf{h}_{so}(t) \tag{8.10}$$

$$= \sum_t \sum_i \frac{1}{2m^2 c^2} \left[\frac{1}{r} \frac{\partial V(r)}{\partial r} \right] \mathbf{l}_{ti} \cdot \mathbf{s}_i = \sum_t \sum_i \xi(r_{ti}) \mathbf{l}_{ti} \cdot \mathbf{s}_i \tag{8.11}$$

where $V(r)$ is the potential energy of an electron in a central field. Usual procedures are followed in calculating the matrix elements $\mathscr{H}'_{jk}$.[10,11] One must first interchange columns in D_j so that D_j and D_k are put into maximum coincidence with each other. It turns out that the number of permutations required for this is $\mu = |j - k| - 1$ if $j \neq k$; $\mu = 0$ if $j = k$. It can then be shown that, for $j \neq k$ *and* $j = k$, the spin-orbit interaction reduces to the one-electron integral

$$\mathcal{H}'_{jk} = (-1)^{\mu} \langle \psi_j | \sum_t \xi_t \mathbf{l}_t \cdot \mathbf{s} | \psi_k \rangle \tag{8.12}$$

For $j \neq k$, this follows directly from the fact that D_j and D_k differ in one spin orbital, with ψ_j entering D_k where ψ_k enters D_j. For diagonal elements $j = k$, it can be shown that spin-orbit interactions for $m_s = +\frac{1}{2}$ and $m_s = -\frac{1}{2}$ electrons in each closed shell cancel each other; the only surviving term is due to the odd electron in the open shell ψ_j.

We now introduce the LCAO (linear combination of atomic orbitals) approximation of the SCF MOs:

$$\phi_i = \sum_{r=1}^{m} c_{ir} \chi_r \tag{8.13}$$

Combining Eqs. 8.3, 8.12, and 8.13, we obtain

$$\mathcal{H}'_{jk} = (-1)^{\mu} \sum_r \sum_s c'_{jr} c'_{ks} \mathbf{F}_{rj,sk} \tag{8.14}$$

in which

$$\mathbf{F}_{rj,sk} = \langle \chi_r \sigma_j | \sum_t \xi_t \mathbf{l}_t \cdot \mathbf{s} | \chi_s \sigma_k \rangle \tag{8.15}$$

$$c'_{lt} = c_{lt} \quad \text{and} \quad \sigma_l = \alpha \quad \text{if} \quad l \leq N$$

$$c'_{lt} = c_{l-N,t} \quad \text{and} \quad \sigma_l = \beta \quad \text{if} \quad l > N \tag{8.16}$$

Because of the r_t^{-3} dependence of ξ_t, we retain only one-center integrals; that is, $\mathbf{F}_{rj,sk} = 0$ unless χ_r and χ_s belong to the same atom ν.[12] In this case, the sum over t appearing in Eq. 8.15 reduces to one term identifying the atom ν to which both χ_r and χ_s belong; at the same time, we express $\mathbf{l}_\nu \cdot \mathbf{s}$ in an equivalent form:

$$\mathbf{F}_{rj,sk} = \langle \chi_r \sigma_j | \xi_\nu [\mathbf{l}_{z\nu} \mathbf{s}_z + \tfrac{1}{2}(\mathbf{l}_{+\nu} \mathbf{s}_- + \mathbf{l}_{-\nu} \mathbf{s}_+)] | \chi_s \sigma_k \rangle$$
$$= G^z_{rj,sk} + G^+_{rj,sk} + G^-_{rj,sk} \tag{8.17}$$

in which

$$\mathbf{l}_{\pm\nu} = \mathbf{l}_{x\nu} \pm i\mathbf{l}_{y\nu}$$
$$\mathbf{s}_\pm = \mathbf{s}_x \pm i\mathbf{s}_y \tag{8.18}$$

are the usual raising and lowering operators.[13]

For spin, we have

$$s_z\alpha = \tfrac{1}{2}\hbar\alpha \qquad s_z\beta = -\tfrac{1}{2}\hbar\beta$$
$$s_+\alpha = 0 \qquad s_+\beta = \hbar\alpha \qquad (8.19)$$
$$s_-\alpha = \hbar\beta \qquad s_-\beta = 0$$

For molecular systems, approximate MO wavefunctions are usually phrased in terms of a real AO basis; the following equations represent application of the angular momentum operators to the set of real p and d orbitals:

$$l_z p_z = 0 \qquad l_z p_x = i\hbar p_y \qquad l_z p_y = -i\hbar p_x$$
$$l_+ p_z = -\hbar(p_x + i p_y) \qquad l_+ p_x = \hbar p_z \qquad l_+ p_y = i\hbar p_z$$
$$l_- p_z = \hbar(p_x - i p_y) \qquad l_- p_x = -\hbar p_z \qquad l_- p_y = i\hbar p_z$$

$$
\begin{array}{l|l}
l_z d_{xy} = -2i\hbar^2 d_{x^2-y^2} & l_z d_{x^2-y^2} = 2i\hbar^2 d_{xy} \\
l_z d_{xz} = i\hbar^2 d_{yz} & l_z d_{z^2} = 0 \qquad\qquad (8.20) \\
l_z d_{yz} = -i\hbar^2 d_{xz} & l_\pm d_{xy} = i\hbar^2 d_{xz} \pm \hbar^2 d_{yz}
\end{array}
$$

$$
\begin{array}{l|l}
l_\pm d_{xz} = -i\hbar^2 d_{xy} \mp \hbar^2 d_{x^2-y^2} \pm \hbar^2 3^{1/2} d_{z^2} & l_\pm d_{x^2-y^2} = -i\hbar^2 d_{yz} \pm \hbar^2 d_{xz} \\
l_\pm d_{yz} = i\hbar^2 d_{x^2-y^2} \mp \hbar^2 i 3^{1/2} d_{z^2} \mp \hbar^2 d_{xy} & l_\pm d_{z^2} = -i\hbar^2 3^{1/2} d_{yz} \mp \hbar^2 3^{1/2} d_{xz}
\end{array}
$$

Equations 8.19 to 8.20 may be used to derive the following expressions for the G integrals needed in Eq. 8.17:

$$
\begin{aligned}
G^z_{rj,sk} = 0 \quad &\text{unless} \quad j \leq N \quad \text{and} \quad k \leq N \quad \text{or} \\
&\text{unless} \quad j > N \quad \text{and} \quad k > N \\
G^+_{rj,sk} = 0 \quad &\text{unless} \quad j > N \quad \text{and} \quad k \leq N \\
G^-_{rj,sk} = 0 \quad &\text{unless} \quad j \leq N \quad \text{and} \quad k > N
\end{aligned}
\qquad (8.21)
$$

Nonzero cases are given by the formulas

$$
\begin{aligned}
G^z_{rj,sk} = {}& i\hbar^2 A_j[\zeta^p_{vjk}(B_{y,x} - B_{x,y}) + \zeta^d_{vjk}(B_{yz,xz} - B_{xz,yz}) \\
& - 2B_{x^2-y^2,xy} + 2B_{xy,x^2-y^2})] \qquad\qquad (8.22) \\
G^\pm_{rj,sk} = {}& \hbar^2 A_j\{\zeta^p_{vjk}[B_{x,z} - B_{z,x} \pm i(B_{y,z} - B_{z,y})] \\
& + \zeta^d_{vjk}[3^{1/2} B_{xz,z^2} - B_{xz,x^2-y^2} - B_{yz,xy} + B_{xy,yz} \\
& + B_{x^2-y^2,xz} - 3^{1/2} B_{z^2,xz} \pm i(3^{1/2} B_{yz,z^2} + B_{yz,x^2-y^2} \\
& + 3^{1/2} B_{z^2,yz} + B_{xy,xz} - B_{xz,xy} - B_{x^2-y^2,yz})]\}
\end{aligned}
$$

The quantities in these expressions are defined as:

$$x = p_x \qquad y = p_y \qquad z = p_z \qquad xy = d_{xy}$$

$$xz = d_{xz} \qquad yz = d_{yz} \qquad z^2 = d_{z^2} \qquad x^2 - y^2 = d_{x^2-y^2}$$

$$A_j = \tfrac{1}{2} \text{ if } j < N \qquad A_j = -\tfrac{1}{2} \text{ if } j > N \tag{8.23}$$

$$B_{p.q} = 1 \text{ if } \chi_r = p \text{ and } \chi_s = q \qquad \text{Otherwise, } B_{p.q} = 0$$

Effective spin-orbit coupling parameters

Equations 8.22 contain an *effective spin-orbit coupling parameter* for atom ν

$$\zeta^p_{\nu jk} = \langle np'_\nu | \, \xi_\nu \, | np''_\nu \rangle \qquad \zeta^d_{\nu jk} = \langle nd'_\nu | \, \xi_\nu \, | nd''_\nu \rangle \tag{8.24}$$

These quantities are appropriate for ejection of an electron from the jth MO of a closed shell molecular ground state with $2N$ electrons. The terms $\zeta^p_{\nu jk}$ and $\zeta^d_{\nu jk}$ are the effective spin-orbit coupling parameters for atom ν in which np'_ν or nd'_ν and np''_ν or nd''_ν are np or nd ν-atom AOs appropriate to the molecular ion states D_j and D_k, respectively. The $\zeta_{\nu jk}$s can be approximated as a mean of the spin-orbit parameters for atom ν in the two molecular ion states D_j and D_k:

$$\zeta_{\nu jk} = \tfrac{1}{2}(\zeta_{\nu j} + \zeta_{\nu k}) \tag{8.25}$$

The $\zeta_{\nu j}$s are characteristic of the valence state of atom ν as it exists in the molecular ion state D_j. These effective spin-orbit coupling parameters have dimensions of energy. They determine the magnitude of the first-order energetic effects caused by spin-orbit coupling. For the Coulomb potential of a bare nucleus with charge $+Ze$, it can be shown[13] that

$$\zeta_{\nu j} = \frac{e^2 \hbar^2}{2m^2 c^2 a_0^3} \cdot \frac{Z^4}{n^3 l(l+\tfrac{1}{2})(l+1)} \tag{8.26}$$

where a_0 is the Bohr radius. In the general central-field many-electron case, $\zeta_{\nu j}$ has been found experimentally to be roughly proportional to Z^2 rather than Z^4. In these calculations $\zeta_{\nu j}$ will be treated as an empirical parameter whose magnitude is determinable from experiment.

8.3. Application of the model using approximate MOs

The LCAO coefficients c_{ir} appearing in Eqs. 8.13 and 8.14 are required in order to apply the above method to any given molecule. In the applications presented here the coefficients are obtained from Mulliken–Wolfsberg–Hemholtz (MWH) MO calculations (Section 4.7). Wavefunctions for the neutral molecule ground states can be described as in Eqs.

8.1 to 8.4. Molecular ion states are obtained by removing one electron from any one of the occupied MOs; corresponding wavefunctions can be written in the form of Eq. 8.5.

We need values for the effective spin-orbit coupling parameters $\zeta_{\nu j}$ as described in Eq. 8.24. It may be expected that this parameter will depend strongly upon the net charge q_ν at the atom ν in the state D_j. For example, $\zeta_{3p} = 318\,\text{cm}^{-1}$ and $180\,\text{cm}^{-1}$ for P^+ and P^-, respectively.[14] The charge appropriate to any given site can be obtained from the MO calculations. These calculations give a final self-consistent charge q_ν for each atom in the neutral molecule ground state. From these same calculations, one can also calculate the electron density $\rho_{\nu j}$ at each atom ν due to the presence of one electron in any given MO ϕ_j. Removal of one electron from the MO ϕ_j to produce the state D_j, or D_{j+N}, results in an increased charge $q_\nu + \rho_{\nu j}$ at the atomic site ν. These charges may then be used to interpolate an effective spin-orbit parameter $\zeta_{\nu j}$ most appropriate to the ion state D_j.[11]

This procedure for determining appropriate parameters $\zeta_{\nu j}$ is rather complicated. A much simpler procedure is to neglect the j dependence of $\zeta_{\nu j}$. Use values of ζ_ν evaluated for a charge $q_\nu + 1/(\text{number of atoms})$; that is, assume that for *all* ion states D_j the charge on atom ν is equal to the charge on that atom in the parent molecule increased by one charge distributed equally over all atoms. These charges are then used to interpolate effective spin-orbit parameters ζ_ν presumed appropriate to *all* molecular ion states D_j. In most cases final results obtained with these ζ_ν differ by negligible amounts from results obtained with the $\zeta_{\nu j}$ using the more complicated procedure. Values of effective spin-orbit coupling parameters $\zeta_{\nu j}^p$ for some common elements are listed in Table 8.1. Spin-orbit parameters appropriate for the atomic charge q_ν in the molecules are obtained by interpolation of the values in Table 8.1 for the positive, neutral, and negative species.

Identification of the symmetry species of each spin-orbit split term in an atom is simple. If a term arises from a configuration with a less than half-filled shell, then it is a *normal multiplet* and after spin-orbit coupling the level with the lowest value of J has the lowest energy. If the configuration has a more than half-filled shell, then the level of highest J has the lowest energy; it is called an *inverted multiplet*. There is no first-order spin-orbit splitting from a half-filled shell. For molecules the procedure is not as straightforward, and, furthermore, interactions between levels of the same symmetry can produce large shifts in the multiplet components.

Solution of the secular determinant, Eq. 8.7, yields theoretical predictions of the molecular-ion-state eigenvalues $E_1, E_2, \cdots, E_j$ including

Table 8.1. Effective Spin-Orbit Coupling Parameters $\zeta_{\nu j}^{p}$ (in eV) for Some Common Elements[a]

	Positive Ions	Neutral Atoms	Negative Ions
B	—	0.0014	0.0005
C	0.0053	0.0037	0.0002
N	0.011	0.0005	0.0077
O	0.0011	0.018	0.015
F	0.040	0.033	—
Ne	0.065	—	—
Al	—	0.0093	0.0047
Si	0.024	0.019	4.3×10^{-5}
P	0.039	0.0001	0.022
S	0.0002	0.047	0.037
Cl	0.080	0.072	—
Ar	0.118	—	—
Sc	0.0067	0.0083	0.0067
Ti	0.011	0.013	0.011
V	0.018	0.019	0.016
Cr	—	0.026	—
Mn	0.031	—	0.037
Fe	0.044	0.052	0.048
Co	0.056	0.068	0.060
Ni	0.075	0.083	—
Cu	—	0.102	—
Ga	—	0.068	0.034
Ge	0.146	0.106	3.1×10^{-5}
As	0.183	7.4×10^{-5}	0.117
Se	0.0011	0.246	0.171
Br	0.389	0.305	—
Kr	0.444	—	—
In	0.287	0.183	—
Sb	0.322	—	—
I	0.801	0.628	—

[a] Spin-orbit coupling parameters for atoms through Kr were obtained from: G. Malli and S. Fraga, *Theor. Chim. Acta*, **7**, 80 (1967); B. W. N. Lo, N. M. S. Saxena, and S. Fraga, ibid., **25**, 97 (1972); J. Thorhallsson, C. Fisk, and S. Fraga, *J. Chem. Phys.*, **48**, 2925 (1968). Constants for the elements heavier than Kr were evaluated from data in C. E. Moore, Nat. Bur. Standards (U.S.) No. 467 (1949, 1952, 1958). For $(\nu l)^*$ configurations, the many-electron spin-orbit coupling constant $\zeta(S, L)$, which is sometimes written as $\lambda(S, L)$, is related to the one-electron spin-orbit parameter ζ_ν by $\zeta(S, L) = \pm \zeta_\nu/(2S)$, the plus sign applying for less-than-half-filled shells and the minus sign for more-than-half-filled shells. The experimental values of $\zeta(S, L)$ have been determined by the relation $\zeta(S, L)J = E(J) - E(J-1)$ using the $E(J)$ values given in Moore's Tables.

spin-orbit interaction. One will normally also obtain the associated eig-envectors represented by the coefficients a_{nj}. In principle, from the transformation properties of each Ω_n under operations of the appropriate *extended point group*, one can determine the symmetry species to which each level belongs. However, this is very complicated in practice. A very simple alternative method for determining these species can be used.

Suppose one is interested in determining the order of the two levels $E_{1/2}$ and $E_{3/2}$ derived from the ejection of a *4e* electron of PI_3. The outer valence electron configuration of PI_3 is $2e^4 3a_1^2 3e^4 4e^4 1a_2^2 4a_1^2$. First per-form a calculation according to Eqs. 8.6 to 8.9 with $M = 4$, mixing only four determinants associated with the electron configuration written as above with one electron deleted from the *4e* orbital. Two levels result, of course, with a splitting of about 0.62 eV. The identity of the $E_{3/2}$ and $E_{1/2}$ levels is still unknown at this point. Next perform a second calculation, this time with $M = 6$, adding two determinants associated with the state $1a_2$, $E_{1/2}$. Three energy levels result from this calculation: one of these, the $E_{3/2}$ level, will be unperturbed by the $1a_2$, $E_{1/2}$ states, and will be identical in the two calculations; the *4e*, $E_{1/2}$ states, on the other hand, will be slightly perturbed upward in energy by the presence of the lower energy $1a$, $E_{1/2}$ states. In this way, one can determine the order of the *4e*, $E_{1/2}$ and *4e*, $E_{3/2}$ companion levels. For the specific case of PI_3 the *4e*, $E_{1/2}$ level lies higher than the *4e*, $E_{3/2}$ level.

States that are split by spin-orbit interaction can, in most cases, be split by either Jahn–Teller or Renner interactions (Chapter 9). It is necessary to differentiate between these different types of interactions. The following observations can be used to indicate whether the dominant effect causing the splitting is spin-orbit coupling: (1) Spin-orbit splittings increase in size as the atomic number of the constituent atoms in the molecule is increased; this increase is unexpected for Jahn–Teller splittings. For example, in a series of methyl halides, an increase in the magnitude of a certain splitting as the atomic number of the halogen increases is substan-tial evidence for spin-orbit splitting. (2) Spin-orbit splittings of various degen-erate states of a given polyatomic molecule may differ drastically, whereas Jahn–Teller splittings should be more uniform in size. (3) Satisfactory agreement between observed and calculated splittings can provide strong support for the type of interaction prevailing. Examples of concomitant Jahn–Teller and spin-orbit interactions are discussed in Section 9.8.

Since the spin-orbit components $E_{3/2}$ and $E_{1/2}$ of a split E or Π state are each doubly degenerate, they are expected to be of comparable intensity on the basis of their total spin-orbital degeneracies. However, since these two spin-orbit components have different energies, the kinetic energies of electrons ejected in forming the states will be different. As a result the cross section per electron can vary considerably for the two components. Such variations have been observed in the $^2P_{3/2} : ^2P_{1/2}$ ratios (statistical ratio 2:1) of the rare gas ions and the $^2D_{5/2} : ^2D_{3/2}$ ratios (statistical ratio 3:2) of Zn^+, Cd^+, and Hg^+, where the deviations from the statistical values are attributed to cross-section variations.[15] Such variations in the cross sections for the spin-orbit components of a molecular ion must certainly occur, making it difficult to predict intensities of molecular spin-orbit components.

8.4. Example: The halogen acids, HX (X=F, Cl, Br, I)

The ground-state electron configuration of the halogen acids is

$$1\sigma^2 2\sigma^2 3\sigma^2 1\pi^4 \tag{8.27}$$

The MOs are written in order of increasing orbital energy, that is, decreasing binding energy. The He I spectra of these molecules are presented in Fig. 8.1 and have been reported by several groups.[16–20] Ionization bands corresponding to ejection of only the 3σ and 1π electrons can be observed below 21 eV. The splitting of the first ionization band increases consistently as the atomic number of the halogen increases. This obvious trend allows easy identification of the first ionization process as ejection of a 1π electron and the splitting as the $^2\Pi_{3/2}$–$^2\Pi_{1/2}$ spin-orbit interval. The band at higher energy corresponds to ejection of the more tightly bound 3σ electrons. From the appearance of these bands we conclude that the 1π electrons are nonbonding, while the

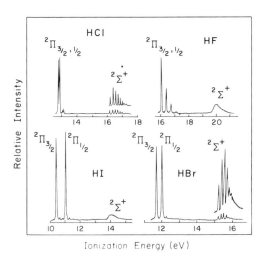

Fig. 8.1. He I spectra of the halogen acids.

3σ electrons are strongly bonding. Vibrational structure can be resolved in all of the ionization bands. The observed and calculated spin-orbit splittings are presented in Table 8.2. The splittings were calculated by means of the spin-orbit model described in the previous section using the self-consistent atomic charges q_ν from MWH calculations and the inter-polated effective spin-orbit coupling parameters ζ_ν^p of Table 8.1. The excellent agreement between calculated and observed splittings vindicates the band assignments and lends support to the spin-orbit model.

8.5. Example: The halogens, X_2 (X=F, Cl, Br, I)

The ground-state electron configuration of the halogens is:

$$1\sigma_g^2 1\sigma_u^2 2\sigma_g^2 2\sigma_u^2 3\sigma_g^2 1\pi_u^4 1\pi_g^4 \tag{8.28}$$

The He I spectra of the diatomic halogens have been reported by several groups;[21-24] selected spectra are presented in Figs. 8.2 and 8.3.Band systems are observed below 21 eV corresponding to ejection of $1\pi_g$, $1\pi_u$, and $3\sigma_g$ electrons. The $(3\sigma_g)$ $^2\Sigma_g^+$ state of F_2 has not been observed, but it is believed to be near 21 eV. Spin-orbit splittings are observed in the $^2\Pi_g$ and $^2\Pi_u$ bands, with the splitting ΔE_{so} increasing as the molecular weight increases. The vibrational stretching frequency ν' in these diatomics exhibits a trend opposite to the spin-orbit splitting; ν' decreases with increasing molecular weight. The magnitudes of these quantities are such that $\Delta E_{so} < \nu'$ for F_2^+, $\Delta E_{so} \simeq \nu'$ for Cl_2^+, $\Delta E_{so} > \nu'$ for Br_2^+ and I_2^+. These features can be observed in the $^2\Pi_g$ bands of Fig. 8.3 and the data in Table 8.2. The $^2\Pi_g$ band in F_2 consists of a vibrational progression with each vibrational band split by a small spin-orbit interaction. The Cl_2 band appears to be anomalous because it resembles a simple vibrational progression. However, comparison with spectra of other chlorine compounds has shown that the magnitudes of the spin-orbit splittings and vibrational frequencies are comparable and that the $^2\Pi_g$ band of Cl_2 consists of two overlapping vibrational progressions with origins separated by ΔE_{so}. The Br_2 bands consists of two well-separated vibrational progressions each associated with a spin-orbit component of the $^2\Pi_g$ state. The I_2 structure consists of two bands for which vibrational structure has recently been resolved.[24] See Fig. 3.4 for hot bands and vibrational structure in the first ionization band $^2\Pi_{3/2,g}$ of I_2. The vibrational frequencies in the ground states of the halogen molecular ions are increased from that of the ground state of the molecule, indicating that the $^2\Pi_g$ state of the ion has a stronger bond than the molecular ground state. This is consistent with the antibonding description of the $1\pi_g$ orbital as obtained from MO theory.

Table 8.2. Observed[a] and Calculated Spin-Orbit Splittings for the Halogen Acids and Halogens

Ionic State	HF E_I	ΔE_{obsd}	ΔE_{calcd}	HCl E_I	ΔE_{obsd}	ΔE_{calcd}	HBr E_I	ΔE_{obsd}	ΔE_{calcd}	HI E_I	ΔE_{obsd}	ΔE_{calcd}
(2π), $^2\Pi_{3/2}$	16.02	0.04	0.034	12.74	0.08	0.076	11.67	0.33	0.333	10.38	0.67	0.671
(2π), $^2\Pi_{1/2}$	16.06			12.82			12.00			11.05		
	F$_2$			Cl$_2$			Br$_2$			I$_2$		
$(1\pi_g)$, $^2\Pi_{3/2,g}$	15.70	0.03	0.035	11.51	0.08	0.067	10.51	0.35	0.309	9.22	0.65	0.607
$(1\pi_g)$, $^2\Pi_{1/2,g}$	15.73			11.59			10.86			9.87		
$(1\pi_u)$, $^2\Pi_{3/2,u}$		—	0.038	13.96	0.08[b]	0.088	12.41	0.34[b]	0.382	10.74	0.80	0.788
$(1\pi_u)$, $^2\Pi_{1/2,u}$	18.39			14.04[b]			12.75[b]			11.54		

[a] Experimental data for HF are from Ref. 17; data for HCl, HBr, and HI are from Ref. 19; data for the halogens are from Ref. 23.

[b] Uncertain value or values obtained from extrapolation.

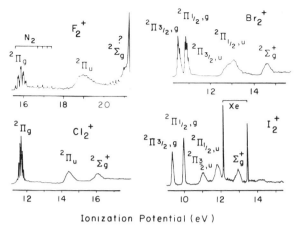

Fig. 8.2. The He I spectra of the halogens. Reproduced with permission from Ref. 21.

The transition to the first excited ionic state $^2\Pi_u$ produces a much broader band than the corresponding $^2\Pi_g$ band. This is indicative of ejection of electrons from a strongly bonding orbital, in agreement with the bonding nature of the $1\pi_u$ orbital. The spin-orbit components of the $^2\Pi_u$ bands have not been resolved in F_2^+ and Cl_2^+, but can be differentiated in Br_2^+ and I_2^+ where the splittings are much larger than in the lighter species. Vibrational structure in this band has been resolved only in Cl_2^+,[22] where the frequency (323 cm^{-1}) is considerably decreased from that of the molecular ground state (565 cm^{-1}). This is another indication of the strong bonding character of the $1\pi_u$ orbital.

The observed and calculated spin-orbit splittings for the molecular halogens are listed in Table 8.2. The differences in the splittings of the $^2\Pi_g$ and $^2\Pi_u$ spin-orbit components of I_2^+ is reproduced by the calculations. These differences can be understood using the simple qualitative spin-orbital scheme of Fig. 8.4.[25] When the symmetry group $D_{\infty h}$ is extended to the double group $D'_{\infty h}$, orbitals with σ_u and σ_g symmetry become $e_{1/2,u}$ and $e_{1/2,g}$, respectively, and orbitals with π_u and π_g symmetry each split into $e_{3/2}$ and $e_{1/2}$ levels of u or g, respectively. As shown in Fig. 8.4, the symmetry allowed mixing of σ character into $\pi_{1/2,g}$ and $\pi_{1/2,u}$ orbitals decreases $\Delta\pi_g$ and increases $\Delta\pi_u$. The good agreement between calculated and experimental spin-orbit splittings places the band assignments on a firm basis.

Spectra of the mixed halides ClF, BrF, ICl, IBr, and so forth, have been reported by several groups.[22,23,26,27] These molecules exhibit spin-orbit

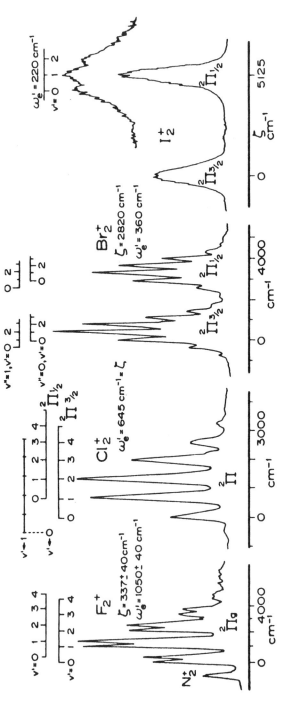

Fig. 8.3. Spin-orbit splittings and vibrational progressions in the ground states $^2\Pi_{3/2,\,1/2,\,g}$ of the halogen molecular ions. Reproduced with permission from Ref. 21.

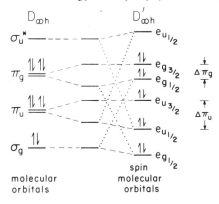

Fig. 8.4. Interpretation of the differences in the splittings of the $^2\Pi_g$ and $^2\Pi_u$ spin-orbit components in terms of spin orbitals. Levels on the left correspond to MOs of the point group, levels in the middle correspond to molecular spin orbitals of the double group, and levels at the right are spin orbitals after interaction (dotted line) between orbitals of the same symmetry. Adapted with permission from Ref. 25.

splittings in their π bands in accordance with the size of the spin-orbit coupling constants appropriate to the particular halogens in the molecule.

8.6.   Example: The PX₃ (X = Cl, Br, I) and PYX₃ (X = Cl, Br; Y = 0, S) series

The outer valence electron configurations of PX_3 and PYX_3 ($X = F$, Cl, Br, I; $Y = 0$, S) may be described[6] as follows:

$$(2e)^4(3a_1)^2(3e)^4(4e)^4(1a_2)^2(4a_1)^2 \quad \text{for} \quad PX_3 \tag{8.29}$$

$$(2e)^4(4a_1)^2(3e)^4(5a_1)^2(4e)^4(1a_2)^2(5e)^4 \quad \text{for} \quad PXY_3 \tag{8.30}$$

The inner na_1 and $1e$ valence electrons are not considered here. The MOs are written in order of increasing orbital energy ε, that is, decreasing binding energy, in accordance with the assignments made in this section. An approximate description of these MOs is presented in Table 8.3. The He I spectra of the PX_3 and PYX_3 series are presented in Figs. 8.5 and 8.6. The spectral peaks outnumber the molecular orbitals in the heavier members of the series. If we accept, for the moment, the ordering of MOs expressed in Eqs. 8.29 and 8.30, the experimental ionization energies $E_j(E_0 = 0$ eV for parent neutral molecule ground state) may be assigned as listed in Table 8.4. For degenerate states the observed splitting ΔE_{obsd} and the splitting ΔE_{calcd} as determined by the spin-orbit model of section 8.3 are also listed. These splittings were calculated through the use of the self-consistent charges q_ν and interpolated effective spin-orbit coupling parameters ζ_ν^p listed in Table 8.5.

The spectra within the PX_3 series and the PYX_3 series are similar if one imagines each band that is split by spin-orbit interaction to be one peak. Using this similarity and the *ab initio* MO orderings for PCl_3 and $POCl_3$

Table 8.3. Localization Properties and Description of the Outer Valence Molecular Orbitals of PCl_3, $POCl_3$ and $PSCl_3$.[a]

MO		Description	Localization	PCl_3	$POCl_3$	$PSCl_3$
PX_3	PYX_3					
—	5e	nonbonding	%P	—	7	1
			%Cl	—	48	86
			%0 or S	—	45	13
$1a_2$	$1a_2$	nonbonding	%P	0	0	0
			%Cl	100	100	100
			%0 or S		0	0
4e	4e	nonbonding	%P	10	14	15
			%Cl	90	78	70
			%0 or S		8	15
$4a_1$	$5a_1$	nonbonding in PCl_3 P—Y bonding in $PYCl_3$	%P	55	19	13
			%Cl	45	36	43
			%0 or S		45	44
3e	3e	strongly bonding	%P	19	15	17
			%Cl	81	74	75
			%0 or S	—	11	8
$3a_1$	$4a_1$	P—Cl bonding	%P	31	17	20
			%Cl	69	64	62
			%0 or S	—	19	18
2e	2e	P—Cl bonding	%P	25	33	25
			%Cl	75	67	34
			%0 or S	—	10	41

[a] From CNDO/2 calculations.

as the basis for a correlation of the bands within each series, the assignments shown in Figs. 8.5 and 8.6 are derived. Confirmation of these assignments is given by the comparison of calculated spin-orbit splittings ΔE_{calcd} and observed splittings ΔE_{obsd}. The observed ionization energies along with ΔE_{obsd} and ΔE_{calcd} are listed in Table 8.4. Quite good agreement is obtained between ΔE_{obsd} and ΔE_{calcd}. In cases where $\Delta E_{calcd} > \Delta E_{obsd}$, the discrepancy may be due to overlap of the $^2E_{1/2}$ and $^2E_{3/2}$ bands (whose natural line-widths can be considerable), resulting in splittings that are beyond the resolution capabilities of the instrument. Cases where $\Delta E_{calcd} < \Delta E_{obsd}$ are the 5e band of $POBr_3$ and the 3e band of PI_3. One possible rationale for this discrepancy lies with the approximate Mulliken–Wolfsberg–Helmholz wavefunctions used in the calculations. For example, the eigenvector for the 5e orbital of $POBr_3$ localizes about 94% of the 5e electron density on the oxygen atom; however, the shift of the corresponding ionization band between $POCl_3$ and $POBr_3$ is 0.79 eV, suggesting that this orbital has substantial halogen character.

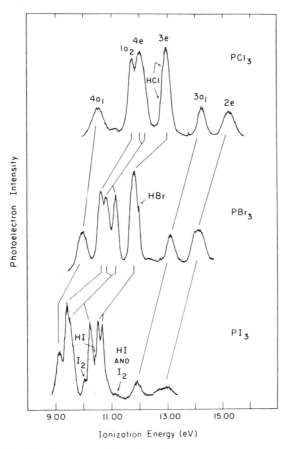

Fig. 8.5.   The He I spectra of the phosphorous trihalides. Lines connect bands associated with molecular orbitals of the same symmetry. Reproduced with permission from Ref. 6.

Examination of the spin-orbit splittings ΔE in Table 8.4 reveals that variations of more than an order of magnitude occur in the splittings of different degenerate electronic states of the same species. This large variation in ΔE is mainly due to two factors: (1) atomic orbital composition of the wavefunction—for example, s orbitals do not couple with spin functions, and certain combinations of p orbitals yield larger splittings than others. (2) atomic charges q_ν in the molecule—the effective spin-orbit coupling parameters ζ_ν are functions of the electron densities at the various atoms; for individual atoms, ζ_ν usually decreases as the atom becomes more negative.

Table 8.4. Vertical Ionization Energies, Spectral Assignments, and Results of Spin-Orbit Interaction Calculations (All Values in eV)[a]

Ionic State	Extended Point Group Species	PCl_3 E_{obsd}	PCl_3 ΔE_{obsd}	PCl_3 ΔE_{calcd}	PBr_3 E_{obsd}	PBr_3 ΔE_{obsd}	PBr_3 ΔE_{calcd}	PI_3 E_{obsd}	PI_3 ΔE_{obsd}	PI_3 ΔE_{calcd}
$(4a_1), {}^2A_1$	$E_{1/2}$	10.51	—	—	9.96	—	—	9.15	—	—
$(1a_2), {}^2A_2$	$E_{1/2}$	11.70	—	—	10.61	—	—	9.42	—	—
$(4e), {}^2E$	$E_{3/2}$	12.00	N.R.	0.06	10.83	0.33	0.30	9.57	0.67	0.63
	$E_{1/2}$				11.16			10.24		
$(3e), {}^2E$	$E_{1/2}$	12.97	N.R.	0.02	11.79	N.R.B.	0.08	10.53	0.15	0.02
	$E_{3/2}$							10.68		
$(3a_1), {}^2A_1$	$E_{1/2}$	14.23	—	—	13.13	—	—	11.80	—	—
$(2e), {}^2E$	$E_{1/2}$	15.20	N.R.	0.02	14.12	N.R.B.	0.13	12.70	N.R.B.	0.35
	$E_{3/2}$							N.R.B.		

Ionic State	Extended Point Group Species	POCl$_3$ E_{obsd}	POCl$_3$ ΔE_{obsd}	POCl$_3$ ΔE_{calcd}	PSCl$_3$ E_{obsd}	PSCl$_3$ ΔE_{obsd}	PSCl$_3$ ΔE_{calcd}	POBr$_3$ E_{obsd}	POBr$_3$ ΔE_{obsd}	POBr$_3$ ΔE_{calcd}	PSBr$_3$ E_{obsd}	PSBr$_3$ ΔE_{obsd}	PSBr$_3$ ΔE_{calcd}
$(5e)$, ^2E	$E_{3/2}$	11.85	—	0.01	10.11	N.R.	0.04	10.99	0.14	0.01	9.82	N.R.	0.04
	$E_{1/2}$							11.13					
$(1a_2)$, $^2A_2$	$E_{1/2}$	12.35	—	—	11.99	—	—	11.36	—	—	10.86	—	—
$(4e)$, ^2E	$E_{3/2}$	12.93	0.05	0.06	12.65	N.R.B.	0.05	11.73	0.24	0.24	11.16	0.22	0.23
	$E_{1/2}$	12.98			12.65			11.97			11.38		
$(5a_1)$, $^2A_1$	$E_{1/2}$	13.48	—	—	12.65	—	—	12.39	—	—	11.80	—	—
$(3e)$, ^2E	$E_{1/2}$	13.85	N.R.	0.01	13.39	N.R.	0.01	12.61	N.R.	0.02	12.15	N.R.	0.01
	$E_{3/2}$												
$(4a_1)$, $^2A_1$	$E_{1/2}$	15.37	—	—	14.78	—	—	14.60	—	—	13.91	—	—
$(2e)$, ^2E	$E_{1/2}$	16.53	N.R.B.	0.02	15.80	N.R.B.	0.01	15.35	N.R.B.	0.11	14.59	N.R.B.	0.10
	$E_{3/2}$												

[a] E_{obsd} = Vertical ionization energy. ΔE_{obsd} = Observed splitting. ΔE_{calcd} = Calculated spin-orbit splitting. N.R. = Spin-orbit component *not resolved*. N.R.B. = Spin-orbit components *not resolved* but band obviously *broad* due to overlapping transitions.

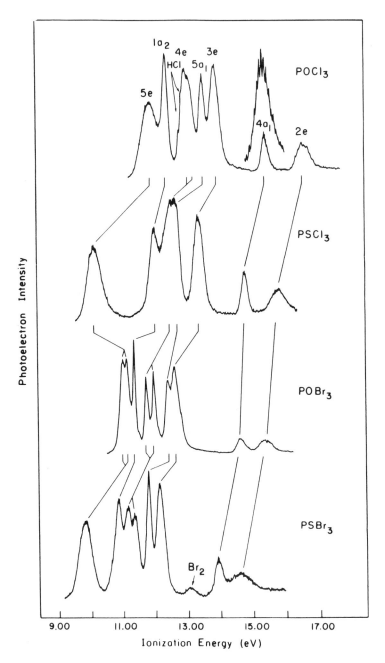

Fig. 8.6. The He I spectra of the phosphorus oxytrihalides and the thiophosphoryl trihalides. Lines connect bands associated with molecular orbitals of the same symmetry. Reproduced with permission from Ref. 6.

Table 8.5. Atom Charges[a] q_ν in Neutral Molecule Ground States and Interpolated Effective Spin-Orbit Coupling Parameters[b]

Molecule PYX₃	q_ν^p			$\zeta_\nu(eV)$		
	P	X	Y	P	X	Y
PCl₃	+0.25	−0.083	—	0.035	0.074	—
PBr₃	+0.18	−0.060	—	0.035	0.321	—
PI₃	+0.043	−0.014	−	0.033	0.669	—
POCl₃	+0.62	−0.033	−0.52	0.038	0.074	0.017
PSCl₃	+0.52	−0.064	−0.33	0.037	0.074	0.045
POBr₃	+0.55	+0.049	−0.70	0.037	0.323	0.017
PSBr₃	+0.47	+0.008	−0.49	0.036	0.320	0.044

[a] Self-consistent charges obtained from Mulliken–Wolfsberg–Helmholtz calculations.

[b] Interpolated from Table 8.1.

The only band exhibiting resolved vibrational structure is the $4a_1$ band of POCl₃. The observed spacing, 242 cm^{-1}, could arise from two possible modes of vibration; the symmetric P—Cl stretch or the symmetric deformation mode that have frequencies of 486 and 267 cm^{-1}, respectively in the neutral molecule.[28]

8.7. Example: The AsX₃ and SbX₃ (X = Cl, Br, I) series with inclusion of d electrons

The role of valence d electrons in spin-orbit interactions can be studied by applying the model calculation of Section 8.3 using minimal basis set (valence ns and np AOs) and extended basis set (valence ns, np, and nd AOs) wavefunctions. The results[7] of such an investigation on the AsX₃ and SbX₃ series are presented here. The outer valence electron configuration of the AsX₃ and SbX₃ series may be described in a manner similar to those of the PX₃ series:

$$(2e)^4(3a_1)^2(3e)^4(4e)^4(1a_2)^2(4a_1)^2 \qquad (8.31)$$

The inner na_1 and $1e$ valence electrons are not considered here. An approximate description of these MOs is presented in Table 8.6. The He I spectra of the AsX₃ and SbX₃ series are presented in Fig. 8.7. The spectral peaks outnumber the molecular orbitals in the heavier members of the series. The atomic charges q_ν in the neutral molecule ground states and interpolated effective spin-orbit coupling parameters ζ_ν are listed in

Table 8.6. Localization[a] Properties and Description of the Outer Valence Molecular Orbitals of the AsX_3 and SbX_3 Series

MO	Description	$AsCl_3$	$AsBr_3$	AsI_3	$SbCl_3$	$SbBr_3$	SbI_3
$4a_1$	nonbonding	76.1	59.2	54.7	77.6	73.2	66.2
$1a_2$	nonbonding	100	100	100	100	100	100
$4e$	nonbonding	100	99	95	100	100	98.2
$3e$	nonbonding	100	100	98.6	100	100	98.7
$3a_1$	bonding	73.2	74.1	83.7	84.1	82.9	81.8
$2e$	bonding	74.1	77.3	78.2	78.5	77.8	77.5

[a] Numbers in the table represent the percent halogen character of the MO. Values are from calculations with minimal basis set.

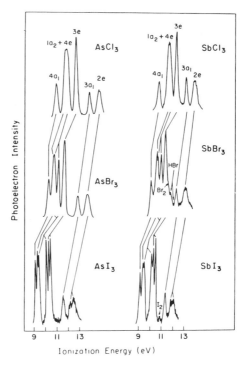

Fig. 8.7. He I spectra of the arsenic and antimony trihalides. Lines connect bands associated with molecular orbitals of the same symmetry. Reproduced with permission from Ref. 7.

Table 8.7. Atom Charges q_ν in Neutral Molecule Ground States and Interpolated Effective Spin-Orbit Coupling Parameters[a] ζ_ν

| Molecule AX$_3$ | Minimal Basis Set | | | | Extended Basis Set on all Atoms | | | | | |
| | q_ν | | ζ_ν^p | | q_ν | | ζ_ν^p | | ζ_ν^d | |
	A	X	A	X	A	X	A	X	A	X
AsCl$_3$	0.32	−0.11	0.169	0.074	0.15	−0.050	0.163	0.074	0.090	0.007
AsBr$_3$	0.21	−0.07	0.165	0.320	0.01	−0.003	0.159	0.326	0.098	0.032
AsI$_3$	0.10	−0.03	0.160	0.666	−0.04	0.013	0.157	0.672	0.101	0.245
SbCl$_3$	0.52	−0.17	0.267	0.073	0.16	−0.054	0.192	0.074	0.154	0.007
SbBr$_3$	0.42	−0.14	0.247	0.314	−0.03	0.010	0.150	0.327	0.162	0.032
SbI$_3$	0.30	−0.10	0.221	0.654	−0.16	0.053	0.124	0.680	0.168	0.205

[a] Values of ζ_ν^p are from Table 8.1. Values of ζ_ν^d are determined from data on the excited states of atoms in C. E. Moore, U.S. Nat. Bur. Standards *467*, (1949, 1952, 1958), Washington, D. C.

Table 8.7. If we accept, for the moment, the ordering of the MOs expressed in Eq. 8.31, the experimental ionization energies E_j ($E_0 = 0$ for parent neutral molecule ground state) may be assigned as listed in Table 8.8. For degenerate ionic states, we list also the observed splitting ΔE_{obsd} and the splitting ΔE_{calcd} as estimated by our model.

The spectra within the AsX$_3$ and SbX$_3$ series are similar if each band that is split by spin-orbit interaction is considered as ionization from one MO. Using this similarity, the MO ordering from the MWH calculation, and the ordering for the PX$_3$ series, the assignments in Fig. 8.7 were derived. Verification of these assignments is given by the agreement between calculated spin-orbit splittings ΔE_{calcd} and observed splittings ΔE_{obsd}.

The assignments suggest that the second band of the chloride and bromide compounds consists of two overlapping transitions, that is, in AsCl$_3$ and SbCl$_3$ the second band consists of transitions to the $^2A_2$ ($1a_2$) and ^2E ($4e$) states; in AsBr$_3$ the second band consists of the $^2A_2$ ($1a_2$) and $^2E_{3/2}$ ($4e$) states; in AsI$_3$ and SbI$_3$ these bands are partially resolved.

It is of interest to compare the spin-orbit splittings calculated from the minimal and extended basis sets (Table 8.8). In most cases the splittings calculated with the minimal basis set, that is, without inclusion of d orbitals, are in better agreement with ΔE_{obsd} than those calculated with the extended basis set, that is, including d orbitals on the central atom and the halogens. This suggests that it is not necessary to include virtual

Table 8.8. Vertical Ionization Energies, Spectral Assignments, and Results of Spin-Orbit Interaction Calculations (All Values in eV)[a]

Ionic State	Extended Point Group Species	AsCl₃				AsBr				AsI₃			
		E_{obsd}	ΔE_{obsd}	ΔE^m_{calcd}	ΔE^e_{calcd}	E_{obsd}	ΔE_{obsd}	ΔE^m_{calcd}	ΔE^e_{calcd}	E_{obsd}	ΔE_{obsd}	ΔE^m_{calcd}	ΔE^e_{calcd}
$(4a_1)$, $^2A_1$	$E_{1/2}$	10.90	—	—	—	10.19	—	—	—	9.11	—	—	—
$(1a_2)$, $^2A_2$	$E_{1/2}$	11.60	—	—	—	10.50	—	—	—	9.27	—	—	—
$(4e)$, ^2E	$E_{3/2}$	11.79	N.R.	0.07	0.07	10.70	0.31	0.28	0.31	9.37	0.72	0.61	0.62
	$E_{1/2}$					11.01				10.09			
$(3e)$, ^2E	$E_{1/2}$	12.59	N.R.	0.01	0.01	11.51	N.R.	0.03	0.09	10.30	0.17	0.12	0.08
	$E_{3/2}$									10.47			
$(3a_1)$, $^2A_1$	$E_{1/2}$	13.75	—	—	—	12.75	—	—	—	11.56	—	—	—
$(2e)$, ^2E	$E_{1/2}$	14.61	N.R.	0.01	0.01	13.60	N.R.B.	0.15	0.19	12.19	0.31	0.36	0.43
	$E_{3/2}$									12.50			

Ionic State	Extended Point Group Species	SbCl$_3$ E_{obs}	ΔE_{obs}	ΔE^m_{calc}	ΔE^e_{calc}	SbBr$_3$ E_{obs}	ΔE_{obs}	ΔE^m_{calc}	ΔE^e_{calc}	SbI$_3$ E_{obs}	ΔE_{obs}	ΔE^m_{calc}	ΔE^e_{calc}
$(4a_1)$, $^2A_1$	$E_{1/2}$	10.70	—	—	—	10.07	—	—	—	9.05	—	—	—
$(1a_2)$, $^2A_2$	$E_{1/2}$	11.57	—	—	—	10.41	—	—	—	9.23	—	—	—
$(4e)$, ^2E	$E_{3/2}$	11.94	N.R.	0.07	0.06	10.65	0.29	0.32	0.27	9.41	0.62	0.65	0.57
	$E_{1/2}$					10.94				10.03			
$(3e)$, ^2E	$E_{1/2}$	12.26	N.R.	0.01	0.01	11.32	N.R.	0.04	0.01	10.16	0.20	0.08	0.06
	$E_{3/2}$									10.36			
$(3a_1)$, $^2A_1$	$E_{1/2}$	13.13	—	—	—	12.26	—	—	—	11.25	—	—	—
$(2e)$, ^2E	$E_{1/2}$	13.88	N.R.	0.02	0.03	13.06	N.R.B.	0.09	0.14	11.72	0.29	0.27	0.51
	$E_{3/2}$									12.01			

[a] E_{obsd} = vertical ionization energy. ΔE_{obsd} = observed splitting. ΔE^m_{calcd} = calculated spin-orbit splitting using minimal basis set. ΔE^e_{calcd} = calculated spin-orbit splitting using extended basis set. N.R. = spin-orbit components *not resolved*. N.R.B. = spin-orbit components *not resolved* but band obviously *broad* due to overlapping transitions.

nd orbitals in spin-orbit interaction calculations of AX_3 type molecules. The consequences of this result must be considered.

The MWH calculations predict only a 5 to 15% involvement of *nd* orbitals in the bonding MOs suggesting that ΔE_{calcd} with the extended basis set will only be slightly perturbed from ΔE_{calcd} with the minimal basis set. One of the most important quantities in this calculation, and the most difficult to obtain accurately, is the molecular wavefunction. The consequences of the choice of wavefunction can be seen from Table 8.7. The extended basis set provides a more uniform molecular charge distribution and, in some cases, inverts the charge by making the central atom negative and the halogens positive (a result that is contrary to chemical intuition). This effect may be the result of (1) an unbalanced basis set[29,30] due to the failure to include in the AO basis set other low-energy unoccupied orbitals besides the *nd* orbitals that are also available for bonding and (2) the overemphasis of *d*-orbital participation by the MWH calculation. The molecules containing iodine atoms provide an extreme example of this, for here there are many low-energy unoccupied AOs that have been neglected in the basis set.

The reasonable agreement between experimental and calculated splittings provided by the minimal basis set computation suggests that the effect on spin-orbit coupling caused by the participation of *d* orbitals in the bonding of the AX_3 series is minimal and can be neglected for such an approximate treatment. It should be interesting to apply the model to molecules containing atoms with occupied valence-shell *d* orbitals in order to investigate their role in spin-orbit interactions.

8.8. Spin-orbit coupling in low-symmetry molecules

Although spin-orbit splittings are a manifestation of the coupling of spin and orbital angular momenta, such splittings have been observed in low-symmetry molecules where the orbital angular momentum is not well defined. Consider the example of the alkyl halides, RX. If the alkyl group R is a methyl or *tert*-butyl group, the halogen X is situated on a threefold axis with well-defined orbital angular momentum. Ejection of the nonbonding electrons on the halogen atom produces $^2E_{3/2}$ and $^2E_{1/2}$ states with a separation of the magnitude expected for spin-orbit coupling.[31-34] If the alkyl group is substituted for one of lower symmetry such that the threefold symmetry axis is destroyed, the splitting persists with approximately the same magnitude.[34,35] Apparently, the local cylindrical symmetry at the halogen atom is effectively unperturbed by the presence of the low-symmetry alkyl group. This is illustrated by the spectrum of ethyl bromide, Fig. 8.8, where the splitting between the bands arising from

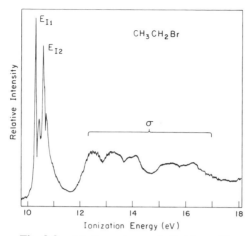

Fig. 8.8.   He I spectrum of ethyl bromide.

ejection of the nonbonding bromine electrons is $\Delta E_{In} = E_{I2} - E_{I1} = 0.31$ eV. It has been shown[35] that the splitting ΔE_{In} is insensitive to the size of the alkyl group and remains constant for alkyl groups with as many as five carbon atoms. Thus, for alkyl halides one expects to find $\Delta E_{In} \simeq 0.07$, 0.31, and 0.63 for $X = Cl$, Br, and I, respectively. However, the mean position of these bands resulting from ejection of nonbonding electrons, $\Delta E_{In}/2$, *is* sensitive to the size of the alkyl group; for $X = Br$, $\Delta E_{In}/2$ is shifted by -0.2 eV per α-methyl substituent and by -0.09 eV per β-methyl substituent.[35]

Despite the sensitivity of the mean position of the nonbonding orbitals to the alkyl substituent R, the influence of R is not strictly polar. The two components of the nonbonding orbitals possess vibrational fine structure indicating that there is conjugative interaction between the groups X and R. Such vibrational structure can be observed in Fig. 8.8. The observed vibrational spacing $\sim 1300 \text{ cm}^{-1}$ is much larger than expected for a vibration whose normal coordinate involves mainly stretching of the C—Br bond $[\bar{\nu}(\text{C—Br}) \simeq 560 \text{ cm}^{-1}]$; the observed spacing is characteristic of CH_2 symmetric bending frequencies $[\bar{\nu}(CH_2 \text{ bend}) \simeq 1400 \text{ cm}^{-1}]$. This indicates that the alkyl group orbital that interacts with the halogen p orbital is predominantly of the C—H bonding type. Ionization of a halogen "nonbonding orbital" in an alkyl halide is expected, therefore, to produce some change in HCH bond angles. Brogli and Heilbronner have shown[34] that the vibrational structure accompanying such nonbonding bands increases in intensity and complexity with increasing size of the alkyl group. In some systems such as cyclopropyl bromide,[35] E_{I1} degenerates into a wide fine-structured band due to the strong conjugation between one of the nonbonding orbitals and the Walsh orbital[36] of appropriate symmetry on the cycloalkyl group.

In cases where the nonbonding orbitals are *not* completely localized on the halogen atom, the splittings ΔE_{In} can vary appreciably. Consider the situation in which the nonbonding orbital must be described as a linear combination of halogen p AOs and orbitals localized on the alkyl group R as

$$\phi_i = c_i(X)p_i(X) + \sum_r c_{ir}\chi_r(R) \qquad (8.32)$$

where $\chi_r(R)$ is an alkyl group orbital of the proper symmetry for mixing with the halogen p orbital. In this situation the effective spin-orbit coupling parameter ζ_{ri} can be expressed as

$$\zeta_{ri} = c_i^2(X)\zeta(X) + \sum_r c_{ir}^2\zeta(R) \qquad (8.33)$$

where $\zeta(X)$ and $\zeta(R)$ are effective spin-orbit parameters for the halogen X and the alkyl group R atoms, respectively. In the simple case where $c_i(X) = 1$ and $c_{ir} = 0$, $\phi_i = p_i(X)$ and $\zeta_{ri} = \zeta(X)$. For cases where $c_{ir} \neq 0$, Eqs. 8.32 adn 8.33 must be retained. This orbital mixing necessarily results in a *decrease* in ζ_{ri} because $\zeta(R) \ll \zeta(X)$ for alkyl groups (C and H atoms). Consequently, the splitting ΔE_{In} is a sensitive probe for the extent of delocalization of the nonbonding halogen atomic orbitals onto the alkyl group, with ΔE_{In} decreasing as delocalization increases. Examples of the dependence of ΔE_{In} on such mixing between $p_i(X)$ and $\chi_r(R)$ orbitals have been reported for haloacetylenes and cyanohalides.[34,37]

8.9. Conclusions concerning spin-orbit coupling

The results of this chapter show that spin-orbit interactions can be of major importance in determining the structure of ionization bands resulting from ejection of electrons from degenerate levels, particularly so for molecules composed of heavy atoms. Although spin-orbit splittings are most prominent in molecules containing heavy atoms, they have been observed in molecules composed of atoms as light as oxygen (for example, see the spectrum of O_2, Fig. 5.3). The semiempirical model developed here for calculation of these spin-orbit interactions is successful in interpreting splittings observed in the PE spectra of the halogen acids, the diatomic halogens, and more complex polyatomic molecules containing halogens. It is evident from these examples that the calculation of spin-orbit interactions and their correlation with observed splittings in photoelectron bands can serve as a useful criterion for confirmation of assignments in PE spectra of molecules composed of heavy atoms.

Spin-orbit splittings have been observed in the PE spectra of many different types of molecules besides the ones discussed here; some of these are GaX_3,[38] InX_3,[38] HgX_2,[39,40] CsX,[41] InX,[42] and XeF_2.[43] The model developed here should be useful in confirming these observed splittings.

References

1. G. Herzberg, *Molecular Spectra and Molecular Structure. III. Electronic Spectra and Electronic Structure of Polyatomic Molecules*, pp. 14–19, 337–339, Van Nostrand, Princeton, N.J., 1966.

2. A. Abragam and M. H. L. Price, *Proc. Roy. Soc., Ser. A*, **205**, 135 (1951); ibid. **206**, 173 (1951); R. H. Pritchard and C. W. Kern, *J. Chem. Phys.*, **57**, 2590 (1972); J. A. Hall, *J. Chem. Phys.*, **58**, 410 (1973).

3. H. Lefebvre and C. M. Moser, *J. Chem. Phys.*, **44**, 2951 (1966); R. L. Ellis, R. Squire, and H. H. Jaffe, *J. Chem. Phys.*, **55**, 3499 (1971); S. D. Augustin, W. H. Miller, R. K. Pearson, and H. F. Schaefer, *J. Chem. Phys.*, **58**, 2845 (1973); B. D. Bird and P. Day, *J. Chem. Phys.*, **49**, 392 (1968).

4. J. C. Green, M. L. H. Green, P. J. Joachim, A. F. Orchard, and D. W. Turner, *Phil. Trans. Roy. Soc. London*, **A268**, 111 (1970).

5. F. A. Grimm, *J. Electron Spectrosc.*, **2**, 475 (1973).

6. J. L. Berkosky, F. O. Ellison, T. H. Lee, and J. W. Rabalais, *J. Chem. Phys.*, **59**, 5342 (1973).

7. T. H. Lee and J. W. Rabalais, *J. Chem. Phys.*, **60**, 1172 (1974).

8. We employ the approximate formulations of H_{so} developed by A. A. Misetich and T. Buch, *J. Chem. Phys.*, **41**, 2524 (1964).

9. R. Lefebvre, in *Modern Quantum Chemistry, Istanbul Lectures*, O. Sinanoglu (Ed.), Part I, p. 125, Academic, New York, 1965; H. C. Longuet-Higgins and J. A. Pople, *Proc. Roy. Soc. London*, **A68**, 591 (1951).

10. In this problem the matrix elements $\mathcal{H}_{jk}$ are complex. We use an eigenvalue subroutine extracted from a program by J. Heinzer, "ESREXN-Simulation of Exchange-Broadened Isotropic ESR Spectra," Program 209, Quantum Chemistry Program Exchange, Indiana University.

11. R. G. Parr, *The Quantum Theory of Molecular Electronic Structure*, p. 24, Benjamin, New York, 1963.

12. R. H. Pritchard, C. W. Kern, O. Zamani-Khamiri, and H. F. Hameka, *J. Chem. Phys.*, **58**, 411 (1973).

13. S. P. McGlynn, L. G. Vanquickenborne, M. Kinoshita, and D. G. Carroll, *Introduction to Applied Quantum Chemistry*, Holt Rinehart, New York, 1972.

14. B. W. N. Lo, K. M. S. Saxena, and S. Fraga, *Theoret. Chim. Acta*, **25**, 97 (1972).

15. T. E. H. Walker, J. Berkowitz, J. L. Dehmer, J. T. Waber, *Phys. Rev. Letters*, **31**, 678 (1973).
16. J. Berkowitz, *Chem. Phys. Letters*, **11**, 21 (1971).
17. C. R. Brundle, *Chem. Phys. Letters*, **7**, 317 (1970).
18. M. J. Weiss, G. M. Lawrence, and R. A. Young, *J. Chem. Phys.*, **52**, 2867 (1970).
19. H. J. Lempka, T. R. Passmore, and W. C. Price, *Proc. Roy. Soc., Ser. A*, **304**, 53 (1968).
20. D. C. Frost, C. A. McDowell, and D. A. Vroom, *J. Chem. Phys.*, **45**, 4255 (1967).
21. A. B. Cornford, D. C. Frost, C. A. McDowell, J. L. Ragle, and I. A. Stenhouse, *J. Chem. Phys.*, **54**, 2651 (1971).
22. S. Evans and A. F. Orchard, *Inorg. Chim. Acta*, **5**, 81 (1971).
23. A. W. Potts and W. C. Price, *Trans. Faraday Soc.*, **67**, 1242 (1971).
24. B. R. Higginson, D. R. Lloyd, and P. J. Roberts, *Chem. Phys. Letters*, **19**, 480 (1973).
25. K. Wittel, *Chem. Phys. Letters*, **15**, 555 (1972).
26. C. P. Anderson, G. Mamantov, W. E. Bull, F. A. Grimm, J. C. Carver, and T. A. Carlson, *Chem. Phys. Letters*, **12**, 137 (1971).
27. R. L. DeKock, B. R. Higginson, D. R. Lloyd, A. Breeze, D. W. J. Cruickshank, and D. R. Armstrong, *Mol. Phys.*, **24**, 1059 (1972).
28. K. Nakamoto, *Infrared Spectra of Inorganic and Coordination Compounds*, Wiley, New York, 1963.
29. P. Politzer and R. S. Mulliken, *J. Chem. Phys.*, **55**, 5135 (1971).
30. R. S. Mulliken, *J. Chem. Phys.*, **36**, 3428 (1962).
31. A. W. Potts, H. J. Lempka, D. G. Streets, and W. C. Price, *Phil. Trans. Roy. Soc. London*, **A268**, 59 (1970).
32. J. L. Ragle, I. A. Stenhouse, D. C. Frost, and C. A. McDowell, *J. Chem. Phys.*, **53**, 178 (1970).
33. T. A. Carlson and R. M. White, *Discussions Faraday Soc.*, **54**, 285 (1972).
34. F. Brogli and E. Heilbronner, *Helv. Chim. Acta*, **54**, 1423 (1971).
35. J. A. Hashmall and E. Heilbronner, *Angew. Chem.*, **9**, 305 (1970).
36. D. A. Walsh, *Nature*, **159**, 167, 712 (1947).
37. E. Heilbronner, V. Hornung, and E. Kloster-Jensen, *Helv. Chim. Acta*, **53**, 331 (1970); E. Heilbronner, V. Hornung, and K. A. Muszkat, ibid., **53**, 347 (1970); H. J. Haink, E. Heilbronner, V. Hornung, and E. Kloster-Jensen, ibid., **53**, 1073 (1970).
38. J. L. Dehmer, J. Berkowitz, L. C. Cusachs, and H. S. Aldrich, *J. Chem. Phys.*, **61**, 594 (1974).
39. J. H. D. Eland, *Int. J. Mass Spectrom. Ion Phys.*, **4**, 37 (1970).
40. J. Berkowitz, *J. Chem. Phys.*, **61**, 407 (1974).
41. J. Berkowitz, J. L. Dehmer, and T. E. H. Walker, *J. Chem. Phys.*, **59**, 3645 (1973).
42. J. Berkowitz and J. L. Dehmer, *J. Chem. Phys.*, **57**, 3194 (1972).
43. C. R. Brundle, M. B. Robin, and G. R. Jones, *J. Chem. Phys.*, **52**, 3383 (1970).

9 Configurational Instability of Molecular Ions

The most intense photoelectron transitions usually occur from the molecular ground state to a point on the ionic potential surface where the nuclear configuration is identical to that of the molecular ground state, that is, a vertical transition. Four possibilities can be envisioned once the molecular ion has been formed in this configuration: (1) The most stable nuclear configuration of the molecule and ion may be identical. In this case the ion vibrates about the equilibrium position, and the PE spectrum appears as a strong 0—0 band followed by a weak vibrational progression. (2) The ion may possess *no* stable nuclear configuration, in which case the atoms fly apart resulting in dissociation. The PE spectrum appears as a broad structureless band. (3) The most stable equilibrium configuration of the ion may be different from that of the molecule, in which case the PE spectrum consists of a prominent vibrational progression in the vibrational mode which leads to the new equilibrium position. (4) If the electronic state of the ion is orbitally degenerate, stability and degeneracy may not be simultaneously possible, and a nuclear displacement will occur that destroys the orbital degeneracy. Such displacements are called Jahn–Teller[1] and Renner[2] distortions and can lead to considerable spectral complexity. This chapter is concerned with the manifestations of Jahn–Teller and Renner effects in PE spectroscopy. The theory of interaction of vibrational and electronic motion is considered in Section 1. This theory is used to demonstrate the Renner effect in Section 2 and the Jahn–Teller effect in Section 3. Examples of the Jahn–Teller effect in the PE spectra of CH_4, NH_3, and C_2H_6 are presented in Sections 3 to 6. Jahn–Teller effects in larger molecules and complications from spin-orbit coupling are considered in Sections 7 and 8, respectively. Section 9 provides a detailed study of the Renner effect in the PE spectrum of H_2O.

9.1. Interaction of electronic and vibrational motion

Consider a system of electrons and nuclei described by a Hamiltonian $\mathcal{H}(q; Q)$ that is a function of electron q and nuclear Q coordinates.

Interaction of the electronic and nuclear motion can be described by a perturbation approach in which the Hamiltonian is expanded in terms of a Taylor series in the normal coordinates of vibration Q_r:

$$\mathcal{H} = \mathcal{H}(q;Q)\big|_{Q_0} + \sum_r^\infty \frac{1}{r!} \frac{\partial^r \mathcal{H}(q;Q)}{\partial Q^r}\bigg|_{Q_0} Q^r + \cdots \qquad (9.1)$$

The first term in Eq. 9.1, $\mathcal{H}(q;Q)\big|_{Q_0}$, is known as the 'zero-order Hamiltonian' and can be expressed as $\mathcal{H}_0$. Operating on Ψ with this expanded Hamiltonian yields

$$\langle \Psi | \mathcal{H} | \Psi \rangle = E_0 + E_1 + E_2 + \cdots \qquad (9.2)$$

where E_1 and E_2 are the perturbation energies resulting from the linear and quadratic expansion terms in Eq. 9.1, respectively. The interaction between two-degenerate functions Ψ_k and Ψ_l in the degenerate state Ψ can be expressed according to the value of the matrix elements $\langle \Psi_k | \mathcal{H} | \Psi_l \rangle$. Expanding these matrix elements with the operator of Eq. 9.1 yields

$$\langle \Psi_k | \mathcal{H} | \Psi_l \rangle = \langle \Psi_k | \mathcal{H}_0 | \Psi_l \rangle + \sum_r Q_r \langle \Psi_k | \frac{\partial \mathcal{H}}{\partial Q_r} | \Psi_l \rangle$$

$$+ \frac{1}{2!} \sum_r Q_r^2 \langle \Psi_k | \frac{\partial^2 \mathcal{H}}{\partial Q_r^2} | \Psi_l \rangle + \frac{1}{2!} \sum_{r,s} Q_r Q_s \langle \Psi_k | \frac{\partial^2 \mathcal{H}}{\partial Q_r \partial Q_s} | \Psi_l \rangle + \cdots \qquad (9.3)$$

The perturbation terms of Eq. 9.3 can cause a splitting of the electronic degeneracy if the normal coordinates Q_r that are active in the displacements are nontotally symmetric. If E_1 is the characteristic value of the linear perturbation term in Eq. 9.3 for a given nontotally symmetric coordinate Q_r, then $-E_1$ is the characteristic value obtained by changing the sign of the Q_r. Any degenerate electronic state cannot be stable in its equilibrium configuration Q_0 if its electronic energy is linearly dependent on asymmetrical vibrational modes. For such a case the electronic potential functions split into components corresponding to a lower molecular symmetry group. These new potentials consist of intersecting paraboloids whose potential minima (if they exist) are not at the symmetrical position Q_0. The new potential surfaces cross each other with equal but opposite slopes at $Q_r = Q_0$. *Such distortions due to the linear perturbation terms in Eq. 9.3 are a manifestation of the Jahn–Teller effect.*[1] *The Jahn–Teller theorem states that if a nonlinear molecule has an orbitally degenerate electronic state when the nuclei are in a symmetrical configuration, then that molecule is unstable with respect to at least one asymmetric displacement of the nuclei that may lift the orbital degeneracy.*

The quadratic perturbation terms in Eq. 9.3 can also lift the degeneracy of an electronic state, but in a manner different from the linear terms. Since the terms are quadratic, changing the sign of the displacement coordinate Q_r to $-Q_r$ has no effect on the perturbation energy E_2. Therefore, even though these terms may lift the degeneracy of an electronic state, *the potential energy surface of each split component will remain symmetric about the equilibrium nuclear configuration Q_0. Such splittings due to the quadratic terms of Eq. 9.3 are a manifestation of the Renner effect.*[2]

Having conceded that electronic degeneracy can be lifted by the perturbation terms of Eq. 9.3, we must consider the conditions necessary for these perturbation terms to be nonzero. The total Hamiltonian $\mathcal{H}$ of Eq. 9.1 necessarily possesses the full symmetry of the molecular point group. This also holds for $\mathcal{H}_0$, the Hamiltonian for the zeroth-order unperturbed molecule. The displacement $\partial\mathcal{H}/\partial Q_r$ transforms as the irreducible representation of the normal coordinate Q_r. The linear perturbation integral in Eq. 9.3 will be finite if the product of the representations of all species within the integral contains the totally symmetric representation, that is, $[\Gamma(\Psi_k) \times \Gamma(\Psi_l)] \times \Gamma(\partial\mathcal{H}/\partial Q_r) \subset \Gamma_1$. This term will cause a splitting of the degeneracy if Q_r is an asymmetric normal mode. The quadratic perturbation integral will be finite for all cases because $\Gamma(\partial^2\mathcal{H}/\partial Q_r^2) \equiv \Gamma_1$ and $[\Gamma(\Psi_k) \times \Gamma(\Psi_l)] \subset \Gamma_1$. The linear terms are usually considerably larger than the quadratic terms; thus the quadratic terms are normally neglected when the linear terms are nonzero.

9.2. Linear molecules—The Renner effect

Consider a linear molecule belonging to the $C_{\infty v}$ point group. The nondegenerate irreducible representations are denoted by A (Σ) and have zero angular momentum about the symmetry axis. The degenerate irreducible representations are all two dimensional, and there is an infinite series of them denoted by $E_1(\Pi)$, $E_2(\Delta)$, $E_3(\Phi)$, $\cdots$, E_k, $\cdots$. The indices k refer to the angular momentum $\pm k$ about the symmetry axis. The representation of the symmetrical product of E_k is given by

$$[E_k^2] = E_{k-k} + E_{k+k} \subset A + E_{2k} \tag{9.4}$$

Now consider the fate of the perturbation terms in Eq. 9.3. All nontotally symmetrical normal modes Q_r of a linear molecule transform as $E_1(\pi)$. In order for the perturbation integrals to be nonzero, the product of the symmetry representations of all species within the integral must contain the symmetric representation A. Such a product for the linear term is

$$[\Gamma(\Psi_k) \times \Gamma(\Psi_l)] \times \Gamma\left(\frac{\partial \mathcal{H}}{\partial Q_r}\right) \subset [E_k \times E_k] \times E_1 = E_1 \times [E_k^2]$$

$$\subset E_1 \times (A + E_{2k}) \subset E_1 + E_{2k+1} + E_{2k-1} \quad (9.5)$$

Hence the integral in the linear perturbation term *does not* contain the symmetric representation A for any asymmetric vibration e_1 and any degenerate state E_k. This linear term is *always zero* in a linear molecule, that is, the Jahn–Teller effect cannot occur in a linear molecule. The corresponding product for the quadratic term is

$$[\Gamma(\Psi_k) \times \Gamma(\Psi_l)] \times \Gamma\left(\frac{\partial^2 \mathcal{H}}{\partial Q_r^2}\right) \subset [E_k \times E_k] \times A$$

$$= A \times [E_k^2] \subset A \times (A + E_{2k}) \subset A + E_{2k} \quad (9.6)$$

The integral in the quadratic term *contains* the symmetric representation A for any asymmetric vibration e_1 and any degenerate state E_k. This quadratic term is *always* permissible in a linear molecule, that is, the Renner effect may occur in any linear molecule if the matrix elements of the quadratic term are large enough to cause a significant splitting. The normal mode Q_r that causes the instability is said to be the *Renner active mode*. The theory of the Renner effect has been reported in considerable detail.[3-5].

9.3. Nonlinear molecules—The Jahn–Teller effect

The validity of the Jahn–Teller theorem is most easily demonstrated qualitatively[6] for a tetraatomic molecule of point group D_{4h}, that is, with all four atoms at the corners of a square. This molecule can have two types of doubly degenerate electronic states, E_g and E_u, and four types of nontotally symmetric normal modes, b_{1g}, b_{1u}, b_{2g}, and e_u. It can be shown that

$$[E_g^2] = [E_u^2] \subset A_{1g} + B_{1g} + B_{2g} \quad \text{and} \quad [B_{1g}^2] = [B_{2g}^2] \subset A_{1g} \quad (9.7)$$

Applying this to a D_{4h} molecule in an E_g state, we see that the integral in the linear perturbation term of Eq. 9.3 is nonzero because

$$\sum_r [\Gamma(\Psi_k) \times \Gamma(\Psi_l)] \times \Gamma\left(\frac{\partial \mathcal{H}}{\partial Q_r}\right) \subset b_{1g} \times [E_g^2] + b_{2g} \times [E_g^2] \subset A_{1g} + \cdots \quad (9.8)$$

Hence the vibrations b_{1g} and b_{2g} are said to be *Jahn–Teller active modes*, for they render the square nuclear configuration unstable for either type of degenerate state E_g or E_u.

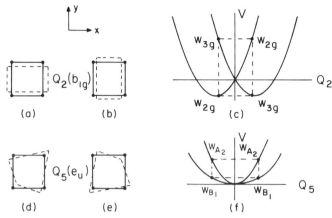

Fig. 9.1. Nontotally symmetric normal modes and potential energy curves for a degenerate electronic state as a function of normal coordinates Q_r for a planar tetraatomic molecule. (a)–(b) and (d)–(e) represent opposite phases of the b_{1g} and e_u vibrations, respectively, with dashed-line quadrangles representing displaced conformations. (c) and (f) represent the splitting of a degenerate state as a result of coupling with the Q_2 and Q_5 normal modes, respectively. Adapted from Ref. 14.

Consider, for example, the splitting of an E_g state by a $Q_2(b_{1g})$ normal mode. In the displaced configuration, the conformation of the nuclei is rectangular (Fig. 9.1a) and there are two electronic states of species B_{2g} and B_{3g} of the D_{2h} point group; B_{2g} is antisymmetric with respect to the xz plane. Assume the energies of these two states to be $W_{2g}(Q_2)$ and $W_{3g}(Q_2)$ with $W_{2g}(Q_2) > W_{3g}(Q_2)$, where we express the energies as a function of the displacement coordinate Q_2. If we change the sign of the normal coordinate displacements to $-Q_2$ (Fig. 9.1b), the energies become $W_{2g}(-Q_2)$ and $W_{3g}(-Q_2)$. Since Fig. 9.1b can be obtained from Fig. 9.1a by a rotation by 90°, it follows that

$$W_{2g}(-Q_2) = W_{3g}(Q_2) \quad \text{and} \quad W_{3g}(-Q_2) = W_{2g}(Q_2) \qquad (9.9)$$

Therefore, the two curves must cross as in Fig. 9.1c. The curves cross at $Q_2 = 0$, and any minima that occur are at $Q_2 \neq 0$. The potential surfaces of Fig. 9.1c correspond to the two Jahn–Teller split components of the E_g state.

If the same procedure is applied to the splitting of an E_g state by the degenerate bending mode $Q_5(e_u)$, the displaced conformation has symmetry C_{2v} and the electronic state splits into an A_2 and B_1 pair (Fig. 9.1d). Let $W_{A_2}(Q_5) > W_{B_1}(Q_5)$ for positive Q_5. Reversing the displacement (Fig. 9.1e) does not change the species of the two electronic component states because the displaced configurations differ only by a rotation of $180°$, hence

$$W_{A_2}(-Q_5) = W_{A_2}(Q_5) \quad \text{and} \quad W_{B_1}(-Q_5) = W_{B_1}(Q_5) \qquad (9.10)$$

The two curves do not cross but have a minimum at $Q_5 = 0$, where they coincide; there is no Jahn–Teller instability (Fig. 9.1f). The same applies for either component of the degenerate $Q_5(e_u)$ normal mode. The splitting into two coaxial paraboloids is similar to the occurrence in linear molecules; the $Q_5(e_u)$ mode is Renner active but not Jahn–Teller active.

The above discussion has qualitatively demonstrated the Jahn–Teller effect for a D_{4h} molecule. The proof of the theorem for the general case[7] and its treatment for more complicated cases are available.[8–15]

General discussion of Jahn–Teller effects

The Jahn–Teller theorem states that a molecule in a degenerate electronic state is unstable toward distortions that lower the molecular symmetry, thereby removing the electronic degeneracy. Part of the difficulty in the search for and measurement of these static distortions lies in the fact that the Jahn–Teller theorem is *permissive only*; a distortion may be very small or dynamic and, in either case, may escape detection. For any given molecule or ion only certain vibrations are Jahn–Teller active and only certain subgroup geometries are accessible by their operation. Jotham and Kettle[16] have presented an analysis of the Jahn–Teller active vibrations and accessible subgroup geometries for degenerate electronic states; data for the most common point groups are reproduced in Appendix VII. One can immediately determine from this Appendix which vibrations can be Jahn–Teller active and which subgroup geometries are possible for a degenerate state of a given molecule. The conditions by which a subgroup of the parent point group is accessible by virtue of a Jahn–Teller active vibration are:[16]

1. A point group is accessible if the irreducible representation of the Jahn–Teller active vibration in the parent point group subduces to the totally symmetric representation of the subgroup.

2. A point group is accessible if there exists no intermediate point group such that a vibration satisfying condition (1) for the final subgroup also satisfies it for the intermediate subgroup. Thus a molecule in a degenerate electronic state will descend in symmetry only as far as the nearest subgroup that will remove its degeneracy.[11] That is, a molecule distorted through Jahn–Teller interactions strives to maintain as much symmetry as possible while simultaneously removing its inherent electronic degeneracy.

The size of Jahn–Teller distortions is ultimately limited by the quasielastic forces that resist it. Jahn–Teller distortions are often small because the restoring potentials are usually large, and the interaction between the electronic and vibrational motions may be very small. The effect is largest when the degenerate orbital concerned is strongly involved in the bonding and is smallest for nonbonding orbitals. In cases where there is a stable distorted configuration, the electronic energy is lowered without shifting the center of mass to first order, so that there is more than one position of equilibrium with equal energy. In other words, the electronic degeneracy is replaced by vibronic degeneracy. Thus a Jahn–Teller distorted molecule has a permanent distortion that lowers its symmetry and produces an observable anisotropy. This is called the *static Jahn–Teller effect*. The *dynamic Jahn–Teller effect* refers to the coupling between the electronic and vibrational angular momenta that results in vibronic energy levels. Jahn–Teller vibronic splittings lift the degeneracy (for example, the sixfold degeneracy of the first quantum of an e vibrational mode excited in a T electronic state of a tetrahedral molecule) of the excited "vibrational" levels, hence splitting the fundamentals into vibronic levels. It is important to note that the overall degeneracy of a level in a Jahn–Teller distorted molecule is not reduced by this effect since the overall Hamiltonian retains its symmetry. Instead of the original electronic degeneracy in which the group operations change the electronic and vibrational wavefunctions independently, we have vibronic degeneracy in which transformations of the electronic wavefunction are inextricably accompanied by transformations of the vibrational wavefunctions.

9.4. Example: The Jahn–Teller effect in methane

The spectrum of methane provides a classical example of the Jahn–Teller effect in a tetrahedral molecule. The MO with lowest binding energy in CH_4 is the triply degenerate t_2 orbital. Ejection of one electron from this orbital leaves the CH_4^+ ion in a $^2T_2$ state that is stable and well known in mass spectrometry. Since the t_2 orbital is strongly bonding, the triply degenerate T_2 state formed by removal of a t_2 electron is subject to Jahn–Teller distortions. Several theoretical investigations[17–22] of the geometry of CH_4^+ have concluded that the Jahn–Teller distortions are large, with the more recent investigations predicting the most stable geometry to be that of D_{2d} symmetry. The spectrum of methane has been investigated by several groups[23–29] with varying degrees of resolution.

The spectral bands of methane resulting from ejection of a t_2 electron are shown in Fig. 9.2, and the vibrational bands below 13.1 eV are expanded in Fig. 9.3. The vibrational frequencies and positions of the

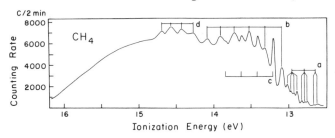

Fig. 9.2. He I spectrum of the $(1a_1^2 2a_1^2 1t_2^5)^2 T_2 \leftarrow (1a_1^2 2a_1^2 1t_2^6)^1 A_1$ transition of methane. Vibrational progressions are indicated by lines in the spectrum. Reproduced with permission from Ref. 23.

bands are listed in Table 9.1. By examination of Figs. 9.2 and 9.3 and Table 9.1, we can make the following observations. (1) The He I PE spectrum of methane is a composite of at least three electronic bands whose respective maxima lie at 13.5, 14.5, and ≈ 15 eV. Thus ejection of a t_2 electron produces an ion that is in one of three different electronic states that are very close in energy. (2) The vibrational structure is not composed of simple progressions. Furthermore, there is a large change in the vibrational structure at ≈ 13.07 eV. Below 13.07 eV, the structure can be assigned as in Fig. 9.3, that is, as three groups of bands with approximately equal spacings between the bands. These groupings and vibrational quantum numbers are explained later in this section. Above 13.07 eV there are three progressions, namely, b, c, and d. The vibrational quantum numbers of progressions b and c can be labeled, for the first vibrational bands observed in these progressions are very intense, well defined, and appear to be the origins of the progressions. The quantum numbers of progression d cannot be labeled because it is not possible to find the origin of the progression due to overlap of the two electronic bands in that region. (3) The vibrational bands appear to be

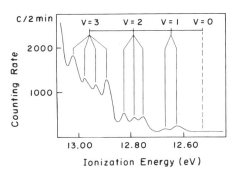

Fig. 9.3. Details of the PE spectrum of methane below 13.1 eV, showing the splittings between vibronic bands. v indicates the vibrational quantum numbers of the progression. The $v = 0$ band is believed to be outside the Franck–Condon region; its approximate position has been determined by extrapolation. Reproduced with permission from Ref. 23.

Table 9.1. Vibrational Progressions in the PE Spectrum of Methane

Interpretation	Energy (eV)	Average Energy
Progression a		
ν_2	12.616	12.644
	12.673	
		} 0.143
	12.753	
$2\nu_2$	12.784	12.787
	12.825	
		} 0.161
	12.887	
$3\nu_2$	12.928	12.948
	12.968	
	13.010	
Progression b		
ν_3	13.08	0.22
$2\nu_3$	13.30	0.21
$3\nu_3$	13.51	0.20
$4\nu_3$	13.71	0.19
$5\nu_3$	13.90	0.18
$6\nu_3$	14.08	
Progression c		
ν_1	13.20	0.21
$2\nu_1$	13.41	0.21
$3\nu_1$	13.62	0.21
$4\nu_1$	13.83	
Progression d		
	14.27	0.16
	14.43	0.14
	14.57	0.13
	14.70	

superimposed on a dissociative continuum. This is most likely a result of dissociation processes such as

$$CH_4^+ \rightarrow CH_3^+ + H \qquad (9.11)$$

and

$$CH_4^+ \rightarrow CH_2^+ + 2H \qquad (9.12)$$

The complex nature of this transition, that is, the anomalous band splittings and peculiar vibrational structure suggests that Jahn–Teller

Table 9.2.   Normal Modes of Vibration of CH_4

$\nu_1(a_1) = 0.361$ eV	$\nu_3(t_2) = 0.374$ eV
$\nu_2(a) = 0.189$ eV	$\nu_4(t_2) = 0.162$ eV

forces are operative. In order to analyze the spectrum it is necessary to invoke Jahn–Teller attitudes.

First consider which of the vibrational modes of methane is active in the Jahn–Teller distortion. The symmetry and frequencies of the normal modes of vibration of CH_4 are given in Table 9.2. Of these vibrations, it is only the asymmetric modes that can produce a distortion. Therefore, the active mode could be either $\nu_2(e)$, $\nu_3(t_2)$, or $\nu_4(t_2)$, and it is not possible to distinguish between these unambiguously by frequency alone. However, it should be possible to distinguish between them by considering the vibronic splittings obtained from excitation of an e or a t_2 mode in a T_2 electronic state, as shown in Fig. 9.4. Excitation of a t_2 mode in a T_2 electronic state produces a multitude of vibronic components, while excitation of an e mode produces the exact pattern of components that is

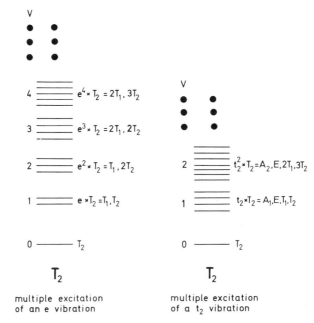

Fig. 9.4.   Vibronic species resulting from multiple excitation of e and t_2 vibrations in a $^2T_2$ electronic state of a tetrahedral molecule. v represents the vibrational quantum number. Reproduced with permission from Ref. 23.

observed in Fig. 9.3 for the vibrational quantum numbers $v = 1$, 2, and 3. Separations between the vibronic peaks are ≈ 0.043 eV. The vibronic splitting is arbitrarily taken to be symmetrical so that the "vibrational" band is chosen as the center of the vibronic bands. Thus the splittings between these "vibrational" bands are 0.143 and 0.161 eV. These, of course, give only an estimate of the size of the vibration. Since the $v_2(e)$ mode is 0.189 eV in the ground state of CH_4, the values obtained are in the expected range for the $v_2(e)$ mode in CH_4^+. With this vibrational analysis, it is possible to elicit a value for the adiabatic ionization energy of methane. The $v = 0$ band must be below the $v = 1$ band at 12.616 eV, and a reasonable extrapolation would place it at ≈ 12.51 eV. No bands have been observed below 12.616 eV, probably because this is outside the Franck–Condon region, and any bands in this area would have extremely weak intensity. However, it is quite certain that the adiabatic ionization energy of methane is below 12.616 eV, and probably lies near 12.51 eV. Unfortunately, there are no theoretical calculations of the vibronic energy levels of CH_4^+ to confirm this analysis.

Next consider the gross change in the vibrational structure of the spectrum that occurs near 13.07 eV. Rather than the complexity of vibronic components that arises from multiple excitation of a degenerate vibration, the structure consists of two interlocking progressions, b and c, of which only progression b shows anharmonicity. It is difficult to assign these progressions on the basis of frequency alone. However, since the vibrational frequencies of the excited ionic states are usually lower than those of the ground state molecule, these progressions having a value of ≈ 0.20 eV probably correspond to $v_1(a_1)$ and/or $v_3(t_2)$, which are 0.361 and 0.374 eV, respectively, in the ground state. We would not expect vibronic splittings upon excitation of v_1 (for it is symmetrical); however, excitation of v_3 could produce such splittings. The bands of progression b at 13.08 and 13.30 eV along with several other bands are not symmetrical and even show small splittings. This is an indication that they are composed of several unresolved components. There is also a noticeable broadening of the vibrational bands of progression b toward higher energy. For these two reasons, progression b is assigned to $v_3(t_2)$. The bands of progression c are much sharper than those of progression b and show no indication of broadening. This progression is thus believed to be $v_1(a_1)$.

The change in vibrational structure near 13.07 eV is a result of a change in the potential surfaces of the excited state. The potential surfaces for a Jahn–Teller distorted tetrahedral molecule have been derived analytically.[11] A two-dimensional cross section in $Q_2 Q_3$ space of the lowest potential energy surface for a T state split by an e vibrational

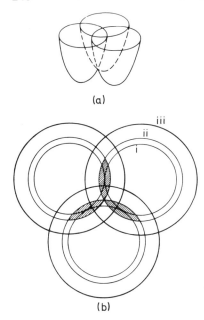

(a)

(b)

Fig. 9.5. (*a*) The lowest potential energy surface in $Q_2 Q_3$ space for a T_2 electronic state interacting with an e vibrational mode. The surface consists of three disjoint paraboloids drawn as surfaces of revolution. In the general case a horizontal section should reveal ellipses rather than circles. (*b*) Cross section through (*a*) as seen from directly above the paraboloids. The inner circles are contour lines whose energies increase as i < ii < iii. The shaded areas represent overlap in the surfaces of the paraboloids between the contour lines (i) and (ii). Reproduced with permission from Ref. 23.

mode appears as in Fig. 9.5*a*. Q_2 and Q_3 are the displacement coordinates of the degenerate e mode. The surface consists of three intersecting disjoint paraboloids, whose minima form the vertices for an equilateral triangle. Figure 9.5*b* is a view of this surface from above showing contour lines within the wells. The contour lines can be listed in order of increasing energy as i < ii < iii. From this diagram, it can be seen that the surface of the three paraboloids begin to overlap at the point (i) that is below the point of intersection at the origin of the coordinate system, (ii). The regions of overlap between contour lines (i) and (ii) are the shaded areas of Fig. 9.5*b*. The change in structure near 13.07 eV should correspond to the point (i) in Fig. 9.5*b* where the surfaces of the three disjoint paraboloids begin to overlap. This intersection promotes the beginning of the two new progressions, b and c, which are much stronger than progression a. Thus, the contour line (i) in Fig. 9.5*b* probably corresponds to the 13.07-eV region of the methane spectrum.

A schematic representatioin of the Jahn–Teller splitting based on the experimental spectrum is shown in Fig. 9.6. This diagram, based on those of Liehr,[11] represents sections through the potential energy hypersurfaces, which have cylindrical symmetry in the harmonic approximation. The initial state of the molecule is shown with the vibrational wavefunction drawn in. At first glance, it may appear that the most probable

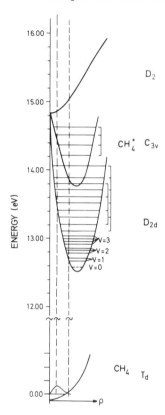

Fig. 9.6. A qualitative configuration-coordinate diagram showing the Jahn–Teller splittings that result from ejection of a t_2 electron in methane. ρ is a distortion coordinate. The positions of the excited electronic states and vibrational levels are as observed in the electron spectrum. Reproduced with permission from Ref. 23.

transition is to an unsplit upper state, that is, where the coordinate ρ is zero. Sturge[30] has shown that since there are two distortion coordinates and the curves of Fig. 9.6 are surfaces of cylindrical symmetry, the maximum in the vibrational wavefunction is not at $\rho = 0$ but at some finite value of ρ. Thus in the Franck–Condon approximation the most probable transition is given by the vertical dashed line at the center of the vibrational wavefunction. The classical turning point for the ground vibrational state is also projected vertically through the upper surfaces. The positions of the three upper surfaces are estimated from the electron spectrum. The Franck–Condon maxima and turning points for all three excited states can be located from the vertical projections. These points are uncertain for the highest excited state due to the diffuseness of the spectrum in that region. The vibrational energy levels of the excited states are shown as observed in the spectrum. The position of the $v = 0$ level is the extrapolated value.

It is of interest to consider the symmetry of the Jahn–Teller distorted

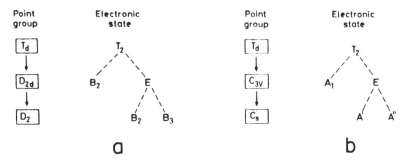

Fig. 9.7. Two probable routes of distortion of the CH_4^+ ion showing the splittings of the T_2 state in the lower symmetry point groups. Reproduced with permission from Ref. 23.

upper states. We can envision two probable routes of distortion of the CH_4^+ ion:

1. The tetrahedral ion is compressed along the z axis with all bond lengths remaining equal so that the symmetry is reduced to D_{2d}; the T_2 state splits into E and B_2 states in the D_{2d} point group. The E state, being degenerate, is subject to further Jahn–Teller distortions. The descent that preserves the most symmetry and, therefore, is the most likely, is the one in which the molecule becomes planar, thus reducing the symmetry to D_2 and splitting the E state into B_2 and B_3 states. This process is illustrated in Fig. 9.7a.
2. One of the C—H bonds is elongated, the three others remaining equal such that the symmetry is reduced to C_{3v}; the T_2 state splits into E and A_1 states in the C_{2v} point group. The degenerate E state, is subject to further distortions. The descent that preserves some symmetry, not considering dissociation, is the one in which two of the three equivalent Hs of the C_{3v} ion move either toward or away from each other to leave only a plane of symmetry. This lowers the symmetry to C_s and splits the E state into A' and A'' states. This process is illustrated in Fig. 9.7b.

These two routes of distortion remove the electronic degeneracy but retain some symmetry in the molecule. If we consider the analytical treatment of Liehr,[11] route (1) should be more probable than route (2) because route (1) preserves the highest symmetry while simultaneously removing the electronic degeneracy. Thus, the first maximum at 13.5 eV should correspond to the vertical transition to the $^2B_2$ state of the D_{2d} ion, and the second maximum at 14.5 eV should correspond to the vertical transition to the $^2A_1$ state of the C_{3v} ion. The third maximum at

≈ 15 eV could be the vertical transition to either the $^2B_2$ or $^2B_3$ state of the D_2 ion, for D_2 is higher symmetry than C_3 and should have lower energy. The symmetries of these electronic states are indicated in Fig. 9.6. These qualitative predictions are in agreement with recent theoretical calculations[17-19] that predict the symmetries of the lowest energy electronic states of the CH$_4^+$ ion to be, in order of increasing energy, $D_{2d} < C_{3v}$.

Conclusions concerning CH$_4^+$

The following conclusions can be drawn from the study of the methane spectrum:

1. The electron spectrum resulting from ejection of a t_2 electron is a composite of three overlapping electronic bands which are separated by several tenths of an electron volt.
2. Four vibrational progressions can be identified in the spectrum. Three of them have been assigned, these assignments being $v_2(e)$ for progression *a*, $v_3(t_2)$ for progression *b*, and $v_1(a_1)$ for progression *c*. Progressions *a* and *b* show vibronic splittings.
3. There are strong Jahn–Teller forces operative in the CH$_4^+$ ion that distort it from regular tetrahedral symmetry.
4. The active vibration that causes the Jahn–Teller splitting is believed to be the $v_2(e)$ mode, which is reduced to ≈ 0.15 eV in the CH$_4^+$ ion.
5. The adiabatic ionization potential of methane is below 12.616 eV; the extrapolated value is ≈ 12.51 eV.
6. The three electronic bands observed in the spectrum of methane are the Jahn–Teller components of the distorted $^2T_2$ state of CH$_4^+$. The lowest energy excited state most likely has D_{2d} symmetry, while the two higher-energy states probably have a symmetry of C_{3v} and D_2 respectively.

9.5. Example: The Jahn–Teller effect in NH$_3^+$

The 14.5 to 18.5-eV region

The $(\cdots 1e^3\ 3a_1^2)\,^2E \leftarrow (\cdots 1e^4\ 3a_1^2)\,^1A_1$ transition of ammonia provides a good example of the Jahn–Teller effect in a trigonal species. Ejection of a $1e$ electron is expected[20] to produce strong Jahn–Teller distortions because the $1e$ orbital has strongly bonding character as a result of its combination of nitrogen $2p_x$ and $2p_y$ orbitals with hydrogen $1s$ orbitals. The spectral bands of ammonia arising from ejection of a $1e$ electron,

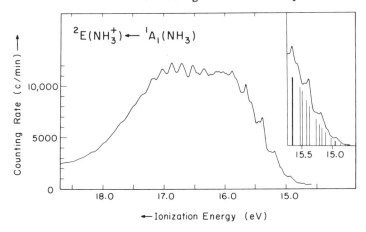

Fig. 9.8. The He I spectrum of the $(1e^3 3a_1^2)^2 E \leftarrow (1e^4 3a_1^2)^1 A_1$ transition of ammonia. An adaption of calculated vibronic levels to the spectrum is shown in the insert. Reproduced with permission from Ref. 31.

Fig. 9.8, have been investigated by several groups.[31–34] The following ovservations can be made:

1. The transition gives rise to a broad electron band that starts at 14.725 eV with maxima at ≈ 15.8 and ≈ 16.8 eV.

2. The electron band shows vibrational structure in two regions. The first region starts at 14.725 eV and reaches up to ≈ 15.9 eV while the latter ranges from ≈ 16.3 to ≈ 17.8 eV. Below 15.9 eV the structure consists of six groups of irregularly spaced bands whose intensity increases rapidly with energy. Above 15.3 eV the bands are considerably broadened and show, in some cases, a doublet structure. In the middle region (15.9–16.3 eV) there are only weak indications of vibrational bands.

3. The electron band appears to consist partly of a continuum. This is most likely a result of dissociation processes such as

$$NH_3^+ \rightarrow NH_2^+ + H \qquad (9.13)$$

and

$$NH_3^+ \rightarrow NH^+ + 2H \qquad (9.14)$$

The appearance potential of NH_2^+ is 15.75 eV.[35]

These observations strongly suggest that Jahn–Teller forces are operative in this ionic state. Ammonia belongs to the C_{3v} point group and has four normal modes of vibration: $\nu_1(a_1) = 413.7\,\text{meV}$, $\nu_2(a_1) = 115.6\,\text{meV}$, $120.0\,\text{meV}$, $\nu_3(e) = 427.0\,\text{meV}$, and $\nu_4(e) = 201.7\,\text{meV}$, where the inversion splitting is indicated for ν_2. According to the Jahn–Teller effect the equilibrium conformation of the ion is distorted from C_{3v} symmetry by the displacements $Q_{2a,b}$ and $Q_{3a,b}$ corresponding to the two doubly degenerate vibrations ν_3 and ν_4, respectively. Q_{2a} and Q_{3a} thereby lower the symmetry to C_s and Q_{2b} and Q_{3b} to C_1. Considering the static distortion, the ^2E state is split into $^2A'$ and $^2A''$ states in C_s symmetry and into ^2A states in C_1 symmetry. The equilibrium conformation with the lowest potential energy should correspond to that of C_s symmetry since this configuration retains the highest symmetry while simultaneously removing the electronic degeneracy.

In the dynamic Jahn–Teller effect the vibronic components of the vibrational levels must be considered. Longuet–Higgins et al.[36] have performed first-order calculations of the vibronic levels of a molecule in a doubly degenerate electronic state, split by a doubly degenerate vibration. The energy levels have been expressed as functions of the parameter $k^2 = 2D$. The coupling parameter D is defined in such a way that $D\hbar\nu$ is the depth of the bottom of the potential well below the undistorted equilibrium position. An adaption of these vibronic energy levels to the spectrum is shown as a bar diagram in the inset of Fig. 9.8. The vibrational energy in the calculation was chosen as that of the ν_4 mode of the neutral molecule. The calculated energy levels corresponding to $D = 2.5$ give the best fitting to the vibrational bands. The heights of the bars in Fig. 9.8 were obtained by fitting a Gaussian function resembling the intensity distribution in the spectrum. The agreement between the bar diagram and the spectrum is remarkably good considering that many approximations and assumptions have been made in the calculations and in the adaptation. Furthermore, there are two doubly degenerate vibrations in the NH_3^+ ion that may couple. However, no indications of any effects of such a coupling have been observed.

Longuet–Higgins et al.[36] have also calculated probabilities for transitions from a nondegenerate state to a doubly degenerate one. These calculations give a double-maximum shape for the electron band in agreement with the observations. For $D = 2.5$ the separation of the maxima is expected to be about five vibrational quanta. The observed separation is $\approx 1\,\text{eV}$, which is equal to five quanta of the ν_4 vibration.

The structure above $15.3\,\text{eV}$ should correspond to transitions to states in the upper branch of the potential. Slonszewski has proposed the existence of such states when $D > 1$.[37] He has also found that the energy-level diagram should consist of series of states that overlap at higher energies and that these states are shortlived. This is in qualitative agreement with the observations of broad bands that appear to form a progression in the region, 16.3 to 17.0 eV. The mean separation is ≈ 165 meV. The fading of the vibrational structure in the middle region might be due to the onset of dissociation at 15.73 eV.

We can now conclude that the $1e$ orbital has an adiabatic ionization energy of 14.725 eV and that the resulting ion shows a Jahn–Teller stabilization $Dh\nu \approx 0.5$ eV. In order to explain the vibrational structure of the electron band in the region of transitions to the lower branch of the potential one has to consider the vibronic levels.

The 10.0 to 12.0- eV region

The outermost orbital of ammonia, $3a_1$, containing the "nonbonding" electrons, is formed from a hybrid of nitrogen $2s$ and $2p_z$ atomic orbitals. Figure 9.9 shows the electron spectrum obtained when an electron is expelled from this a_1 orbital by He I resonance radiation, that is, the $\cdots 1e^4 3a_1^1, {}^2A_1 \leftarrow \cdots 1e^4 3a_1^2, {}^1A_1$ transition. It consists of a long progression of vibrational bands starting at 10.073 eV. The progression shows negative anharmonicity with spacings ranging from ≈ 111 to ≈ 140 meV. The vibrations are of the ν_2 bending mode, which produces the inversion of ammonia. The origins of the band at the lowest energy are uncertain. It may be the $0 \leftarrow 0$ band or a hot band enhanced in intensity by a high

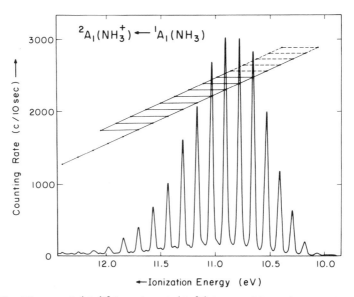

Fig. 9.9. The $\cdots 1e^4 3a_1^1)^2 A_1 \leftarrow (\cdots 1e^4 3a_1^2)^1 A_1$ transition of ammonia excited by He I resonance radiation. Reproduced with permission from Ref. 31.

Franck–Condon factor. A similar progression shifted by $\approx 340\,meV$ towards higher binding energy is also observed. This structure is probably due to simultaneous excitation of $n\nu_2$ vibrations and a ν_1 vibration, giving a value of 340 meV for the ν_1 vibrational energy in the $^2A_1$ state of NH_3^+. Observation of such a long vibrational progression in the ν_2 mode indicates that the NH_3^+ ion is very nearly planar in its ground ionic state. Therefore, the $3a_1$ electrons are certainly structure determining, even though they are generally considered to be "nonbonding."

9.6. Example: The Jahn–Teller effect in $C_2H_6^+$

Molecular orbital calculations on the ground state of neutral ethane predict a staggered geometry (D_{3d} symmetry) to be energetically preferred with an electron configuration of

$$1a_{1g}^2 1a_{2u}^2 2a_{1g}^2 2a_{2u}^2 1e_u^4 3a_{1g}^2 1e_g^4 \qquad (9.15)$$

This electronic configuration along with Koopmans' theorem suggests a ^2E_g ground state for the ethane cation $C_2H_6^+$, although *ab initio* calculations[38–40] are not in complete agreement about the exact nature of the ground state. All of the experimental studies on ethane predict a ^2E_g ground state[28,41,42] for the molecular ion.

The PE spectrum of ethane is shown in Fig. 9.10, and the bands with resolved vibrational structure are shown on an expanded energy scale in

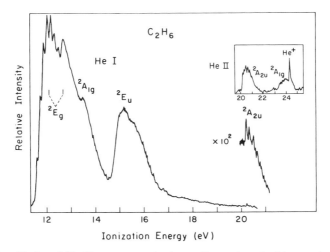

Fig. 9.10. He I and He II spectrum of ethane. Reproduced with permission from Ref. 41.

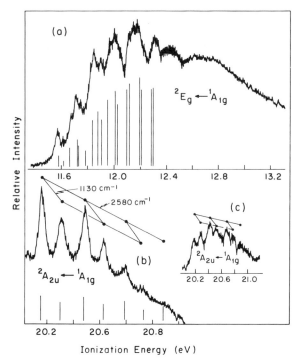

Fig. 9.11. (a) The He I spectrum of the $^2E_g \leftarrow {}^1A_{1g}$ transition of ethane. An adaption of calculated vibronic levels to the spectrum is shown as a bar diagram. (b) The He I spectrum of the $^2A_{2u} \leftarrow {}^1A_{1g}$ transition of ethane. The bar diagram beneath the spectrum represents the intensities of the vibrational bands corrected for analyzer response. (c) The He II spectrum of the $^2A_{2u} \leftarrow {}^1A_{1g}$ transition of ethane. Reproduced with permission from Ref. 41.

Fig. 9.11. The intensities of the higher vibrational levels of the ~ 20 eV band are distorted in the He I spectrum due to the discrimination of the electrostatic analyzer against low-kinetic-energy electrons. This spectrum has been corrected to a first approximation, Fig. 9.11b, by dividing the observed intensities by the corresponding kinetic energy of the photoelectrons. The following observations can be made from Figs. 9.10 and 9.11.

1. The spectrum can be subdivided into three ionization regions:
    a. 11.5 to 14.5 eV—There are three band maxima in this region occurring at 12.00, 12.72, and 13.5 eV.
    b. 14.5 to 16.5 eV—This region contains a broad asymmetric band (adiabatic $E_I \sim 14.7$ eV) that appears to be composed of two overlapping transitions centered at ~ 15.0 and 15.8 eV.

   c. 20.0 to 26.0 eV—Two bands are observed in this region with adiabatic E_Is at 20.6 and ~ 23.6 eV.

2. A vibrational progression with irregular spacings begins at 11.57 eV and submerges beneath a continuum at ~ 12.55 eV. The spacings between the main peaks vary between 1050 and 1280 cm^{-1} with an average spacing of ≈ 1110 cm^{-1}. Irregular shoulders and smaller peaks accompany this main progression.

3. The band near 20 eV contains evidence for two interlocking vibrational progressions of ~ 2580 cm^{-1} that are offset by ~ 1130 cm^{-1}.

Molecular orbitals of ethane

The molecular orbitals of ethane can be described as follows:

$1e_g$: Pseudo-π orbitals consisting of C $2p_x$ and $2p_y$ and H $1s$ AOs; C—H bonding and C—C antibonding.

$3a_{1g}$: σ Orbital composed of C $2p_z$ and H $2s$ AOs; strongly C—C and C—H bonding.

$1e_u$; Pseudo-π orbitals consisting of C $2p_x$ and $2p_y$ and H $1s$ AOs; C—C and C—H bonding.

$2a_{2u}$: σ Orbital composed of predominantly C $2s$ AOs with small admixtures of C $2p_z$ and $1s$ and H $1s$ AOs; C—C antibonding and C—H bonding.

$2a_{1g}$: σ Orbital composed of predominantly C $2s$ AOs with small admixtures of C $2p_z$ and $1s$ and H $1s$ AOs; C—C bonding and C—H antibonding.

$1a_{1g}$ and $1a_{2u}$: Core orbitals of carbon ($1s$) arranged in bonding and antibonding combinations.

The five outermost orbitals are expected to be in the valence region (binding energy <30 eV) with the core orbitals being more tightly bound.

12.5 to 14.5-eV region

Applying Koopmans' theorem, the ionization bands expected in this region are ^2E_g and $^2A_{1g}$ due to ejection of $1e_g$ and $1a_{1g}$ electrons. The irregular vibrational progression, with its numerous shoulders and small peaks, observed below 12.55 eV strongly suggests that vibronic interaction is occurring in this transition. Interactions such as this, that is, Jahn-Teller interactions, occur when degenerate vibrational modes are excited in a degenerate electronic state. According to the theorem, the $C_2H_6^+$ ion in a ^2E_g ionic state is subject to first-order perturbations that remove the degeneracy by lowering the symmetry of the ion. Considering the static effect, the equilibrium conformation of the ion is distorted from

D_{3d} symmetry by displacements $Q_{a,b}$ corresponding to the doubly degenerate vibrations. These displacements can split the degeneracy of the ^2E_g state into ^2A_g and ^2B_g states of C_{2h} symmetry or two ^2A_g states of C_i symmetry. The equilibrium conformation should correspond to that of C_{2h} symmetry since this configuration retains the highest symmetry while simultaneously removing the electronic degeneracy.

The irregularity of the vibrational progression and its numerous shoulders are manifestations of the dynamic Jahn–Teller effect. The ethane molecule possesses twelve normal modes of vibration, six of which are degenerate (Table 9.3). The Jahn–Teller active vibration must be either ν_8, ν_{11}, ν_{12}, or some combination of these because only degenerate vibrations can produce this instability and these are the only degenerate modes with frequencies near that observed. Since the $1e_g$ MO is C—H bonding and C—C antibonding, it is likely that the active vibration is a CH_3 deformation mode, that is, ν_8 or ν_{11}. A reduction from their ground-state value to the observed spacing of $\sim 1110\ cm^{-1}$ is expected for this type of transition.

Table 9.3. Normal Modes of Vibration of Ethane

Normal Mode	Symmetry	Wavenumber/(cm^{-1})	Approximate Description
ν_1	a_{1g}	2899	CH stretching
ν_2	a_{1g}	1375	CH_3 deformation
ν_3	a_{1g}	993	C—C stretching
ν_4	a_{1u}	275	torsion
ν_5	a_{2u}	2954	CH stretching
ν_6	a_{2u}	1379	CH_3 deformation
ν_7	e_u	2994	CH stretching
ν_8	e_u	1486(IR); 1491(R)	CH_3 deformation
ν_9	e_u	821(IR); 831(R)	bending
ν_{10}	e_g	2963	CH stretching
ν_{11}	e_g	1460	CH_3 deformation
ν_{12}	e_g	1155	bending

The vibronic energy levels of Longuet–Higgens et al.[36] have been adapted to the spectrum as shown in the bar diagram of Fig. 9.11a. The vibrational energy in the calculation was chosen at $1110\ cm^{-1}$, the average observed spacing. The heights of the bars in Fig. 9.11a have been obtained by fitting a gaussian function resembling the intensity distribution in the spectrum. The calculated energy levels agree best with the vibronic bands when $D \sim 2.5$.

The calculated transition probabilities[36] from a nondegenerate state to a doubly degenerate one indicate that the band will have two maxima separated by about five vibrational quanta for $D \sim 2.5$. The observed separation of the first two band maxima is ~ 0.72 eV, which is equal to five quanta of the observed vibrational spacing. The structure above 12.55 eV should correspond to transitions to states in the upper branch of the potential. The existence of a series of overlapping short-lived states in this upper potential has been proposed by Slonzewski[37] when $D > 1$ in qualitative agreement with the observation of a broad structureless band in this region implying the onset of dissociation.

It can now be concluded that the bands at 12.00 and 12.72 eV are the Jahn–Teller components of the split $^2E_g \leftarrow {}^1A_{1g}$ transition with the $^2A_{1g} \leftarrow {}^1A_{3g}$ transition appearing at 13.5 eV. The ground-state ion shows a Jahn–Teller stabilization of $D\hbar\nu \sim 0.35$ eV due to coupling of the degenerate vibrational mode, ν_8 or ν_{11}. The vibrational structure of the electron band in the region of transitions to the lower branch of the potential is best explained in terms of vibronic energy levels.

The energy of the dissociation process

$$C_2H_6 \rightarrow CH_3^+ + CH_3 + e^- \tag{9.16}$$

is 13.6 eV. Lathan et al.[38] have shown that the energy of the dissociation process

$$C_2H_6^+ \rightarrow CH_3^+ + CH_3 \tag{9.17}$$

represents the energy of a C—C one-electron bond between a methyl radical and a methyl cation, that is, ~ 37 to 45 kcal/mole. Since the $3a_{1g}$ orbital is the main C—C bonding orbital, the above considerations suggest that ejection of a $3a_{1g}$ electron will lead to a continuous spectrum beyond 13.6 eV with an origin for the $^2A_{1g} \leftarrow {}^1A_{1g}$ transition in the region, 11.6 to 12.0 eV. Thus, the origin of the $^2A_{1g} \leftarrow {}^1A_{1g}$ transition almost certainly underlies the $^2E_g \leftarrow {}^1A_{1g}$ band, adding to the complexity of the spectrum.

14.5- to 16.5-eV region

Ionization of the third occupied orbital of ethane, $1e_u$, leaves the ion in a doubly degenerate electronic state that is subject to Jahn–Teller perturbations. The band observed in this region, being broad and asymmetric, fits this pretension. The maxima at 15.0 and 15.8 eV appear to be the two Jahn–Teller split components of the ^2E_u band.

20- to 26-eV region

The transitions resulting from ejection of $2a_{2u}$ and $2a_{1g}$ electrons are expected in this region and can be confidently assigned to the bands at 20.16 and 23.6 eV, respectively. Vibrational structure is resolved in the former but not the latter. The lack of resolved structure in the latter may be due to the low resolution obtained with He II excitation. The structure observed in the $^2A_{2u}$ band (Fig. 9.11b) consists of vibrational peaks of alternating intensity and spacing. This suggests that there are two interlocking progressions of $\sim 2580\,\text{cm}^{-1}$ displaced by $\sim 1130\,\text{cm}^{-1}$. Totally symmetric vibrational modes are expected to be excited in such a transition to a nondegenerate electronic state in which geometrical changes are expected to be small. The high frequency mode is most likely ν_1, the C—H stretching mode reduced from its ground state value, and the low frequency mode is most likely ν_3, the C—C stretching mode increased from its ground-state value. These changes are expected from the C—H bonding and C—C antibonding properties of the $2a_{2u}$ orbital.

9.7. Jahn–Teller effects in larger molecules

The effects of Jahn–Teller interactions have been observed or implied in the interpretation of the PE spectra of many molecules with degenerate orbitals. Some of these cases are mentioned below.

Cyclopropane

The outer electron configuration[43] of C_3H_6 is $\cdots 3e''^4 1e'^4$, hence ejection of an electron from either one of these outer orbitals produces an ion in a degenerate electronic state that is susceptible to Jahn–Teller distortions. The $1e'$ orbital is mainly of the σ C—C bonding type, while the $3e''$ orbital is of π type and involved only in out-of-plane C—H bonding. Consequently, Jahn–Teller interactions in the $^2E'$ state are expected to be much larger than those in the $^2E''$ state. The transition to the ground state of the ion $^2E' \leftarrow {}^1A_1'$ exhibits two distinct maxima separated by $\sim$ 0.77 eV.[43] The lowest-energy component of these two bands possesses an apparent vibrational progression with a spacing of $\sim 480\,\text{cm}^{-1}$. Only a vibration of e' symmetry is capable of splitting the degeneracy of the $^2E'$ state; the $\nu_{11}(e')$ normal mode is suggested[43] as the most likely candidate. The e' vibration splits the $^2E'$ state into A_1 and B_2 states of C_{2v} symmetry. The transition to the first excited ionic state $^2E'' \leftarrow {}^1A_1$ is very broad, however, it shows little evidence of Jahn–Teller splitting in agreement with theoretical predictions.[44,45]

Cyclobutane

The outer molecular orbital of $C_4H_8(D_{2d})$ is expected to be the degenerate $4e$ σ-type orbital analogous to that of cyclopropane. The transition to the ionic ground state $^2E \leftarrow ^1A_1$ consists of two maxima separated by about 0.6 eV.[46,47] The degeneracy of the ^2E state of the D_{2d} ion can be split by b_1 or b_2 vibrations to produce D_2 or C_{2v} ions, respectively. The two maxima observed in the spectrum indicate that Jahn–Teller forces are active in the ion.

Allene

Ionization of the outer molecular orbital, *2e*, of C_3H_4 produces a band with two maxima separated by ~ 0.6 eV[48] and an apparent vibrational progression of ~ 720 cm^{-1}. The D_{2d} symmetry of this ^2E ground state can be split into D_2 or C_{2v} symmetry by b_1 and b_2 vibrations. Theoretical studies[44] indicate that the most favorable distortion is from D_{2d} to D_2 symmetry and that the active vibration is most likely ν_4 (b_1), the torsional mode of allene with a frequency of 865 cm^{-1}.

Methyl fluoride

The outermost orbital of CH_3F is the degenerate $2e$ orbital; it exhibits the expected Jahn–Teller splitting upon ionization. A complex vibrational structure with spacings of ~ 1050 cm^{-1} and ~ 950 cm^{-1} has been observed in this transition.[49]

9.8. Concomitant Jahn–Teller and spin-orbit interactions

The degenerate states that are subject to Jahn–Teller distortions are also subject to spin-orbit interactions. When both of these interactions occur in an electronic state, usually one is considerably more dominant than the other, and the spectrum can be interpreted by consideration of the combined effects of the the the two interactions. For example, in the series CH_3F, CH_3Cl, CH_3Br, and CH_3I, progressive increases in the magnitude of the spin-orbit parameters of the halogens results in the dominance of spin-orbit splittings in the heavier molecules and the dominance of Jahn–Teller splittings in CH_3F, although both types of interactions are operative throughout the series. Guides for distinguishing between spin-orbit and Jahn–Teller splittings are discussed in Section 8.3.

The combined effects of Jahn–Teller and spin-orbit interactions are amply demonstrated by the series of Group IV tetramethyl compounds, $C(CH_3)_4$, $Si(CH_3)_4$, $Ge(CH_3)_4$, $Sn(CH_3)_4$, and $Pb(CH_4)_4$.[50,51] Semiempirical calculations for these molecules are in agreement that the outermost

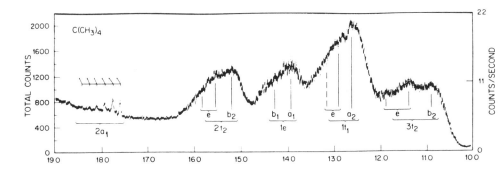

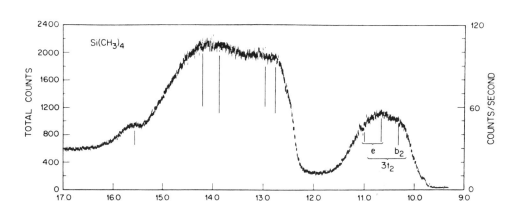

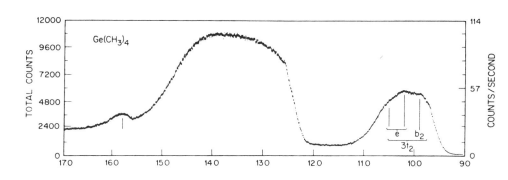

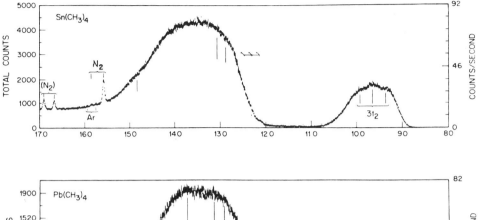

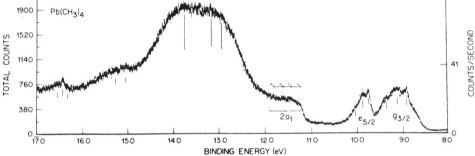

Fig. 12. The He I spectra of the Group IV tetramethyl series $C(CH_3)_4$, $Si(CH_3)_4$, $Ge(CH_3)_4$, $Sn(CH_3)_4$, and $Pb(CH_3)_4$. Reproduced with permission from Ref. 50.

orbital is the triply degenerate $3t_2$ MO. The spectra[50] of these Group IV tetramethyl compounds are presented in Fig. 9.12. The structure observed in the outermost band of the C, Si, Ge, and Sn compounds is easily interpreted in terms of Jahn–Teller interactions. Jahn–Teller forces split the $^2T_2$ state into E and B_2 states, with the E state exhibiting a further splitting into two components. This results in three closely spaced electronic states and produces an ionization band with three maxima as in the case of methane (Section 9.4).

The spectrum of $Pb(CH_3)_4$ is considerably different from the other members of the series; the low-energy region consists of two bands with $\sim 2:1$ intensity distribution whose separation is much larger than that expected from Jahn–Teller splitting. It appears that spin-orbit interaction has become dominant as a result of the large spin-orbit parameter of the Pb atom and must be considered in the interpretation of the $Pb(CH_3)_4$ spectrum. Following the treatment of Carlson et al.,[50] a $^2T_2$ state in the T_d point group can undergo spin-orbit interaction yielding two new states, the twofold degenerate $^2E_{5/2}$ state and the fourfold degenerate $^2G_{3/2}$ state, where the symbolism is now that of *spinor* groups. If the molecule during the course of its vibrations distorts to the D_{2d} point group, then the symmetry of the twofold degenerate state changes to $^2E_{3/2}$ and the fourfold degenerate state can undergo Jahn–Teller splitting to $^2E_{3/2}$ and $^2E_{1/2}$ states. From a consideration of the relative areas of the two low-energy bands, it seems reasonable that the one centered at 9.10 eV corresponds to the transition to the fourfold degenerate $^2G_{3/2}$ state, while the one centered at 9.75 eV corresponds to the transition to the twofold degenerate $^2E_{5/2}$ state.

Vibronic structure resulting from spin-orbit and Jahn–Teller effects is difficult to interpret because the electronic states are subject to several types of perturbations, all of which tend to produce rather complicated vibronic patterns. In the case of $Pb(CH_3)_4$ both transitions in question have a low-energy shoulder and high-energy structure that could result from a mixing of the vibronic eigenfunctions of the two $^2E_{3/2}$ levels such that the higher vibronic levels of both states are mutually enhanced. This implies that the $^2G_{3/2}$ *spinor* state is undergoing Jahn–Teller splitting resulting in the formation of a low-energy $^2E_{3/2}$ state and a higher-energy $^2E_{1/2}$ state. Since both of these states have twofold orbital degeneracy, the $^2G_{3/2}$ band can be assumed to result from the overlap of these two symmetric subbands and, in that way, obtain an estimate of the magnitude of the Jahn–Teller effect as one-half the total band width at half-maximum height; this amounts to ~ 0.4 eV. Since Jahn–Teller splitting occurs about a center of symmetry, the magnitude of the spin-orbit splitting can be estimated by comparing the midpoint energy of the $^2G_{3/2}$ band with the vertical transition energy of the $^2E_{5/2}$ state; this amounts to ~ 0.65 eV. Figure 9.13 represents a one-dimensional cross section through the potential function of a $^2T_2$ state that is under the influence of both Jahn-Teller and spin-orbit interactions. The potentials change as a function of a bending distortion that changes the molecular symmetry from T_d^* to D_{2d}^*, where the asterisk represents the spinor description. Part (*b*), representing large spin-orbit and small Jahn–Teller effects, seems to correspond best to the situation in $Pb(CH_3)_4^+$.

9.9. Example: The Renner effect in H_2O^+

The energy levels and structure of H_2O^+ are important from the standpoint of ionic structure and from that of astrophysics, for it has recently been observed in emission from the comet Kohoutek.[52] The emission spectra of H_2O^+ and D_2O^+ have been studied in the laboratory by Lew and Heiber,[53] although emission has not been observed from the lowest

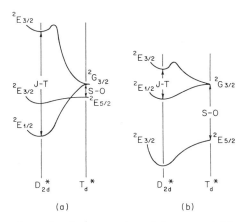

Fig. 9.13.   One-dimensional cross section of the potential functions of a $^2T_2$ state which is under the influence of both Jahn–Teller and spin-orbit interactions. (*a*) Large Jahn–Teller and small spin-orbit effects. (*b*) Large spin-orbit and small Jahn–Teller effects. Reproduced with permission from Ref. 50.

vibronic levels of the $^2A_1$ state nor from any of the levels of the $^2B_2$ state. In order to understand the geometry and properties of the ion in these states, it is therefore necessary to analyze the high resolution PE spectra.

The results of several *ab initio* SCF-MO calculations have established the molecular orbital structure of water as $1a_1^2\ 2a_1^2\ 1b_2^2\ 3a_1^2\ 1b_1^2$ as described in Figs. 1.2 and 1.3. The computational results indicate that the three valence orbitals that are accessible by He I radiation have the following characteristics: the $1b_1$ orbital consists of a nonbonding $2p_x$ AO localized on the oxygen atom; the $3a_1$ orbital, which is strongly H—H bonding, is a combination of H_{1s} and O_{2p_z} AOs; and the $1b_2$ orbital is a combination of O_{2p}, and H_{1s} AOs arranged in an H—H antibonding and O—H bonding configuration. The early PE spectra of water[32,54] (Fig. 1.2) were not of sufficiently high resolution to resolve the vibronic structure. The high-resolution PE spectra of H_2O, HDO, and D_2O are presented in this section, and analysis of the bands is discussed.[55] The analysis of the transitions to the ground state $^2B_1$ and the first excited state $^2A_1$ of the ion indicate that these states are derived from a $^2\Pi_u$ state of the linear ion by means of a strong Renner effect. Discussion of this Renner effect requires a detailed study of the fine structure of the bands. We proceed in this vein.

The ionic ground state, $^2B_1$

The photoelectron transition to the ground state of the ion $\cdots\ 1b_1^1$, $^2B_1 \leftarrow \cdots\ 1b_1^2,\ ^1A_1$, is shown in Figs. 9.14 and 9.15 for H_2O and D_2O.

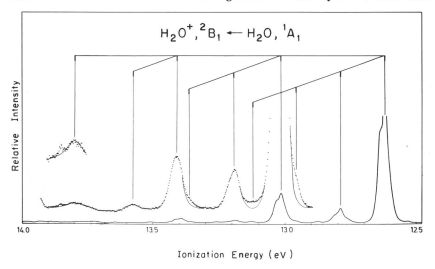

$$H_2O^+, {}^2B_1 \leftarrow H_2O, {}^1A_1$$

Relative Intensity

Ionization Energy (eV)

Fig. 9.14. The Photoelectron transition to the ground state of H_2O^+, $\cdots 1b^1$, ${}^2B_1 \leftarrow \cdots 1b_1^2$, ${}^1A_1$, produced by He I (584 Å) radiation. In the upper spectrum the contribution from the He 537-Å line is subtracted point by point. The small peak at 13.85 eV is interpreted as the first menber of the second band system. Reproduced with permission from Ref. 55.

The ionization energies and vibrational assignments of the observed bands are listed in Table 9.4. The vibrational analysis of this band is exceedingly straightforward since the vibrational frequencies are very similar to those of the molecular ground state. In the C_{2v} molecules only the symmetric vibrational modes, ν_1 and ν_2, are active, while in the C_s molecule, all three modes are symmetric and active. The overwhelming intensity of the (0, 0, 0) band and the similarity of the vibrational frequencies of the ion to those in the molecular ground state and Rydberg states leading to this ionic state indicate that the electron is expelled from an essentially nonbonding orbital of the oxygen atom. An asymmetry in the band shapes is recognized upon close observation of the vibrational bands, especially those of H_2O. This asymmetry is due to partially resolved rotational structure. The (0, 0, 0) band of the ${}^2B_1 \leftarrow {}^1A_1$ transition of H_2O is compared[56] with the first member of the $\cdots 2b_1^1 \, 3a_1^1$, ${}^1B_1 \leftarrow 2b_1^2$, ${}^1A_1$ Rydberg series reported by Johns[57] in Fig. 9.16. The similarity between the rotational structure observed by Johns and the contour of the (0, 0, 0) photoelectron band indicates that the molecule in the Rydberg state is very similar to the positive ion and that the rotational selection rules are similar (Section 3.10) in both spectra.

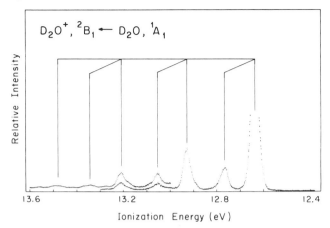

Fig. 9.15 The photoelectron transition to the ground state of $D_2O^+, \cdots 1b_1^1$, $^1B_1 \leftarrow \cdots 1b_1^2, {}^1A_1$, produced by He I radiation. Reproduced with permission from Ref. 55.

The peak intensities of Table 9.4 are determined from the relative areas of the bands for the various vibrational states of the ion. A Franck–Condon analysis, performed according to the method of Coon et al.,[58] allows one to determine the difference in geometry between the $^1A_1$ state of H_2O and the $^2B_1$ state of H_2O^+. In order to use this method, the L matrices relating the internal coordinates to the normal coordinates must be calculated. Computing the L'' matrix for the $^1A_1$ state and the L'

Table 9.4. Vibrational Analysis of the Bands in the $^2B_1 \leftarrow {}^1A_1$ Transition of H_2O and D_2O^a

| | H_2O | | D_2O | |
$V_1 V_2 V_3$	Relative Intensity	Peak Position (eV)	Relative Intensity	Peak Position (eV)
0 0 0	1000	12.624	1000	12.637
0 1 0	118.0	12.797	150.0	12.768
0 2 0	16.1	12.960	—	—
1 0 0	239.0	13.021	224.0	12.928
0 3 0	3.6	13.121	—	—
1 1 0	25.7	13.191	25.0	13.053
1 2 0	2.9	13.354	—	—
2 0 0	33.9	13.406	41.0	13.210
2 1 0	6.4	13.567	6.0	13.339
3 0 0	8.9	13.784	4.0	13.486

^{a} Reprinted with permission from Ref. 55.

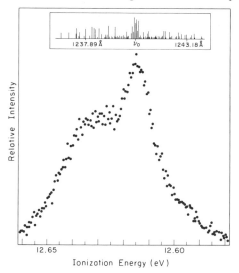

Fig. 9.16.   The $(0, 0, 0) \leftarrow (0, 0, 0)$ transition of the $\cdots 1b_1^1, \, ^2B_1 \leftarrow \cdots 1b_1^2, \, ^1A_1$ photoelectron transition excited with Ne I radiation. The asymmetry of the band is due to partially resolved rotational structure. The upper inset shows the rotational structure in the 1240-Å Rydberg band of H_2O as calculated by Johns.[57] Reproduced with permission from Ref. 56.

matrix for the $^2B_1$ state yields the results

$$\Delta r_{O-H} = +0.08 \text{ Å}, \quad \Delta \text{HOH} = 4.4°$$

or

$$\Delta r_{O-H} = -0.10 \text{ Å}, \quad \Delta \text{HOH} = 5.9°$$

or the negatives of either of these. The first of these models is chosen, as the sign of both bond lengths and bond angle changes in this model are in agreement with those derived from rotational constants.[53] Use of the ground-state zero-point values of HOH = 105.0° and $r_{O-H} = 0.956$ Å leads to estimates of HOH = 109.4° and $r_{O-H} = 1.03$Å in the $^2B_1$ state. These values may be compared with the values of 110.5° and 0.999 Å obtained from rotational analysis of the emission spectrum.[53] The small change in bond lengths and bond angle upon ionization indicates that the $1b_1$ electrons are largely nonbonding.

The first excited ionic state, $^2A_1$

The photoelectron transition to the first excited state of the ion, $\cdots 3a_1^1$ $1b_1^2, \, ^2A_1 \leftarrow \cdots 3a_1^2 \, 1b_1^2, \, ^1A_1$, is shown in Figs. 9.17 to 9.19 for H_2O,

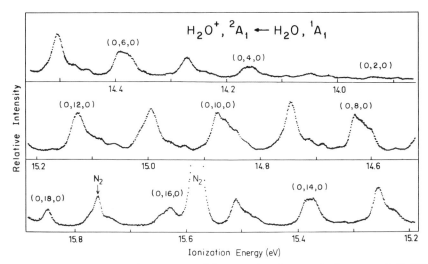

Fig. 9.17. The photoelectron transition to the first excited state of H_2O^+, $\cdots 3a_1^1 \, 1b_1^2, \, ^2A_1 \leftarrow \cdots 3a_1^2 \, 1b_1^2, \, ^1A_1$. Reproduced with permission from Ref. 55.

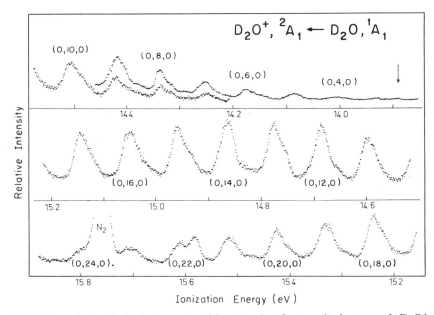

Fig. 9.18. The photoelectron transition to the first excited state of D_2O^+, $\cdots 3a_1^1 \, 1b_1^2, \, ^2A_1 \leftarrow \cdots 3a_1^2 \, 1b_1^2, \, ^1A_1$, showing some of the observed vibronic bands. The arrow indicates Ar ionized by 537-Å radiation. Reproduced with permission from Ref. 55.

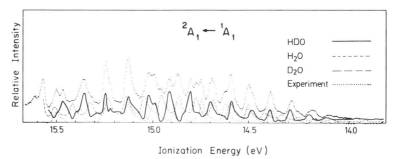

Fig. 9.19. The photoelectron transition to the first excited state of HDO^+, $\cdots 4a''^1 1a''^2$, $^2A' \leftarrow \cdots 4a'^2 1a''^2$, $^1A'$, obtained by irradiating a mixture of 50% H_2O and 50% D_2O with Ne 584-Å radiation. Reproduced with permission from Ref. 55.

D_2O, and HDO. The energy levels and assignments of these transitions for H_2O, D_2O, and HDO are listed in Table 9.5. The vibrational structure of the H_2O and D_2O bands appears to consist of a single progression in the bending mode, ν_2. In HDO this band system has a more complex structure, which is interpreted in terms of the excitation of ν_1 as well as ν_2.

ISOTOPE SHIFTS. The isotope shifts of the vibrational bands in the H_2O and D_2O spectra are shown in Fig. 9.20 plotted against the vibrational energy of the H_2O bands. The points form a straight line with a slope of 0.289. The value of the slope is very important, since from this number it is possible to deduce that the equilibrium configuration in the $^2A_1$ ionic state is very close to linear or perhaps quasilinear. Assuming that the strong lines in the spectrum originate in the (0, 0, 0) vibrational level of the ground state and neglecting the rotational contribution to the isotope shift, the change in the isotope shift throughout the band system is due primarily to the change in the vibrational quantum numbers of the ionic state. The principal factors determining the ratio of the isotope shift, $\Delta E(H_2O{-}D_2O)$, to the vibrational frequency, E_ν, in the ionic state are (1) the nuclear masses, (2) the HOH angle, and (3) the particular vibration excited, that is, ν_1, ν_2, or ν_3. The value of $\Delta E/E_\nu$ will also be dependent on the force constants of the ionic state, but only in a secondary manner. For the purpose of calculating $\Delta E/E_\nu$, the stretching and bending force constants for the ground state of the water molecule were used.[59] The calculated values are listed in Table 9.6 as a function of the bond angle and the particular vibration involved. The observed value of $\Delta E/E_\nu$ from Fig. 9.20 can only be explained if it is assumed that the ion is linear, or

Table 9.5. Vibrational Assignments, Vibronic Subbands, and Energies for the $^2A_1 \leftarrow {}^1A_1$ Transition (eV)

v_2	Subband	H_2O (PE spectrum)	H_2O (Emission spectrum[a])	D_2O (PE spectrum)	HDO (PE spectrum)
1	Σ	13.838	—	—	—
2	Π	13.933	—	—	—
3	$\Sigma\Delta$	14.050	—	13.933	—
4	Π	14.168	—	14.007	14.09
5	$\Sigma\Delta$	14.271	14.267	14.089	14.18
6	Π	14.385	14.385	14.180	14.385
7	$\Sigma\Delta$	14.504	14.501	14.259	14.48
8	Π	14.629	14.623	14.345	14.585
9	$\Sigma\Delta$	14.745	14.745	14.426	14.695
10	Π	14.875	14.869	14.507	14.80
11	$\Sigma\Delta$	14.997	14.997	14.596	14.91
12	Π	15.127	—	14.686	—
13	$\Sigma\Delta$	15.253	15.252	14.775	—
14	Π	15.381	—	14.864	—
15	$\Sigma\Delta$	15.510	15.510	14.959	—
16	Π	15.634	—	15.049	—
17	$\Sigma\Delta$	15.766	—	15.143	—
18	Π	15.897	—	15.239	—
19	$\Sigma\Delta$	16.034	—	15.331	—
20	Π	16.159	—	15.421	—
21	$\Sigma\Delta$	16.295	—	15.517	—
22	Π	—	—	15.610	—
23	$\Sigma\Delta$	—	—	15.710	—
24	Π	—	—	15.803	—
25	$\Sigma\Delta$	—	—	15.902	—
26	Π	—	—	15.995	—
27	$\Sigma\Delta$	—	—	16.095	—

[a] From Ref. 53.

nearly linear, in the $^2A_1$ state and that the spectrum involves a progression of v_2. v_3 can be ruled out simply on the basis of it being an antisymmetric mode. This information leads to the conclusion that the origin of the band is near 13.8 eV, that the spectrum consists principally of a long progression in v_2, and that the ion is nearly linear or quasilinear in the $^2A_1$ state.

VIBRONIC STRUCTURE. Linear triatomic molecules have a doubly degenerate vibrational mode that produces an angular momentum, $lh/2\pi$, about the internuclear axis. If v_2 quanta of the bending vibration are excited, the quantum number l takes the values $v_2, v_2-2, v_2-4, \cdots, 1$ or 0 depending on whether v_2 is odd or even. For a $D_{\infty h}$ molecule in a

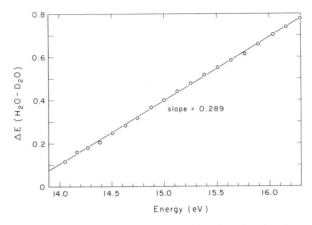

Fig. 9.20. H_2O–D_2O isotope shifts plotted against the vibrational energies of the H_2O bands for the $^2A_1 \leftarrow ^1A_1$ transition.

symmetric electronic state, levels with $l = 0, 1, 2, 3, \cdots$ correspond to Σ_g^+, Π_u, Φ_u, $\cdots$ vibronic states. The vibronic symmetries of the levels alternate between even and odd l values for successive quanta of the bending vibration. Levels with the same value of v_2 but different values of l have the same energy in an harmonic force field. A splitting of the l degeneracy is produced by higher-order terms in the potential function. The nature of the splitting has been observed to be quadratic in those molecules where the effects of l-type splitting have been studied experimentally.[3]

Relating the above to the water spectrum requires careful analysis of the fine structure associated with the transition. The low-resolution spectrum shows that the vibrational bands in the $^2A_1 \leftarrow ^1A_1$ transition have alternating widths, that is, alternating bands are broad or sharp. The high resolution spectrum shows that these vibrational bands are actually composed of several components with widely differing intensities, as

Table 9.6. Calculated Isotope Factors ($\Delta E/E_\nu$) for H_2O and D_2O

Apex Angle	$\Delta E_{\nu_1}/E_{\nu_1}$	$\Delta E_{\nu_2}/E_{\nu_1}$	$\Delta E_{\nu_3}/E_{\nu_3}$
180°	0.000	0.276	0.117
150°	0.025	0.251	0.118
120°	0.075	0.201	0.120
90°	0.118	0.158	0.123

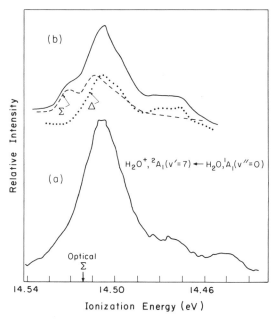

Fig. 9.21. Observed (a) and calculated (b) contours for the $v' = 7 \leftarrow v'' = 0$ band of the $^2A_1 \leftarrow {}^1A_1$ transition of H_2O. The selection rules employed are $\Delta K = 0$, ± 1, the line halfwidth is 50 cm^{-1}, the temperature is taken to be 300°K, and the Σ–Δ separation is 60 cm^{-1}. The last peak in the observed contour is probably due to the Γ subband. Reproduced with permission from Ref. 55.

illustrated in Figs. 9.21a and 9.22a. The assignments of the bands with a strong peak at the high energy side to a combination of Σ, Δ, $\cdots$ subbands, and the more flat topped bands to a combination of Π, Φ, $\cdots$ subbands, are made by comparison with the emission spectrum[53] and by the synthetic contours for Σ, Δ and Π, Φ shown in Figs. 9.21b and 9.22b, respectively. The contributions from the various subbands are shown by the dashed lines. There is reasonable agreement between the calculated contours, and also between the Σ–Δ separation obtained by contour matching with the emission spectrum. The dashed line curves show that when the rotational quantum number $K' > 2$, the $\Delta K = +1$ subband is fairly sharp, whereas the $\Delta K = -1$ subband is weaker and broader. This suggests that the selection rules of Dixon et al.[60] (Section 3.10) are obeyed for this band. The calculated subband contours were obtained by assuming equal transition probabilities for $\Delta K = 0$ and $\Delta K = \pm 1$. The discrepancies in the contours indicate that this intensity ratio is probably incorrect.

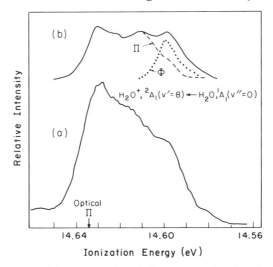

Fig. 9.22.   Observed (a) and calculated (b) contours for the $v' = 8 \leftarrow v'' = 0$ band of the $^2A_1 \leftarrow {}^1A_1$ transition of H_2O. The selection rules employed are $\Delta K = 0$, ± 1, the line halfwidth is 60 cm^{-1}, the temperature is taken to be 300°K, and the Π–Φ separation is 113 cm^{-1}. Reproduced with permission from Ref. 55.

THE EQUILIBRIUM CONFIGURATION OF THE $^2A_1$ ION. The observation of a long progression in ν_2 with a reduction in frequency from its ground-state value indicates that an electron has been removed from a strongly bonding orbital and that the major dimensional change upon ionization must be in the bond angle. This is also consistent with the strong H—H bonding character of the $3a_1$ orbital. By use of the Franck–Condon principle, Brundle and Turner[54] and Potts and Price[32] concluded that the $^2A_1$ state of H_2O^+ is probably very nearly linear. The subbands observed by Lew and Heiber[53] resemble those of a bent to linear transition; however, the lowest vibrational level observed in this study was $v_2' = 5$. It is necessary to find a criterion to determine whether the ion is linear in the first few vibrational levels or if a small potential hump exists.

Dixon[61] has shown that for quasilinear triatomics with small potential humps at the linear configuration, the vibrational intervals decrease uniformly below the potential maximum and then begin to increase when the molecule has sufficient energy to overcome the barrier to linearity. Since the $^2B_1$ and $^2A_1$ states are components of a $^2\Pi_u$ state with a large Renner (electrostatic) splitting, then the only levels of the upper state that are determined entirely from the form of the $^2A_1$ potential curve are the Σ levels. All other levels are subject both to shifts and to possible

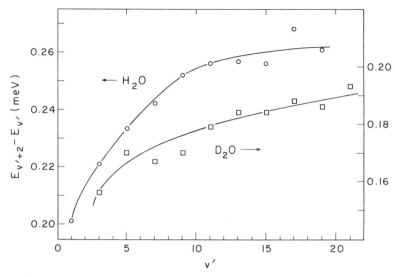

Fig. 9.23. The variation with vibrational quantum number of the observed Σ vibrational spacings in the $^2A_1$ band of H_2O^+ and D_2O^+.

resonances with higher levels of the $^2B_1$ state. The Σ levels are, therefore, of crucial importance in determining whether or not a small potential hump exists at the linear configuration of the $^2A_1$ state. Assuming that the higher energy (stronger) part of each Σ, Δ pair can be taken as the Σ energy, the intervals between the Σ levels can be plotted against the vibrational quantum number as in Fig. 9.23. The intervals for H_2O^+ and D_2O^+ increase with increasing v_2' beginning with the first intervals observed. This implies that there is no hump at the bottom of the $^2A_1$ potential well.

The bending potential function for the $^2A_1$ state can be calculated[55] by using

$$V(\theta) = \frac{Hf_m(\theta^2 - \theta_m^2)^2}{[f_m\theta_m^4 + (8H - f_m\theta_m^2)\theta^2]} \tag{9.18}$$

when the potential has a hump or a flat bottom and

$$V(\theta) = f_h\theta^2 + f_q\theta^4 \tag{9.19}$$

when the potential corresponds to an anharmonic linear state. In these equations H is the barrier height, f_m is the force constant at the potential minimum, and θ_m is the equilibrium angle. A Hamiltonian describing large vibrational amplitudes as defined by Barrow et al.[62] is appropriate

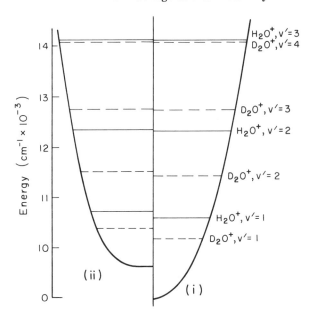

Fig. 9.24. Calclulated potential curves and Σ levels for the $^2A_1$ state of H_2O^+ and D_2O^+ using (i) an almost harmonic minimum and (ii) a potential hump of $1\ cm^{-1}$ at the linear configuration. Solid lines represent H_2O levels, and dashed lines represent D_2O levels. Either of these potentials can be used to fit the experimental data. Reproduced with permission from Ref. 55.

for calculating the Σ vibrational levels. The lowest Σ vibrational levels of the $^2A_1$ state have been calculated[55] by using three different potentials, (1) a harmonic potential perturbed by a small quartic term, (2) a barrier of $1\ cm^{-1}$ at the linear configuration (essentially a flat minimum), and (3) a barrier of $300\ cm^{-1}$ at the linear configuration. Model (1) gives the best agreement with the experimental vibrational levels, while model (3) gives the worse agreement. The experimental Σ vibrational levels and those levels calculated from models (1) and (2) are shown in Fig. 9.24. The agreement obtained with experiment suggests that the true potential shape is between that of models (1) and (2), that is, a very anharmonic potential with a linear equilibrium configuration.

From Fig. 9.23 it is noted that large deviations from the asymptote occur near $v_2' = 15$ and 17 in H_2O^+. This is probably caused by Fermi resonance between vibrational levels of the same symmetry. Since $\nu_1 \sim \nu_3 \sim 3\nu_2$, possibilities for the interacting levels are $(2, v_2, 0)$ with $(0, v_2+6, 0)$ or $(0, v_2, 2)$ with $(0, v_2+6, 0)$. Similar effects are observed for D_2O.

THE RENNER EFFECT. It has been shown[2,3] that, for linear molecules in degenerate electronic states, a strong coupling exists between the vibrational angular momentum, $lh/2\pi$, of the bending vibration, v_2, and the electronic orbital angular momentum, $\Lambda h/2\pi$. As a result of this coupling, l and Λ are no longer good quantum numbers, but $K = |l \pm \Lambda|$ is a good quantum number for all degrees of interaction. For a given value of v_2, the possible values of K are $v + \Lambda$, $v + \Lambda - 2$, $v + \Lambda - 4$, $\cdots$, 1 or 0 depending on whether $v + \Lambda$ is odd or even. Because of this interaction, the degenerate electronic state is split into two components in the nonlinear conformation. The Born–Oppenheimer approximation is no longer valid here, for the vibronic wavefunctions are dependent upon the potential functions of both component states and are essentially inseparable. Three distinct cases of the Renner effect are possible corresponding to various shapes of the potential curves: (1) both curves have a minimum at the linear configuration, (2) the upper curve has a minimum and the lower curve has a maximum at the linear configuration, and (3) both curves have a maximum in the linear configuration.

In the case of H_2O^+ the first excited ionic state, $^2A_1$, and the ground ionic state, $^2B_1$, are assigned to the upper and lower components, respectively, of the split $^2\Pi_u$ state of the linear configuration. It has been shown that the $^2A_1$ state has a linear equilibrium configuration, thus corresponding to case (2) of the Renner effect. The energy of the lowest stable vibronic level of the $^2A_1$ state is unusual. Dressler and Ramsay[3] have presented a detailed correlation between the levels of a Π_u state and its component B_1 and A_1 states. Renner[2] has shown that the energies of the Σ vibronic levels associated with the upper potential well remain finite, while the energies of the Σ vibronic levels associated with the lower potential curve approach zero. Thus in correlating the levels between these states, the total angular momentum about the axis of the linear molecule correlates with the rotational angular momentum about the figure axis of the bent molecule. The lowest vibronic level of the $^2\Pi_u$ state with $v_2 = 0$ correlates with the $K = 0$, $v_2 = 0$ level of the $^2B_1$ state. The lowest stable levels of the $^2A_1$ state with even values of K are assigned odd vibrational quantum numbers and vice versa. Thus the lowest level of the $^2A_1$ state corresponds to $v_2 = 1$ and has an energy contribution from two quanta of the bending vibration.

The second excited ionic state, $^2B_2$

The photoelectron transition to the second excited ionic state of H_2O^+, $\cdots 1b_2^1\ 3a_1^2\ 1b_1^2,\ ^2B_2 \leftarrow \cdots 1b_2^2\ 3a_1^2\ 1b_1^2,\ ^1A_1$, is shown in Fig. 9.25. The origin of this band system has been the subject of some speculation,[32,54]

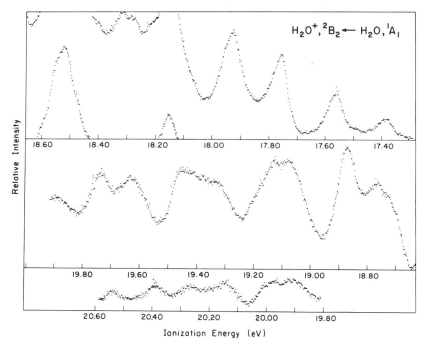

Fig. 9.25. The photoelectron transition to the second excited state of H_2O^+, $\cdots 1b_2^1 3a_1^2 1b_1^2$, $^2B_2 \leftarrow \cdots 1b_2^2 3a_1^1 1b_1^2$, $^1A_1$. Reproduced with permission from Ref. 55.

but subtraction of the contribution of spectra arising from 537 Å radiation has shown that the origin is probably at 17.301 eV, as deduced by Bergmark et al.[63] The pattern stems mainly from progressions in ν_1 and ν_2, the symmetric stretching and bending vibrations. The spectrum of H_2O^+ consists of progressions of progressions, that is, $(v, 0, 0)$, $(v, 1, 0)$, $(v, 2, 0)$, $\cdots$. Comparisons of the height of the first three peaks leads to the conclusion that the relative change in the normal coordinate is approximately the same for the bending and stretching motions. Using the method of Coon et al.[58], it is found that $\Delta r_{O-H} \sim 0.2$ Å and $\Delta HOH \sim 18°$. Since the $1b_2$ orbital is O—H bonding but H—H antibonding, the removal of a $1b_2$ electron should lead to an increase in bond length and a decrease in bond angle, leading to an estimate of the geometry of the $^2B_2$ state of the ion of $r_{OH} \sim 1.16$ Å and HOH $\approx 86°$. In the higher homologs there is progressively less angle change, so that in H_2Te the progression in ν_2 is not observed.

Since the $^2B_2$ state is produced by the removal of a strongly bonding electron, both the stretching and the bending vibrations are probably rather anharmonic. This causes the different superimposed progressions to become out of step, leading to the marked increase in complexity in the region of 18.6 eV. It is also likely that, as in water itself, the ν_3 vibrations may be excited owing to stretch-stretch and stretch-bend interaction caused by a rather anharmonic potential. Unfortunately, the resolution in this region is such that it is difficult to give an unambiguous analysis.

Conclusions about the ionic states of water

A qualitative configuration-coordinate diagram of the ground state of H_2O and the three lowest ionic states of H_2O^+ is shown in Fig. 9.26. The most probable transition is indicated by the solid vertical line that corresponds to the Franck–Condon maximum. The classical Franck–Condon turning points are indicated by two dashed vertical lines. The ionic potential surfaces were drawn by considering the energy of the observed Franck-Condon maxima and turning points along with the calculated equilibrium angles. The $^2B_1$ and $^2A_1$ states coalesce into a

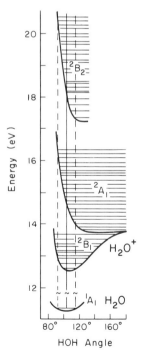

Fig. 9.26.   A qualitative configuration-coordinate diagram of the ground state of H_2O and the three lowest energy states of H_2O^+.

degenerate $^2\Pi_u$ state in the linear configuration. The observed vibrational levels in all three of the ionic states are inserted in the diagram.

We conclude that characteristic rotational envelopes can be observed in the PE spectra of polyatomic molecules and assigned to particular combinations of vibronic subbands. The selection rules necessary for the production of synthetic contours agree with the theory for the selection rules in photoelectron transitions given by Dixon et al. (Section 3.10).

References

1. H. A. Jahn and E. Teller, *Proc. Roy. Soc., Ser. A*, **161**, 220 (1937).
2. G. Herzberg and E. Teller, *Z. Phys. Chem.*, **B21**, 410 (1933); R. Renner, *Z. Phys.* **92**, 172 (1934).
3. K. Dressler and D. A. Ramsay, *Phil. Trans. Roy. Soc. London*, **A251**, 553 (1959).
4. J. A. Pople and H. C. Longuet–Higgins, *Mol. Phys.*, **1**, 372 (1958).
5. R. N. Dixon, *Trans. Faraday Soc.*, **60**, 1363 (1964); ibid., *Mol. Phys.*, **9**, 357 (1965).
6. H. Sponer and E. Teller, *Rev. Mod. Phys.*, **13**, 75 (1941).
7. W. Moffitt and A. D. Liehr, *Phys. Rev.*, **106**, 1195 (1957); E. I. Blount, *J. Math. Phys.*, **12**, 1890 (1971).
8. J. T. Hougen, *J. Mol. Spectrosc.*, **13**, 149 (1964).
9. J. H. Van Vleck, *J. Chem. Phys.*, **7**, 72 (1939).
10. W. Moffitt and W. Thorson, *Phys. Rev.*, **108**, 1251 (1957).
11. A. D. Liehr, *J. Phys. Chem.*, **67**, 389 (1963).
12. H. A. Jahn, *Proc. Roy. Soc., Ser. A*, **164**, 117 (1938).
13. R. Englman, *The Jahn–Teller Effect in Molecules and Crystals*, Wiley, New York, 1972.
14. G. Herzberg, *Electronic Spectra and Electronic Structure of Polyatomic Molecules*, Van Nostrand, Princeton, N.J., 1967.
15. R. S. Mulliken and E. Teller, *Phys. Rev.*, **61**, 283 (1942).
16. R. W. Jotham and S. F. A. Kettle, *Inorg. Chim. Acta*, **5**, 183 (1971).
17. R. N. Dixon, *Mol. Phys.*, **20**, 113 (1971).
18. F. A. Grimm and J. Godoy, *Chem. Phys. Letters*, **6**, 336 (1970).
19. J. Arents and L. C. Allen. *J. Chem. Phys.*, **53**, 73 (1970).
20. C. A. Coulson and H. L. Strauss, *Proc. Roy. Soc., Ser. A*, **269**, 443 (1962).
21. R. G. Pearson, *J. Amer. Chem. Soc.*, **91**, 4947 (1969).
22. G. S. Handler and H. W. Joy, *Int. J. Quantum Chem.*, **3S**, 529 (1970).
23. J. W. Rabalais, T. Bergmark, L. O. Werme, L. Karlsson, and K. Siegbahn, *Phys. Scripta*, **3**, 13 (1971).
24. A. W. Potts and W. C. Price, *Proc. Roy. Soc., Ser. A*, **326**, 165 (1972).
25. B. P. Pullen, T. A. Carlson, W. E. Moddeman, G. K. Schweitzer, W. E. Bull, and F. A. Grimm, *J. Chem. Phys.*, **53**, 768 (1970).

26. D. W. Turner, C. Baker, A. D. Baker, C. R. Brundle, *Molecular Photoelectron Spectroscopy*, Wiley-Interscience, New York, 1970.
27. A. W. Potts, H. J. Lempka, D. G. Streets, and W. C. Price, *Phil. Trans. Roy. Soc. London*, **A268**, 59 (1970).
28. A. D. Baker, C. Baker, C. R. Brundle, and D. W. Turner, *Int. J. Mass Spectrom. Ion Phys.*, **1**, 285 (1968).
29. A. W. Potts, T. A. Williams, and W. C. Price, *Faraday Discussions Chem. Soc.*, **54**, 104 (1972).
30. M. D. Sturge, *Solid State Phys.*, **20**, 92 (1967).
31. J. W. Rabalais, L. Karlsson, L. O. Werme, T. Bergmark, and K. Siegbahn, *J. Chem. Phys.*, **58**, 3370 (1973).
32. A. W. Potts and W. C. Price, *Proc. Roy. Soc., Ser. A*, **326**, 181 (1972).
33. M. J. Weiss and G. M. Lawrence, *J. Chem. Phys.*, **53**, 214 (1970).
34. G. R. Branton, D. C. Frost, F. G. Herring, C. A. McDowell, and I. A. Stenhouse, *Chem. Phys. Letters*, **3**, 581 (1969).
35. V. H. Dibeler, A. Walker, and H. M. Rosenstock, *J. Res. Nat. Bur. Stand. A*, **70**, 459 (1966).
36. H. C. Longuet–Higgens, U. Öpik, M. H. L. Pryce, and R. A. Sack, *Proc. Roy. Soc., Ser. A*, **244**, 1 (1958).
37. J. C. Slonczewski, *Phys. Rev.*, **131** 1596 (1963).
38. W. A. Lathan, L. A. Curtiss, and J. A. Pople, *Mol. Phys.*, **22**, 1081 (1971).
39. J. N. Murrell and W. Schmidt, *J. C. S. Faraday Trans.*, **68**, 1709 (1972).
40. E. Clementi and H. Popkie, *J. Chem. Phys.*, **49**, 2382 (1968).
41. J. W. Rabalais and A. Katrib, *Mol. Phys.*, **27**, 923 (1974).
42. B. Narayan, *Mol. Phys.*, **23**, 281 (1972).
43. H. Basch, M. B. Robin, N. A. Kuebler, C. Baker, and D. W. Turner, *J. Chem. Phys.*, **51**, 52 (1969).
44. E. Haselbach, *Chem. Phys. Letters*, **7**, 428 (1970).
45. C. G. Rowland, *Chem. Phys. Letters*, **9**, 169 (1971).
46. P. Bischof, E. Hasselbach, and E. Heilbronner, *Angew. Chem. (Int. Edit.)*, **9**, 953 (1970).
47. D. W. Turner, *Proc. Roy. Soc., Ser. A*, **307**, 15 (1968).
48. C. Baker and D. W. Turner, *Chem. Commun.*, 480 (1969); R. K. Thomas and H. Thompson, *Proc. Roy. Soc., Ser. A*, **339**, 29 (1974).
49. C. R. Brundle, M. B. Robin, and H. Basch, *J. Chem. Phys.*, **53**, 2196 (1970).
50. A. E. Jonas, G. K. Schweitzer, F. A. Grimm, and T. A. Carlson, *J. Electron Spectrosc.*, **1**, 29 (1972/73).
51. S. Evans, J. C. Green, P. J. Joachim, A. F. Orchard, D. W. Turner, and J. P. Maier, *J. C. S. Faraday II*, **68**, 905 (1972).
52. P. A. Wehinger, S. Wyckoff, C. H. Herbig, G. Herzberg, and H. Lew, *Astrophys. J.*, **190**, L43 (1974).
53. H. Lew and J. Heiber, *J. Chem. Phys.*, **58**, 1246 (1973).
54. C. R. Brundle and D. W. Turner, *Proc. Roy. Soc., Ser. A*, **307**, 27 (1968).
55. R. N. Dixon, G. Duxbury, J. W. Rabalais, and L. Åsbrink, *Mol. Phys.*, **31**, 423 (1976).

56. L. Åsbrink and J. W. Rabalais, *Chem. Phys. Letters*, **12**, 182 (1971).
57. J. W. C. Johns, *Can. J. Phys.*, **41**, 209 (1963).
58. J. B. Coon, R. E. deWames, and C. M. Loyd, *J. Mol. Spectrosc.*, **8**, 285 (1962).
59. G. Herzberg, *Infrared and Raman Spectra of Polyatomic Molecules*, Van Nostrand, New York, 1945.
60. R. N. Dixon, G. Duxbury, M. Horani, and J. Rostas, *Mol. Phys.*, **22**, 977 (1971).
61. R. N. Dixon, *Trans. Faraday Soc.*, **60**, 1363 (1964).
62. T. Barrow, R. N. Dixon, and G. Duxbury, *Mol. Phys.*, **27**, 1217 (1974).
63. T. Bergmark, L. Karlsson, R. Jadrny, L. Mattsson, R. G. Albridge, and K. Siegbahn, *J. Electron Spectrosc.*, **4**, 85 (1974).

10 Simplifying Approaches for Complex Molecules

In many large molecules it is not possible to resolve vibrational fine structure, spin-orbit splittings, or multiplet structure as in small molecules. For such species the assignment of spectral bands can become very complicated, for there are inherently a large number of ionization bands, and the availability and reliability of theoretical calculations decreases as the molecular size increases. We are left in a dilemma that can only be resolved by reverting to various approximation schemes that simplify the problem to a tractable level. The judicious use of "chemical intuition" is often an asset in such approximate schemes. The purpose of this chapter is to develop various approximations that simplify interpretation of the spectra of large molecules and to apply these approximations to some representative examples. The first section describes the LCAO-MO or *atoms-in-molecules* approach coupled with experimental data for the interpretation of spectra. This is followed in Section 2 by a discussion of the *molecules-in-molecules* or *composite molecule* approach to spectral interpretation. Since many molecules can be treated as composite species, numerous examples of this approach are provided in Sections 3 to 5. The spectrum of benzene is discussed in detail because it is a prototypal aromatic system and an important subunit for composite molecule schemes. Sections 6 to 11 deal with interactions between specific groups and their consequences and various approximations and schemes that provide some understanding and simplifications of complex spectra.

10.1. Atoms in molecules approach (LCAO)

Molecular orbital calculations that use linear combinations of atomic orbitals (LCAO) for construction of wavefunctions are in a sense *atoms-in-molecules* methods. In such schemes the properties of a molecule are derived from those of its constituent atoms and from interactions between those atoms. The calculations are often parameterized according to spectroscopic information concerning the constituent atoms and their various stages of ionization. Such calculations can vary in complexity from the simple Hückel scheme to the sophisticated *ab initio* methods. In applying the *atoms-in-molecules* method to the elucidation of spectra, it is often advantageous to use the combined information obtained from LCAO-MO computations and spectral data. The following example illustrates the use of computational and spectroscopic data to elucidate the electronic structure of SCl_2.

An example; SCl_2

The spectrum[1] of SCl_2 is shown in Fig. 10.1 along with a correlation of the MOs of SCl_2 with the AOs of sulfur and chlorine. The correlation diagram is constructed as follows. The valence-orbital ionization energies[2] of sulfur and chlorine are used for the AO energy levels, and the MOs of SCl_2 are determined from a CNDO/2 calculation. The energies of the occupied MOs are obtained from the PE spectrum with the exception of $1b_1$ and $1a_1$, which are obtained from the CNDO/2 calculation. The energies of the virtual MOs $4b_1$ and $5a_1$ are obtained from the electronic absorption spectrum.[1] The correlation lines connecting the atomic and molecular levels are drawn according to the results of the MO computations. The lines connect the MOs to the AOs, which form the largest part of their constitution. Molecular orbitals that are mainly of the chlorine nonbonding type $3b_1$ and $1a_2$ are not connected to the sulfur AOs.

The identification of the PE bands of SCl_2 with specific MOs needs some clarification. The combined $3p$ AOs of chlorine transform as a_1, a_2, b_1, and b_2 in C_{2v} symmetry, while those of sulfur transform as a_1, b_1, and b_2. The MO calculations indicate a strong interaction between the b_2 orbitals of sulfur and chlorine to form the $2b_2$ (antibonding) and $1b_2$ (bonding) MOs. The remaining chlorine orbitals a_1, a_2, and b_1 remain largely nonbonding and relatively unshifted from their positions in free chlorine. From this interaction the scheme emerges: the first and fifth MOs are of b_2 symmetry with three largely chlorine nonbonding orbitals situated between them. We can now fit these MOs to the spectrum. The first band exhibits extensive vibrational structuring with a spacing of $548\,\text{cm}^{-1}$. Since the only symmetric ground state fundamentals of SCl_2 are $\nu_1 = 514\,\text{cm}^{-1}$ (S—Cl stretch) and $\nu_2 = 208\,\text{cm}^{-1}$ (bending), the observed progression must be ν_1 increased from its ground state value. This is in agreement with the S—Cl antibonding nature of the $2b_2$ MO, suggesting that the vibrational force constant of the ionic ground state is larger than that of the molecular ground state. The strongly bonding $1b_2$ orbital is widely split from the $2b_2$ orbital and is identified with the

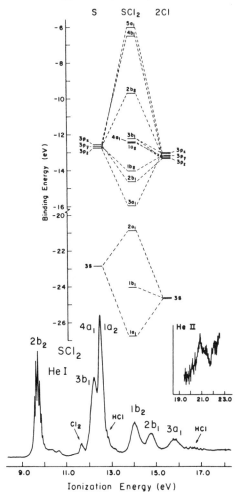

Fig. 10.1. Photoelectron spectrum of SCl_2 and correlation of the valence molecular orbitals of SCl_2 with the atomic orbitals of sulfur and chlorine. Reproduced with permission from Ref. 1.

structureless band at 14.05 eV. The remaining three orbitals are in the 12.0- to 12.8-eV region. The sharp intense band at 12.46 eV is typical for ejection of nonbonding electrons. However, its unusually large width (full width at half maximum of 0.23 eV) as compared to similar bands in other chlorine compounds (for example, $S_2Cl_2^1 \sim 0.14$ eV; $Cl_2O^3 \sim 0.07$ eV; $C_6H_5Cl^4 \sim 0.11$ eV) suggests that there are two overlapping transitions in this band. Thus the band at 12.24 eV probably consists of a single transition, while the one at 12.46 eV consists of two overlapping transitions. The ordering of the three levels is taken as that predicted from the CNDO/2 calculations $3b_1 > 4a_1 > 1a_2$. The remaining bands above 14.5 eV fit clearly into the scheme of strongly bonding orbitals.

10.2.   Composite molecule approach (LCMO)

The *composite molecule* or *molecules-in-molecules* approach is based on the assumption that a large molecule may be hypothetically fragmented into smaller molecular subunits and that information concerning these subunits may be utilized to develop a quantum-chemical picture of the entire molecule. This approach can be applied to many molecules; for example, butadiene can be described in terms of two ethylene moieties, styrene in terms of ethylene and benzene, nitroamide H_2NNO_2 in terms of ammonia and nitrogen dioxide, and so forth. This method is particularly useful in deciphering the spectra of large molecules and is developed here in considerable detail. The purpose of the composite molecule approach is to provide a simple description of the MOs of a complex molecule. The synthesis of the *component* information into *whole-molecule* information is presented at the one-electron MO level. For open-shell molecules it is necessary to carry out this synthesis at the many-electron state or configuration level. The value of the composite molecule approach is determined by its accuracy and the number of basis functions required to produce this accuracy; the fewer this number for a given precision, the greater the utility of the approach.

Linear combination of molecular orbitals (LCMO)

Methods that use linear combinations of molecular subunit MOs to construct the MOs of the whole molecule are called LCMO methods.[2,5] These methods are usually restricted to the π subset of MOs or the interaction of π and n (nonbonding) orbitals of similar symmetry. The LCMO theory can be developed in a manner similar to Hückel theory. Consider a molecule that can be divided into two subunits P and Q by scission of a bond connecting atom p on subunit P with atom q on subunit Q. The MOs of subunit P are denoted ϕ_m^P and the Hückel Hamiltonian on subunit P by

$$\mathcal{H}^P = \mathcal{T}^P + \mathcal{V}^P \tag{10.1}$$

where $\mathcal{H}^P$ is operative only in the region P of the whole molecule. The

Table 10.1. LCMO notation[2] for molecule PQ and subunits P and Q. ($\mathcal{H}^P$ acts only in the region P; $\mathcal{H}^Q$ acts only in region Q; $\mathcal{H}'$ acts only in the bond region pq)

	Subunit P	Subunit Q	Molecule PQ
MO Wavefunction	$\phi_m^P = \sum_\mu c_{m\mu}^P \chi_\mu^P$	$\phi_m^Q = \sum_\mu c_{m\mu}^Q \chi_\mu^Q$	$\phi_m = \sum_\mu c_{m\mu} \chi_\mu$
MO Energy	ε_m^P	ε_m^Q	ε_m
Hamiltonian	$\mathcal{H}^P = \mathcal{T}^P + \mathcal{V}^P$	$\mathcal{H}^Q = \mathcal{T}^Q + \mathcal{V}^Q$	$\mathcal{H} = \mathcal{T}^P + \mathcal{T}^Q + \mathcal{V}^P + \mathcal{V}^Q + \mathcal{H}'$ $= \mathcal{H}^P + \mathcal{H}^Q + \mathcal{H}'$
Elements	$\mathcal{H}_{mn}^{PP} = \mathcal{H}_{mn}^P = \varepsilon_m^P \delta_{m,n}$	$\mathcal{H}_{mn}^{QQ} = \mathcal{H}_{mn}^Q = \varepsilon_m^Q \delta_{m,n}$	$\mathcal{H}_{mn}^{PQ} = (\mathcal{H}_{mn}^{PQ})'$ $= c_{mp}^P c_{nq}^Q \beta_{pq}$
β_{pq}	—	—	$\beta_{pq} = \langle \chi_p^P \vert \mathcal{H}' \vert \chi_q^Q \rangle$

Hamiltonian for the entire molecule is

$$\mathcal{H} = \mathcal{T}^P + \mathcal{T}^Q + \mathcal{V}^P + \mathcal{V}^Q + \mathcal{H}' = \mathcal{H}^P + \mathcal{H}^Q + \mathcal{H}' \qquad (10.2)$$

where $\mathcal{H}'$ represents the interaction of subunits P and Q. Table 10.1 summarizes the notation applied. In keeping with the simplifications of Hückel theory, assume that $\mathcal{H}'$ is localized in the region of the pq bond and is otherwise inoperative in regions P or Q; hence we can define

$$\mathcal{H}' = \mathcal{H}_{pq} \qquad (10.3)$$

The matrix elements of the one-electron Hamiltonian $\mathcal{H}$ are given by

$$\langle \phi_m^P \vert \mathcal{H} \vert \phi_n^P \rangle = \langle \phi_m^P \vert \mathcal{H}^P \vert \phi_n^P \rangle = \mathcal{H}_{mn}^P = \varepsilon_m^P \delta_{m,n} \qquad (10.4)$$

since ϕ_m^P is, by definition, an eigenvector of $\mathcal{H}^P$ with eigenvalue ε_m^P. Similarly, one finds

$$\langle \phi_m^Q \vert \mathcal{H} \vert \phi_n^Q \rangle = \langle \phi_m^Q \vert \mathcal{H}^Q \vert \phi_n^Q \rangle = \mathcal{H}_{mn}^Q = \varepsilon_m^Q \delta_{m,n} \qquad (10.5)$$

and

$$\langle \phi_m^P \vert \mathcal{H} \vert \phi_n^Q \rangle = \left\langle \sum_\mu c_{m\mu}^P \chi^P \middle\vert \mathcal{H}' \middle\vert \sum_\nu c_{n\nu}^Q \chi_\nu^Q \right\rangle$$
$$= \langle c_{mp}^P \chi_p^P \vert \mathcal{H}' \vert c_{nq}^Q \chi_q^Q \rangle = c_{mp}^P c_{nq}^Q \beta_{pq} \qquad (10.6)$$

where β_{pq} is a standard Hückel resonance integral.

The whole-molecule MO eigenvectors and energies that result from interaction of the ϕ_m^P and ϕ_n^Q subunit MOs across the bond pq are obtained from the solutions of the secular determinant

$$\begin{vmatrix} \varepsilon_m^P - \varepsilon & c_{mp}^P c_{nq}^Q k_{pq}\beta \\ c_{nq}^Q c_{mp}^P k_{pq}\beta & \varepsilon_n^Q - \varepsilon \end{vmatrix} = 0 \qquad (10.7)$$

where $\beta_{pq} = k_{pq}\beta$. This method can be generalized to the case in which all MOs of subunit P are allowed to interact with all MOs of subunit Q.

The LCMO method provides an approximate description of the MOs of a composite molecule. Before proceeding with applications of this approximate scheme, we present a method for measuring the interaction between subunits and a qualitative discussion of the consequences resulting from interaction.

Measuring the interaction between subunits

PE spectroscopy provides a direct probe into the interaction between orbitals or subunits of a composite molecule. For a measure of this interaction it is desirable to have a "before-after" dichotomy of the molecular orbitals involved. The extent of interaction of two orbitals will be measured by *the magnitude of the one-electron level splitting in the presence of interaction compared to the theoretical splitting in the absence of such interaction.* On an MO diagram this would be defined according to Fig. 10.2a, where $\Delta\varepsilon_b$ is the splitting between two orbitals before interaction, $\Delta\varepsilon_a$ is the splitting after interaction, and $\Delta\varepsilon_a - \Delta\varepsilon_b$ is the measure of the interaction. This definition is consistent with simple bonding ideas. For example, Fig. 10.2b shows the splitting of the *1s* levels of two

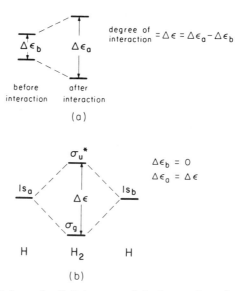

Fig. 10.2. (a) "Before-after" dichotomy of the interaction of two levels and the measure ΔE of the degree of interaction. (b) Splitting of the *1s* levels of two hydrogen atoms to form molecular hydrogen.

hydrogen atoms interacting to form molecular hydrogen. In this case, $\Delta\varepsilon_b = 0$ and $\Delta\varepsilon_a = \Delta\varepsilon$ is approximately proportional to bond strength. It should be noted that this definition is within a one-electron framework where the total energy is the sum of individual one-electron energy levels. This is not strictly correct in SCF theory because one-electron energies do not yield term values by simple addition.

Energy levels and wavefunctions of the composite molecule

Having defined the measure of interaction, we want to know what general conclusions can be drawn about the energy levels and wavefunctions of the composite molecule. Some simple conclusions from perturbation theory are particularly useful here.[6] For this purpose we define $\mathcal{H}^\circ$ for a composite molecule with the corresponding wavefunctions ϕ_n° and energies ε_n°. The interaction between the two subunits of the molecule is described by the perturbation $\mathcal{H}'$. After "turning on" the interaction between the ith level and all of the other levels, it can be shown by perturbation theory that the energy of the ith level is

$$\varepsilon_i = \varepsilon_i^\circ + \sum_{j \neq i} \frac{|H_{ij}'|^2}{\varepsilon_i^\circ - \varepsilon_j^\circ} \tag{10.8}$$

where $H_{ij}' = \langle \phi_i^\circ | \mathcal{H}' | \phi_j^\circ \rangle$. The perturbed wavefunctions are

$$\phi_i = \phi_i^\circ + \sum_{j \neq i} \frac{H_{ij}'}{\varepsilon_i^\circ - \varepsilon_j^\circ} \phi_j^\circ \tag{10.9}$$

These formulas apply to nondegenerate levels in which the overlap between interacting wavefunctions is neglected.

A number of important conclusions can be reached through these simple formulas. The summations in Eqs. 10.8 and 10.9 indicate that (a) *changes in energy levels and wavefunctions are pair-wise additive.* In order to consider how level i is affected by the perturbation we first consider how it interacts with level $i+1$, then linearly add its interaction with level $i+2$, and so forth.

Consider the interaction between levels i and j of which i has lower energy. The energies of these levels after interaction are, according to Eq. 10.8

$$\varepsilon_i = \varepsilon_i^\circ + \frac{|H_{ij}'|^2}{\varepsilon_i^\circ - \varepsilon_j^\circ} \quad \text{and} \quad \varepsilon_j = \varepsilon_j^\circ + \frac{|H_{ij}'|^2}{\varepsilon_j^\circ - \varepsilon_i^\circ} \tag{10.10}$$

Since we have assumed that $\varepsilon_i^\circ < \varepsilon_j^\circ$ and since $|H_{ij}'|^2$ is positive, it follows that $\varepsilon_i < \varepsilon_i^\circ$ and $\varepsilon_j > \varepsilon_j^\circ$ or (b) *when two levels interact the lower level is stabilized and the upper level is destabilized.* This is the well-known conclusion that two interacting levels repel each other and that the magnitude of the perturbation is inversely proportional to the energy difference between the interacting levels.

Now consider the wavefunctions of these two interacting levels

$$\phi_i = \phi_i^\circ + \frac{H'_{ij}}{\varepsilon_i^\circ - \varepsilon_j^\circ}\,\phi_j^\circ \quad \text{and} \quad \phi_j = \phi_j^\circ + \frac{H'_{ij}}{\varepsilon_j^\circ - \varepsilon_i^\circ}\,\phi_i^\circ \qquad (10.11)$$

In general, it is found consistently that H'_{ij} has the opposite sign to the overlap S_{ij}; that is, H'_{ij} is negative for positive S_{ij} and positive for negative S_{ij}. This simply implies that positive overlap indicates stabilization or bonding. Consider a given interaction where S_{ij} is positive. Then H'_{ij} is negative, and, since $\varepsilon_i^\circ - \varepsilon_j^\circ$ is also negative, ϕ_j° mixes into ϕ_i° with a plus sign as $\phi_i = \phi_i^\circ + |c|\phi_j^\circ$; similarly $\phi_j = \phi_j^\circ - |c|\phi_i^\circ$. This defines the shape of the orbitals after interaction: (c) *If two orbitals interact, the orbital of lower-energy mixes into itself the higher-energy orbital in a bonding manner, while the higher-energy orbital mixes into itself the lower-energy orbital in an antibonding manner.* This means that the higher-energy component takes the node.

The extent of these interactions increases with the proximity of the two interacting orbitals; the most prominent interactions occurring from first-order mixing of two levels with identical energies ε_i°. In this case one uses degenerate perturbation theory, although the general features of conclusions (b) and (c) are preserved. The two degenerate orbitals χ_i and χ_j originally at ε_0 mix to form two new orbitals $\chi_i + \chi_j$ and $\chi_i - \chi_j$, with energies $\varepsilon_0 + H'_{ij}$ and $\varepsilon_0 - H'_{ij}$, respectively.

10.3. Benzene—A prototypal aromatic molecule and subunit for composite molecules

Since benzene is a prototypal aromatic system and a subunit to be used in construction of composite molecules in the following sections, it seems appropriate to consider, in some detail, the interpretation of the benzene spectrum. Although the spectrum of benzene has been studied since the beginning of PE spectroscopy,[7] the assignment of its ionization bands has been a subject of much controversy. The assignments are now believed to be correct, largely due to the vast amount of experimental and theoretical evidence accumulated by various researchers. The evidence used in interpretation of the benzene spectrum includes vibrational analysis of the high-resolution PE spectrum;[8-12] comparisons with Rydberg series,[8] energy loss spectra,[13] and photoionization mass spectra;[8,14] angular distributions of photoelectrons;[15] MO models ranging from simple Hückel theory to *ab initio* SCF calculations;[8,16,17] and studies of intensity

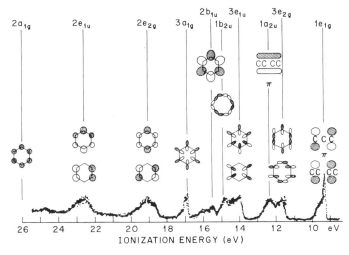

Fig. 10.3. The He II spectrum of benzene with schematic molecular orbitals and assignments. Reproduced with permission from Ref. 11.

changes resulting from Ne I, He I, and He II radiation.[12] The benzene spectrum is presented in Fig. 10.3 along with the band assignments and schematic MOs as suggested by Lindholm.[11] There are ten valence orbitals in benzene, and they all occur below 27 eV. The ionization energies, assignments, and relative intensities of the bands are listed in Table 10.2. The recent MO calculations[16,17] are in agreement with these assignments. The low-energy bands correspond to ionization of MOs that

Table 10.2. Analysis of the Benzene Spectrum[9]

Orbital Assignment	Ionization Energy (eV)	Band Intensity[a]	Number of Electrons	Intensity per Electron
$1e_{1g}(\pi)$	9.241 (A)	10.0	4	8.8
$3e_{2g}$	11.490 (A)	17.1	6	10.0
$1a_{2u}(\pi)$	12.1			
$3e_{1u}$	13.8	18.6	6	10.8
$1b_{2u}$	14.7			
$2b_{1u}$	15.4	6.3	2	11.0
$3a_{1g}$	16.9	5.1	2	9.0
$2e_{2g}$	19.2	13.4	4	11.8
$2e_{1u}$	22.5	11.7	4	10.2
$2a_{1g}$	25.9	—	—	—

[a] Measured as total area of photoelectron band.

are composed of carbon-$2p$ and hydrogen-$1s$ AOs, while the bands above 18 eV result from MOs composed of predominantly carbon-$2s$ AOs. The analysis of the individual photoelectron bands of benzene is now discussed in some detail.

The first photoelectron band at 9.241 eV exhibits[10] complicated vibrational structure with several different fundamentals excited. This complex vibrational structure indicates that the molecule changes its geometry upon ionization. Ejection of the outermost π electron produces a $^2E_{1g}$ ionic state; this state is subject to Jahn–Teller distortions through the e_{2g} vibrational modes to produce an ion of D_{2h} symmetry. Åsbrink et al.[10] have studied pyrazine in order to obtain an understanding of the complex

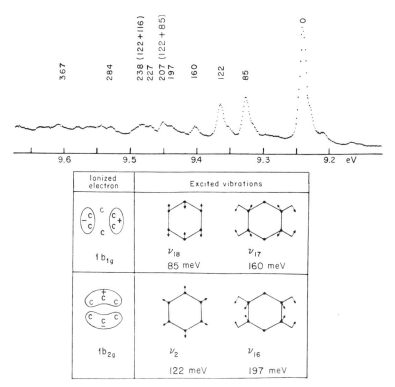

Fig. 10.4. Vibrational structure of the 9.241-eV ionization band of benzene excited with the 1215-Å hydrogen line. The vibrational energies are given in milli-electron volts. Hot bands can be seen at -11, -33, and -76 meV, and with lower intensity at -20, -45, and -87 meV. The lower part of the figure illustrates the connection between the ejected electrons and the vibrations excited in the transition. Benzene is represented in D_{2h} symmetry for this purpose. Reproduced with permission from Ref. 10.

structure of benzene. It is well known that the vibrational frequencies of benzene and pyrazine are approximately the same. The conclusions of this study are that the $^2E_{1g}$ ion is distorted by Jahn–Teller forces to produce $^2B_{1g}$ and $^2B_{2g}$ states of the D_{2h} ion. However, the Jahn–Teller splitting is very small because the $1e_{1g}$ electrons are essentially nonbonding in the benzene molecule. For ejection of such electrons, the Jahn–Teller forces are comparable to, if not weaker than, the restoring forces in a vibration of e symmetry; hence the vibronic wavefunction cannot be separated into an electronic and vibrational part. The two minima on the single potential surface are so close together that they are spanned by the zero-point energy contributions of the e_{2g} vibrations. The first ionization band of benzene is shown on an expanded scale in Fig. 10.4, and the correlation between the ejected electrons and the excited vibrations is illustrated in the lower part of the diagram. Comparison of this structure with that of pyrazine indicates that the 85- and 160-meV vibrations result from ionization of the b_{1g} (D_{2h}) component of the e_{1g} orbital and that the 122- and 197-meV vibrations result from ionization of the b_{2g} (D_{2h}) component of the e_{1g} orbital. These vibrations correspond to the ν_{18} (e_{2g}), ν_{17} (e_{2g}), ν_2 (a_{1g}), and ν_{16} (e_{2g}) modes of benzene (Fig. 10.4), respectively.

The assignment of the second and third ionization bands of benzene in the region 11.4 to 13 eV has remained unsettled until recently.[10–12,16] It has been shown[10] that the structure in the 11.490-eV band is very similar to that of the 9.241-eV band with the exception of the ν_2 breathing vibration, which is absent in the 11.940 eV band (Fig. 10.5). This implies that the ejected electron must come from a degenerate orbital, and hence it is assigned to ionization of the $3e_{2g}(\sigma)$ orbital. It seems highly unlikely that the complex structure at 11.49 eV could arise from ionization of the

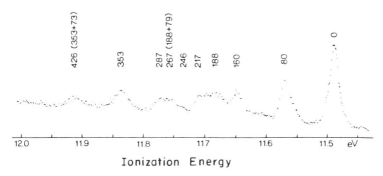

Fig. 10.5. Vibrational structure of the 11.49 eV ionization band of benzene excited with the He I line. Reproduced with permission from Ref. 10.

nondegenerate $1a_{2u}(\pi)$ orbital, for ionization of such a π orbital should produce a progression in the ring breathing mode. The third band at 12.1 eV exhibits very little structure, possibly because of its overlap with the band at 11.49 eV and the strongly bonding nature of the orbital ionized. The 12.1-eV band is assigned to ionization of the $1a_{2u}(\pi)$ orbital.

Analysis of Rydberg series in the absorption spectrum of benzene supports the above assignments. There are three Rydberg series in benzene with the same quantum defect, 0.46, but converging towards three different limits, 9.3, 11.5, and 16.9 eV. It is reasonable to assume that the nature of the Rydberg orbitals is similar in all three cases. It is known from symmetry considerations[11] that the Rydberg orbitals at 9.3 and 16.9 eV are p type; hence the one at 11.5 eV is also likely to be of p type. Excitations from e_{2g}-type orbitals to p-type Rydberg orbitals are allowed, but those from a_{2u} type orbitals are forbidden. This indicates that the 11.5-eV band cannot be produced by ionization of an a_{2u} orbital. These considerations lead to the outer orbital assignment $\cdots 1a_2^2(\pi)$, $3e_{2g}^4(\sigma)$, $1e_{1g}^4(\pi)$ for benzene.

The remaining bands in the benzene spectrum are assigned as indicated in Fig. 10.3 The bands at 13.8 and 14.7 eV are interpreted as ionizations of $3e_{1u}$ and $1b_{2u}$ orbitals, respectively. The 13.8-eV band exhibits the double-maximum characteristic of a transition that is perturbed by strong Jahn–Teller forces. The $^2E_{1u}$ state is subject to Jahn–Teller interactions, and the $3e_{1u}$ orbital is strongly bonding; hence the observed splitting is as expected for such a transition.

The vibrational structure of the 15.4- and 16.9-eV bands exhibits progressions in only the symmetric stretching modes, a_{1g}. This indicates that the ejected electrons result from nondegenerate orbitals, in agreement with the MO calculations that suggest that these orbitals are $2b_{1u}$ and $3a_{1g}$, respectively.

The bands at higher energies, 19.2, 22.5, and 25.9 eV, arise from MOs composed mainly of carbon $2s$ AOs. The MO calculations are in general agreement in assigning these to the $2e_{2g}$, $2e_{1u}$, and $2a_{1g}$ ionizations, respectively. This assignment of the ten valence orbitals of benzene is the one most widely accepted.

10.4.   Composite molecules from homologous subunits

Composite molecules that are derived from homologous subunits provide classic examples for applying the LCMO method. Such molecules include butadiene—two ethylene subunits, biphenyl—two benzene subunits, ethane—two methyl subunits, and so forth. The π-orbital interactions in several composite molecules are now discussed:

Butadiene

Butadiene can be considered as two ethylene subunits, P and Q, with energies and wavefunctions for these subunits given as:

Region P	Region Q

$$\varepsilon_{-1} = \alpha - \beta_{12} \qquad \phi_{-1} = \frac{1}{2^{1/2}}(\chi_1 - \chi_2) \qquad \varepsilon'_{-1} = \alpha - \beta_{34} \qquad \phi'_{-1} = \frac{1}{2^{1/2}}(\chi_3 - \chi_4)$$

$$\varepsilon_1 = \alpha + \beta_{12} \qquad \phi_1 = \frac{1}{2^{1/2}}(\chi_1 + \chi_2) \qquad \varepsilon'_1 = \alpha + \beta_{34} \qquad \phi'_1 = \frac{1}{2^{1/2}}(\chi_3 + \chi_4)$$

$$(10.12)$$

The bond notation is denoted in Fig. 10.6, where the energy levels of two

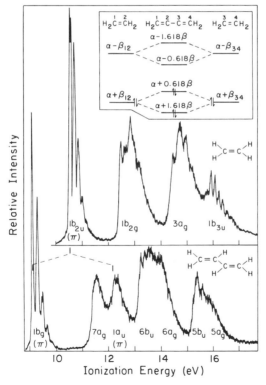

Fig. 10.6. He I spectra of ethylene and butadiene along with a diagram of the energy levels of two interacting ethylene molecules.

interacting ethylenes are correlated with those of butadiene and the spectra of ethylene and butadiene are presented. The secular equation for butadiene, from Section 10.2, is

$$
\begin{array}{cccc}
 & \phi_{-1} & \phi_1 & \phi'_{-1} & \phi'_1 \\
\phi_{-1} & \alpha - \beta_{12} - \varepsilon & 0 & -\dfrac{\beta_{23}}{2} & -\dfrac{\beta_{23}}{2} \\[2mm]
\phi_1 & 0 & \alpha + \beta_{12} - \varepsilon & \dfrac{\beta_{23}}{2} & \dfrac{\beta_{23}}{2} \\[2mm]
\phi'_{-1} & -\dfrac{\beta_{23}}{2} & \dfrac{\beta_{23}}{2} & \alpha - \beta_{34} - \varepsilon & 0 \\[2mm]
\phi'_1 & -\dfrac{\beta_{23}}{2} & \dfrac{\beta_{23}}{2} & 0 & \alpha + \beta_{34} - \varepsilon
\end{array} = 0
$$

$$(10.13)$$

A particular matrix element in the preceding determinant has the value

$$
H_{23} = \left\langle \frac{1}{2^{1/2}}(\chi_1 + \chi_2) \left| \mathcal{H} \right| \frac{1}{2^{1/2}}(\chi_3 - \chi_4) \right\rangle
$$

$$
= \tfrac{1}{2}\langle (\chi_1 + \chi_2) | \mathcal{H} | (\chi_3 - \chi_4) \rangle = \tfrac{1}{2}\langle \chi_2 | H_{23} | \chi_3 \rangle = \tfrac{1}{2}\beta_{23} \quad (10.14)
$$

The solutions to the secular determinant of Eq. 10.13 with $\beta_{12} = \beta_{23} = \beta_{34}$ and $\beta_{13} = \beta_{14} = \beta_{24} = 0$ are $\varepsilon = \alpha \pm 1.618\beta$ and $\alpha \pm 0.618\beta$ (Fig. 10.6). The alteration in bond lengths of butadiene can be taken into account by varying the value of β_{ij}; for example, let $\beta_{12} = \beta_{34} = 1.1\beta$ and $\beta_{23} = 0.87\beta$. The π orbitals in the spectra can be correlated as follows. The first band of ethylene at 10.51 eV results from ionization of the $1b_{2u}(\pi)$ orbital as confirmed by vibrational analysis.[18] The outermost bands of butadiene are $1b_g(\pi)$ at 9.06 eV, $7a_g(\sigma)$ at 11.47V and 11.2A eV, and $1a_u(\pi)$ at 12.23V and ~12.00A eV. The π, σ, π ordering has been established by use of the perfluoro effect (Section 10.6) and vibrational analysis.[18] The 9.06 eV band contains a strong vibrational progression of 1500 cm^{-1} and two weak progressions of 1200 and 520 cm^{-1}. These are identified with the molecular ground state symmetric C_1—C_2 stretching mode (1643 cm^{-1}), the C_2—C_3 stretching mode (1205 cm^{-1}), and the in-plane C=C—C bending mode (513 cm^{-1}), respectively. That C_1—C_2 stretching should be strongly excited, C_2—C_3 stretching should be weakly excited, and C=C—C bending should be weakly excited and increased in frequency are in agreement with the strongly C_1—C_2 bonding and C_2—C_3 antibonding nature of the $1b_g(\pi)$ orbital of butadiene as predicted by conclusion (c) of Section 10.2.

The Hückel parameters α and β can be determined from the spectrum.

First we see that the value of β (the resonance integral) can be determined from the splitting of the π levels of butadiene as $2.94 \, \text{eV} = -(1.618\beta - 0.618\beta)$, hence $\beta = -2.94 \, \text{eV}$. Next the parameter α (the coulomb integral) can be determined as $-9.06 \, \text{eV} = \alpha + 0.618\beta$ or $-12.00 \, \text{eV} = \alpha + 1.618\beta$; with $\beta = -2.94 \, \text{eV}$, we obtain $\alpha = -7.24 \, \text{eV}$.

Diacetylene

The energy levels and wavefunctions of diacetylene can be derived from those of two interacting acetylenes. The procedure is analogous to that of butadiene. In diacetylene the degenerate π_u orbitals of the interacting acetylenes at $11.40 \, \text{eV}$ split symmetrically into two degenerate orbitals, π_u and π_g, at 10.17 and $12.62 \, \text{eV}$, respectively.[18] The spectrum provides a measure of the Hückel parameter β of $-2.45 \, \text{eV}$.

Biphenyl

Biphenyl can be treated as two coplanar interacting benzene subunits with energies and wavefunctions derived from benzene. Applying a Hückel treatment according to the procedure followed for butadiene yields the π energy level scheme and wavefunctions of Fig. 10.7. The degenerate $e_{1g}(\pi)$ orbitals of the benzene subunits at $9.24 \, \text{eV}$ interact to form $\phi_6(b_{2g})$ at $-8.32 \, \text{eV}$, $\phi_5(a_{1u})$ and $\phi_4(b_{1g})$ at $\sim -9.25 \, \text{eV}$, and $\phi_3(b_{3u})$ at $-9.80 \, \text{eV}$, while the $a_{2u}(\pi)$ orbitals of benzene form $\phi_2(b_{2g})$ at $-11.20 \, \text{eV}$ and $\phi_1(b_{3u})$ at $\sim -13.5 \, \text{eV}$. Two of these orbitals, ϕ_4 and ϕ_5, are not detectably split in the spectrum, in agreement with the Hückel predictions. Orbitals ϕ_6 and ϕ_3 acquire antibonding and bonding character, respectively, as a result of the interaction. The spectral identification of the bands from ϕ_2 and ϕ_1 orbitals is more difficult because of the intervention of σ orbitals in the same region. The position of ϕ_2 has been established by studying the effects of substituents on the levels. The location of the ϕ_1 band is uncertain, but can be predicted as $\sim 13.5 \, \text{eV}$ from the Hückel calculations as parameterized for biphenyl. Such parameterization is obtained by equating the experimental energy differences between the bands to the differences in the corresponding Hückel eigenvalues of Fig. 10.7. Using those experimental energy gaps that are known to greatest precision, α and β have the values of -6.60 and $-2.43 \, \text{eV}$, respectively.

The values of α and β for several conjugated hydrocarbons determined in this manner are listed in Table 10.3. These parameters could obviously be extended to include heteroatom interactions in hydrocarbons. Such parameterization of α and β using experimental values is useful for

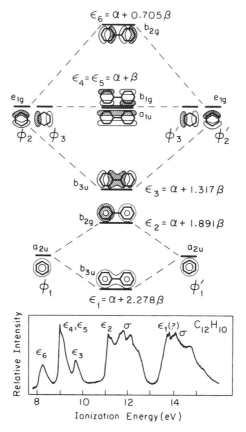

Fig. 10.7.   He I spectrum of biphenyl along with a diagram of the energy levels and wavefunctions of two interacting benzene molecules.

interpretation of π bands in more complicated molecules. An example of the use of such parameterizations is presented below for the acenes.

Acenes

The acenes, benzene, naphthalene, anthracene, naphthacene, and so forth, are planar aromatic hydrocarbons whose π orbital energies can be calculated easily using Hückel theory. The physical and chemical properties, including the π orbital energies, of these molecules are smooth, monotonic funcitons of the number of annexed rings in the species. One would expect that the parameters α and β derived for benzene could be used to fit the outer π bands of the acenes in Fig. 10.8 to a Hückel energy level scheme. Such a scheme is shown in Fig. 10.9, where orbitals

Table 10.3. Hückel Parameters α and β Obtained from Experimental Spectra[a]

Molecule	Hückel Parameters (eV)	
	β	α
Butadiene	−2.94	−7.24
Diacetylene	−2.45	−8.65
Biphenyl	−2.43	−6.60
Styrene	−2.63	−6.72
Phenylacetylene	−2.12	−7.31
Benzene	−3.02	−6.24

[a] In most cases the value of β can be determined from differences between several π energy levels. In such cases, the β reported here is the mean value of all these determinations.

belonging to the same irreducible representation are connected by lines. The experimental E_Is of the outer bands of the acenes are also presented in this diagram. Assuming that Koopmans' theorem holds, we can connect the experimental E_Is with lines in a manner similar to those in the Hückel scheme. It is not possible to follow these correlations above 11.0 eV because of the onset of σ-orbital ionizations. From this diagram we see that the experimental E_Is are similar to those in the Hückel scheme. This provides evidence for the assignment of the outermost energy levels of the acenes as π types. Hückel theory predicts two degeneracies in anthracene. Experimentally these levels are nondegenerate; however, the bands of anthracene at $\sim$10.2 eV are overlapping and near degenerate.

The success of such a parameterization as the one in Fig. 10.9 illustrates the utility of simple correlative attitudes in deciphering spectra. Heilbronner et al.[20] have used a perturbation treatment on the Hückel model to account for bond length changes accompanying the ionization process. The perturbations on the energy levels $\delta\varepsilon_i$ are estimated from the changes in bond order $p_{\mu\nu}$ upon removal of an electron from orbital ϕ_i by the formula

$$\delta\varepsilon_i = by_i = b \sum_{\mu,\nu} (p^+_{\mu\nu,i} - p_{\mu\nu})(p_0 - p_{\mu\nu}) \tag{10.15}$$

In the preceding equation, $p_{\mu\nu}$ and $p^+_{\mu\nu,i}$ are the bond orders of the $\mu\nu$ bond in the molecule and in the molecular ion, respectively, and p_0 is the

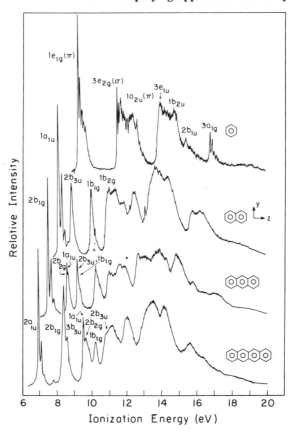

Fig. 10.8. He I spectra of the acenes benzene, naphthalane, anthracene, and naphthacene. The irreducible representations for the outer orbitals are indicated on the spectra.

bond order for a π bond of standard length (in benzene $r_0 = 1.39$ Å) having the value 2.3. The constant b depends on the force constants of the bonds and the derivative $d\beta/dr$ of the resonance integral. In practice b is handled as an adjustable parameter. The perturbation is introduced on the Hückel eigenvalues as

$$\varepsilon_i = \varepsilon_i + \delta\varepsilon_i = \alpha + x_i\beta + by_i \qquad (10.16)$$

where $\varepsilon_i = \alpha + x_i\beta$ is the Hückel eigenvalue. Using this perturbative approach, Heilbronner et al.[20] derive the parameters $\alpha = -5.864 \pm 0.110$, $\beta = -3.196 \pm 0.107$, and $b = -7.859 \pm 1.475$ by a least-squares fit appropriate to acene molecules. Excellent agreement between calculated and experimental E_Is is obtained from such a parameterization.

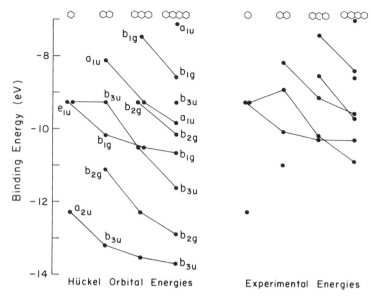

Fig. 10.9. Correlation diagram (left) of the acene orbital energies with $\beta = -3.0$ and $\alpha = -6.24$ as derived from the benzene spectrum. Lines connect orbitals of the same symmetry. The correlation of the experimental E_Is (right) is drawn according to those in the Hückel scheme.

10.5. Composite molecules from heterologous subunits

There are many examples of composite molecules derived from heterologous subunits; in this section we consider some of the best studied examples of this type of molecule, that is, substituted benzenes and ethylenes.

Substituted benzenes

Electronic interaction between the phenyl group and its substituents is of utmost importance in predicting and understanding the reactivities of substituted benzene compounds. The perturbation on the electronic structure supplied by a substituent results in a general redistribution of all electrons in the system. The most extensive redistribution of electronic charge occurs in the outer MOs, for the electrons occupying these orbitals are easily polarized. These outer electrons are the π electrons of the phenyl moiety and the π or n (nonbonding) electrons of the substituent. It has long been recognized[4,21–23] that the first E_I of substituted benzenes, relative to benzene, can be shifted to lower or higher binding energy by the action of an electron releasing or an electron withdrawing substituent, respectively. The presence of the substituent also lifts the degeneracy of the $1e_{1g}\pi$ orbital of benzene. One of the best methods of investigating the interaction between the phenyl ring and a substituent is through PE spectroscopy, because this allows measurement and identification of the ionization energies of the perturbed π electrons of the phenyl group as well as the π and n electrons of the substituent. The splittings between these molecular orbitals are a direct result of conjugative and inductive effects between the components of the composite molecule.

Substitution of a hydrogen atom in benzene by some other species results in a lowering of the molecular symmetry and a splitting of the doubly degenerate $1e_{1g}$ orbital. This is evident from an examination of the symmetry properties of the outer π orbitals of substituted benzenes in Fig. 10.10. Treating the composite molecule as one of C_{2v} symmetry, we see that at the point of substitution one of the orbitals (b_1 type) has its maximum electron density, while the other (a_2 type) has a node and thus zero electron density. The substituent in Fig. 10.10 has a π-type orbital that interacts with the b_1 phenyl orbital. The splitting occurs with the consequences described as in Section 10.2. The a_2 orbital remains largely unshifted from its position of 9.24 eV in benzene. The magnitude of the splitting between the b_1-type orbitals is dependent upon the electronegativity (through inductive effects) of the substituent and the ability of the substituent π electrons to conjugate with the phenyl π electrons (resonance effect.) The orbitals of the phenyl moiety and the substituent generally become mixed in the composite system so that localization on one subunit no longer exists. Even though the orbitals of the composite molecule are delocalized over the entire system, it is helpful to refer to them in terms of their parentage. The split $e_{1g}(\pi)$ orbitals of the phenyl group are usually the least tightly bound orbitals of the composite molecules. The ring $1a_{2u}(\pi)$ and $3e_{2g}(\sigma)$ orbitals and the substituent n and π orbitals are all possible candidates for the penultimate position,

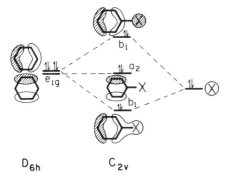

D_{6h} C_{2v}

Fig. 10.10. Interaction between the $1e_{1g}(\pi)$ orbitals of benzene and the π-type orbital of a substituent X.

that is, the next to the last occupied orbital. It is convenient to label this region of the spectrum as the "penultimate region" and the orbitals occupying this region as assuming a "penultimate role." The identities of the $a_{2u}(\pi)$ and the substituent n and π orbitals are marred by the presence of σ orbitals that begin to appear in this region. An additional complication occurs in the composite molecules due to the change in vibrational structure and general shape of these bands that may appear as a result of partial delocalization. The properties and reactivities of substituted benzenes are generally considered to be determined to a large extent by the electronic properties of the substituent. It is, therefore, important to investigate the extent of delocalization between the phenyl group and its substituent and, in particular, the nature of the so-called "nonbonding" electrons in the composite molecule.

ISOELECTRONIC SERIES WITH SEVEN-VALENCE-ELECTRON SUBSTITUENTS. The spectra of the isoelectronic series, toluene, aniline, phenol, and fluorobenzene (Fig. 10.11), form an interesting group of molecules for studying substituent effects.[24] All of these substituents except the methyl group contain nonbonding $p\pi$-type electrons. The low-energy ionization bands of these molecules result from the split $1e_{1g}$ orbitals and the n substituent orbital. The lowest energy band corresponds to the antibonding combination between π and n, that is, the b_1 orbital, while the second band corresponds to the unperturbed a_2 orbital. This assignment is based on the energy and vibrational structure of these outer bands. Molecular orbital calculations on these systems place the b_1 and a_2 π orbitals at lowest binding energy.

The details of the vibrational structure of these two bands are shown on expanded scales in Fig. 10.12. The vibrational structure observed in the first band of all four molecules is very similar. In each case it consists of two progressions, one of ~ 1540 to ~ 1690 cm^{-1} and the other of ~ 520 to ~ 560 cm^{-1}. Reference to the infrared and Raman spectra of these molecules shows that there is always a symmetric vibrational mode of the ground state that has a frequency close to the intervals observed in the spectra. The vibrational frequencies of such multielectron systems are not expected to change by a large amount in the positive ions. On this basis, the higher-frequency interval is identified as a symmetric "X-sensitive" mode. "X-sensitive" modes are those vibrations in which the substituent, X, moves with appreciable amplitude. The second band for each molecule in Fig. 10.12 contains a single short progression of ~ 810 cm^{-1} in phenol and fluorobenzene and ~ 485 cm^{-1} in aniline. Both of these frequencies can be identified with X-sensitive modes of the ground state molecule. Vibrations excited in PE spectra usually correspond to motions of the atoms in that part of the molecule where the ejected electron is most

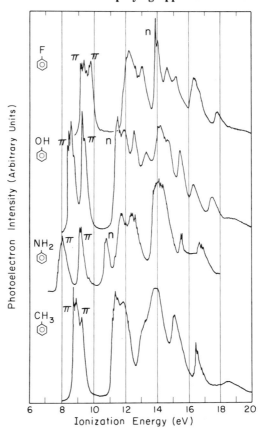

Fig. 10.11. He I spectra of substituted benzenes with seven-valence-electron substituents: fluorobenzene, phenol, aniline, and toluene. Reproduced with permission from Ref. 24.

heavily localized. Excitation of ring C—C stretching and X-sensitive modes in the two outer bands of these molecules establishes the ejected electrons as delocalized over the composite molecule systems.

Delocalization of these mobile π electrons causes large shifts in the outer orbitals of the composite molecule systems. The splittings between the e_{1g} orbitals of the composite molecules toluene, aniline, phenol, and fluorobenzene are 0.52, 1.23, 0.91, and 0.71 eV, respectively. These splittings are a direct result of conjugative and inductive effects of the substituent. The NH_2 group is the best electron donor and splits the phenyl π orbitals widely by its conjugative action. The first E_I of fluorobenzene is lower than that of benzene, and its outer π orbitals are

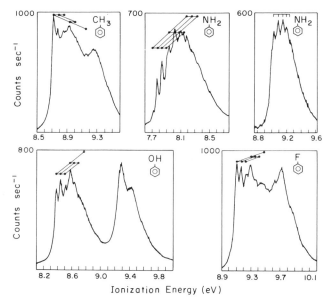

Fig. 10.12. Details of the vibrational structure in the two lowest energy ionization bands of toluene, aniline, phenol, and fluorobenzene. Vibrational progressions are indicated by lines above the spectra. Reproduced with permission from Ref. 24.

well separated as a result of the resonance effect of fluorine. Although the inductive effect of the fluorine atom withdraws electron density from the ring, its $p\pi$ atomic orbital participates in the phenyl $\pi(b_1)$ orbital allowing delocalization and lowering of the energy of this orbital. The resonance effect of the fluorine atom appears to be about the same order of magnitude as that of the methyl group. This is borne out in chemical reactions, for it has been shown that the methyl group is only weakly activating towards electrophilic aromatic substitution.

The b_1 and a_2 π orbitals of substituted benzenes appear consistently in the region of ~9 eV (Fig. 10.13). Unlike these π orbitals, the nonbonding orbitals of the substituents vary over a wide range, their energy being strongly dependent upon the electronegativity of the substituent. The nonbonding orbitals of fluorobenzene appear as two sharp peaks at 13.85 and 13.98 eV. These two observed peaks are assigned to transitions from two different nonbonding orbitals for the following reasons. The nonbonding electrons of Cl, Br, and I have been shown[4] to exist in two different orbitals when the halogen is a substituent on an aromatic ring. One of these orbitals is parallel to the plane of the ring, while the other is

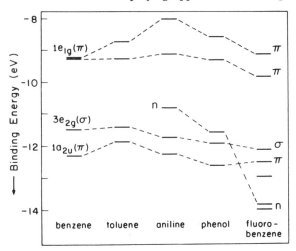

Fig. 10.13. Energy level diagram for the outer molecular orbitals of benzene and the isoelectronic series toluene, aniline, phenol, and fluorobenzene. Vertical ionization energies are used throughout the diagram. Reproduced with permission from Ref. 24.

perpendicular to the plane of the ring. The in-plane orbital remains essentially nonbonding, and ejection of its electrons produces a very sharp and intense peak. The out-of-plane orbital conjugates with the phenyl π orbitals, thereby assuming some bonding character and producing a spectral band that is not as sharp as the former. The out-of-plane orbital appears at higher binding energy because of stabilization by the π electrons. The two bands observed in fluorobenzene fit this pattern, the lower energy band being much sharper and more intense than the high-energy band. The presence of these two expected bands is confirmed if one plots the binding energy of the nonbonding orbitals of halogen substituted benzenes versus the electronegativity of the halogen (Fig. 10.14). Allred electronegativities[25] are used in this plot since they are empirical in nature and are derived for atoms in their most common oxidation states. The points form two straight lines that include the two values for the fluorobenzene nonbonding orbitals with the points corresponding to the in-plane orbital forming the line of steepest slope. The splitting between the halogen n orbitals increases as the E_I of the n orbital approaches that of the phenyl π orbitals. This indicates a greater interaction due to the closer proximity in energy (Section 10.2) of the interacting orbitals.

The third E_I of aniline at 10.78 eV corresponds to ionization of the b_1 bonding combination between the π and n orbitals. Although MO

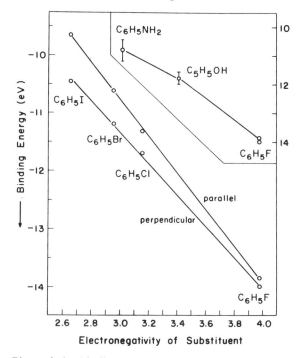

Fig. 10.14. Plots of the binding energies of the nonbonding orbitals of some substituted benzenes versus the Allred electronegativity of the substituent. For $C_6H_5NH_2$ and C_6H_5OH the vertical E_is are indicated by circles, and the lower and upper extents of the bands are indicated by lines. These nonbonding orbitals are not as sharp as those of the halobenzenes. Reproduced with permission from Ref. 24.

calculations show that these electrons become delocalized onto the phenyl group, the band is commonly called a nitrogen n orbital because of the large electron density of this b_1 orbital on the nitrogen atom. The position of this band is unshifted from its 10.8-eV value in ammonia. One would have expected a shift in this orbital simply from comparison with the large splitting in the $e_{1g}(\pi)$ orbitals and the computational results that show significant mixing of the n and $b_1(\pi)$ orbitals. The constancy of the position of the n orbital of aniline is attributed to the cancellation of opposing resonance and inductive effects as follows. The n orbital of ammonia is inherently of higher binding energy than the $e_{1g}(\pi)$ orbitals of benzene. Resonance interaction between them results in destabilization of the $b_1(\pi)$ orbital and stabilization of the n orbital. In the inductive effect the substituent interacts with the σ orbitals. In this case the nitrogen atom is more electronegative than the ring carbon atoms; therefore, electron density is drawn from the ring to the substituent, resulting in a destabilization of the n orbital. These two opposing effects in aniline are evidently of the same order of magnitude and cancel each other out. In fluorobenzene the inductive effect is dominant, resulting in the nonbonding orbitals being destabilized by ~ 2.1 eV from their position in hydrogen fluoride. The inductive effect in phenol again predominates over the resonance effect, shifting the n orbital to lower binding energy by ~ 1.05 eV from its position in water. The decrease in the shift of this nonbonding orbital in the series fluorobenzene, phenol, and aniline reflects the electronegativity differences between the fluorine, oxygen, and nitrogen atoms (Fig. 10.14).

MO CALCULATIONS ON SUBSTITUTED BENZENES. The eigenvalues and localization properties of the outer MOs of several substituted benzenes are listed in Table 10.4. Only the π, n, and lowest-energy σ orbitals are listed. A serious deficiency of these calculations is the improper positioning of the π and σ manifolds relative to each other. The computations place the lowest energy σ orbital between or very near the split $e_{1g}(\pi)$ orbitals with several additional σ orbitals in the penultimate region. In addition to this, the neglect of electron delocalization and correlation effects in Koopmans' theorem results in eigenvalues that are usually higher than experimental E_Is. In order to compensate for these deficiences, the eigenvalues of the π and n orbitals in Table 10.4 are empirically reduced from the computed values by 20%, while the σ eigenvalues remain untouched. Despite this attempted correction, the ordering of the π and σ eigenvalues relative to each other remains uncertain.

The calculations are useful, however; the reorganization of electron density that takes place upon substitution onto the benzene ring is best understood by studying the localization properties of the MOs in the composite molecules. Considering the phenyl $\pi(e_{1g})$ orbitals, the b_1 component becomes partially delocalized on the substituent, while the a_2 component, with a node at the point of substitution, remains localized on the phenyl ring. The nonplanarity of toluene and aniline removes the clear $\sigma-\pi$ separability that is maintained in the planar molecules. The $\pi(b_1)$ orbital in these molecules contains considerable C—H and N—H σ character, respectively. The innermost $\pi(b_1)$ orbital of the phenyl group remains highly localized on the ring. The n orbital of the substituent becomes delocalized in the composite molecule; however, it always maintains more than 50% of its electron density on the substituent.

ISOELECTRONIC SERIES WITH UNSATURATED SUBSTITUENTS. The spectra of substituted benzenes with eleven-valence-electron substituents[26]

Table 10.4. Results of INDO Calculations on Some Substituted Benzenes

Molecule	Ionization Energy (eV)	MO Eigen-value (eV)	Electron Density[a]	MO Type
Toluene	8.72	8.81	0.49 ring C 0.51 CH_3	$\pi(b_1)$
	10.90	10.28	0.72 ring 0.28 CH_3	σ
	9.24	10.54	1.00 ring C	$\pi(a_2)$
	11.87	12.27	0.96 ring C 0.04 CH_3	$\pi(b_1)$
Aniline	7.71	9.08	0.79 ring C 0.21 NH_2	$\pi(b_1)$
	8.94	10.66	1.00 ring C	$\pi(a_2)$
	11.88	12.21	0.95 ring 0.05 NH_2	σ
	10.45	13.15	0.42 ring C 0.58 NH_2	n
	12.48	18.66	0.86 ring C 0.14 NH_2	$\pi(b_1)$
Phenol	8.37	8.79	0.70 ring C 0.30 OH	$\pi(b_1)$
	9.28	10.70	1.00 ring C	$\pi(a_2)$
	11.91	12.61	0.73 ring C 0.27 OH	σ
	11.22	13.71	0.45 ring C 0.55 OH	n
	12.61	18.90	0.89 ring C 0.11 OH	$\pi(b_1)$
Fluorobenzene	9.11	10.22	0.77 ring C 0.23 F	$\pi(b_1)$
	9.82	11.10	1.00 ring C	$\pi(a_2)$
	11.75	13.63	0.94 ring C 0.06 F	σ
	13.85	15.70	0.37 ring C 0.63 F	n
	13.98	16.23	0.47 ring C 0.53 F	n

Table 10.4. (*continued*)

Molecule	Ionization Energy (eV)	MO Eigen-value (eV)	Electron Density[a]	MO Type
Fluorobenzene (*continued*)	12.5	19.51	0.78 ring C 0.22 F	$\pi(b_1)$
Styrene	8.42	9.30	0.65 ring C 0.11 X 0.24 Y	$b_1(\pi)$
	9.13	10.71	1.00 ring C	$a_2(\pi)$
	10.55	13.64	0.45 ring C 0.32 X 0.23 Y	$b_1(\pi)$
	—	18.96	0.91 ring C 0.07 X 0.02 Y	$b_1(\pi)$
Benzaldehyde	9.69	9.72	0.47 ring C 0.08 X 0.45 Y	n (nonbonding p_x and p_y orbital on O)
	9.40	10.23	0.75 ring C 0.04 X 0.21 Y	$b_1(\pi)$
	9.81	10.90	1.00 ring C	$a_2(\pi)$
	—	14.87	0.88 ring C 0.08 X 0.04 Y	$b_1(\pi)$
Phenylacetylene	8.75	9.61	0.72 ring C 0.08 X 0.20 Y	$b_1(\pi)$
	9.34	10.74	1.00 ring C	$a_2(\pi)$
	10.36	12.58	0.39 ring C 0.28 X 0.33 Y	σ (in-plane acetylenic p orbital)
	11.03	14.32	0.38 ring C 0.34 X 0.28 Y	b_1 (out-of-plane acetylenic π orbital)
	—	19.10	0.88 ring C 0.09 X 0.03 Y	$b_1(\pi)$

Table 10.4. (*continued*)

Molecule	Ionization Energy (eV)	MO Eigen- value (eV)	Electron Density[a]	MO Type
Nitrosobenzene	8.09	8.59	0.20 ring C 0.46 X 0.34 Y	n (nonbonding p_x and p_y orbitals on N and O)
	9.49	10.3	0.75 ring C 0.04 X 0.21 Y	$b_1(\pi)$
	9.90	11.00	1.00 ring C	$a_2(\pi)$
	—	14.6	0.33 ring C 0.33 X 0.34 Y	$b_1(\pi)$
	—	19.2	0.90 ring C 0.07 X 0.03 Y	$b_1(\pi)$
Benzonitrile	9.62	10.13	0.75 ring C 0.06 X 0.19 Y	$b_1(\pi)$
	10.15	10.96	1.00 ring C	$a_2(\pi)$
	11.93	13.10	0.33 ring C 0.28 X 0.39 Y	σ (in-plane nitrile p orbital)
	12.18	14.7	0.37 ring C 0.33 X 0.30 Y	b_1 (out-of-plane nitrile π orbital)
	—	19.4	0.87 ring C 0.09 X 0.04 Y	$b_1(\pi)$

[a] The localization properties of the MOs are represented by the fraction of electron density in various regions. The "ring C" designates the carbon atoms of the phenyl group. X and Y designate the two atoms of the substituent group that are derived from the second row of the periodic table. X is the atom adjacent to the phenyl group, whereas Y is one atom removed from the phenyl group.

styrene, benzaldehyde, and nitrosobenzene are shown in Fig. 10.15. The bands with binding energy less than 11 eV are of primary interest since these arise from interaction between the $e_{1g}(\pi)$ orbital and the outermost π or n orbital of the substituent. The bands above 11 eV arise from σ orbitals of the substituent. The severe overlap of bands above 11 eV

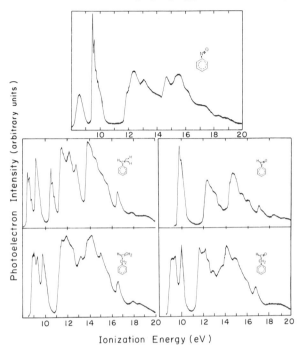

Fig. 10.15. He I spectra of nitrosobenzene, styrene, allylbenzene, benzaldehyde, and phenylacetaldehyde. Reproduced with permission from Ref. 26.

makes identification and assignment of the bands in this region uncertain. The outermost π, n, and σ orbitals of these composite molecules are correlated in Fig. 10.16. Vertical E_Is are employed in order to maintain consistency, for in many cases it is not possible to locate the adiabatic E_Is. It should be emphasized that this correlation does not imply that the MOs of the composite molecule are localized on the grouping indicated; it merely connects the MOs of the composite molecule with those of the constituent molecules from which they are derived.

Delocalization of the mobile π electrons causes large shifts in the outer orbitals of these composite molecule systems. In styrene the conjugation between the ethylenic and phenyl π electrons splits the degeneracy of the $e_{1g}(\pi)$ orbitals, destabilizing the $b_1(\pi)$ phenyl orbital and stabilizing the $b_1(\pi)$ ethylenic orbital. In allylbenzene the CH_2 group serves to "insulate" the two conjugated groups from each other, thus reducing the interaction. This is indicated by the reduction in the splitting of the b_1 and a_2 π orbitals of the phenyl moiety. The inductive effect of the CH_2 group shifts the b_1 phenyl orbital to a position that corresponds to that in

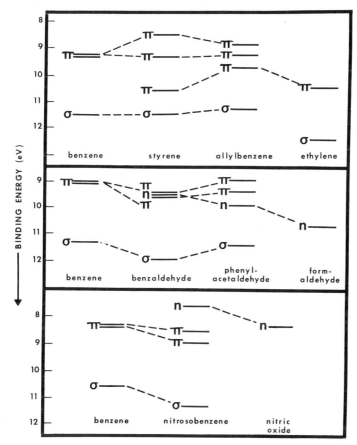

Fig. 10.16. Energy level diagrams for the outer molecular orbitals of substituted benzenes with eleven-valence-electron substituents. The molecular orbitals of the composite molecules are correlated with those of the constituent molecules from which they are derived. Reproduced with permission from Ref. 26.

toluene, while the ethylene π orbital is shifted to a position near the value of the propylene π orbital.

In benzaldehyde the n orbital of the aldehyde group occurs in the energy region of the $e_{1g}(\pi)$ orbitals, while the CO π orbital occurs at much higher energy (14.09 eV in formaldehyde). The aldehyde group is electron withdrawing, and its inductive effect draws electron density from the ring, shifting the $1e_{1g}(\pi)$ orbitals to higher binding energy. The n orbital is near degenerate with these $1e_{1g}$ orbitals. In phenylacetaldehyde these two groups are decoupled, and the electron-releasing effect of the methylene group destabilizes the $e_{1g}(\pi)$ orbitals and stabilizes the n

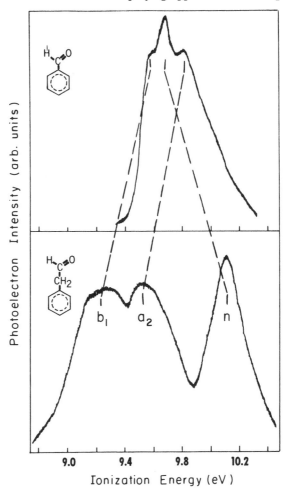

Fig. 10.17. The ionization bands in the low-binding-energy region of the benzaldehyde and phenylacetaldehyde spectra shown on an expanded scale. The correlations between the bands are indicated by dashed lines. Reproduced with permission from Ref. 26.

orbital to a position near that of acetaldehyde. These shifts are indicated on the expanded scale spectra of Fig. 10.17.

The n orbital of nitrosobenzene is delocalized over both the N and O atoms, and, as a result, its energy is lower than that of the $1e_{1g}(\pi)$ orbitals. This is one of the few substituted benzenes in which an orbital largely localized on the substituent is at lower binding energy than the $1e_{1g}(\pi)$ orbitals. (Another example is iodobenzene.)

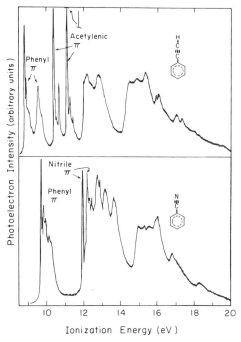

Fig. 10.18. He I spectra of phenylacetylene and benzonitrile. Reproduced with permission from Ref. 26.

The spectra of substituted benzenes with nine-valence-electron substituents[26] phenylacetylene and benzonitrile are shown in Fig. 10.18. The acetylene and cyanide groups have degenerate π orbitals; only one member of these degenerate sets can interact with the $1e_{1g}$ orbitals. The other in-plane π orbital can interact with the ring σ orbitals (Table 10.4). In phenylacetylene the $1e_{1g}$ and acetylenic π orbitals are both widely split, Fig. 10.19, as a result of conjugation between the π electrons. The first E_I is not as low as that of styrene, reflecting the higher electronegativity of the acetylene group with respect to the ethylene group. In benzonitrile the $1e_{1g}$ orbitals are stabilized and the degenerate nitrile π orbitals are destabilized from their positions in the constituent molecules. Also, the splittings between these two sets of degenerate π orbitals are not as large as those of phenylacetylene. The reason for these pecularities in the spectrum of benzonitrile is that the inductive effect is stronger than the resonance effect in this molecule. The resonance effect is weak because of the large energy difference between the interacting phenyl e_{1g} and CN π orbitals. The inductive effect is strong because the CN group is strongly electron withdrawing. This results in a flow of electron density

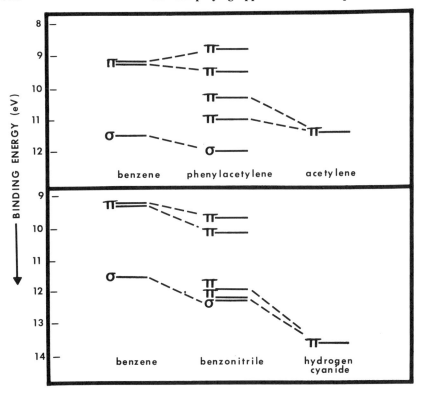

Fig. 10.19. Energy level diagrams for the outer molecular orbitals of substituted benzenes with nine-valence-electron substituents. The molecular orbitals of the composite molecules are correlated with those of the constituent molecules from which they are derived. Reproduced with permission from Ref. 26.

from the ring to the CN group, increasing the binding energy of the split $1e_{1g}$ orbitals and decreasing the binding energy of the split CN π orbitals.

The spectra of most of these benzene derivatives with unsaturated substituents exhibit vibrational structure similar to the structure in the seven-valence-electron substituent case. In most cases the vibrations correspond to stretching modes of the phenyl ring and unsaturated portions of the substituent and very often to stretching of the C—R bond connecting the substituent to the phenyl group.

Nitrobenzene is another example[27] of a substituted benzene in which the unsaturated NO_2 group is strongly electron withdrawing. The first E_I, Fig. 10.20, is ~0.75 eV higher than that of benzene, and the $1e_{1g}$ orbitals are split by only ~0.36 eV. Thus the strong inductive effect of the nitro group serves to stabilize the $1e_{1g}$ orbitals. The bands centered near

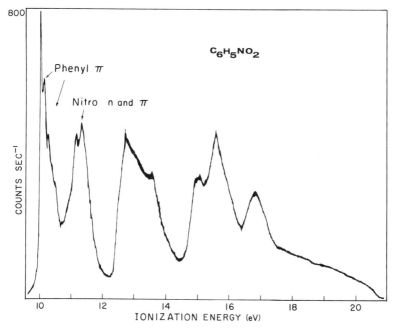

Fig. 10.20. He I spectrum of nitrobenzene. Reproduced with permission from Ref. 27.

11.2 eV correspond to the outer n and π orbitals of the nitro group. These have been identified by correlation with the spectrum of nitromethane.[27]

Substituted ethylenes

Substituted ethylenes provide an interesting example of the interaction between the vinyl π orbital and the π or n orbitals of substituents. Since the molecules are relatively small, it is possible to resolve vibrational structure in the bands and derive information on resonance and inductive effects that is otherwise unobtainable. Two types of molecules are considered in this section, (1) vinyl halides and (2) propene and substituted propenes.

VINYL HALIDES. Interpretation of the spectra of vinyl halides requires the consideration of resonance and inductive effects between the vinyl group and the halogen atom. As a halogen atom approaches a vinyl group, the p orbital of the halogen pointing in the direction of the vinyl group is used to form a C—X σ bond. The two remaining p orbitals are

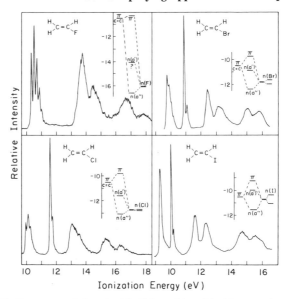

Fig. 10.21   He I spectra of the vinyl halides with orbital interaction diagrams for the $n(X)$ and vinyl π orbitals as derived from their respective orbitals in ethylene and the halogen acids, HX. The position of the n halogen orbitals is taken as the mean value of its spin-orbit split components.

oriented perpendicular (a'') and parallel (a'), respectively, to the molecular plane. The a' orbital remains largely noninteracting, thus a halogen nonbonding orbital, while the a'' orbital interacts with the vinyl $\pi(a'')$ orbital. This resonance interaction between the a'' orbitals produces an antibonding and bonding combination of a'' MOs as described in Section 10.2. The inductive effect caused by the differences in electronegativity of the carbon and halogen atoms also influences the positions of the bands—however, only by shifts and not by splitting of degenerate orbitals.

These effects are best described by reference to the vinyl halide spectra in Fig. 10.21. Orbital interaction diagrams are included in the figure with the vinyl halides treated as composite molecules that are dichotomized into ethylene and halogen atoms. First, consider the case of vinyl chloride: the n Cl orbitals are inherently more stable than the vinyl π orbital. The strong resonance interaction between these two orbitals produces an antibonding combination of predominantly vinyl π character and a bonding combination of predominantly Cl n character. The vibrational structure of the outermost band confirms its bonding character in the C=C region and antibonding character in the C—Cl region; two progressions with mean spacings of 1320 and 650 cm^{-1} are observed corresponding to a decrease in the C=C stretching mode ($1620\ \mathrm{cm}^{-1}$) and an increase in the C—Cl stretching mode ($605\ \mathrm{cm}^{-1}$) from the ground state values. The bonding $n(a'')$ combination is strongly stabilized and has the broad structureless appearance of a bonding orbital. The noninteracting $n(a')$ orbital retains the sharp appearance of a nonbonding orbital with only a short progression in the C—Cl stretching mode. The large electronegativity difference between the chlorine and carbon atoms leads to an inductive effect that reduces the shielding of the carbon atoms, resulting in an increase in the binding energy of the vinyl π electrons. Conversely, the shielding of the chlorine atom is increased, resulting in a decrease in the binding energy of the Cl electrons. Thus the effects of resonance and induction are opposed. The resonance effect is stronger than the inductive effect, as evidenced by the destabilization of the π orbital of vinyl chloride relative to that of ethylene. Since the $n(a')$ orbital is unaffected by conjugation with the vinyl π orbital, its destabilization by $\sim 1.0\ \mathrm{eV}$ from its position in HCl is a direct measure of the inductive effect. The consequences of these effects in vinyl chloride are to provide some π character to the C—Cl bond, making it shorter and stronger than an ordinary C—Cl single bond.

The remaining spectra in Fig. 10.21 can be interpreted in a similar manner. In vinyl bromide the situation is almost identical to that of vinyl chloride with the exception of the weaker inductive effect of the bromine atom; the $n(a')$ orbital of vinyl bromide is destabilized by only $\sim 0.8\ \mathrm{eV}$ from its position in HBr. In vinyl iodide the interacting π and n orbitals are near degenerate. As a result, the split a'' orbitals contain almost equal mixtures of vinyl π and iodine $n(a'')$ character, and the resulting bands do not resemble those of vinyl chloride and bromide. The weak inductive effect of the iodine atom results in a destabilization of the $n(a')$ orbital by $\sim 0.5\ \mathrm{eV}$. In vinyl fluoride the large energy difference between the interacting π and $n(a'')$ orbitals reduces the resonance interaction such that the π orbital is destabilized by only $\sim 0.2\ \mathrm{eV}$. The $n(a')$ orbital of fluorine does not appear as a sharp nonbonding band as in the other halides; this is true for most fluorine substituted hydrocarbons. The fluorine orbitals are approximately the same size as those of carbon, providing for considerable overlap with a resulting loss of their nonbonding character. The $n(a')$ orbital of fluorine is evidently mixed into the C—C and C—H orbitals in the 13 to 15-eV region.

PROPYLENE.   *The vinyl π orbital of propylene is destabilized from its position in ethylene by 0.79 eV. This shift results from combined resonance and inductive effects which both tend to destabilize the π orbital. The vinyl π orbital acquires antibonding character through resonance*

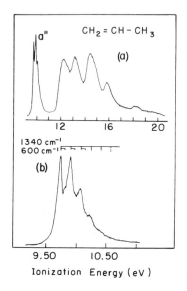

Fig. 10.22. (*a*) He I spectrum of propylene. (*b*) Expansion of first band. Reproduced with permission from Ref. 28.

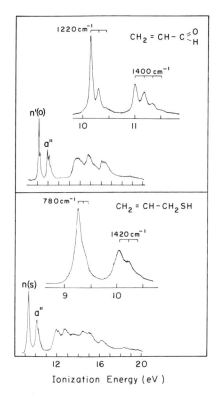

Fig. 10.23. He I spectra of acrolein and allylmercaptan. Reproduced with permission from Ref. 28.

320

interaction with the methyl group orbitals (hyperconjugation) and is further destabilized by the electron donating inductive effect of the methyl group. The vibrational structure, Fig. 10.22, consists of stretching and bending modes;[28] the former decreased and the latter increased in frequency from the ground-state values, in keeping with the properties of the orbital.

ALLYL COMPOUNDS. The spectra of some substituted propylene or allyl compounds are shown in Figs. 10.23 and 10.24 along with expanded spectra for some of the outer π and n orbitals. Although the resonance effect is operative in these molecules by means of hyperconjugation

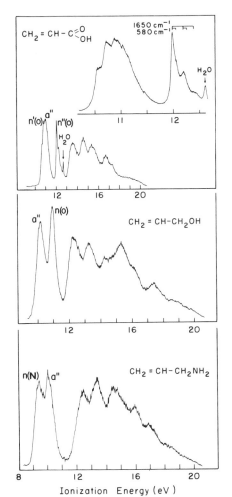

Fig. 10.24. He I spectra of acrylic acid, allylalcohol, and allylamine. Reproduced with permission from Ref. 28.

through the coupling methylene group, its magnitude is decreased because of the large spatial separation of the interacting π type orbitals. In
allylmercaptan and allylamine the vinyl π orbital is inherently more
stable than the sulfur or nitrogen n orbitals. In this case the n orbitals
must be correlated to those of methylmercaptan and methylamine. As a
result resonance interaction destabilizes the $n(a'')$ orbital and stabilizes
the π orbital. The situation is reversed in allylalcohol, where the $n(a'')$
oxygen orbital is more stable than the vinyl π orbital. In acrolein and
acrylic acid the dominant force is the large inductive effect of the oxygen
atoms, which results in a stabilization of the vinyl π orbital. There are
two types of n orbitals in acrylic acid, that is, $n(a')$ and $n(a'')$.

An interesting example of the separation of interacting groups occurs in
the spectra of the three isomeric bromopropenes in Fig. 10.25. In 1- and
2-bromopropene, where the bromine atom is adjacent to the double
bond, the expected interaction occurs—strong resonance interaction splits
the π and $n(a'')$ orbitals, while the $n(a')$ orbital remains nonbonding. In
3-bromopropene the bromine atom is far enough removed from the vinyl
group such that the resonance interaction between the π type orbitals is
very weak. The free rotation of the bromine atom provides local high

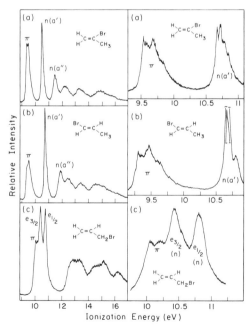

Fig. 10.25.   He I spectra of (a) 2-bromopropene, (b) 1-bromopropene, and (c)
3-bromopropene with expansion of the low-energy bands.

symmetry such that the two spin-orbit components, $e_{3/2}$ and $e_{1/2}$, of the bromine n orbital are observed. The vinyl π orbital assumes a position near that of the π orbital of propylene.

10.6. The perfluoro effect

It is well known that a substituent fluorine atom exerts a strong inductive effect throughout the molecule, drawing electrons toward itself and thereby binding them more tightly to the molecule. This effect is specific, because its influence on π and σ electrons is different. It has been shown that *perfluorination of planar molecules has a much larger stabilization effect on σ MOs than on π MOs*; Brundle et al.[29] have called this the *perfluoro effect. In several nonaromatic planar molecules, it has been found that the π E_Is are shifted by 0 to 0.5 eV upon perfluorination, whereas the σ E_Is increase uniformly by 2 to 3 eV, as illustrated in Figs. 10.26 and*

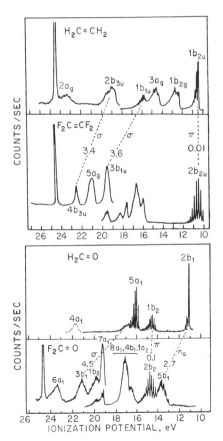

Fig. 10.26. Perfluoro effect in the ethylene-tetrafluoroethylene and the formaldehyde-difluoroformaldehyde pair. Reproduced with permission from Ref. 29.

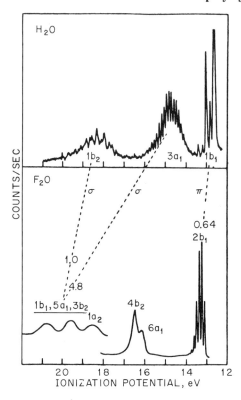

Fig. 10.27. Perfluoro effect in the water-oxygen difluoride pair. Reproduced with permission from Ref. 29.

10.27. The cause of the perfluoro effect can be rationalized through MO theory.[29] In planar nonaromatic molecules perfluorination results in a heavy mixing of the fluorine σ AOs with the σ MOs of the molecule, but the π orbitals remain largely segregated. For example, in tetrafluoroethylene the vinyl π MO remains predominantly localized on the two carbon atoms, whereas the σ MOs acquire large coefficients on the fluorine atoms. Since the σ MOs are more delocalized onto the fluorine atoms than the π MOs, the σ MOs will be more strongly stabilized due to the high E_I of the fluorine atom (17.42 eV). Although the strong inductive effect of the fluorine atoms tends to stabilize the π MO also, this effect is opposed by the acquisition of C—F antibonding character in the π MO which tends to destabilize it. Thus the perfluoro effect can be summarized as a strong stabilization of the σ MOs due to extensive mixing with the σ orbitals of the highly electronegative fluorine atoms and a relatively small stabilization of the π MOs due to the less extensive mixing with the fluorine orbitals and the countereffect of the C–F antibonding character. The extent of the π MO shift will vary from molecule to molecule depending upon their opposing delocalization and antibonding character, but it should remain consistently smaller than the σ shift.

Nonaromatic molecules

The ethylene-tetrafluoroethylene pair provides a good example of the perfluoro effect; a gap of ~6 eV is found between the π and σ orbitals of C_2F_4. It is very difficult to correlate the bands in the 16 to 18-eV region of the spectrum of the perfluorinated analog, because this is the region where the fluorine n orbitals appear. It is, however, possible to correlate some of the bands beyond this highly overlapped region. The water-oxygen difluoride pair also illustrates the main features of the effect: the nonbonding $1b_1$ π orbital suffers only a small shift compared to the $3a_1$ and $1b_2$ σ orbitals. The fluorine n orbitals are clearly situated in the 16 to 17-eV region. The acquisition of O–F and C–F antibonding character in the $2b_1$ and $2b_{2u}$ orbitals of OF_2 and C_2F_4, respectively, is verified by the change in vibrational structure of these bands from the corresponding bands of the perhydronated analogs, that is, the vertical and adiabatic E_Is of these bands are now widely separated, and O–F and C–F stretching modes, respectively, are excited. The effect works surprisingly well on the formaldehyde-difluoroformaldehyde pair, because in this case the oxygen n orbital (σ .type) is very strongly stabilized despite the fact that the substitution occurs at the carbon atom. Molecular orbital calculations show that this n orbital is not completely localized on the oxygen atom; it is partially delocalized into the σ framework of the molecule. This is verified by its drastic change in appearance from H_2CO to F_2CO, indicating considerable interaction with the fluorine atoms and acquisition of bonding character. The second band of the H_2CO-F_2CO pair remains unshifted upon perfluorination, confirming its assignment as a π orbital.

Aromatic molecules

The general rules governing the perfluoro effects of π and σ MOs can also be applied to aromatic molecules, although in this case the π shift is generally larger than in nonaromatics. In the spectra of benzene, naphthalene, and pyridine, Figs. 10.28 to 10.30, all of the E_Is are increased upon perfluorination; some specificity still remains, however, for the π levels are shifted less than the σ levels. In the case of benzene and hexafluorobenzene, Brundle et al.[29] indicate that the $\cdots 1a_{2u}^2(\pi)3e_{2g}^4(\sigma)1e_{1g}^4(\pi)$ ordering for the outer orbitals leads to a reasonable set of perfluoro shifts. The 15 to 19-eV region of the C_6F_6 spectrum is very complex due to the mixing of fluorine n and carbon σ orbitals; this region has not been interpreted. There are five occupied π

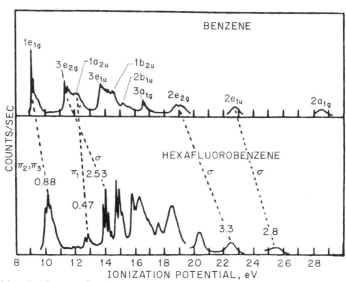

Fig. 10.28. Perfluoro effect in the benzene-hexafluorobenzene pair. Reproduced with permission from Ref. 29.

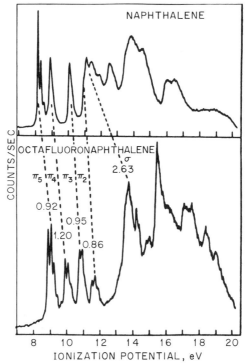

Fig. 10.29. Perfluoro effect in the naphthalene-octafluoronaphthalene pair. Reproduced with permission from Ref. 29.

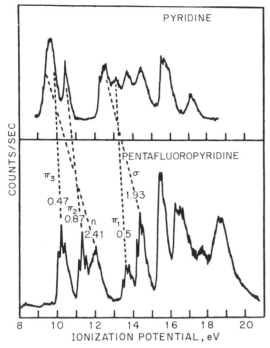

Fig. 10.30. Perfluoro effect in the pyridine-pentafluoropyridine pair. Reproduced with permission from Ref. 29.

orbitals in naphthalene, four of which are expected to lie above the σ orbitals.[30] This assignment is in agreement with the perfluoro shifts of Fig. 10.29 with the π shifts being ~0.4 to 1.0 eV and the σ shifts being 2.3 eV or greater. For pyridine, Fig. 10.30, the situation is more uncertain than the previous cases. In this case, the introduction of a nitrogen atom into the aromatic ring in place of one of the ring carbons causes the doubly degenerate orbitals of benzene to split into two orbitals. However, the presence of the nitrogen n orbital in the same energy region of the spectrum obscures their identities. The pyridine spectrum exhibits two low-energy bands, the first of which is broadened and obviously contains two transitions.[29] The outer π levels are expected[31] to be ~1 eV apart; hence the first band is expected to be an overlap of the n and one of the π orbitals, while the second band corresponds to the remaining π orbital. This provides the ordering $\pi_2 > \pi_3 \sim n$, which gives reasonable perfluoro shifts. Brundle et al.[29] use the order $\pi_2 > \pi_3 > n$ in Fig. 10.30; however, this exact ordering is immaterial since the first two overlapping bands are separated by only ~0.13 eV.

The success of the perfluoro effect lies in the ability of fluorine atoms to produce such a large selective change in binding energies without adding too much complexity to the spectrum. The fluorine atoms add extra bands in the region $\sim$16 to 18 eV; therefore, only bands below and above this region can be correlated. It has been found that the trifluoromethyl group produces an effect similar to that of the fluorine atom; this has been demonstrated[29] in various methyl-trifluoromethyl pairs. The limitations of the perfluoro effect are: (1) The molecule must be planar so that there is clear σ–π separability. (2) The hydrogen and perfluoro pairs must have very nearly the same geometries for the relations between the bands to persist. (3) It is desirable to have MO calculations on the pair of molecules investigated in order to evaluate the extent of delocalization into the fluorine orbitals and provide an independent measure of the perfluoro shifts. (4) The technique is of little value for bands between $\sim$15 and $\sim$19 eV due to the complexity of the structure in this region.

10.7. Substituent additivity effects

Introduction of a given substituent into a molecule of a class of *closely related* molecules produces a change ΔE_I^i in the ionization energy E_I^i that is sensibly constant and, for multiple substitution, roughly additive. McGlynn et al.[32] have shown that the ionization energy $E_I^i(NX)$, where i is an MO index, X is a substituent index, and N is the number of substituents, can be expressed as

$$E_I^i(NX) = E_I^i + N\Delta E_I^i(X) \tag{10.17}$$

where E_I^i is the ionization energy of type i MO in some specified parent, unsubstituted molecule and $\Delta E_I^i(X)$ is a constant for a given substituent X within a class of *closely related parent molecules*. The catch in the terminology resides in the phrase "closely related parent molecules"; the limits of this phrase are yet to be defined. This method can be applied according to the following example.[32] Consider the tentative assignment[29] of the 13.4-eV band of acetone to E_I^π, the C=O π MO. The assignment may be validated by considering the series formaldehyde (H_2CO), acetaldehyde (CH_3CHO), and acetone (($CH_3)_2CO$). The π band of formaldehyde, whose assignment is well established, has E_I^π at 14.09A or 14.51V eV and may be correlated with either the second or third bands of acetaldehyde that occur at E_I (second) = 12.61A or 13.2V eV and E_I (third) = 13.51A or 14.19V eV, respectively. The π band of formaldehyde may also be correlated with either the second or third bands of acetone that occur at E_I (second) = 11.99A or 12.6V eV and E_I (third) = 12.79A or 13.4V eV, respectively. Evaluating the three sets of data simultaneously, the adiabatic ionization energy differences for correlation of the π band of H_2CO with E_I (second) of CH_3CHO and $(CH_3)_2CO$ are

$$H_2CO \xrightarrow{-1.48} CH_3CHO \xrightarrow{-0.62} (CH_3)_2CO$$

whereas an assumed correlation of the π band of H_2CO with E_I (third) yields

$$H_2CO \xrightarrow{-0.59} CH_3CHO \xrightarrow{-0.71} (CH_3)_2CO$$

The latter scheme exemplifies the view expressed in Eq. 10.17 and supports the acetone assignment $E_I(\pi) = 13.4$ eV.

There are a considerable number of systems that support this substituent additivity approach. (1) Consider the $E_I(\pi)$ values for methylated ethylenes: $CH_2{=}CH_2(10.51$ eV); $CH_2{=}CHCH_3$ (9.73 eV); $CH_2{=}C(CH_3)_2(9.23$ eV); $(CH_3)HC{=}C(CH_3)_2(8.67$ eV); $(CH_3)_2$ $C{=}C(CH_3)_2(8.30$ eV). The $CH_2{=}C{=}$ and $(CH_3)_2C{=}C{=}$ groups are pseudoisoelectronic with the $={=}C{=}O$ group, and the average values of $\Delta E_I^\pi(CH_3)$ for the ethylenic groups, -0.64 and -0.47 eV, respectively, are comparable to the values of -0.59 and -0.71 eV found for the carbonyl group. (2) The ionization energies corresponding to the second band of CH_3CHO and $(CH_3)_2CO$, E_I (second), correspond to the first band of methane at 12.5A or 13.5V eV and undoubtedly represent ejection of an electron that has significant amplitude on the $—CH_3$ group. McGlynn et al.[32] have used this substituent additivity method to assign the n and πE_Is of various monocarbonyls and α-dicarbonyls. Some of their examples for the monocarbonyls are presented below.

n Ionization

The effects of $—CH_3$ and $—OH$ substitution on E_I^n (the n orbital of O) for formaldehyde are

$$
\begin{array}{ccc}
\underset{(10.88)}{\overset{\displaystyle O}{\underset{\displaystyle \|}{H\diagdown C \diagup H}}}
\xrightarrow{-0.68}
\underset{(10.20)}{\overset{\displaystyle O}{\underset{\displaystyle \|}{CH_3\diagdown C \diagup H}}}
\xrightarrow{-0.54}
\underset{(9.66)}{\overset{\displaystyle O}{\underset{\displaystyle \|}{CH_3\diagdown C \diagup CH_3}}}
\end{array}
$$

$\downarrow +0.63$ $\downarrow +0.62$

$$
\begin{array}{cc}
\underset{(11.51)}{\overset{\displaystyle O}{\underset{\displaystyle \|}{H\diagdown C \diagup OH}}}
\xrightarrow{-0.69}
\underset{(10.82)}{\overset{\displaystyle O}{\underset{\displaystyle \|}{CH_3\diagdown C \diagup OH}}}
\end{array}
$$

where $\Delta E_{\mathrm{I}}^{n}(CH_3)$, in electron volts, is indicated on the horizontal arrows, $\Delta E_{\mathrm{I}}^{n}(OH)$ on the vertical arrows, and E_{I}^{n} is in parentheses below the molecular representation. Methylation of either formaldehyde or formic acid yields similar $\Delta E_{\mathrm{I}}^{n}(CH_3)$ values. Substitution of an —OH group in either formaldehyde or acetaldehyde yields identical $\Delta E_{\mathrm{I}}^{n}(OH)$ values. It also appears that multiple substitution is only slightly saturative, as indicated by the small change from -0.68 to -0.54 eV caused by the series methylation that yields acetone.

Similar additivity effects are found for the n orbital of oxygen in the acrolein-acrylic acid cycle:

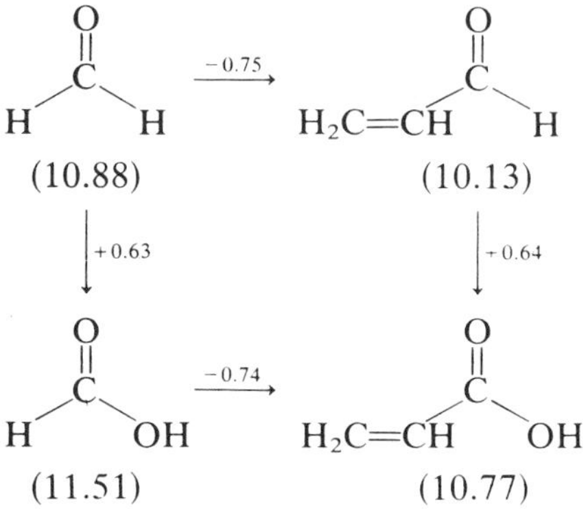

The effects of —CH_3 and —NR_2 (R = H, CH_3) substitution on E_{I}^{n} (the n orbital of O) for formamide are illustrated for the formamide-N,N-dimethylacetamide series in the cycle:

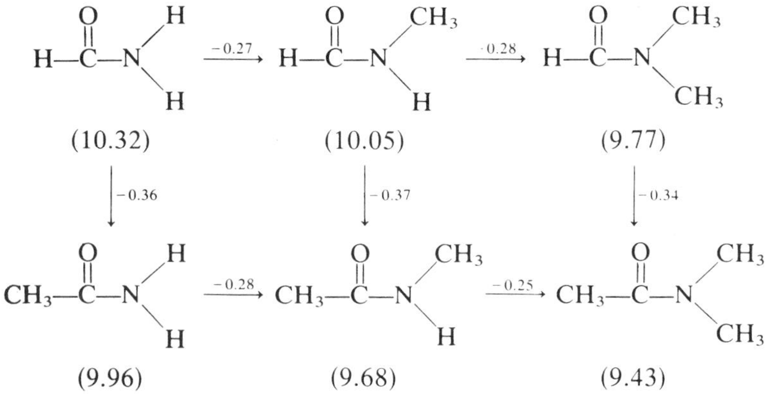

It appears that N-methylation decreases E_{I}^{n} by 0.27 eV, whereas C-methylation produces a decrease of 0.36 eV.

The effects of methylation of the simplest homologs are

$$H_2CO \text{————} CH_3CHO \qquad \Delta E^n = -0.68$$
$$HCOOH \text{————} CH_3COOH \qquad \Delta E_I^n = -0.69$$
$$HCONH_2 \text{————} CH_3CONH_2 \qquad \Delta E_I^n = -0.36$$

The small magnitude of the last value relative to the first two demands explanation. In formaldehyde-acetaldehyde, the adiabatic and vertical E_Is are coincident, while in formic-acetic acids the vertical E_Is are coincident with the second vibrational peaks of the coupled progression in the C=O stretching mode. Hence, the ΔE_I^n values are identical for these two couples whether one uses vertical or adiabatic values. In formamide-acetamide, the vertical energies occur at the second and third vibrational peaks, respectively, of the coupled vibrational progression in the C=O stretching mode. Hence the adiabatic ΔE_I^n for this couple differs from the vertical ΔE_I^n by one quantum of a C=O stretching vibration. In this fashion, it is found for formamide-acetamide that ΔE_I^n is $-0.55A$ eV and $-0.36V$ eV. Hence the apparent discrepancy in the above tabulation is resolved to within the error of experiment. At the same time, this example points to a limitation intrinsic to the use of vertical ΔE_I quantities.

π Ionization

There are two types of π orbitals to consider in these molecules. The most stable of the pair, π_1, is a symmetric combination of the C=O π orbital and the $p\pi$-type orbital of the substituent nitrogen atom; it is predominantly C=O π type. The other member, π_2, is an antisymmetric combination of the C=O π orbital and the $p\pi$-type orbital of the substituent atom; it is predominantly substituent $p\pi$-type.

The effects of methylation on $E_I^{\pi_2}$ are illustrated by the following cycle:

The decrements for N-methylation are surprisingly constant at $\sim$0.6 eV and for C-methylation at $\sim$0.2 eV. A comparable cycle for the π_2 orbital that is largely localized on the hydroxyl oxygen atom in the formic acid-methyl acetate series is:

$$
\begin{array}{ccc}
\underset{(12.51)}{\overset{O}{\underset{H}{\parallel}}\!\!\!\overset{}{\underset{}{}}\!\!\!C\!\!-\!\!OH} & \xrightarrow{\;-0.46\;} & \underset{(12.05)}{\overset{O}{\underset{CH_3}{\parallel}}\!\!\!C\!\!-\!\!OH} \\[2pt]
\Big\downarrow{\scriptstyle-0.96} & & \Big\downarrow{\scriptstyle-0.89} \\[2pt]
\underset{(11.55)}{\overset{O}{\underset{H}{\parallel}}\!\!\!C\!\!-\!\!OCH_3} & \xrightarrow{\;-0.39\;} & \underset{(11.16)}{\overset{O}{\underset{CH_3}{\parallel}}\!\!\!C\!\!-\!\!OCH_3}
\end{array}
$$

Similar cycles for the C=O π_1 orbital pertinent to the formaldehyde-N,N-dimethylacetamide series are:

$$
\begin{array}{cccc}
(14.5) & (14.75) & (14.30) & (13.70) \\[4pt]
\underset{H}{\overset{H}{}}\!C\!\!=\!\!O & \xrightarrow{+0.25} \underset{NH_2}{\overset{H}{}}\!C\!\!=\!\!O & \xrightarrow{-0.45} \underset{NHCH_3}{\overset{H}{}}\!C\!\!=\!\!O & \xrightarrow{-0.60} \underset{N(CH_3)_2}{\overset{H}{}}\!C\!\!=\!\!O \\[4pt]
\Big\downarrow{\scriptstyle-0.31} & \Big\downarrow{\scriptstyle-0.55} & \Big\downarrow{\scriptstyle-0.60} & \Big\downarrow{\scriptstyle-0.60} \\[4pt]
\underset{H}{\overset{CH_3}{}}\!C\!\!=\!\!O & \xrightarrow{+0.01} \underset{NH_2}{\overset{CH_3}{}}\!C\!\!=\!\!O & \xrightarrow{-0.50} \underset{NHCH_3}{\overset{CH_3}{}}\!C\!\!=\!\!O & \xrightarrow{-0.60} \underset{N(CH_3)_2}{\overset{CH_3}{}}\!C\!\!=\!\!O \\[4pt]
(14.19) & (14.20) & (13.7) & (13.10)
\end{array}
$$

Some comment on the left-most cycle, where obvious discrepancies occur, is required. These discrepancies are associated with the fact that the vertical E_Is differ from the adiabatic E_Is by varying numbers of vibrational quanta. To be specific, in the formaldehyde-acetaldehyde couple, the vertical E_I of formaldehyde occurs at the third vibrational peak, whereas in acetaldehyde it must, in order to bring the value -0.31 into accord with the other values for C-methylation, occur at the fifth vibrational peak. If $1210\,\mathrm{cm}^{-1}$ is used for the coupled vibrational quantum, this correction yields $\Delta E_I^{\pi_1} = -0.61$ eV for the formaldehyde-acetaldehyde couple. In addition, this same assertion yields $\Delta E_I^{n} = 0.31$ eV for the acetaldehyde-acetamide couple and removes the discrepancy that existed with respect to the formaldehyde-formamide couple. Finally, this same supposition leads to an adiabatic E_I of 14.59 eV for $E_I^{\pi_2}$ of acetaldehyde; a value that is in excellent agreement with the value 14.5 eV found by Gaussian analysis of the PE spectrum.

In terms of the cited cycle and the analysis just given, it is found that $\Delta E_I^{\pi_2}$ (CH$_3$) for C-methylation is remarkably constant at approximately -0.58 eV, $\Delta E_I^{\pi_1}$ (NH$_2$) at $+0.25$ eV, $\Delta E_I^{\pi_1}$ (NHCH$_3$) at -0.20 eV, $\Delta E_I^{\pi_1}$ (N(CH$_3$)$_2$) at -0.80 eV, and $\Delta E_I^{\pi_1}$ (CH$_3$) for N-methylation at -0.55 eV.

The totality of ΔE_Is, referred to formaldehyde as a parent molecule, are listed for E_I^n and E_I^π in Table 10.5. The $E_I^{\pi_2}$ column refers to ionization of the π orbital that is heavily localized on the substituent group. Since this π_2 orbital is introduced by the substituent itself, the $\Delta E_I^{\pi_2}$ values of Table 10.5 are defined as $\Delta E_I^{\pi_2} = E_{I,HCOX}^\pi - E_{I,HX}^\pi$, where $E_{I,HCOX}^{\pi_2}$ is the ionization energy of the π_2 orbital that is heavily localized on group X of the molecule HCOX and $E_{I,HX}^\pi$ is the ionization energy of the π orbital of X in the HX molecule. Thus the number listed in column $\Delta E_I^{\pi_2}$ for —NHCH$_3$ is $E_I^{\pi_2}$ for HCONHCH$_3$ minus E_I^π of the first band in NH$_2$CH$_3$. In view of the small values of $\Delta E_I^{\pi_2}$ found in this manner, the π_2 identifications given appear to be relatively secure. That is, all π_2 bands of HCOX molecules are energetically very similar to the related bands of HX molecules.

Table 10.5. Effects of Substitution on Ionization Energies[a] (in eV) of Formaldehyde[32]

Substituent, X	ΔE_I^n	$\Delta E_I^{\pi_1}$	$\Delta E_I^{\pi_2}$
—H	—[b]	—[b]	—[b]
—OH	$+0.63$	—	-0.10
—OCH$_3$	$+0.14$	—	-0.28
—NH$_2$	-0.36	$+0.25$	-0.33
—CH$_3$	-0.68	-0.55^c	-0.30
—CH=CH$_2$	-0.75	—	—
—NHCH$_3$	-0.85	-0.20	$+0.17$
—N(CH$_3$)$_2$	-1.12	-0.80	$+0.35$

[a] ΔE_I^n refers to the shift of the oxygen n orbital of the carbonyl group, $\Delta E_I^{\pi_1}$ to the shift of the C=O π_1 orbital, and $\Delta E_I^{\pi_2}$ to the shift in the substituent π_2 orbital (see text for further details).
[b] The shifts listed below are with reference to the hydrogen substituent.
[c] The vertical E_1 of acetaldehyde has been adjusted by two vibrational quanta to obtain this value.

Effects of extended alkylation

The additive substituent effects described above pertain to certain model systems with a selected limited number of substituents. One might expect that these effects would reach a "saturation" level with increasing number and/or size of the substituents. Indeed, recent studies[33,34] on alkyl substitution have shown that there is a rapid nonlinear decrease in the E_1 with increasing number and size of the substituents. These trends are apparent in Fig. 10.31, where the first E_1s of various molecules are shown to be nonlinear functions of the degree of alkyl substitution. Figure 10.31a shows[33] that the decrease in E_1^π of ethylene as the degree of substitution increases is not linear but progressively dampened. The decrease is more pronounced for the larger substituents (t-Bu > Et > Me). As the number of carbon atoms in the substituent increases, Fig. 10.31b shows that the E_1s of n orbitals are decreased, but this decrease becomes progressively smaller as the substituent size increases.

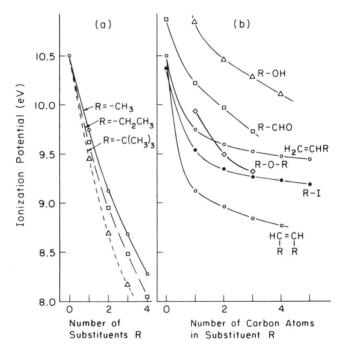

Fig. 10.31. (*a*) The π E_1s of substituted ethylenes versus number of substituents for methyl, ethyl, and *tert*-butyl substituents. (*b*) The first E_1 (n orbital ionization) of alkyl alcohols, aldehydes, ethers, iodides, and ethylenes versus the number of carbon atoms in the n-alkyl group. Data obtained from Refs. 33 and 34.

The decrease of the E_I with increasing degree of alkyl substitution indicates that the introduction of an alkyl group stabilizes the ground state of the ion more than that of the molecule. This large stabilization of the molecular ion is principally the result of delocalization of the positive charge: the higher the electron density in the vicinity of the ionized orbital, the more easily the electron deficiency in the ion can be delocalized, thus leading to a greater stabilization of the ion and consequent reduction in the E_I. The greater stabilization of the ion is attributed to hyperconjugation and electrostatic induction caused by the alkyl substituents. These effects are enhanced by the number of σ electrons in the alkyl groups and hence the number of carbon atoms in the substituent. Studies[33] of various isomers of alkene molecules indicates that the E_Is of cis and trans alkenes are identical, within experimental error, and that E_Is of gem isomers are slightly higher than the corresponding cis and trans forms. In general, the structural effects displayed by the n or π orbitals upon alkyl substitution are, in order of increasing importance: the degree (number of substituents), the nature (size) of the substituents, and the relative positions of the substituents.

10.8. He I–He II intensity effects

The differential cross section for the $1s$ orbital of hydrogen and the valence np orbitals of some nonmetals are plotted in Fig. 10.32 as a function of the photoelectron kinetic energy. For elements of the same period, the maxima in these curves are shifted towards higher kinetic energy as the nuclear charge Z increases. This can be explained by the following qualitative concepts: The spatial extension of AOs diminishes

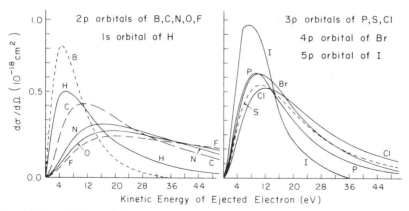

Fig. 10.32. Differential photoionization cross section for various AOs as a function of kinetic energy of the photoelectrons.

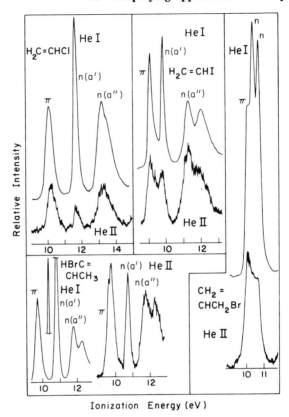

Fig. 10.33.   He I and He II spectra of halogen substituted alkenes.

as nuclear charge increases so that the optimum overlap or match-up of the plane wave and AO occur at shorter wavelengths or higher photoelectron energies for increasing Z. The highest probability for photoionization is determined by the best match-up of plane wave and AO. Also, because of the higher binding energy of electrons in atoms with high nuclear charges, the momentum distribution in the AO is shifted towards higher momenta, resulting in an increased probability for the ejection of high energy photoelectrons.

For many large molecules it is not possible or not practical to perform extended cross-section calculations as in Chapter 6; in these cases the qualitative variations in relative band intensities as a function of incident photon energy can be used as an aid in spectral assignments. For example, consider the spectra of the halogen substituted alkenes in Fig. 10.33. It is possible to identify the vinyl π band and the n bands of the halogens through use of Fig. 10.32 as follows: The ionization bands with which we are concerned in Fig. 10.33 are in the region of ~ 10 eV. Using He I and He II radiation for excitation provides photoelectron kinetic energies of ~ 11 eV and ~ 30 eV, respectively. Reference to Fig. 10.32 shows that the intensity of the $2p$ carbon AO diminishes slightly in going from 11 to 30 eV; however, the outer p orbitals of P, S, Cl, Br, and I suffer a considerable reduction in intensity at higher kinetic energies. Thus we expect the intensities of $3p$, $4p$, and $5p$ AOs to be suppressed relative to $2p$ AOs at high kinetic energies. This effect is observed experimentally in the chlorine and bromine compounds of Fig. 10.33 and the compounds of Figs. 7.1 and 7.2. In these figures we see that the π orbital, which is predominantly carbon $2p$ AOs, retains its intensity under He II excitation while the halogen n orbitals suffer a reduction in intensity. An exception is vinyl iodide where the $n(a'')$ band remains strong under He II radiation. This is due to the significant mixing of the carbon $2p$ and iodine $5p$ orbitals to form the π and $n(a'')$ MOs. Another useful feature exhibited by Fig. 10.32 is that the intensities of the $2p$ orbitals of nitrogen, oxygen, and fluorine are slightly higher with He II than with He I radiation. This opposite effect of He II radiation on $2p$ orbitals as opposed to higher np orbitals is useful for distinguishing nonbonding orbitals of nitrogen, oxygen, and fluorine. However, it should be kept in mind that these intensity schemes are only qualitative, and their gross oversimplification must be recognized.

10.9. "Through-space," "through-bond," and "hyperconjugative" interactions

"Through-space" interactions

The substituted benzene and ethylene molecules discussed in previous sections are examples of *classical conjugation*, that is, the interaction of π-type orbitals that are immediately adjacent and parallel to one another. The splitting of energy levels in these classically conjugated systems is a special case of "through-space" interactions. In general, "through-space" interactions refer to *interactions resulting from the direct overlap of orbitals, including those between nonneighboring centers*.

As an example of these interactions, consider the case of norbornadiene with its π orbitals arranged in symmetric $\pi_+(a_1)$ and antisymmetric configurations $\pi_-(b_2)$ with respect to a mirror plane between the π orbitals, Fig. 10.34a. The conditions for interaction of these orbitals are satisfied: (1) Symmetry-adapted linear combinations π_+ and π_- can be formed. (2) The energies are well matched. (3) The orbitals are in spatial

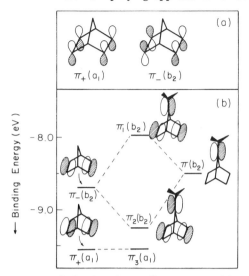

Fig. 10.34. (*a*) π-type orbitals of norbornadiene arranged in symmetry adapted linear combinations $\pi_+(a_1)$ and $\pi_-(b)_2$) with respect to a mirror plane containing the bridgehead carbons. (*b*) Correlation of the π E_1s of norbornadiene, 7-isopropylidene norbornane, and 7-isopropylidene norbornadiene.

positions such that significant overlap between them can occur. The magnitude of the splitting $\Delta E = E(\pi_-) - E(\pi_+)$ is determined by the overlap between the π orbitals. This overlap is inversely proportional to the distance between the interacting orbitals. The energy levels of these symmetry adapted combinations are illustrated on the left side of Fig. 10.34*b*. The splitting[35,36] ΔE is 0.85 eV, and the interacting π orbitals are on atomic centers separated by 2.5 Å. This can be compared to the interacting π orbitals of butadiene where $\Delta E = 2.94$ eV and the nearest π atomic centers are 1.48 Å apart.

In the case of norbornadiene, Helibronner et al.[35] have been able to prove that the antibonding combination of orbitals π_- is indeed the orbital with lowest E_1, as expected from rule (c) of section 10.2. It is possible to prove this by using the selective perturbation of an ethylenic π orbital attached to the bridgehead position. The exocyclic π bond of 7-isopropylidene norbornane is illustrated on the right side of Fig. 10.34*b*; it is antisymmetric with respect to the plane containing the two bridgehead carbons and is designated as $\pi(b_2)$. In the molecule 7-isopropylidene norbornadiene, the $\pi(b_2)$ orbital selectively interacts with the $\pi_-(b_2)$ orbital of norbornadiene while leaving the $\pi_+(a_1)$ orbital unperturbed; as a result, the two new π orbitals are widely split. This is illustrated in the center of Fig. 10.34*b*.

An unusual example of "through-space" interaction is the case of *trans*-azomethane.[37]

One would expect that the overlap between the nitrogen n orbitals is negligible. The actual overlap between these n orbitals has been shown[37,38] to be sizable, negative, and probably larger in magnitude than the corresponding vicinal overlap for *cis*-azomethane. Since the overlap between ϕ_{n_1} and ϕ_{n_2} is negative, $H_{12} = \langle \phi_{n_1} | H' | \phi_{n_2} \rangle$ is positive (see Eq. 10.11 and discussion thereof), and the antisymmetric combination $\phi_- = \frac{1}{2}(\phi_{n_1} - \phi_{n_2})$ is more stable than the symmetric combination $\phi_+ = \frac{1}{2}(\phi_{n_1} + \phi_{n_2})$; the observed splitting is $\Delta E = E_- - E_+ = 3.3 \ eV$.[37]

"Through-bond" interactions

"Through-bond" interactions refer to the *coupling between two orbitals by virtue of their mutual symmetry allowed interaction with a third semilocalized orbital.* Such qualitative considerations are necessary for the interpretation of the spectra of large complex molecules where the results of direct "through-space" coupling do not provide an adequate spectral interpretation. The splitting between the nitrogen n orbitals of diazabicyclo[2.2.2]octane (DABCO) is a classic example of "through-bond" coupling. Consider the "zero-order" case of DABCO, where the ϕ_n orbitals of the nitrogens and the σ MOs are *not* interacting. The situation is depicted in Fig. 10.35*a*. The "through-space" interaction between the ϕ_n orbitals is expected to be small due to the 3-Å separation between the nitrogen centers (consider the small splitting in norbornadiene). The symmetric bonding combination ϕ_+ is expected to lie below the antisymmetric antibonding combination ϕ_- (Fig. 10.35*b*). Next, we "turn on" the interaction of ϕ_+ and ϕ_- with the carbon σ MOs. The ϕ_- orbital is unperturbed because there are no carbon σ MOs of the appropriate symmetry for mixing. However, the ϕ_+ orbital and the carbon σ orbitals do have the same symmetry; their interaction is large enough, Fig. 10.35*c*, to destabilize the ϕ_+ orbital to the extent that the ϕ_+ and ϕ_- orderings are reversed from the conventional models. The splitting observed[39] in the PE spectrum is $\Delta E = E_- - E_+ = 2.13 \, eV$. The reversal of the conventional symmetric-antisymmetric level ordering according to the nodal rule is predicted by MO calculations.[39] This reversed level ordering in DABCO and *trans*-azomethane gives these molecules specific properties

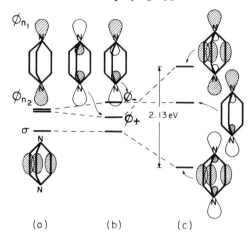

Fig. 10.35. (a) "Zero order" case of noninteracting nitrogen ϕ_{n_1} and ϕ_{n_2} orbitals and carbon σ orbitals in diazabicyclo[2.2.2]octane(DABCO). (b) "Through space" interaction between ϕ_{n_1} and ϕ_{n_2} orbitals to form the combinations $\phi_+ = (1/2^{1/2})(\phi_{n_1} + \phi_{n_2})$ and $\phi_- = (1/2^{1/2})(\phi_{n_1} - \phi_{n_2})$. ($c$) "Through bond" interaction between ϕ_+ and σ orbitals.

and results in electronic absorption spectra that are considerably different from those with the conventional ordering.

"Hyperconjugative" interactions

The term *hyperconjugation*[40,41] *refers to the delocalization that may occur when an alkyl* (CH_3 *or* CH_2) *group is conjugated with an unsaturated* (*ethylenic, acetylenic, or phenyl*) *group.* One of the simplest examples of hyperconjugation is in the methylacetylene molecule, which may be written

$$H_3 \equiv C\text{---}C \equiv CH$$

such that the three C—H bonds of the methyl group are conjugated with the C≡C triple bond. In the molecular point group C_{3v}, the CH group orbitals transform as a_1 and e and the π orbitals of the C≡C bond transform as e. Each component of e has a nodal plane through the molecular axis just as π orbitals. These e orbitals interact to produce an antibonding and a bonding combination of H≡C and C≡C orbitals. As a result of this delocalization (1) the energy of the ground state is slightly lowered; (2) the C—C single bond takes on multiple bond characteristics, that is, it is 1.459 Å as compared to 1.536 Å in ethane; and (3) the E_I of the C≡C π orbital is 10.37 eV as compared to 11.40 eV in acetylene.[4]

However, these results are not due to hyperconjugation alone, as the inductive effect of the methyl group is also operative.

A separation of these hyperconjugative and inductive effects has been accomplished for allyl halides $CH_2=CH-CH_2X$ (X = F, Cl, Br, I) by Schmidt and Schweig.[42] Allyl halides exist predominantly in the *gauche* configuration such that the vinyl π orbital can conjugate with both a C—X and a C—H bond orbital. The hyperconjugative effect destabilizes the π orbital (antibonding combination between vinyl π and —CH_2X), and the inductive effect stabilizes the π orbital (the halogens are more electronegative than hydrogen). The experimental results of Schmidt and Schweig[42] show that the π orbital of allyl chloride, bromide, and iodide is destabilized relative that of ethylene (10.51 eV) due to the predominant hyperconjugative interaction, while the π orbital of allyl fluoride is stabilized due to the predominant inductive effect of the fluorine atom. These two interactions cannot be separated in allyl compounds. However, the inductive effect of the halogen atoms can be estimated from the phenyl-halide spectra by observing the shift of the $\pi(a_2)$ MO from its position in benzene; these shifts are[4] F = −0.58, Cl = −0.45, Br = −0.43, and I = −0.18 eV. Subtracting these inductive effects from the observed total shifts of the π orbitals in the allyl compounds $CH_2=CH-CH_2X$ (F = −0.05, Cl = 0.17, Br = 0.33, I = 0.76 eV) yields the bare hyperconjugative effects ε of the CH_2X substituents as F = 0.53, Cl = 0.62, Br = 0.76, and I = 0.94 eV.

Hyperconjugation in *gauche* allyl halides occurs between the vinyl π orbital and a C—X and C—H bond orbital. The hyperconjugative effect in propylene involves two C—H bond orbitals and destabilizes the π orbital by 0.63 eV. Thus a value of 0.31 eV can be ascribed to the hyperconjugative effect of a *gauche* C—H bond orbital. Subtracting this C—H contribution from the hyperconjugative effect of the CH_2X substituents yields an approximation to the hyperconjugative effect ε of a *gauche* C—H bond as F = 0.22, Cl = 0.31, Br = 0.45, and I = 0.63.

Treating hyperconjugation as ordinary forms of conjugation by means of perturbation theory, one expects ε to be indirectly proportional to the energy difference ΔE between the interacting MOs. As basis MOs for this test, Schmidt and Schweig[42] select the π MO of ethylene and the $\sigma(a_1)$ MOs (C—X bonding) of the methyl halides CH_3X. They find that a plot of ε versus $1/\Delta E$ gives a straight line that passes through the origin as it must. On the assumption that the interaction integrals between various C—X bonding MOs and the π MO of ethylene do not differ by large magnitudes, this result provides a confirmation of hyperconjugation as a conjugation mechanism that can be described by the same model as classical π conjugation. The hyperconjugative ability of this series has been extracted from the treatment[42] as C—F < C—Cl ≈ C—H < C—Br < C—I.

10.10.   Correlation of spectra in related series of molecules

Interpretation of the spectrum of a complex molecule can be facilitated by correlation of the ionization bands with those of a series of related molecules. In such correlations the symmetries of the molecules may change, causing certain orbitals to become degenerate or lifting the degeneracy of other orbitals. These effects are usually observable in PE spectroscopy and provide clues for assignment of bands. For example, the correlation of the ionization bands of the isoelectronic second-row hydrides CH_4, NH_3, H_2O, HF, and Ne, as illustrated in Fig. 4.2, allows one to follow the formation and destruction of orbital degeneracies. This correlation, as discussed in Chapter 4, provides a certain level of confidence that the interpretation of the spectral bands in any one of the molecules of the series is reasonable, if not correct.

Another slightly more complex example is that of the bromomethane series (Fig. 10.36). The "simplist" member of the series, methane, exhibits only two bands corresponding to the two occupied valence MOs $1t_2$ (14 eV) and $2a_1$ (23 eV). In bromomethane the tetrahedral symmetry is reduced to C_{3v}, splitting the $1t_2$ orbital into $5a_1$ (C—Br bonding) and $1e$ (C—H bonding) orbitals. The $1e$ band shows evidence for Jahn–Teller

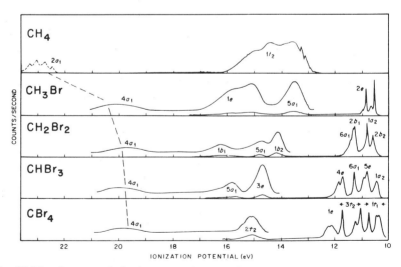

Fig. 10.36. Spectra of the bromomethane series. Reproduced with permission from Ref. 43.

interactions. Substitution of a bromine atom also introduces nonbonding orbitals ($2e$) in the 10.5- to 11-eV region. Spin-orbit splittings are clearly observed in this $2e$ band.

All of the degeneracies are removed in methylene bromide (C_{2v}); hence the orbital angular momentum is quenched, and spin-orbit effects disappear from the spectrum. The bromine n orbitals are all inequivalent, giving four bands in the 10- to 12-eV region. The σ-bond region also splits into three nondegenerate bands corresponding to C—H and C—Br bonding orbitals.

In bromoform, C_{3v} symmetry, the six bromine n orbitals form two degenerate, $4e$ and $5e$, and two nondegenerate, $6a_1$ and $1a_2$, MOs. The bands in the 10- to 12-eV region are in agreement with this deduction; the two sharp bands correspond to the nondegenerate levels, while the two split bands correspond to the spin-orbit components of the degenerate levels. The individual assignments are tentative; however, all six expected components are observed. The σ-bonding region coalesces to $2e$ (C—Br bonding) and $5a_1$ (C—H bonding) orbitals.

The tetrahedral symmetry is restored in carbon tetrabromide and the eight bromine n orbitals become $3t_2$, $1t_1$ and $1e$ MOs. Each of these bands is expected to be split by spin-orbit coupling and may even be further complicated by Jahn–Teller distortions. The eight maxima observed in the low-energy region are tentatively assigned in Fig. 10.36. Some of the maxima may be a result of vibrational or Jahn-Teller structure. The σ orbitals of CBr_4 coalesce into a single $2t_2$ orbital with no evidence for Jahn–Teller or spin-orbit splittings. The $2a_1$ orbital of methane is destabilized by approximately 3 eV upon substitution of a bromine for a hydrogen, however, further substitution of bromine atoms has little effect on it.

The correlation of spectral bands as presented in the above discussion generates some understanding of the spectra of complex molecules. This level of understanding is difficult to achieve on the basis of a single isolated spectrum.

10.11. Electronegativity correlations

Electronegativity is a measure of the attraction of an atom in a molecule for electrons. The concept of electronegativity is useful for interpreting spectra[44] and gaining "chemical intuition" concerning electronic effects within a molecule. There are two practical ways of using this concept.

1. Since nonbonding electrons are generally highly localized on a specific atom in the molecule, their energies are determined, to a large extent, by the electronegativity of that atom. The approximate

Table 10.6. Approximate Range of E_Is for Nonbonding (p-type) Orbitals of Selected Atoms

Atom	Electro-negativity	Type of n Orbital	E_I^n of Methyl Derivative (eV)	Approximate Range of E_I^n (eV)
F	4.0	$2p$	CH_3F^a	13.5–16.0
Cl	3.0	$3p$	CH_3Cl (11.3)	10.5–12.8
Br	2.8	$4p$	CH_2Br (10.5, 10.8)	10.0–12.0
I	2.5	$5p$	CH_3I (9.6, 10.2)	9.2–11.0
O	3.0	$2p$	CH_3OH (10.8)	9.2–12.6
N	3.5	$2p$	CH_3NH_2 (9.0)	7.8–10.2
S	2.5	$3p$	CH_3SH (9.4)	8.7–10.5

^{a} The fluorine $2p$ AOs are strongly involved in the bonding, and, as a result, no fluorine n orbital exists in CH_3F. A fluorine n orbital does exist in fluorobenzene (Section 10.5) at 13.9 eV.

energy ranges of n-type MOs for several atoms are listed in Table 10.6. These bands are usually easy to identify because they are sharp, prominent, and occur in specific spectral regions which are dependent upon the type of atom involved. These $n E_I$s yield approximately linear correlations when plotted against the electronegativity of the atom upon which n is localized. This is illustrated in Fig. 10.14 for the phenyl halide series. Such correlations are possible for many halogen containing molecules such as HX, alkyl halides, inorganic halides, and so forth.

2. Substitution of an atom in a molecule by one of higher electronegativity usually results in an increase in the E_Is of the MOs associated with that atom and the atoms adjacent to it. The more electronegative atom tends to draw electron density toward itself, binding these electrons more tightly and thus increasing the E_Is. The reverse effect is expected when the replacement atom is of lower electronegativity. The perfluoro effect is an extreme example of such electronegativity effects.

References

1. R. J. Colton and J. W. Rabalais, *J. Electron Spectrosc.*, **3,** 345 (1974).
2. S. P. McGlynn, L. G. Vanquickenborne, M. Kinoshita, and D. G. Carroll, *Introduction to Applied Quantum Chemistry*, Holt Rinehart, New York, 1972.
3. A. B. Cronford, D. C. Frost, F. G. Herring, and C. A. McDowell, *J. Chem. Phys.*, **55,** 2820 (1971).

4. D. W. Turner, C. A. Baker, A. D. Baker, and C. R. Brundle, *Molecular Photoelectron Spectroscopy*, Wiley-Interscience, London, 1970.

5. M. J. S. Dewar, *Proc. Cambridge Phil. Soc.*, **45,** 639 (1949); idem., *J. Chem. Soc.*, 2329 (1950); idem., *J. Amer. Chem. Soc.*, **74,** 3341, 3345, 3350, 3353, 3355, 3357 (1952).

6. R. Hoffmann, *Account. Chem. Res.*, **4,** 1 (1971).

7. A. D. Baker, C. R. Brundle, and D. W. Turner, *Int. J. Mass Spectrom. Ion Phys.*, **1,** 443 (1968).

8. B. O. Jonsson and E. Lindholm, *Ark. Fys.*, **39,** 65 (1969); E. Lindholm and B. O. Jonsson, *Chem. Phys. Letters*, **1,** 501 (1967).

9. L. Åsbrink, O. Edqvist, E. Lindholm, and L. E. Selin, *Chem. Phys. Letters*, **5,** 192 (1970).

10. L. Åsbrink, E. Lindholm, and O. Edqvist, *Chem. Phys. Letters*, **5,** 609 (1970).

11. E. Lindholm, *Discussions Faraday Soc.*, **54,** 200 (1973).

12. A. W. Potts, W. C. Price, D. G. Streets, and T. A. Williams, *Discussions Faraday Soc.*, **54,** 168 (1973).

13. E. N. Lassettre, A. Skerbele, M. A. Dillon, and K. H. Ross, *J. Chem. Phys.*, **48,** 5066 (1968); A. Skerbele and E. N. Lassettre, *J. Chem. Phys.*, **42,** 395 (1965).

14. V. H. Dibeler and R. M. Reese, *J. Res. Nat. Bur. Stand.*, A, **68,** 409 (1964).

15. T. A. Carlson and C. P. Anderson, *Chem. Phys. Letters*, **10,** 561 (1971).

16. E. Lindholm, C. Fridh, and L. Åsbrink, *Discussions Faraday Soc.*, **54,** 127 (1973).

17. L. Prand, P. Millie, and G. Berthier, *Theor. Chim. Acta*, **11,** 169 (1968); J. M. Schulman and J. W. Moskowitz, *J. Chem. Phys.*, **47,** 3491 (1967); ibid., **43,** 3287 (1965); P. A. Clark and J. L. Ragle *J. Chem. Phys.*, **46,** 4235 (1967); M. D. Newton, F. P. Boer, and W. N. Lipscomb, *J. Amer. Chem. Soc.*, **88,** 2353 (1966).

18. C. R. Brundle, M. B. Robin, H. Basch, M. Pinsky, and A. Bond, *J. Amer. Chem. Soc.*, **92,** 3863 (1970).

19. C. Baker and D. W. Turner, *Proc. Roy. Soc.*, Ser. A, **308,** 19 (1968).

20. P. A. Clark, F. Brogli, and E. Heilbronner, *Helv. Chim. Acta*, **55,** 1415 (1972).

21. M. J. S. Dewar and S. D. Worley, *J. Chem. Phys.*, **50,** 654 (1969).

22. K. Watanabe, T. Nakayama, and J. Mottl, *J. Quant. Spectrosc. Radiat. Transfer.*, **2,** 369 (1962).

23. A. D. Baker, D. P. May, and D. W. Turner, *J. Chem. Soc. B*, 22, (1968).

24. T. P. Debies and J. W. Rabalais, *J. Electron Spectrosc.*, **1,** 355 (1972/73).

25. A. L. Allred, *J. Inorg. Nucl. Chem.*, **17,** 215 (1961).

26. R. J. Colton and J. W. Rabalais, *J. Electron Spectrosc.*, **1,** 83 (1972/73).

27. J. W. Rabalais, *J. Chem. Phys.*, **57,** 960 (1972).

28. A. Katrib and J. W. Rabalais, *J. Phys. Chem.*, 77, 2358 (1973).

29. C. R. Brundle, M. B. Robin, N. A. Kuebler, and H. Basch, *J. Amer. Chem. Soc.*, **94,** 1451 (1972); C. R. Brundle, M. B. Robin, and N. A. Kuebler, *J. Amer. Chem. Soc.*, **94,** 1466 (1972).

30. R. J. Buenker and S. D. Peyerimhoff, *Chem. Phys. Letters*, **3**, 37 (1969).
31. A. D. Baker, D. Betteridge, N. R. Kemp, and R. E. Kirby, *Chem. Commun.*, 286 (1970).
32. J. L. Meeks, H. J. Maria, P. Brint, and S. P. McGlynn, *Chem. Rev.*, **75**, 603 (1975).
33. P. Masclet, D. Grosjean, G. Mouvier, and J. Dubois, *J. Electron Spectrosc.*, **2**, 225 (1973).
34. B. J. Cocksey, J. H. D. Eland, and C. J. Danby, *J. Chem. Soc. B*, 790 (1971).
35. R. Hoffman, E. Heilbronner, and R. Gleiter, *J. Amer. Chem. Soc.*, **92**, 706 (1970); F. Brogli, E. Heilbronner, and J. Ipaktschi, *Helv. Chim. Acta*, **55**, 2447 (1972); E. Heilbronner and H. D. Martin, *Helv. Chim. Acta*, **55**, 1490 (1972).
36. H. Bock and P. D. Mollere, *J. Chem. Educ.*, **51**, 506 (1974).
37. E. Haselbach, J. A. Hashmall, E. Heilbronner, and V. Hornung, *Angew. Chem. (Int. Edit.)*, **8**, 878 (1969); E. Haselbach and E. Heilbronner, *Helv. Chem. Acta*, **53**, 684 (1970).
38. M. B. Robin and W. T. Simpson, *J. Chem. Phys.*, **36**, 580 (1962); R. Hochstrasser and S. Lower, ibid., **36**, 3505 (1962); M. B. Robin, R. R. Hart, and N. A. Kuebler, *J. Amer. Chem. Soc.*, **89**, 1564 (1967).
39. E. Heilbronner and K. Muszkat, *J. Amer. Chem. Soc.*, **92**, 3818 (1970); R. Hoffmann, A. Imamura, and W. J. Hehre, *J. Amer. Chem. Soc.*, **90**, 1499 (1968).
40. R. S. Mulliken, *J. Chem. Phys.*, **7**, 339 (1939).
41. M. J. S. Dewar, *Hyperconjugation*, Ronald, New York, 1962.
42. H. Schmidt and A. Schweig, *Angew. Chem. (Int. Edit.)*, **12**, 307 (1973).
43. A. W. Potts, H. J. Lempka, D. G. Streets, and W. C. Price, *Phil. Trans. Roy. Soc. London*, **A268**, 59 (1970).
44. A. D. Baker, D. Betteridge, N. R. Kemp, and R. E. Kirby, *Int. J. Mass Spectrom. Ion Phys.*, **4**, 90 (1970).

11 Spectral Interpretation and Application

This chapter contains a summary of the methods available for interpretation of PE spectra and a discussion of some of the specific applications of the technique. *The most important application of PE spectroscopy is to determine accurate values of the E_Is of electrons in various molecular energy levels and to use these E_Is as well as information obtained from the spectral fine structure to elucidate the electronic structure of molecules and ions.* Photoelectron spectroscopy has evolved as one of the best methods of obtaining information such as bond lengths, geometries, vibrational frequencies, and nature of the electronic states of molecular ions. The preceding chapters have been devoted to just that ; E_Is and spectral fine structure have been used to obtain information about electronic structures. The remainder of this chapter deals with more specific, nevertheless, important applications of PE spectroscopy. Related topics such as charge-exchange mass spectroscopy and Rydberg spectra are also discussed.

11.1. Summary of methods for interpreting PE spectra

Various aspects of molecular electronic structure that are manifest in PE spectra have been discussed in previous chapters. In such discussions the spectral band assignments have been accepted as already known or have been deduced from the specific features of the phenomenon under investigation. A spectroscopist who is armed with a knowledge and understanding of these phenomena has all of the tools necessary for interpretation of PE spectra. The following is a summary of the approaches that can be used for interpretation and assignment of photoelectron bands.

1. *Quantum Chemical Calculations*—Calculations of electronic structure, whether they be rigorous or crude in their approximations, are

highly desirable and almost indispensable in interpreting PE spectra. No matter how crude the calculation, the important point is that it provides an ever essential *model* to be used in interpretation. Naturally, the more rigorous calculations usually provide more realistic models than the very approximate schemes. Such realistic models should contain the capacity for identifying and assigning all of the observed spectral bands and features. To this end, molecular orbital calculations have proven most valuable, and approaches such as *ab initio* SCF methods, CNDO, INDO, MINDO, SPINDO, Xα-method, extended-Hückel and Hückel methods, and LCMO methods are all useful and worthy in that they provide interpretative models, albeit with varying degrees of sophistication. (See Section 4.7 for a discussion of computational details.)

2. *Spectral Fine Structure*—The observation of fine structure in spectral bands, such as vibrational progressions (Sections 3.3 to 3.6), rotational structure (Section 3.10), spin-orbit splittings (Chapter 8), and Jahn–Teller and Renner effects (Chapter 9) can lead to positive identification of specific spectral bands provided that the features are interpreted properly. Indeed, the analysis of spectral fine structure, if used correctly and with a knowledge of selection rules (Sections 3.7 to 3.10), can be one of the most conclusive means of identifying transitions. Full use of such spectral information should be employed whenever it is present.

3. *Band Intensities*—The observation of different intensities in various spectral bands or a change in band intensities as a function of excitation energy (or kinetic energy of ejected electrons) can lead to band assignments if these intensities can be reproduced by cross-section calculations such as those of Chapters 6, 7, and 10. The use of different excitation sources, such as Ne I, He I, He II, or X-rays, can produce large changes in spectral intensities that can be reproduced by calculations, or even rationalized, in an attempt at assignment.

4. *Angular Distributions*—Studies of the angular distributions of photoelectrons lead to experimental determination of the asymmetry parameter β. The β values for specific MOs can be obtained theoretically (Chapters 6 and 7) and used as a criterion for spectral assignment. Problems arise in the difficulty and accuracy of such calculations for large molecules and in the practical experimental problems.

5. *Perfluoro Effect*—The large shifts in the energies of spectral bands that occur upon perfluorination can be used as a criterion for distinguishing between π- and σ-orbital ionizations of planar

molecules. Such shifts can provide powerful evidence (Section 10.6) for the elucidation of π and σ manifolds.

6. *Correlation with a Series of Related Compounds*—In complex molecules where many of the ionization bands overlap or contain no fine structure, correlation of the bands with a series of related compounds as in Section 10.10 may be a most useful clue to band assignments. The energies of MOs are expected to vary smoothly and in predictable manners as atoms are substituted within a molecule; such *trends* can usually be reproduced by MO computations.

7. *Electronegativity Trends*—Molecular orbitals that are highly localized on specfic atoms, such as nonbonding orbitals, inner valence orbitals, or core orbitals, usually occur in specific spectral regions that are determined mainly by the electronegativity of the specific atom (Section 10.11). The halogen, oxygen, or sulfur nonbonding orbitals or metal d orbitals generally occur in predictable spectral regions regardless of the nature of the molecule within which the atom is contained.

8. *Band Shapes*—Even though fine structure is not resolved, it is sometimes possible to obtain qualitative information from band shapes. For example, nonbonding orbitals produce sharp bands with identical adiabatic and vertical E_Is, whereas strongly bonding orbitals are expected to produce a more rounded spectral band and may lead to very broad bands if dissociation occurs.

9. *Orbital Interactions*—Symmetry restrictions allow only certain orbitals to interact in symmetrical molecules. Such "through-space" interactions, as between the π orbitals of benzene and ethylene and their substituents (Sections 10.4 and 10.5), or "through-bond" interactions (Section 10.9) often serve to distinguish specific spectral bands when a composite molecule is compared to its separate subunits.

10. *Charge-Exchange Mass Spectra*—Charge-exchange mass spectra (Section 11.7) can be used to determine the excitation energy required to produce certain fragment ions. Since fragment ions are produced by ejection of electrons that are strongly bonding in the region of fragmentation, correlation with PE spectral bands provides information on the bonding and localization properties of the ejected electrons.

11. *Rydberg Spectra*—Rydberg series converge to molecular ionization energies, the bands exhibit fine structure similar to that of PE spectral bands, and *high* resolution conditions can be obtained in optical Rydberg spectra. Hence the analysis of these bands can

prove useful in identification of PE spectral bands, and furthermore, the combined use (Section 11.8) of Rydberg and PE spectra allow accurate determination of quantum defects δ. The δ values can lead to convincing assignments for specific bands through application of selection rules. Also, the use of fluorescence spectra of molecular ions can establish an assignment on the basis of the stringent selection rules of optical spectroscopy.

12. *Atoms-in-Molecules and Molecules-in-Molecules Approaches*— Simplified, although extremely useful, MO diagrams can be constructed (Sections 10.1 and 10.2) by using the valence-state ionization energies of the constituent atoms of a molecule or the energies of the subunits of a composite molecule and allowing those orbitals of similar symmetry to interact. Although these schemes are qualitative, they can provide helpful first approximations for interpretation of complex spectra.

13. *Substituent Additivity Effects*—Shifts in the energies of certain orbitals caused by addition of specific substituents (Section 10.7) onto a molecule can provide a unique means of identifying a certain orbital within a series of similar molecules.

Correct band assignments and MO-ordering schemes are, in principle, verifiable using all of the thirteen assigment criteria listed above. The best assignments and those that are most firmly established use all or at least several of these approaches. The question of what establishes a spectral band assignment and MO ordering as conclusive is difficult, if not impossible, to answer for the general case. As any reader of the PE spectroscopic literature will quickly realize, some researchers expend considerable effort establishing a firm and conclusive assignment and MO ordering for a single molecule, whereas others assign the spectra of entire families of molecules with minimal effort; obviously, the question is one of personal taste and attitudes. It is generally believed that the more assignment approaches that are consistent with a given interpretation, the more firmly established is the assignment.

11.2.   Molecular structure determinations

Ultraviolet PE spectroscopy in its present status is not a general technique for molecular structure determinations; the more established classical methods remain superior. There are some specific cases, however, where it can provide quick evidence for gross molecular structure with less work then the classical methods. Some of these examples, discussed in this section are hexafluorobutadiene, substituted biphenyls, and phenyl ethylenes. Aside from these determinations of molecular structures, PE spectroscopy is often very helpful in determinating *ionic structures*. The geometry of a molecular ion in its ground or excited states can often be deduced from an analysis of the vibrational and rotational structure of high-resolution spectra; for example, see Sections 3.3 and 3.4, and particularly Eq. 3.25, Chapter 7, and Chapter 9. Photoelectron spectroscopy has made significant contributions to our knowledge of the geometrical structures of small molecular ions. Now we focus attention on the determination of molecular structures.

Hexafluorobutadiene

The low-energy E_1s of the planar cis and trans forms of butadiene differ by only ~ 0.1 eV according to recent calculations. The reason for this small difference is that the lowest E_1s of the two isomers differ only in the extent of the weak interaction between the carbon atoms at the extremities of the chain, C_1 and C_4. However, from Section 10.2 we know that the two occupied π orbitals of butadiene, π_1 and π_2, are delocalized over the entire molecule with π_1 bonding between all atoms and π_2 antibonding between C_2 and C_3. If the molecule is twisted about the C_2—C_3 bond, conjugation between $C_2(p\pi)$ and $C_3(p\pi)$ diminishes progressively until the angle between the planes of the two ethylene groups reaches 90°. In this position $C_2(p\pi)$ and $C_3(p\pi)$ cannot interact, and the π_1 and π_2 eigenvalues approach each other. Therefore, one expects that in nonplanar butadienes the π_1–π_2 splitting will collapse as the angle between the planes of the two ethylene subunits approaches 90°.

Brundle and Robin[1,2] have effectively demonstrated this phenomenon using the fluoro derivatives of butadiene. From comparison of the spectra of butadiene and 1,1,4,4-tetrafluorobutadiene, Fig. 11.1, we see that the perfluoro effect (Section 10.6) is operative; that is, π_1 and π_2 are relatively unchanged, while the first σ band is shifted by ~ 2.5 eV. The relative ordering of the first σ and inner π orbital of butadiene has been discussed in detail.[2] The perfluoro effect is valid in this case because it is known that both butadiene and 1,1,4,4-tetrafluorobutadiene have planar trans structures in the gas phase. Comparison of these spectra with that of hexafluorobutadiene reveals that the 3-eV splitting between the π_1 and π_2 MOs of butadiene collapses to only 1 eV in the perfluorinated derivative. This is indicative of a decreased coupling between the two olefin π orbitals caused by twisting about the C_2—C_3 bond. With this clue, hexafluorobutadiene was investigated by optical spectroscopy and electron diffraction and was found[2] to have a cis nonplanar structure with a dihedral angle of 42 to 48° between the planes of the ethylene groups.

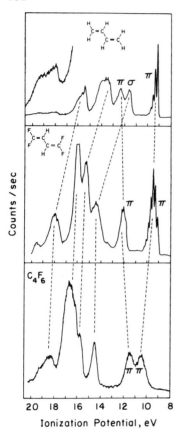

Fig. 11.1. Correlation of the He I spectra of butadiene, 1,1,4,4-tetrafluorobutadiene, and hexafluorobutadiene. Reproduced with permission from Ref. 2.

Steric inhibition of resonance

Photoelectron spectroscopy has been used to investigate *steric inhibition of resonance* by Maier and Turner in a number of conjugated organic molecules.[3,4] In the case of biphenyl compounds,[3] the maximum resonance interaction between the π electrons of the two phenyl groups occurs when the rings are coplanar. If coplanarity is prevented by steric factors, the resonance interaction is reduced according to the degree of nonplanarity between the two rings. This twisting about the central bond is reflected in the energies of the occupied π orbitals. The resonance interactions between the π orbitals of the two rings tend to keep the molecule coplanar, while the nonbonded repulsions tend to be minimal in the perpendicular orientation. The introduction of substituents into the ortho positions of biphenyl increases these repulsions, thus enhancing the tendency towards nonplanarity. The effect of rotation about the central bond on the π

levels of biphenyl can be visualized by referring to Fig. 10.7. The π_4 and π_5 orbitals are predominantly invariant to rotation, while the π_3 and π_6 orbitals reflect changes in conjugation. Treating biphenyl as a composite molecule, it can be shown within the framework of a Hückel LCMO calculation that the splitting between the π_3 and π_6 orbitals can be written as

$$\Delta\varepsilon_{3,6} = \varepsilon(\pi_6) - \varepsilon(\pi_3) = -\tfrac{2}{3}\beta \cos\theta \qquad (11.1)$$

where β is the standard Hückel resonance integral and θ is the dihedral angle between the planes of the two phenyl groups. The maximum interaction is in the coplanar case where the dihedral angle is zero. In Eq. 11.1 the $\pi-\sigma$ interactions in the nonplanar molecules are ignored. If β is assumed to be effectively constant for the interaction of the phenyl π orbitals across the central bond of biphenyl, a plot of $\Delta\varepsilon_{3,6}$ versus $\cos\theta$ should give a straight line. Using θ determined from electron diffraction studies, a plot of $\Delta\varepsilon_{3,6}$ versus $\cos\theta$ yields a reasonable linear regression.[3] However, the gradient is ~ 1.7, and extrapolation to $\theta = 90°$ yields a finite intercept. These effects must certainly be a reflection on the simplicity of the method, because it neglects $\sigma-\pi$ interactions, interactions with antibonding orbitals, and changes in β. Such an empirical relationship is useful, however, to determine the conformation of other substituted biphenyls, and has indeed been applied with success.

Similar studies have been carried out on a variety of complex molecules. Maier and Turner[4] have also studied phenylethylenes, anilines, and phenols; however, in this case the lack of electron diffraction data required the assumption of a linear regression between $\Delta\varepsilon$ and $\cos\theta$ as defined by only two coordinates, at $\theta = 0°$ and $90°$. Cowling and Johnstone[5] have used PE spectroscopy to estimate the dihedral angles in a series of substituted anilines. In this case the lack of experimental data on the dihedral angles was overcome by the following two methods: (1) Resonance interaction between the nitrogen n orbital and the phenyl π orbitals is maximum when $\theta = 0°$ and decreases as θ increases until $\theta = 90°$, where there is no conjugation between the n and π orbitals. Assuming a linear relationship between these two extremes and using the energy of the n orbital of aniline to represent the case $\theta = 0°$, it is possible to use E_Is of n orbitals of substituted anilines, corrected for the inductive effect of the substituent, to determine the unknown dihedral angles. (2) In the second method, no external standard is used; the separation between the two phenyl π orbitals, $\pi(b_1)-\pi(a_2)$, corrected for the inductive effect of the substitutent group, is used to measure the interaction. The two extreme cases of interaction, maximum resonance and no conjugation, are chosen as aniline and N,N-dimethyl-2-methylaniline. Values obtained from both of these methods are used in plots of $\Delta\varepsilon$ versus $\cos\theta$ or $\cos^2\theta$. The two methods yield reasonably consistent values of θ.

Heilbronner et al.[6] have investigated the relationship between angle of twist and the position of π bands for diene and biphenyl type compounds. Approximately linear relationships between $\Delta\varepsilon$ and $\cos\theta$ were found; however, the authors emphasize that nonplanar deformations of a conjugated system can be effected not only by a pure twisting about a bond, but also by out-of-plane bending deformations. Thus, characterization of deformation requires defining both the angle of twist and out-of-plane bending. The simple models as used in biphenyl, aniline, and so forth are, therefore, applicable within very narrow limits.

Berkowitz et al.[7] have examined the spectrum of the hydrocarbon (α-napthyl)-$(CH_2)_4$-(α-napthyl) in order to determine if it exists in the open-chain or cyclic form in the gas phase at $\sim 200°C$. The determination was made by comparing the spectrum of (α-naphthyl)-$(CH_2)_4$-(α-naphthyl) with that of n-Bu-α-naphthalene. If the former compound exists in the open-chain form, the two spectra should be very similar, whereas if it exists in the cyclic form, interaction between the naphthyl groups should cause substantial shifts in the outer π orbitals. It was found that the E_Is of these two compounds are identical, within the experimental precision, requiring that $\geq 90\%$ of (α-naphthyl)-$(CH_2)_4$-(α-naphthyl) exists in the open-chain form at 200°C.

11.3. Qualitative analysis

Photoelectron spectroscopy is a useful technique for qualitative identification of a substance or qualitative analysis of a mixture of gases or volatile liquids. The technique has not reached the popularity of the more established classical methods of qualitative analysis, although this is probably due to its tender age. The PE spectrum of a molecule provides a "fingerprint" that can be used for qualitative identification. The analysis of mixtures is complicated by the inevitable overlapping of bands; however, mixtures of only two to three components are generally identifiable. When mixtures of liquids are analyzed, those components with the highest vapor pressures naturally appear with highest intensities. These different volatilities may permit resolution of the mixture into its components by continuously pumping on the mixture while simultaneously sampling the spectrum.

Photoelectron spectroscopy provides differentiation between geometrical isomers that is sometimes difficult by other means. For example, in Fig. 11.2 the spectrum of 3-bromothiophene is distinctly different from that of 2-bromothiophene, and both are easily distinguished from that of the

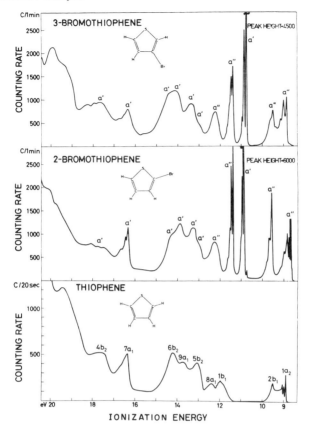

Fig. 11.2. He I spectra of 3-bromothiophene, 2-bromothiophene, and thiophene. Reproduced with permission form Ref. 8.

parent molecule thiophene.[8] The sharp n-type bands between 10.7 and 11.8 eV are unmistakable evidence for a bromine substitutent. Corresponding "fingerprints" of four chloroalkene compounds are presented in Fig. 11.3. Consider a comparison of the 1-chloropropene and 1-chloro-2-methylpropene spectra. The vinyl π band and the chlorine n-bands are clearly distinguished in the regions ~ 9.2 and ~ 11.3 eV, respectively. Although these spectral regions are quite similar, the presence of an additional methyl group in one compound provides considerable difference in the 12- to 16-eV region and allows easy differentiation between the two. The spectra of the 1,1-dichloropropene and 1,3-dichloropropene pair are considerably different, even in the lower-energy region. The remarkable difference in the chlorine n bands reflects the influence of the

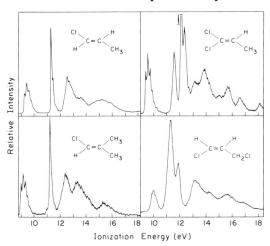

Fig. 11.3. He I spectra of 1-chloropropene, 1-chloro-2-methylpropene, 1,1-dichloropropene, and 1,3-dichloropropene.

position of the chlorine atom on the interaction between the chlorine n orbitals and the vinyl π orbital. It should also be noted that the presence of one or two chlorine atoms in these molecules is easy to detect from the different intensities in the ~ 11.0- to 12.4-eV region. Other examples of the PE spectra of isomers that can be used for their identification are *cis*- and *trans*-1,3, dichloropropene,[9] 2-methylpropane and *n*-butane,[10] *n*-pentane and neopentane,[10] and 2-methylbutane and *n*-pentane[10] pairs.

The spectral regions observed in PE spectroscopy can be divided into a low- and high-energy region. We define the low-energy region as < 17 eV and the high-energy region as ~ 17 to 40 eV. The low-energy region contains bands resulting from ionization of p-type and hydrogen-$1s$ orbitals. These are the bands that most PE spectroscopic investigations are concerned with, for they represent the most loosely bound electrons in the molecule and are most significant in terms of chemical reactivity. However, the investigation of the high-energy region using He II radiation can be very enlightening in terms of elemental analysis. Potts et al.[11] have shown that s-type orbitals of molecules have E_Is in this region and that they possess considerable atomic character. For example, molecules containing C, N, or O always have s-type bands around 22, 27, or 32 eV, respectively. It is apparent that the *inner valence shells show some of the atomic properties characteristic of the X-ray spectra of core orbitals*. The positions of the valence s-type bands in various elements are listed in Table 11.1.[11] From this table we see that analysis of the nonmetal

Table 11.1. Vertical $E_I s^{11}$ of Valence s-type Bands of the Simple Hydride XH_n of the Element X

Element	s-type E_I(eV)	Element	s-type E_I(eV)
F	39.0	As	19.0
Cl	25.7	Sb	17.3
Br	24.4	C in CH_3	22.3
I	21.7	C in CH_2	21.5
O	32.2	C in CH	20.3˙
S	22.2	C	19.4
Se	21.0	Si	18.2
Te	18.6	Ge	18.4
N	27.0	Sn	16.9
P	19.0		

elements in a molecule is feasible providing that good He II spectra are available.

11.4. Investigations of transient species

Photoelectron spectroscopy is applicable to studies of transient species provided that high enough concentrations of these unstable species can be generated. There are practical problems involved, because generally the transient species is a minor component in a mixture of more stable species. The unstable species is generally produced by an electrodeless discharge through the parent gas or parent-gas–rare-gas mixture or by pyrolysis of the parent molecules. The problem of maintaining a high steady-state concentration of the transient is critical; it is effected by differential pumping and, in some cases, condensing the parent gas in a suitable cold trap. Several research groups have investigated transient or unstable species; some of these investigations are discussed below.

Methyl radical

The spectral band corresponding to the first E_I of CH_3 has been observed[12] in the pyrolysis products of azomethane. The spectrum consists of two vibrational components separated by 2720 ± 30 cm^{-1}. The 0–0 band at 9.837 ± 0.005 eV dominates the spectrum, indicating that the ejected electron is nonbonding and that the geometry of CH_3 and CH_3^+ are nearly identical. The most reasonable assignment for the vibration is the totally symmetric ν_1 mode of CH_3^+, although experimental infrared values of these vibrations are not available for comparison.

Atoms

The spectra of atomic hydrogen, nitrogen, oxygen, fluorine, chlorine, and bromine have been obtained by Jonathan and coworkers.[13,14] Atoms are generated by microwave discharges through the parent gas. The spectrum of atomic hydrogen shows a single E_I at 13.61 eV due to the $^1S \leftarrow ^2S$ transition and that of atomic nitrogen shows a single E_I at 14.55 eV due to the $^3P \leftarrow ^4S$ transition. In atomic oxygen, transitions are found at 13.62, 16.96, and 18.63 eV corresponding to the transitions from the ^3P groundstate to the ^4S, ^2D, and ^2P ionic states, respectively.

Other molecules

Some of the transient or unstable molecules that have been observed by PE spectroscopy are: $O_2(^1\Delta_g)$,[14] SO,[15] CS,[16–18] HCP,[19] S_2O,[20] ClO_2,[21] O_3,[22,23] 2,2,6,6-tetramethylpiperidine-*N*-oxyl and di-*tert*-butylnitroxide radicals,[24] NF_2,[25] $(CF_2)_2NO$,[25] SO_3F,[25] and H_2CS.[26] Excited oxygen O_2 ($^1\Delta_g$) was produced[14] at an estimated 10 to 12% concentration by a microwave discharge in molecular oxygen. Photoionization of valence shell electrons in O_2 ($^1\Delta_g$) produces $^2\Pi_g$, $^2\Pi_u$, $^2\Phi_u$, and $^2\Delta_g$ states of O_2^+. The transition $O_2^+(^2\Pi_g) \leftarrow O_2(^1\Delta_g)$ to the ground state of the ion occurs at 11.09 eV. Transitions to excited ionic states have been observed;[14] however, firm assignments are not yet possible. Many of the transient species listed above contain unpaired electrons in their ground states; hence multiplet splittings are observed in many of the spectra. These splittings, together with the superposition of the parent molecule spectrum, make interpretation difficult. Despite these difficulties, the application of PE spectroscopy to transient species has been fruitful, and vigorous activity in this area will certainly continue.

11.5.   *d*-Orbital participation in bonding

The influence of virtual *d*-orbitals on occupied bonding orbitals is discussed in this section. It is well known that partially filled *d* orbitals can be very important in the bonding properties of an element, for example, the transition metal complexes. The present discussion is limited to the question of whether or not virtual *d* orbitals are important in formulating models for describing the bonding of molecules containing the nonmetals of Groups IVA to VIIA.

Sharp contrasts exist when the chemical properties of the second-row elements C, N, O, and F are compared to those of the corresponding elements of the other rows. One of the reasons for these differences is the capability of expansion of the valence shells of the third- to sixth-row elements in order to accomodate extra electron density. The second row elements have only $2s$ and $2p$ AOs available for bonding while the heavier elements have ns, np, and nd AOs available. The question of *if* and *when* these nd orbitals partake in bonding has been studied by many techniques.

It is possible to use PE spectroscopy to investigate the degree of d-orbital participation through shifts in the binding energies of occupied MOs of a homologous series. Proceeding in this manner, several researchers have claimed evidence for virtual d-orbital participation in molecular bonding. Some of the most convincing examples are: XF_4(X = C, Si, Ge);[27] $X(CH_3)_4$(X = C, Si, Ge, Sn, Pb);[27] $C_6H_5XH_2$, $(C_6H_5)_2XH$, $(C_6H_5)_3X$(X = N, P, As, Sb);[28] RQ (R = H, CH_3, SiH_3, GeH_3; Q = NCO, NCS, N_3);[29] SiH_3X, SiH_2X_2, GeH_2X_2 (X = F, Cl, Br, I);[30,31] $(MH_3)_2Y$, MH_3SH (M = C, Si, Ge; Y = O, S, Se, Te);[32] $(CH_3)_3Si—C≡C—X$ (X = H, F, Cl, Br, I).[33]

The work of Carlson et al.[27] on the XF_4 and $X(CH_3)_4$ (X = C, Si, Ge, Sn, Pb) series provides convincing evidence that backbonding into vacant d orbitals causes stabilization of certain MOs and that this effect increases in importance as the atomic number of the central atom increases. Consider the spectra of the tetrafluoride compounds in Fig. 11.4. Silicon and germanium have the potentiality of backbonding through their unfilled valence-shell d orbitals. Using MO calculations it was found that[27] all orbitals are stabilized to some extent by the expansion of the Si and Ge valence shells to include d orbitals, but in particular the nonbonding fluorine orbitals, $3t_2$ and $1e$, *should* be preferentially stabilized. This stabilization arises from a delocalization of the high valence shell electron density into the central atom's vacant d orbitals, thus providing a shift of electronic density into the spatial region between the central atom and the surrounding fluorines. The delocalization reduces inter-electron repulsions, and molecular bonding becomes more energetically favorable. The assignment scheme for these molecules, as predicted by Carlson et al.,[27] is presented in Fig. 11.5. The assignments can be summarized as follows. As expected for an orbital that remains totally ligand nonbonding, the $1t_1$ MO remains essentially constant in energy. The $1e(n)$ MO of SiF_4 and GeF_4 experiences significant stabilization due to participation of the central atom's unfilled $3d$ and $4d$ AOs, respectively, in the molecular bonding. The effect of d-orbital stabilization on the $3t_2$ MO is overridden by the commensurate destabilization of the $2t_2$ MO, primarily induced by its increasing fluorine character. In the case of GeF_4 the filled $3d$ AOs are of an appropriate energy for mixing with e or t_2 MOs. Such a mixing would cause destabilization of the MOs; however, it is not expected to be large. The $2a_1$ orbital is significantly destabilized as the atomic number of

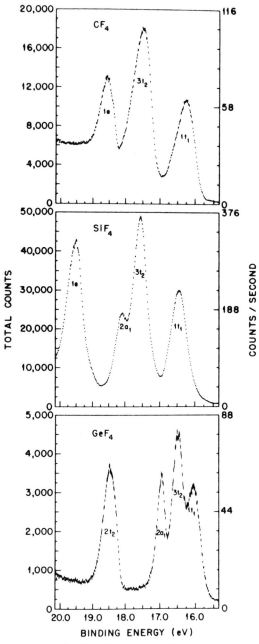

Fig. 11.4. He I spectra of CF_4, SiF_4, and GeF_4. Reproduced with permission from Ref. 27.

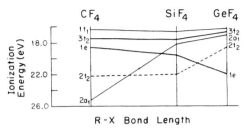

Fig. 11.5. Plot of experimental ionization energies versus central atom-fluorine bond lengths for CF_4, SiF_4, and GeF_4. (C—F = 1.32 Å, Si—F = 1.54 Å, Ge—F = 1.67 Å). Reproduced with permission from Ref. 27.

the central atom increases due to its decreasing F $2s$ and increasing F $2p$ character, while simultaneously retaining a significant contribution from the central atom's valence s orbital (which tends towards lower potential energy with increasing atomic number).

The tetramethyl compounds $X(CH_3)_4$ are isoelectronic with the tetrafluoride compounds and, as expected, exhibit rather similar trends.[27] The various MOs exhibit logical trends as a function of increasingly heavier central atoms, with the ligand-nonbonding orbitals remaining essentially constant in energy (except for the $1e$ case) and the orbitals with central-atom character progressively decreasing in binding energy.

Frost et al.[31] have studied the XH_2Y_2, XHY_3, and XY_4 (X = C, Si; Y = Cl, F) series in order to determine if $p \rightarrow d$ π bonding (between the halogen p orbitals and Si d orbitals) was significant. In these systems the nonbonding halogen orbitals are symmetry related to the d orbitals of Si, but not symmetry related to any of the molecular bonding orbitals. Frost et al.[31] emphasize that mere shifts in the energy of the n orbitals is inconclusive evidence for $p-d$ π bonding because such comparisons are invalidated by the fact that in the silicon compounds the halogen atoms are further away from each other than in the carbon analogs; therefore, interactions between the n orbitals will be decreased anyway. They suggest that the weighted average E_I of the n orbitals should be used, because this will be independent of halogen-halogen p_π–p_π interaction. For example, of the three n-orbitals of $SiCl_4$, t_1, e, and t_2, t_1 is stabilized, e is relatively unchanged, and t_2 is destabilized relative to their counterparts in CCl_4. The weighted average energy of the n orbitals is

$$\varepsilon_{n, \text{Av}} = \frac{3\varepsilon_{t_1} + 2\varepsilon_e + 3\varepsilon_{t_2}}{8} \tag{11.2}$$

It can only be influenced by the effective electronegativity of the adjacent atom and by interaction with bonding orbitals at higher binding energy. The electronegativity effect is expected to cause a decrease in the n E_Is proceeding from CCl_4 to the analogs with heavier central atoms. The interaction with bonding orbitals is expected to remain relatively constant for the pairs CX_4 and SiX_4. Therefore, using Eq. 11.2, one obtains a stabilization for $\varepsilon_{n, Av}$ of 0.26 eV from CCl_4 to $SiCl_4$, which must be caused mainly by $p \rightarrow d \pi$ bonding.

Other strong evidence for d-orbital participation comes from the observation of changes in an n-type band. The sharp characteristic shape of n-type bands is expected to assume broader profiles corresponding to bonding orbitals if $p \rightarrow d$ or $\pi \rightarrow d$ type interactions are operative. The observation of such changes upon substitution of Si or Ge for C is considered rather conclusive evidence for d-orbital participation in the bonding. Cradock et al.[29] have used such observations in XH_3NCO and XH_3NCS (X = C, Si, Ge) in order to postulate $\pi \rightarrow d$ bonding.

From the large amount of evidence acquired, it is evident that virtual d orbitals are important in the bonding of some molecules containing elements beyond the second row.

11.6. Applications to solids and surfaces

The study of surfaces by means of PE spectroscopy is becoming increasingly more common. It is evident that the range of application[34,35] to solids is broad and considerable vigor in this rapidly expanding area is expected. Such studies provide new means of investigating adsorption, chemisorption, valence band structure of solids, band gaps in semiconductors, catalysis, and so forth. In surface studies it is essential to work under conditions of ultrahigh vacuum, $\sim 10^{-10}$ to 10^{-11} torr, in order to maintain clean surfaces. Such high-vacuum conditions present problems because most He resonance discharges are windowless, allowing helium to leak into the target chamber. A suitable differentially pumped system has been described,[36] and a source with an aluminum window has been used,[37] although the window reduces photon flux considerably. Another problem is that of calibration of the energy scale. Mixing or coating the sample with a standard compound for calibration, as in X-ray PE spectroscopy, is undesirable because the valence bands of the calibrant are usually not well separated from those of the sample resulting in considerable overlap of the two spectra. Also, the addition of a standard may shift the Fermi level, which is the usual reference level for solids. Two approaches to calibration are (1) to take the observed onset of the spectrum as the Fermi level and consider that it is unshifted or (2) to calibrate the spectrometer by measuring its transmission or work function. Poole et al.[38] have described the latter procedure and have suggested that the zinc peak at 9.46 eV serves as a useful calibrant. If the instrument used has

X-ray capabilities also, the calibration can be extended from levels to the outer levels.

The necessity for ultrahigh vacuum and clean surfaces stems shallow depths sampled by the technique. This range is roughly ∴ depending upon the type of sample and kinetic energy of the ph⌐trons. The penetration of the exciting radiation is greater than the escape depth of the photoelectrons. The escape depth for organic compounds is greater than that of heavy dense metals.

Despite the difficulties involved, PE spectroscopy has been used successfully for many surface studies. For example, the spectra of adsorbed molecules such as N_2, NO, CO, H_2, C_2H_6, C_2H_4, C_2H_2, C_6H_6, and CH_3OH on metal surfaces have been reported,[39] and various processes such as bond dissociation in the adsorbed species have been observed. In the case of benzene adsorbed on a Ni surface, both π and σ levels of benzene can be observed in the chemisorbed layer; however, the outermost π level is preferentially shifted upon chemisorption. This indicates that the π orbitals of benzene are particularly important in the chemisorption mechanism.

In the case of CH_3OH adsorbed on tungsten, studies[40] have shown that dissociation occurs at low coverage to produce CO(ads) + H(ads). As the coverage increases, sites capable of adsorbing more complex species, such as CH_3OH(ads) or CH_3O(ads), are filled as indicated by the appearance of new spectral features.

Brundle et al.[41,42] have reported studies of adsorbed species using ultraclean vacuums, $\sim 10^{-12}$ torr. They find that the limit of detection of adsorbed water is at least 10% of a monolayer, while the detection limit of metals is about 2% of a monolayer. An interesting discovery is that there is a chemical shift between species adsorbed at room temperature and the further adsorption at lower temperature. Close investigation of the valence band structure of CO adsorbed on molybdenum at room temperature and at 77°K indicate that: (1) The CO adsorbed at 77°K has retained its MO identity, merely suffering an upward shift of the levels. This suggests that CO is present as an individual electronic entity with rather weak bonding to the surface. (2) At room temperature, the adsorbed CO structure loses its resemblance to the MO structure of free CO. The observed structural features at 4 and 6 eV are ubiquitous in that they appear in about the same positions for CO on W, CO on Ti, CO_2 on W, and O_2 on Ni. The results suggest that the observed structures correspond to individual C and O atoms at the surface, hence a dissociated CO state. By "dissociated" it is meant that the individual bonding, C–metal and O–metal, is stronger than the direct C—O bonding.

Many basic studies have been applied to metals,[43-45] metal oxides,[46,47] and chalogenides.[48] Spectra of condensed organic molecules usually resemble the gas-phase spectra shifted by a constant amount and broadened such that vibrational structure is obliveated. The spectra of naphthalene[49] and perylene[50] crystals are representative of condensed organic molecules.

11.7.   Charge-exchange mass spectroscopy and PE spectroscopy

Mass spectroscopy and PE spectroscopy are both concerned with the production of ions, and each depends upon the distribution function for that ion to determine the appearance of the resultant spectrum. A correlation exists between these two techniques that supplies information concerning the fragmentation processes of dissociative ionic states; this information can be helpful in assigning PE spectral bands. Figure 11.6 shows the relationship between the ionization processes for a mass spectrum and a PE spectrum of a simple diatomic system. The energy shift ΔE_s between the appearance potential (AP) of ions in the mass spectrometer and the E_I in the PE spectrometer is dependent upon the degree of vibrational excitation (for AB^+) and the depth of the potential well (for daughter ions). A measurement of the difference $AP - E_I$, where AP corresponds to the daughter ions, can provide information on the depth of the potential well for the parent ion AB^+.

Mass spectra used in such studies are of the charge-exchange type[51,52] In a charge-exchange mass spectrometer, a monatomic gaseous ion (for example, rare gas ions) transfers its positive charge to a neutral molecule (takes an electron from the molecule) and in so doing forms a molecular

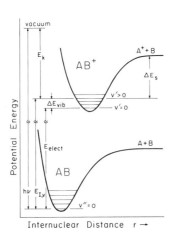

Fig. 11.6 Relationship between the excitation processes for PE spectroscopy and mass spectrometry. The quantities are defined as follows: $h\nu$ = excitation energy, E_k = kinetic energy of ejected electron, $E_{k,\nu}$ = energy required to eject an electron from molecule AB and leave the ion AB^+, in the νth vibrational state, E_{elect} = energy difference between the ground electronic state of AB and the ground ionic state of AB^+, that is, the adiabatic E_I, ΔE_s is the maximum energy shift between PE spectra and mass spectra. The maximum energy shift between the two methods applys to the daughter ions, but any energy shift of the parent ion only has this maximum as an upper limit.

ion in its ground or excited states. The new molecular ion can decompose to form daughter ions in a variety of ways depending upon its particular electronic state. The advantage of using charge exchange rather than conventional mass spectrometry is that in the charge-exchange method, the excitation energy E_{exc} is known precisely to be

$$E_{exc} = E_{RE} - E_{I,\nu} \qquad (11.3)$$

where E_{RE} is the recombination energy of the atomic ion and $E_{I,v}$ is the energy required to eject an electron from molecule AB and form the ion AB^+ in the vth vibrational state. In conventional mass spectrometry using electron bombardment for ionization, the excitation energy transferred cannot be accurately determined because the scattering angles and energies of the primary and secondary electrons are not known. Using the charge-exchange method, no molecular ions are observed until E_{RE} is greater than or equal to the first E_I; as E_{RE} increases, it becomes possible to form the molecular ion in various excited ionic states with different degrees of vibrational excitation. As mentioned earlier, the primary ions are usually rare gas ions; however, small molecular ions with accurately known E_{RE}s, such as H_2O, H_2S, CO, N_2, CO_2, and so forth, can also be used.

The charge-exchange mass spectrum consists of a plot of relative ion abundances versus E_{RE} of the primary ion. The most abundant molecular ion is usually the parent ion, with lower distributions of the daughter ions. Each distribution has its maximum at various values of E_{RE}. The decomposition processes of excited molecular ions appears to be independent of how the ions are excited—whether by electron bombardment, photon bombardment, or charge transfer. Correlation of the mass spectrum with the PE spectrum allows an assessment of which states of the molecular ion lead to dissociation and the nature of these dissociation products. A knowledge of the dissociation products from a particular ionic state can be correlated with the bonding characteristics of the electrons ejected in forming that state. This is effectively illustrated by the following example.

Example: Benzene

The charge-exchange mass spectrum of benzene[53] is correlated with its PE spectrum in Fig. 11.7. This mass spectrum provides the following information.* Excitation of the mass spectrum with energies less than

---

* For a complete discussion of the benzene spectrum, see Section 10.3.

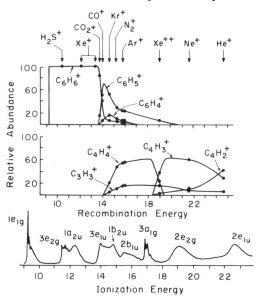

Fig. 11.7. Charge-exchange mass spectrum of benzene as a function of the recombination energy of the primary ions. The primary ions are listed at the positions of their recombination energies, and the secondary molecular ions and fragment distributions are indicated by the curves. The PE spectrum of benzene is drawn at the lower part for comparison. Reproduced with permission from Ref. 53.

~14 eV produces mainly the parent ion $C_6H_6^+$. The three photoelectron bands in this region arise from ejection of $1e_{1g}(\pi)$, $3e_{2g}(\sigma)$, and $1a_{2u}(\pi)$ electrons. The corresponding molecular ions formed by ejection of these electrons are evidently quite stable with little or no decomposition occurring. In charge exchange with CO^+ ($\dot{E}_{RE} = 14.0\,eV$), the electron corresponding to the photoelectron band at 13.8 eV will be preferentially ejected. The graph shows a rapid increase in $C_6H_5^+$ at this energy, suggesting that the electron ejected must be strongly C—H bonding, that is, $3e_{1u}$. In charge exchange at ~14.7 eV, the electrons corresponding to the 14.7-eV PE spectral band will be ejected. The graph indicates a rapid increase of $C_4H_4^+$ and $C_3H_3^+$ at this energy, suggesting that the electron ejected is strongly C—C bonding, that is, $1b_{2u}$. The mass spectrum between 15 and 18 eV reveals no new features. The rapid rise in the $C_4H_3^+$ fragment at ~19 eV indicates that a strongly C—C bonding electron of probably a different type, that is, s rather than p, has been removed. The corresponding PE spectral band at 19.2 eV is assigned to $2e_{2g}$ s-type electrons.

11.8. Rydberg spectra and PE spectra

Molecular Rydberg states are generated by exciting a valence electron to an orbital that is so large and diffuse that it engulfs the whole molecule, that is, a Rydberg orbital. The energy of this type of transition is given as

$$E_n = E_I - \frac{R}{(n-\delta)^2} \tag{11.4}$$

where E_I is the valence electron ionization energy, R is the Rydberg constant (13.60 eV), n is the principal quantum number of the AO that dominates the constitution of the Rydberg orbital, and δ is the quantum defect. Rydberg transitions are observed as a series of bands whose energies can be fit according to Eq. 11.4 using consecutively increasing n values. At high n values the series converges to the E_I of the valence electron that has been excited, permitting the determination of accurate E_Is. In practice it is not always easy to carry out this process, because Rydberg series originating from different valence orbitals will be inevitability overlapped, making it difficult to find the convergence limit; also, δ is unknown and must be chosen to fit the series.

Since the Rydberg orbital is large and diffuse and extends far beyond the periphery of the molecule, an electron residing in this orbital will contribute very little to the bonding in the molecule. In fact, the Rydberg electron is so far removed from the molecule that the Rydberg state is not unlike the ionic state corresponding to ejection of that same electron. It has been noted that the vibrational and rotational structure of Rydberg transitions is similar to that observed in the corresponding photoelectron bands.[54,55] This relationship between Rydberg spectra and PE spectra can be mutually beneficial: The highly accurate E_Is determined from PE spectra can be used in Eq. 11.4 to locate the positions E_n of the Rydberg bands. From a knowledge of E_n and E_I, the quantum defect δ can be accurately determined. From the size of the quantum defect, one can determine the symmetry of the Rydberg orbital, and from the selection rules for electronic absorption[56] one can sometimes correlate the transition with a certain valence MO. Such a determination of the nature of the valence MO involved in the transition provides an independent method for assigning photoelectron bands. This procedure has been used with success by E. Lindholm et al.[53,57,58] Successful application of the method requires some knowledge of the nature of quantum defects, upon which we now elaborate.

Quantum defects δ

The origin of the quantum defect is the penetration of the Rydberg electron into the interior of the molecule. The deeper this penetration, the closer the Rydberg electron is to the nuclei, and hence the larger the effective core charge experienced by the electron. Electrons in penetrating orbitals will naturally be more tightly bound than those in nonpenetrating orbitals, resulting in deviations from the formula $E_n = E_I - R/n^2$; these deviations are taken into account by introducing the quantum defect δ. Evidently, δ is large if the probability is high that the electron is near the nuclei (such as ns electrons) and smaller if the probability function near the nuclei is small (such as np and nd electrons). For molecules composed of second-row atoms, it has been found[56] that δ is ~ 0.1 for nd-, ~ 0.3 to 0.5 for np-, and ~ 0.9 to 1.2 for ns-type Rydberg MOs. Since core shielding is less effective in the third-row atoms, the corresponding quantum defects for molecules composed of such atoms are larger and are given by ~ 0.4 for nd-, ~ 1.5 for np-, and ~ 2.0 for ns-type Rydberg MOs.[59] Comparison of δ values of Rydberg series in different molecules is legitimate, for the δs depend mainly only on the nature of the initial valence orbital from which the electron is excited.

Determining the magnitude of molecular δ values from the nature of a Rydberg MO is not as straightforward as in atoms, because in molecules, the center of the Rydberg orbital may not coincide with any of the nuclei in the molecule. Two procedures can be used for determining the magnitude of molecular δ: (1) The AO composition of the Rydberg MO can be determined from MO calculations, and the molecular δ can be evaluated as a linear combination of atomic δs, each scaled according to the coefficient of the AO in the Rydberg MO.[59,60] (2) The probability of the electron being near the nuclei can be estimated as the square of the wavefunction representing the Rydberg MO and the δ values can be scaled according to these probability distributions.[53,55] The use of such quantum defects in assigning PE spectra is illustrated in the following example.

Example: Benzene

Lindholm and his group[53,57] have used Rydberg spectra effectively in interpretation of the PE spectrum of benzene. Their Rydberg spectra were determined by ultraviolet-absorption spectroscopy and electron-impact energy-loss spectroscopy. The electron-impact energy-loss method of Lassettre et al.[61] and Boersch et al.[62] is particularly useful in this study. This method allows reliable determination of the intensities of absorption bands, and by varying the impact velocity and the scattering angle, one can distinguish between allowed and forbidden transitions. Only allowed transitions are observed with high-energy incident electrons (~ 100 eV) and small scattering angles ($0°$).

The analysis of the benzene system by Lindholm et al.[53] is summarized as follows. The excited states of benzene below ~ 7 eV are molecular transitions arising from the $1e_{1g}(\pi) \rightarrow 1e_{2u}(\pi^*)$ excitation. The higher excited states (>7 eV) appear to be mainly Rydberg states. Only transitions to $^1E_{1u}$- and $^1A_{2u}$-type states are allowed upon excitation of a $1e_{1g}(\pi)$ electron. The electron-impact energy-loss curve of Lassettre et al.[61] has one extremely strong and broad band at 7.0 eV corresponding to the $\pi \rightarrow \pi^*$ transition, followed by many strong and weak transitions extending to ~ 16 eV. These higher energy peaks correspond to Rydberg series whose convergence values correspond to the E_Is. The s-, p-, and d-Rydberg orbitals of benzene have been examined, and the δ values have been estimated from the penetration of these different orbitals.[53] Two strong Rydberg series converging to 9.3 eV with $\delta = 0.46$ and 0.16 are traceable. The orbital ionized at 9.3 eV (Fig. 11.7) is the $1e_{1g}(\pi)$ orbital, hence the allowed Rydberg transitions are $1e_{1g} \rightarrow npe_{1u}$, npa_{2u}. The measured quantum defects agree with those estimated from theory. Next the photoelectron band at 11.4 eV (Fig. 11.7) must be interpreted as either $3e_{2g}(\sigma)$ or $1a_{2u}(\pi)$. The first assignment gives the Rydberg transition $3e_{2g} \rightarrow npe_{1u}$, which fits an observed series with the expected $\delta = 0.46$. Interpretation of the 11.4 eV band as $1a_{2u}$ gives allowed Rydberg transitions to nsa_{1g}-, nda_{1g}-, and nde_{1g}-type orbitals. This latter assignment is in disagreement with the observed Rydberg structure and quantum defects. The photoelectron band at 12.1 eV is interpreted as $1a_{2u}(\pi)$. Excitation of this electron produces the Rydberg transitions indicated above with $\delta = 0.8$, 0.5, and 0.04, respectively, converging to 12.1 eV as observed. The 13.8-eV photoelectron band is interpreted as $3e_{1u}$; corresponding Rydberg transitions $3e_{1u} \rightarrow nsa_{1g}$, nda_{1g}, nde_{1g}, and nde_{2g} are all observed with $\delta = 0.8$, 0.5, 0.4, and 0.11, respectively. The δ values for the nda_{1g} and nde_{1g} Rydberg series are higher than expected for ordinary d orbitals because the wavefunctions for these orbitals have large amplitudes in the plane of the benzene molecule, and, therefore, the penetration is large. In this manner, Lindholm et al. are able to assign all of the bands in the electron-impact energy-loss spectrum and simultaneously provide external vindication of the assignment of the benzene PE spectrum.

11.9. Novel applications

Rate of H–D exchange

Hodges et al.[63] have used PE spectra of halogen-substituted hydrocarbons and hydrocarbons containing quaternary carbon atoms to interpret the nature of platinum(II)-catalyzed hydrogen-deuterium exchange. Representative halogen-substituted alkanes and aromatic compounds and branched-chain hydrocarbons containing an ethyl group adjacent to a quaternary carbon atom were studied by PE spectroscopy and in H–D exchange reactions catalyzed by $PtCl_4^{2-}$ in 50 mole% acetic$[^2H_1]$acid-deuterium oxide. For the halogen-substituted compounds, H–D exchange occurs in those molecules for which the first E_I is associated with an MO centered on carbon-hydrogen bonds. For these compounds, an exponential relationship exists between the rate constant for exchange and the E_I. When the first E_I is associated with electrons of the halogen atom, displacement of the halogen occurs. For the branched-chain hydrocarbons, the rate constant for exchange in the ethyl group correlates with the rate constants for the linear alkanes, provided an E_I attributable to electrons of the ethyl group is used.

Intramolecular hydrogen bonding

Franklin et al.[64] have examined the PE spectra of a series of intramolecularly hydrogen-bonded compounds (amine and iodine substituted alcohols) along with their non-hydrogen-bonded analogs (methylated species) at temperatures ranging between 25 and 200°C. In the hydrogen-bonded compounds the E_I of the electron donor, that is, nitrogen or iodine, was temperature dependent and decreased with increasing temperature. The E_Is of these compounds were markedly greater than the values for the non-hydrogen-bonded analogs. This difference is interpreted as due to the stabilization energy of the "lone-pair" electrons in hydrogen bonding. The destabilization of the nitrogen or iodine "lone-pair" E_I with increasing temperature is evidence for breaking the hydrogen bond. The data are consistent with classical models for hydrogen bonding; for example, in 2-aminoethanol the structure is believed to be

$$CH_2\text{---}CH_2$$
$$H_2N \qquad O$$
$$H$$

where the hydrogen bond is indicated by a dotted line. Franklin et al.[64] were able to estimate the energies of these hydrogen bonds by using the observed shifts in the E_Is and thermochemical data.

References

1. C. R. Brundle and M. B. Robin, *Determination of Organic Structures by Physical Methods*, F. C. Nachod and J. J. Zuckerman (Eds.), Vol. 3, Academic, New York, 1971.

2. C. R. Brundle and M. B. Robin, *J. Amer. Chem. Soc.*, **92,** 5550 (1970).

3. J. P. Maier and D. W. Turner, *Discussions Faraday Soc.*, **54,** 149 (1972).

4. J. Maier and D. W. Turner, *J. C. S. Faraday Trans.*, **2,** 69, 196, 521 (1973).

5. S. A. Cowling and R. A. W. Johnstone, *J. Electron Spectrosc.*, **2,** 161 (1973).

6. C. Batich, O. Ermer, E. Heilbronner, and J. R. Wiseman, *Angew. Chem. (Int. Edit.))*, **12,** 312 (1973).

7. J. Berkowitz, J. L. Dehmer, K. Shimada, and M. Szwarc, *J. Electron Spectrosc.*, **2,** 211 (1973).

8. J. W. Rabalais, L. O. Werme, T. Bermark, L. Karlsson, and K. Siegbahn, *Int. J. Mass Spectrom. Ion Phys.*, **9,** 185 (1972).

9. D. Betteridge and A. D. Baker, *Anal. Chem.*, **42,** 43 (1970).

10. M. J. S. Dewar and S. D. Worley, *J. Chem. Phys.*, **40,** 654 (1969).

11. A. W. Potts, T. A. Williams, and W. C. Price, *Faraday Discussions Chem. Soc.*, **54,** 104 (1972).

12. L. Golob, N. Jonathan, A. Morris, M. Okuda, and K. J. Ross. *J. Electron Spectrosc.*, **1,** 506 (1972/73).

13. N. Jonathan, A. Morris, D. J. Smith, and K. J. Ross, *Chem. Phys. Letters* **7,** 497 (1970); N. Jonathan, A. Morris, M. Okuda, and D. J. Smith, *Electron Spectroscopy*, D. A. Shirley (Ed.), p. 345, North-Holland, Amsterdam, 1972.

14. N. Jonathan, D. J. Smith, and K. J. Ross, *J. Chem. Phys.*, 53, 3758 (1970); N. Jonathan, A. Morris, K. J. Ross, and D. J. Smith, *J. Chem. Phys.*, 54, 4954 (1971).

15. N. Jonathan, D. J. Smith, and K. J. Ross, *Chem. Phys. Letters*, **9,** 217 (1971).

16. N. Jonathan, A. Morris, M. Okuda, D. J. Smith, and K. J. Ross, *Chem. Phys. Letters*, **13,** 334 (1972).

17. D. C. Frost, S. T. Lee, and C. A. McDowell, *Chem. Phys. Letters*, **17,** 153 (1972).

18. G. H. King, H. W. Kroto, and R. J. Suffolk, *Chem. Phys. Letters*, **13,** 45 (1972).

19. D. C. Frost, S. T. Lee, and C. A. McDowell, *Chem. Phys. Letters*, **23,** 472 (1973).

20. D. C. Frost, S. T. Lee, and C. A. McDowell, *Chem. Phys. Letters*, **22,** 243 (1973).

21. A. B. Cornford, D. C. Frost, F. G. Herring, and C. A. McDowell, *Chem. Phys Letters*, **10,** 345 (1971).

22. C. R. Brundle, *Chem. Phys. Letters*, **26,** 25 (1974).

23. D. C. Frost, S. T. Lee, and C. A. McDowell, *Chem. Phys. Letters*, **24,** 149 (1974).

24. I. Morishima, K. Yoshikawa, T. Wonezawa, and H. Matsumoto, *Chem. Phys. Letters*, **16,** 336 (1972).

25. A. B. Cornford, D. C. Frost, F. G. Herring, and C. A. McDowell, *Discussions Faraday Soc.*, **54,** 56 (1972).

26. H. W. Kroto and R. J. Suffolk, *Chem. Phys. Letters*, **15,** 545 (1972).

27. A. Jonas, G. K. Schweitzer, F. A. Grimm, and T. A. Carlson, *J. Electron Spectrosc.*, **1,** 29 (1972/1973).

28. T. P. Debies and J. W. Rabalais, *Inorg. Chem.*, **13,** 308 (1974).

29. S. Cradock, E. A. V. Edsworth, and J. D. Murdock, *J. C. S. Faraday Trans.*, **68**, 86 (1971).
30. S. Cradock and R. A. Whiteford, *Trans. Faraday Soc.*, **67**, 3425 (1971); ibid., **68**, 281 (1972).
31. D. C. Frost, F. G. Herring, A. Katrib, R. A. N. McLean, J. E. Drake, and N. P. C. Westwood, *Can. J. Chem.*, **49**, 4033 (1971); idem., *Chem. Phys Letters.* **10**, 347 (1971).
32. P. Mollere, H. Bock, G. Becker, and G. Fritz, *J. Organometal. Chem.*, **46**, 89 (1972).
33. G. Bieri, F. Brogli, E. Heilbronner, and E. Kloster–Jensen, *J. Electron Spectrosc.*, **1**, 67 (1972/73).
34. H. D. Hagstrom, *Science*, **178**, 275 (1972).
35. J. T. Yates, Jr., *Chem. Eng. News*, **52**, 19 (1974).
36. J. K. Cashion, J. L. Mees, D. E. Eastman, J. A. Simpson, and C. E. Kuyatt, *Rev. Sci. Instrum.*, **42**, 1670 (1971).
37. J. A. Kinsinger, W. L. Stebbings, R. A. Valenzi, and J. W. Taylor, *Anal. Chem.*, **44**, 773 (1972).
38. R. T. Poole, R. C. G. Leckey, J. G. Jenkin, and J. Liesegang, *J. Phys. E*, **6**, 201, 226 (1973).
39. D. E. Eastman and J. Demuth, *Phys. Rev Letters*, **32**, 1123 (1974).
40. W. F. Egelhoff and J. W. Linnett, Physical Electronics Conference, Feb. 25–27, 1974.
41. C. R. Brundle and M. W. Roberts, *Proc. Roy. Soc. London*, **A331**, 382 (1972); idem., *Surface Sci.*, **38**, 234 (1973).
42. S. J. Atkinson, C. R. Brundle, and M. W. Roberts, *J. Electron Spectrosc.*, **2**, 105 (1973). idem., *Chem. Phys. Letters*, **24**, 175 (1974).
43. D. E. Eastman, *Electron Spectroscopy*, D. A. Shirley (Ed.), p. 487, North-Holland, Amsterdam, 1972.
44. N. V. Smith, *Phys. Rev. B*, **3**, 1862 (1971).
45. N. V. Smith and M. M. Traum, *Electron Spectroscopy*, D. A. Shirley, (Ed.), p. 541 North-Holland, Amsterdam, 1972; J. C. Tracy and J. E. Rowe, ibid., p. 587.
46. C. R. Helms and W. E. Spicer, *Appl. Phys. Letters*, **21**, 237 (1972).
47. R. A. Powell, W. E. Spicer, and J. C. McMenamin, *Electron Spectroscopy*, D. A. Shirley (Ed.), p. 575, North-Holland, Amsterdam, 1972.
48. S. J. Veseley, D. W. Langer, and R. L. Hengehold, *Electron Spectroscopy*, D. A. Shirley (Ed.), p. 535, North-Holland, Amsterdam, 1972.
49. T. Hirooka, K. Tanaka, K. Kuchitsu, M. Fujihara, H. Inokuchi, and Y. Harada, *Chem. Phys. Letters*, **18**, 390 (1973).
50. K. Seki, H. Inokuci, and Y. Harada, *Chem. Phys. Letters*, **20**, 197 (1973).
51. V. Cermak and Z. Herman, *Nucleonics*, **19**, 106 (1961).
52. E. Lindholm, *Advan. Chem.*, **58**, 1 (1966).
53. B. Ö. Jonsson and E. Lindholm, *Ark. Fys.* **39**, 65 (1969).
54. L. Åsbrink and J. W. Rabalais, *Chem. Phys. Letters*, **12**, 182 (1971).
55. E. Lindholm, *Ark. Fys.*, 40, 97 (1969).

56. G. Herzberg, *Electronic Spectra of Polyatomic Molecules*, Van Nostrand, Princeton, N. J., 1966.

57. E. Lindholm and B. O. Jonsson, *Chem., Phys. Letters*, **1**, 501 (1967).

58. C. Fridh, L. Åsbrink, B. O. Jonsson, and E. Lindholm, *Int. J. Mass Spectrom. Ion Phys.*, **7**, 85, 101, 215, 229, 485, (1972).

59. J. W. Rabalais, J. R. McDonald, V. Scherr, and S. P. McGlynn, *Chem. Rev.*, **71**, 73 (1971).

60. J. W. Rabalais, *Molecular Electronic Spectroscopy of Isoelectronic Series: Linear Triatomic Groupings Containing Sixteen Valence Electrons*, Ph.D. Dissertation, Louisiana State University, Baton Rouge, La., 1970.

61. E. N. Lassettre and S. A. Francis, *J. Chem. Phys.*, **40**, 1208 (1964); A. Skerbele and E. N. Lassettre, *J. Chem. Phys.*, **42**, 395 (1965); and later papers.

62. H. Boersch, J. Geiger, and W. Stickel, *Z. Phys.*, **180**, 415 (1964); J. Geiger and W. Stickel, *J. Chem. Phys.*, **43**, 4535 (1965).

63. R. J. Hodges, D. E. Webster, and P. B. Wells, *J. C. S. Dalton Trans.*, 2577 (1972).

64. S. Leavell, J. Steichen, and J. L. Franklin, *J. Chem. Phys.*, **59**, 4343 (1973).

Appendix I. Relations Involving Rotation Matrices, Spherical Harmonics, and Clebsch–Gordon Coefficients

We follow the Condon–Shortley phase convention in defining the spherical harmonics[1]

$$Y_{l,m}(\theta, \phi) = (-1)^{(|m|+m)/2} N_{l,m} P_l^{|m|}(\cos \theta) e^{im\phi} \tag{I.1}$$

in which

$$N_{l,m} = \left[\frac{(2l+1)}{4\pi} \cdot \frac{(l-|m|)!}{(l+|m|)!} \right]^{1/2} \tag{I.2}$$

and $P_l^{|m|}(\cos \theta)$ is the usual associated Legendre function. We note that

$$Y_{l,m}^*(\theta, \phi) = (-1)^m Y_{l,-m}(\theta, \phi) \tag{I.3}$$

The coupling equation for spherical harmonics is

$$Y_{l(p), m(p)}(\theta, \phi) Y_{l(q), m(q)}(\theta, \phi) = \sum_l Q_{l(p), l(q), l} C(l_p l_q l; m_p, m_q)$$
$$\times C(l_p l_q l; 0, 0) Y_{l, m(p)+m(q)}(\theta, \phi) \tag{I.4}$$

in which $l(p) = l_p$, $m(p) = m_p$,

$$Q_{l(p), l(q), l} = \left[\frac{(2l_p+1)(2l_q+1)}{4\pi(2l+1)} \right]^{1/2} \tag{I.5}$$

and $C(l_p l_q l; m_p, m_q)$ is the Clebsch–Gordon coefficient.[2] More generally,

$C(l_p l_q l; m_p, m_q, m) = 0$ unless $m = m_p + m_q$. The addition theorem is

$$P_l(\cos \theta) = 4\pi(2l+1)^{-1} \sum_m Y^*_{l, m}(\theta_1, \phi_1) Y_{l, m}(\theta_2, \phi_2) \tag{I.6}$$

where θ is the angle between two lines passing through the origin with polar and azimuthal angles (θ_1, ϕ_1) and (θ_2, ϕ_2).

Rayleigh's expansion of a plane wave

$$e^{i\mathbf{k} \cdot \mathbf{r}} = \sum_{l=0}^{\infty} i^l(2l+1)j_l(kr)P_l(\cos \theta) \tag{I.7}$$

in which $j_l(kr)$ is the spherical Bessel function, may be rewritten

$$e^{i\mathbf{k} \cdot \mathbf{r}} = 4\pi \sum_{l=0}^{\infty} i^l j_l(kr) \sum_{m=-l}^{l} Y_{l, m}(\theta'_k, \phi'_k) Y^*_{l, m}(\theta'_r, \phi'_r) \tag{I.8}$$

using Eq. I.6; here, (θ'_k, ϕ'_k) and (θ'_r, ϕ'_r) are spherical polar coordinates that define the directions of $\mathbf{k}$ and $\mathbf{r}$, respectively, and we elect the molecular coordinate system for defining the polar axis and zero meridian.

We have already given the relation Eq. 6.1, which expresses spherical harmonics in the molecular coordinate system in terms of spherical harmonics in the laboratory system. The inverse is

$$Y_{l, m}(\theta, \phi) = \sum_{m'} \mathbf{D}^{l*}_{m, m'}(\boldsymbol{\beta}) Y_{l, m'}(\theta', \phi') \tag{I.9}$$

The coupling rule for $\mathbf{D}$ matrices is

$$\mathbf{D}^{\lambda}_{\mu, \nu} \mathbf{D}^{l}_{m, n} = \sum_j C(\lambda k j; \mu, m) C(\lambda l j; \nu, n) \mathbf{D}^{j}_{\mu+m, \nu+n} \tag{I.10}$$

A useful complex conjugate relation is

$$\mathbf{D}^{l*}_{m, n} = (-1)^{m-n} \mathbf{D}^{l}_{-m, -n} \tag{I.11}$$

It is necessary to average over all molecular orientations specified by the Euler angles $\boldsymbol{\beta} = (\alpha, \beta, \gamma)$. The integral

$$\int\int\int \sin \beta \, d\beta \, d\alpha \, d\gamma \mathbf{D}^{\lambda*}_{\mu, \nu}(\boldsymbol{\beta}) \mathbf{D}^{l}_{m, n}(\boldsymbol{\beta}) = 8\pi^2(2l+1)^{-1}\delta_{l, \lambda}\delta_{\mu, m}\delta_{\nu, n} \tag{I.12}$$

will often be useful. Note that when the average value is taken, one divides by the integral

$$\int\int \sin \beta \, d\beta \, d\alpha \, d\gamma = 8\pi^2 \tag{I.13}$$

so that the factor $8\pi^2$ is canceled.

In general, the Clebsch–Gordon coefficient

$$C(l_p l_q l; m_p, m_q) = 0$$

unless

$$l = l_p + l_q, \; l_p + l_q - 1, \cdots, |l_p - l_q| \\ l_p \geq |m_p|; \quad l_q \geq |m_q|; \quad l \geq |m_p + m_q| \tag{I.14}$$

Furthermore,

$$C(l_p l_q l; 0, 0) = 0$$

unless

$$l = l_p + l_q, \; l_p + l_q - 2, \cdots, |l_p - l_q| \tag{I.15}$$

that is, $l + l_p + l_q$ must be even. Finally,

$$C(\lambda l j; \mu, \nu) = (-1)^{\lambda - \mu} \left(\frac{2j+1}{2l+1} \right)^{1/2} C(\lambda j l; \mu, -\mu - \nu) \tag{I.16}$$

is a useful symmetry rule. This can be combined with the relation

$$C(\lambda 0 j; \mu, 0) = \delta_{\lambda, j} \tag{I.17}$$

to yield

$$C(\lambda \lambda' 0; \mu, -\mu) = (-1)^{\lambda - \mu} (2\lambda' + 1)^{-1/2} C(\lambda 0 \lambda'; \mu, 0) \\ = (-1)^{\lambda - \mu} (2\lambda + 1)^{-1/2} \delta_{\lambda, \lambda'} \tag{I.18}$$

a useful expression for evaluating certain C coefficients.

References

1. E. U. Condon and G. H. Shortley, *The Theory of Atomic Spectra*, Cambridge U. P., London 1951.
2. M. E. Rose, *Elementary Theory of Angular Momentum*, Wiley, New York, 1957.

Appendix II.   Atomic Orbitals

We first define the AO χ as follows:

$$\chi = f_n(r)\,Y_{l,\,m}(\theta_r',\,\phi_r') \tag{II.1}$$

in which the function $f(r)$ depends upon the particular type of AO basis employed. Slater-type orbitals (STO) and Gaussian-type orbitals (GTO) are most commonly used:

$$f_n^{\text{STO}}(r) = \mathcal{N}_n^{\text{STO}}\, r^{n-1}\exp\left(-\zeta r\right)$$

$$f_n^{\text{GTO}}(r) = \mathcal{N}_n^{\text{GTO}}\, r^{n-1}\exp\left(-\zeta^2 r^2\right)$$

$$\mathcal{N}_n^{\text{STO}} = \frac{(2\zeta)^{n+1/2}}{[(2n)!]^{1/2}} \tag{II.2}$$

$$\mathcal{N}_n^{\text{GTO}} = \left(\frac{(2\zeta)^{2n+1}}{(2n-1)!!}\right)^{1/2}\left(\frac{2}{\pi}\right)^{1/4}$$

Here, $n!! = n(n-2)(n-4)\cdots 1$. For a particular AO χ_p, we would inject the parameters $\zeta(p)$, $n(p)$, $l(p)$, and $m(p)$ into Eqs. II.1 and II.2, and we would specify r as r_p, the electron distance in Bohr units (1 Bohr = 0.5292 Å) measured from the nucleus at which χ_p is centered; the polar coordinate angles $(\theta_r',\,\phi_r')$ are in the molecular coordinate system defined in Chapter 6.

Two different approaches using GTOs are quite common. In the first, Cartesian Gaussians as we have defined them in Eqs. II.1 and II.2 are used. In the second method, GTOs are limited to those for which $n(p) = 1$ and $l(p) = 0$. Linear combinations of these simple *1s* Gaussians are often centered at appropriate points away from nuclear centers to represent lobes of AOs like *2p* and *3d*; hence, they are called Gaussian-lobe functions.

Gaussian-lobe functions are necessarily real functions. On the other hand, STOs and Cartesian GTOs are complex for $l > 0$ and $m \neq 0$

according to Eq. II.1. The corresponding *real* AOs may be written as follows:

$$\begin{pmatrix} 2p_x \\ 2p_y \end{pmatrix} = \mathbf{a}_1 \begin{pmatrix} 2p_{+1} \\ 2p_{-1} \end{pmatrix} \qquad \begin{pmatrix} 3d_{xz} \\ 3d_{yz} \end{pmatrix} = \mathbf{a}_1 \begin{pmatrix} 3d_{+1} \\ 3d_{-1} \end{pmatrix}$$

$$\begin{pmatrix} 3d_{x^2-y^2} \\ 3d_{xy} \end{pmatrix} = \mathbf{a}_2 \begin{pmatrix} 3d_2 \\ 3d_{-2} \end{pmatrix} \tag{II.2}$$

where

$$\mathbf{a}_m = \begin{pmatrix} \dfrac{(-1)^m}{2^{1/2}} & \dfrac{1}{2^{1/2}} \\[2ex] \dfrac{-i(-1)^m}{2^{1/2}} & \dfrac{i}{2^{1/2}} \end{pmatrix} \tag{II.3}$$

Expressions for *4f* AOs, and so forth are similar to these.

Appendix III. Evaluation of Overlap Integral Between Atomic Orbital and Plane Wave

We now prove Eq. 6.24, which gives the overlap integral, Eq. 6.14, between the AO χ_p and the plane wave, as expressed by Eq. 6.5:

$$F_{pk} = L^{-3/2}\langle\chi_p|\,e^{i\mathbf{k}\cdot\mathbf{r}}\rangle \tag{III.1}$$

Here, $\mathbf{r}$ is the vector distance from the laboratory origin to the electron position. Since .the AO χ_p is expressed in terms of electron coordinates relative to the nucleus at which χ_p is centered, we introduce the relation $\mathbf{r} = \mathbf{r}_p + \mathbf{R}_p$, in which the vector $\mathbf{R}_p$ defines the position of nucleus p relative to the laboratory origin, and the vector $\mathbf{r}_p$ defines the position of the electron relative the nuclear center p. Thus,

$$F_{pk} = L^{-3/2}\exp{(i\mathbf{k}\cdot\mathbf{R}_p)}\langle\chi_p|\exp{(i\mathbf{k}\cdot\mathbf{r}_p)}\rangle \tag{III.2}$$

Now, we replace r by r_p in Eqs. I.8 and II.1, combine these two equations with Eq. III.2, and integrate over $0 \leq \theta_{r(p)} \leq \pi$ and $0 \leq \phi_{r(p)} \leq 2\pi$.

The result is, in fact, Eq. 6.24 with

$$T = 4\pi L^{-3/2}\int f_n(r)j_l(kr)r^2\,dr \tag{III.3}$$

in which we have now suppressed the subscript p on T, n, l, and r. Comparison of Eqs. 6.8, 6.18, 6.19, 6.24, and III.3 now reveals that the factor L^3 in Eq. 6.8 and the factor $L^{-3/2}$ in Eq. III.3 may be omitted for all practical purposes because of their effective cancellation.

Noting the relationship between spherical Bessel functions and ordinary Bessel functions of the first kind

$$j_l(kr) = \left(\frac{2kr}{\pi}\right)^{-1/2} J_{l+1/2}(kr) \tag{III.4}$$

we express the integral appearing in Eq. III.3 as follows:

$$\int f_n(r) j_l(kr) r^2 \, dr = \mathcal{N}_n \left(\frac{2k}{\pi}\right)^{-1/2} I_{l+1/2, \, n-l}(k, \zeta) \tag{III.5}$$

where

$$I^{\text{STO}}_{\lambda, \nu}(k, \zeta) = \int \exp\left(-\zeta r\right) J_\lambda(kr) r^{\lambda+\nu} \, dr \qquad \mathcal{N}_n = \mathcal{N}^{\text{STO}}_n \tag{III.6}$$

if $f_n(r) = f^{\text{STO}}_n(r)$ and

$$I^{\text{GTO}}_{\lambda, \nu}(k, \zeta) = \int \exp\left(-\zeta^2 r^2\right) J_\lambda(kr) r^{\lambda+\nu} \, dr \qquad \mathcal{N}_n = \mathcal{N}^{\text{GTO}}_n \tag{III.7}$$

if $f_n(r) = f^{\text{GTO}}_n(r)$; see Eqs. II.2. Integration by parts yields the following useful recursion formulae (arguments of $I_{\lambda, \, \nu}$ are suppressed):

$$I^{\text{STO}}_{\lambda, \nu} = \zeta^{-1}[(2\lambda + \nu) I^{\text{STO}}_{\lambda, \nu-1} - k I^{\text{STO}}_{\lambda+1, \, \nu-1}] \tag{III.8}$$

$$I^{\text{GTO}}_{\lambda, \nu} = 2\zeta^{-2}[(2\lambda + \nu - 1) I^{\text{GTO}}_{\lambda, \nu-2} - k I^{\text{GTO}}_{\lambda+1, \, \nu-2}] \tag{III.9}$$

For STOs, the subscript ν can thus be reduced to zero, for which[1]

$$I^{\text{STO}}_{\lambda,0} = \frac{(2k)^\lambda \Gamma(\lambda + \frac{1}{2})}{\Gamma(\frac{1}{2})(k^2 + \zeta^2)^{\lambda+1/2}} \tag{III.10}$$

For GTOs, we may make direct use of the general equation

$$I^{\text{GTO}}_{\lambda, \nu} = \frac{k^\lambda \Gamma(\lambda + \frac{1}{2}\nu + \frac{1}{2})}{2^{\lambda+1} \zeta^{2\lambda+\nu+1} \Gamma(\lambda + 1)} \cdot {}_1F_1\left(\lambda + \frac{1}{2}\nu + \frac{1}{2}; \lambda + 1; -\frac{k^2}{4\zeta^2}\right) \tag{III.11}$$

where ${}_1F_1(\alpha; \beta; x)$ is the confluent hypergeometric function. Or, for GTOs, we may first make use of Eq. III.9 to reduce subscript ν to zero or one. Since

$${}_1F_1(\alpha; \alpha; x) = e^x \tag{III.12}$$

the special case of Eq. III.11

$$I^{\text{GTO}}_{\lambda, 1} = k^\lambda (2\zeta^2)^{-\lambda-1} \exp\left(\frac{-k^2}{4\zeta^2}\right) \tag{III.13}$$

is very simple. On the other hand, for $\nu = 0$, we must first use Krummer's relation[1]

$$_1F_1(\alpha; \beta; x) = e^x {}_1F_1(\beta - \alpha; \beta; -x) \qquad \text{(III.14)}$$

which gives

$$I^{\text{GTO}}_{\lambda, 0} = \frac{k^\lambda \Gamma(\lambda + \frac{1}{2})}{2^\lambda \zeta^{2\lambda+1} \Gamma(\lambda + 1)} \cdot {}_1F_1\left(\frac{1}{2}; \lambda + 1; \frac{k^2}{4\zeta^2}\right) \exp\left(\frac{-k^2}{4\zeta^2}\right) \qquad \text{(III.15)}$$

We then use the standard recursion formula[2]

$$(\beta - \alpha)x\,{}_1F_1(\alpha; \beta + 1; x) = \beta(x + \beta - 1) \cdot {}_1F_1(\alpha; \beta; x)$$
$$+ \beta(1 - \beta) \cdot {}_1F_1(\alpha; \beta - 1; x) \qquad \text{(III.16)}$$

to reduce the second argument to $-\frac{1}{2}$ or $+\frac{1}{2}$, for which

$$_1F_1(\tfrac{1}{2}; -\tfrac{1}{2}; x) = (1 - 2x)e^x \qquad \text{(III.17)}$$

and Eq. III.12 applys respectively.

It turns out that for $1s$, $2p$, $3s$, and $3d$ GTOs, all integrals reduce to Eq. III.13 via Eq. III.9. For $2s$ and $3p$ GTOs, one is led to the more complicated Eqs. III.14 to III.17.

We have thus established analytical formulas for calculating the factors T_p using a basis of either Gaussian-lobe functions, STOs, or Cartesian GTOs as defined in Appendix II.

Very often, Schmidt-orthogonalized or Löwdin-orthogonalized STOs are used as a basis.[3] It has been shown that when photoionization cross sections are calculated with the semiempirical CNDO–MO wavefunctions, it is very important to utilize an orthogonalized AO basis.[4] Clearly, calculation of the T_p or F_{pk} over such an orthogonalized AO basis is readily accomplished. For example, for the common orthogonalized $2s$ STO,

$$2s^\circ = (1 - S^2_{12})^{-1/2}(2s^{\text{STO}} - S_{12} 1s^{\text{STO}}) \qquad \text{(III.18)}$$

in which $S_{12} = \langle 1s^{\text{STO}} \mid 2s^{\text{STO}} \rangle$, we can immediately write

$$T_{2s^\circ} = (1 - S^2_{12})^{-1/2}(T_{2s} - S_{12} T_{1s}) \qquad \text{(III.19)}$$

where T_{1s} and T_{2s} are evaluated over STOs as described above.

Finally, of course, one may elect to use numerical integration techniques to evaluate the integral in Eq. III.3 for AOs other than Gaussian-lobe functions, Cartesian GTOs, STOs, or orthogonalized STOs.

References

1. I. N. Sneddon, *Special Functions of Mathematical Physics and Chemistry*, Oliver and Boyd, London, 1956.
2. H. Jeffreys and B. S. Jeffreys, *Methods of Mathematical Physics*, p. 627, Cambridge U. P., London, 1950.
3. P. O. Löwdin, *Advan. Quantum Chem.*, **5,** 185 (1970).
4. J. J. Huang, F. O. Ellison, and J. W. Rabalais, *J. Electron Spectrosc.*, **3,** 339 (1974).

Appendix IV. Atomic Units

Practical calculations are carried out most easily in atomic units (a.u.). We use the Bohr unit of length, $1 \ \mathrm{b} = a_0 = \hbar^2/(me^2)$, and the Hartree unit of energy, $1 \ \mathrm{H} = e^2/a_0 = me^4/\hbar^2$. Thus in simplified form,

$$r = r'a_0 = r'(0.5292 \ \text{Å}/\mathrm{b}) \tag{IV.1}$$

$$E = \frac{E'e^2}{a_0} = E'(27.21 \ \mathrm{eV}/\mathrm{H})$$

where primed quantities are in atomic units and unprimed in conventional units.

The wave vector $\mathbf{k}$ has dimension length^{-1}; thus

$$k = \frac{k'}{a_0} \tag{IV.2}$$

Let the photon energy $E = \hbar\omega$ and the ionization energy $= E_1^j$. Conversion of Eq. 6.7 yields

$$k' = [2(E' - E_1^{i'})]^{1/2} \tag{IV.3}$$

Equations 6.8 and 6.9 likewise are simplified in atomic units. The gradient matrix elements, Eq. 6.10, are dimensionally identical to k:

$$G_{jl} = \frac{G_{jl}'}{a_0} \tag{IV.4}$$

The overlap integrals F_{pk}, Eq. 6.14, are actually dimensionless. However, we noted in Appendix III that the factor L^3 in Eq. 6.8 is effectively canceled by the square of $L^{-3/2}$ appearing in Eq. III.3. If, as suggested, $L^{-3/2}$ is omitted in practical evaluation of the T_p, these integrals will then be of dimension (length)$^{3/2}$. The F_{pk} will be of the same dimension by virtue of Eq. 6.24. Thus

$$F_{pk} = F_{pk}'a_0^{3/2} \tag{IV.5}$$

We can now convert Eqs. 6.8 and 6.20 using Eqs. IV.1 and IV.2, IV.4 and IV.5, and the dimensionless fine-structure constant $\alpha = e^2/(\hbar c) = 1/137.036$:

$$\frac{d\sigma}{d\Omega} = \left(\frac{d\sigma}{d\Omega}\right)' a_0^2 \tag{IV.6}$$

$$\left(\frac{d\sigma}{d\Omega}\right)' = \frac{\alpha k' |u \cdot P'_{0j}|^2}{(2E'\pi)}$$

and $|u \cdot P'_{0j}|^2$ is given by Eq. 6.20 excluding, however, the factor $\hbar^2$. Finally, the total cross section σ given by Eq. 6.54 is

$$\sigma = \sigma' a_0^2$$

$$\sigma' = \frac{4\alpha k'(3a' + b')}{3E'} \tag{IV.7}$$

in which a' and b' are calculated according to Eq. 6.52, all quantities being expressed in atomic units.

Appendix V. Gradient Integrals in Terms of Gaussian-Lobe Expansions

To compute the gradient integrals over STOs, Eq. 6.8, we utilize expansions of the real STOs in terms of simple $1s$-Gaussian-type orbitals. Let such an expansion of an STO with orbital exponent $\zeta_\mu = 1$ be written[*]

$$\bar{\chi}_\mu^{(r)} = \sum_i \bar{\beta}_{\mu i} \exp\left(-\bar{\alpha}_{\mu i} r_{\mu i}^2\right) \tag{V.1}$$

If one is expanding an ns STO, all of the GTOs could be centered at the origin, and all $r_{\mu i} = r_{ns, i}$ would represent the electron distance from the origin. For np STOs, we use the Gaussian-lobe approach. Two identical but oppositely signed expansions of the two lobes of the np STO are developed; for example, the $r_{\mu i} = r_{np_x, i}$ would represent the electron distance from a point $(-R_{np_x, i}, 0, 0)$ for odd i and $(+R_{np_x, i}, 0, 0)$ for even i, $\beta_{np_x, 2i} = -\beta_{np, 2i+1}$ and $\alpha_{np_x, 2i} = \alpha_{np^x, 2i-1}$. The Gaussian-lobe expansion method eliminates the need for performing complicated rotations to obtain values of the gradient elements over STOs; all two-center gradient integrals between $1s$ GTOs depend only upon the distance between the centers.

If one is given the expansion parameters $\bar{\beta}_{\mu i}$, $\bar{\alpha}_{\mu i}$, and $\bar{R}_{\mu i}$ for an STO $\bar{\chi}_\mu^{(r)}$ with $\zeta_\mu = 1$, it is easily shown by scaling that the general expansion for $\zeta_\mu \neq 1$ is:

$$\chi_\mu^{(r)} = \sum \beta_{\mu i} \exp\left(-\alpha_{\mu i} r_{\mu i}^2\right) \tag{V.2}$$

$$\beta_{\mu i} = \bar{\beta}_{\mu i} \zeta_\mu^{3/2} \tag{V.3}$$

$$\alpha_{\mu i} = \bar{\alpha}_{\mu i} \zeta_\mu^2 \tag{V.4}$$

$$R_{\mu i} = \frac{\bar{R}_{\mu i}}{\zeta_\mu} \tag{V.5}$$

[*] Note that our definition here of the $1s$ GTO differs slightly from that given by Eq. II.2 and used in Appendix III.

The gradient integrals over real STOs thus become

$$\gamma_{\mu\nu x'} = \sum_i \sum_j \beta_{\mu i}\beta_{\nu j}g_{\mu i,\,\nu j,\,x'} \tag{V.6}$$

Using the techniques described by Shavitt,[1] it can be shown that

$$g_{\mu i,\,\nu j,\,x'} = 2\Delta X'_{\mu i,\,\nu j}\left(\frac{\alpha^3\pi}{\alpha_{\mu i}\alpha_{\nu j}}\right)^{1/2}\exp\left(-\alpha\Delta R_{\mu i,\,\nu j}\right) \tag{V.7}$$

in which

$$\alpha = \frac{\alpha_{\mu i}\alpha_{\nu j}}{\alpha_{\mu i} + \alpha_{\nu j}}$$

$$\Delta X'_{\mu i,\,\nu j} = X'_{\mu i} - X'_{\nu j} \tag{V.8}$$

$$\Delta R_{\mu i,\,\nu j} = |\mathbf{R}_{\mu i} = \mathbf{R}_{\nu j}|$$

Here, $X'_{\mu i}$ is the x coordinate of the ith GTO in the expansion of the $\chi_\mu^{(r)}$ and $\Delta R_{\mu i,\,\nu j}$ is the distance between the two $1s$ GTOs.

We employ the expansion parameters $\alpha_{\mu i}$ and $\beta_{\mu i}$ for $1s$ and $2s$ STOs in terms of $1s$ GTOs as developed by R. L. Stewart.[2] For $2p$ STOs, we use parameters computed by Sambe[3] for the $2p$ hydrogen-atom eigenfunctions. In this case, the basic parameters are appropriate to an STO with $\zeta = \frac{1}{2}$ rather than $\zeta = 1$ as we have described beginning with Eq. 6.12. To account for this difference, we must replace Eqs. 6.14 to 6.16 by the following

$$\beta_{2p,\,i} = \bar\beta_{2p,\,i}(2\zeta_\mu)^{3/2}$$

$$\alpha_{2p,\,i} = \bar\alpha_{2p,\,i}(2\zeta_\mu)^2 \tag{V.9}$$

$$R_{2p,\,i} = \frac{\bar R_{2p,\,i}}{2\zeta_\mu}$$

in which the parameters with bars are taken from Ref. 3.

The ultimate accuracy of the gradient integrals G_{jl} are limited by the accuracy of the GTO expansions of the STOs. For example, for the H_2O molecule we also performed an exact calculation of all gradient matrix elements $\gamma_{\mu\nu}$ over STOs and elements G_{jl} over MOs. Results obtained with six-term GTO expansions of the STOs were identical in at least three significant figures. Finally, the $\bar S_3 + \bar S_4 + \bar S_5$, which depend upon the G_{jl}, were obtained with two-significant-figure accuracy using the GTO expansion of STOs.

References

1. I. Shavitt, *Methods in Computational Physics*, B. Alder (Ed.), Vol. 2, Academic, New York, 1963.
2. R. L. Stewart, *J. Chem. Phys.*, **52**, 431 (1970).
3. H. Sambe, *J. Chem. Phys.*, **42**, 1732 (1965).

Appendix VI. Relative Magnitude of OPW Corrections on PW $\sigma_\perp$ and β

Equations 6.25 to 6.54 and 6.59 may be used to give the following expressions

$$\beta = \frac{2b}{3a+b}$$

$$\sigma = C(3a+b) \qquad\qquad (\text{VI.1})$$

$$\sigma_\perp = \frac{3C(2a+b)}{8\pi}$$

where $C = 4e^2 kL^3/(3mc\omega)$. In the PW approximation, $b = b^0$ and $a = 0$; thus

$$\beta^{\text{PW}} = 2$$

$$\sigma^{\text{PW}} = Cb^0 \qquad\qquad (\text{VI.2})$$

$$\sigma_\perp^{\text{PW}} = \frac{3Cb^0}{8\pi}$$

In the OPW approximation, $b = b^0 + b'$; thus

$$\beta^{\text{OPW}} = \frac{2(b^0 + b')}{3a + b^0 + b'}$$

$$\sigma^{\text{OPW}} = C(3a + b^0 + b') \qquad\qquad (\text{VI.3})$$

$$\sigma_\perp^{\text{OPW}} = \frac{3C(2a + b^0 + b')}{8\pi}$$

The fractional differences between the two approximations can be written

$$\frac{\delta\beta}{\beta^{PW}} = \frac{-3a}{3a + b^0 + b'}$$

$$= \frac{-3\varepsilon}{3\varepsilon + 1 + \varepsilon'}$$

$$\frac{\delta\sigma}{\sigma^{PW}} = 3\varepsilon + \varepsilon'$$

$$\frac{\delta\sigma_\perp}{\sigma_\perp^{PW}} = 2\varepsilon + \varepsilon' \tag{VI.4}$$

where $\delta\beta = \beta^{OPW} - \beta^{PW}$, and so forth, $\varepsilon = a/b^0$, and $\varepsilon' = b'/b^0$.

It appears from all numerical results that we have obtained so far that $2\varepsilon + \varepsilon' \to 0$ for high electron energy (that is, large k). Thus in the *limit of high k*

$$\frac{\delta\beta}{\beta^{PW}} = -\frac{3\varepsilon}{\varepsilon + 1}$$

$$\frac{\delta\sigma}{\sigma^{PW}} = \varepsilon$$

$$\frac{\delta\sigma_\perp}{\sigma_\perp^{PW}} = 0 \tag{VI.5}$$

For atomic ns electrons and for certain types of molecular photoionizations, ε also approaches zero for large k; for photoionization of $2p$-like electrons, however, values of $\varepsilon \sim \frac{1}{2}\varepsilon' \sim 0.5$ at large k are not unusual.

For large k, we thus see that the PW and OPW approximations are in substantial agreement with respect to the normal differential cross section $\sigma_\perp$; however, the OPW corrections may actually be very significant for the asymmetry parameter β and for the total cross section σ.

Appendix VII. Jahn–Teller Active Vibrations and Accessible Subgroup Geometries for Degenerate States of Molecules Belonging to the Most Common Point Groups[a]

Parent Point Group	Jahn–Teller Active vibrations	Electronic States Split	Accessible Subgroup Geometries
D_{2d}	b_1	E	D_2
	b_2	E	C_{2v}
C_3	e	E	C_1
C_{3v}	e	E	C_1, C_s
C_{3h}	e'	E', E''	C_s
D_3	e	E	C_1, C_2
D_{3h}	e'	E', E''	C_s, C_{2v}
D_{3d}	e_g	E_g, E_u	C_i, C_{2h}
C_4	$2b$	E	C_2
C_{4v}	b_1	E	C_{2v}
	b_2	E	C_{2v}
C_{4h}	$2b_g$	E_g, E_u	C_{2h}
D_4	b_1	E	D_2
	b_2	E	D_2

Parent Point Group	Jahn–Teller Active vibrations	Electronic States Split	Accessible Subgroup Geometries
D_{4h}	b_{1g}	E_g, E_u	D_{2h}(rhombus)
	b_{2g}	E_g, E_u	D_{2h}(rectangle)
D_{4d}	b_1	E_2	D_4
	b_2	E_2	C_{6v}
	e	E_1, E_3	D_2, C_{2v}, C_2
C_5	e_1	E_2	C_1
	e_2	E_1	C_1
C_{5v}	e_1	E_2	C_s, C_1
	e_1	E_1	C_s, C_1
C_{5h}	e_1'	E_2', E_2''	C_s
	e_2'	E_1', E_1''	C_s
D_5	e_1	E_2	C_2, C_1
	e_2	E_1	C_2, C_1
D_{5h}	e_1'	E_2', E_2''	C_{2v}, C_s
	e_2'	E_1', E_1''	C_{2v}, C_s
D_{5d}	e_{1g}	E_{2g}, E_{2u}	C_{2h}, C_i
	e_{2g}	E_{1g}, E_{1u}	C_{2h}, C_i
C_6	e_2	E_1, E_2	C_2
C_{6v}	e_2	E_1, E_2	C_{2v}, C_2
C_{6h}	e_{2g}	$E_{1g}, E_{1u}, E_{2g}, E_{2u}$	C_{2h}
D_6	e_2	E_1, E_2	D_2, C_2
D_{6h}	e_{2g}	$E_{1g}, E_{2g}, E_{1u}, E_{2u}$	D_{2h}, C_{2h}
D_{6d}	b_1	E_3	D_6
	b_2	E_3	C_{6v}
	e_2	E_1, E_5	D_2, C_{2v}, C_2
	e_4	E_2, E_4	D_{2d}, S_4
T	e	$E, T, G_{3/2}$	D_2
	t	$T, G_{3/2}$	C_3, C_2, C_1
T_h	e_g	$E_g, E_u, T_g, T_u, G_{3/2g}, G_{3/2u}$	D_{2h}
	t_g	$T_g, T_u, G_{3/2g}, G_{3/2u}$	C_{2h}, S_6, C_i
T_d	e	$E, T_1, T_2, G_{3/2}$	D_{2d}, D_2
	t_2	$T_1, T_2, G_{3/2}$	C_{3v}, C_{2v}, C_s, C_i
0	e	$E, T_1, T_2, G_{3/2}$	D_4, D_2
	t_2	$T_1, T_2, G_{3/2}$	D_3, D_2, C_2, C_1
0_h	e_g	$E_g, E_u, T_{1g}, T_{1u}, T_{2g}, T_{2u}, G_{3/2g}, G_{3/2u}$	D_{4h}, D_{2h}
	t_{2g}	$T_{1g}, T_{1u}, T_{2g}, T_{2u}, G_{3/2g}, G_{3/2u}$	$D_{3d}, D_{2h}, C_{2h}, C_i$
S_4	$2b$	E	C_2
S_6	e_g	E_g, E_u	C_i
$C_{\infty v}, D_{\infty h}$	none	—	—

[a] R. W. Jotham and S. F. Kettle, *Inorg. Chim. Acta*, **5,** 183 (1973).

Appendix VIII.   Photoelectron Spectra of a Variety of Molecules

This appendix contains PE spectra and some energy-level diagrams, Figs. VIII.1 to VIII.28, for a variety of molecules that have not been discussed in the text. Some of these spectra have not been published previously, and firm assignments of the spectral bands do not exist. For cases in which the spectra have already been published, the reference in the figure caption refers to the bibliography of Appendix IX and allows one to go back to the original paper.

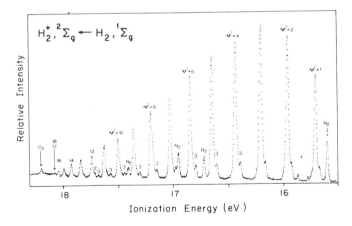

Fig. VIII.I   He I spectrum of H_2. The $v' = 0$ band is outside the figure. 3 indicates the $J'' = 3 \rightarrow J' = 3$ transitions. The transitions $J'' = 0 \rightarrow J' = 0$, $J'' = 1 \rightarrow J' = 1$, and $J'' = 2 \rightarrow J' = 2$ are not resolved. Reproduced with permission from Ref. 9.

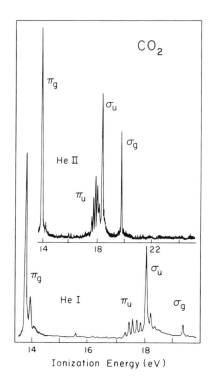

Fig. VIII.2. He I and He II spectra of CO_2.

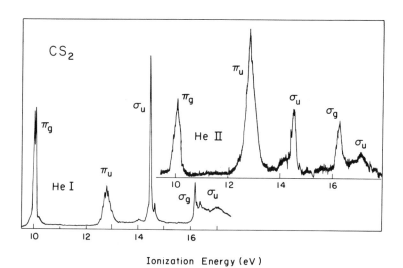

Fig. VIII.3. He I and He II spectra of CS_2.

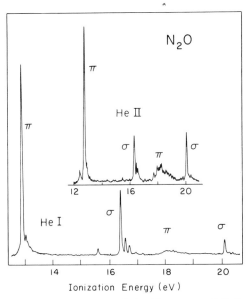

Fig. VIII.4. He I and He II spectra of N_2O.

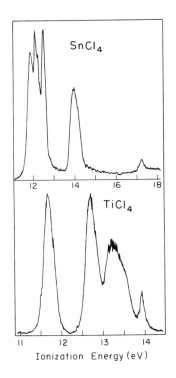

Fig. VIII.5. He I spectra of $SnCl_4$ and $TiCl_4$.

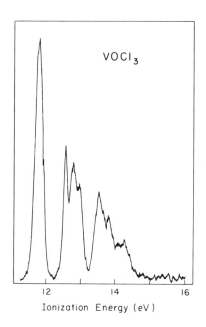

Ionization Energy (eV)

Fig. VIII.6. He I spectrum of VOCl₃.

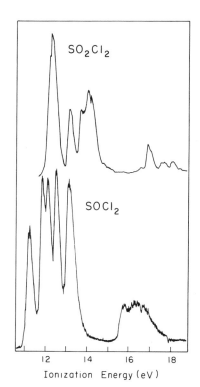

Ionization Energy (eV)

Fig. VIII.7. He I spectra of SO₂Cl₂ and
SOCl₂

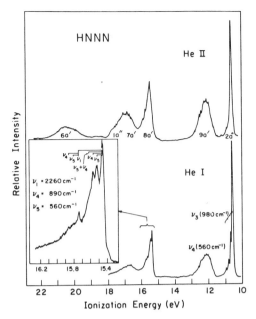

Fig. VIII.8. He I and He II spectra of NH₃. The 15.2- to 16.3-eV region is expanded in the insert. Reproduced with permission form Ref. 529.

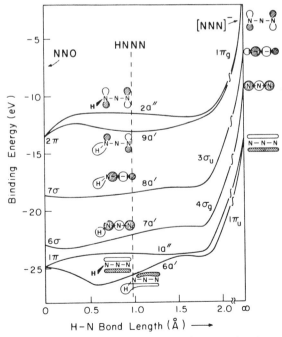

Fig. VIII.9. INDO eigenvalues for the outer valence molecular orbitals of HN₃ as a function of H—N bond length with N₂O and N₂⁻ as the extreme limits. Reproduced with permission from Ref. 529.

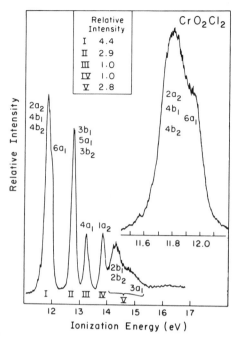

Fig. VIII.10. He I spectrum of chromyl chloride. No bands were observed beyond 15.5 eV. The 11.5- to 12.1-eV region is shown on an expanded scale. Relative integrated band intensities are shown in the insert. The molecular orbital assignments are listed near the respective ionization bands. Reproduced with permission from Ref. 530.

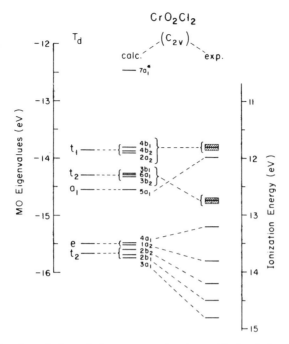

Fig. VIII.11. Correlation of the outer molecular orbitals of the CrO_2Cl_2 C_{2v} structure with the orbitals of a hypothetical tetrahedral sturcture and with the expermimental ionization bands. Reproduced with permission from Ref. 530.

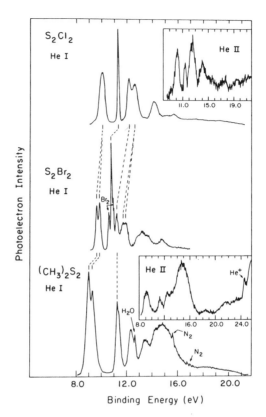

Fig. VIII.12. PE spectra of S_2Cl_2, S_2Br_2, and $(CH_3)_2S_2$. Reproduced with permission from Ref. 216.

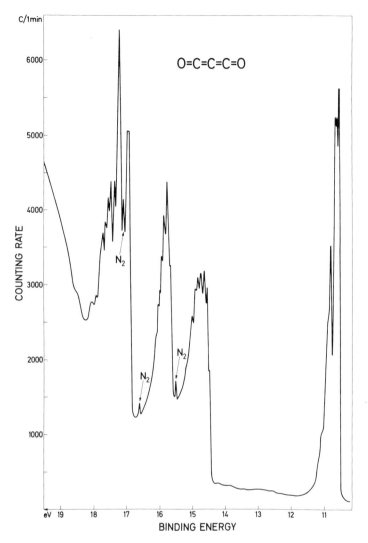

Fig. VIII.13. He I spectrum of carbon suboxide. Reproduced with permission from Ref. 648.

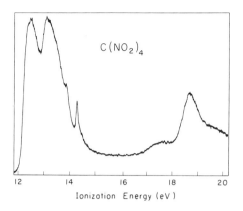

Fig. VIII.14. He I spectrum of tetranitromethane.

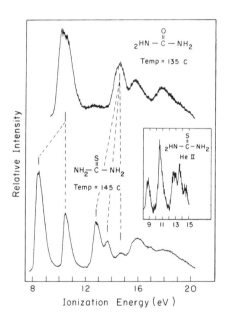

Fig. VIII.15. PE spectra of urea and thiourea with correlations between the bands. Reproduced with permission from Ref. 263.

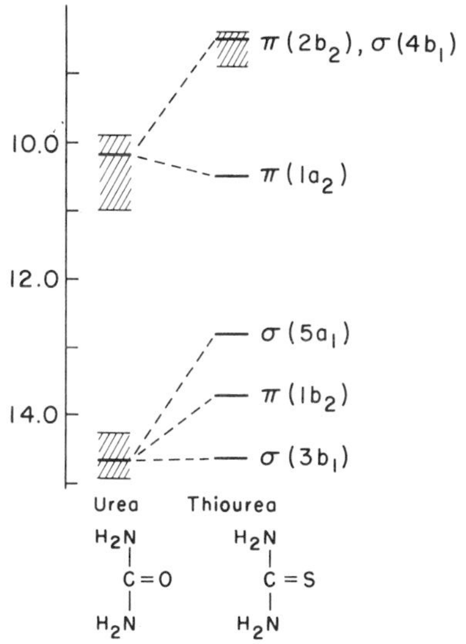

10.0
12.0
14.0
π (2b₂), σ (4b₁)
π (1a₂)
σ (5a₁)
π (1b₂)
σ (3b₁)
Urea
Thiourea
H2N
C=O
H2N
H2N
C=S
H2N

Fig. VIII.16. Correlation of the outer electronic energy levels of urea and thiourea. Experimental energies are used. Reproduced with permission from Ref. 263.

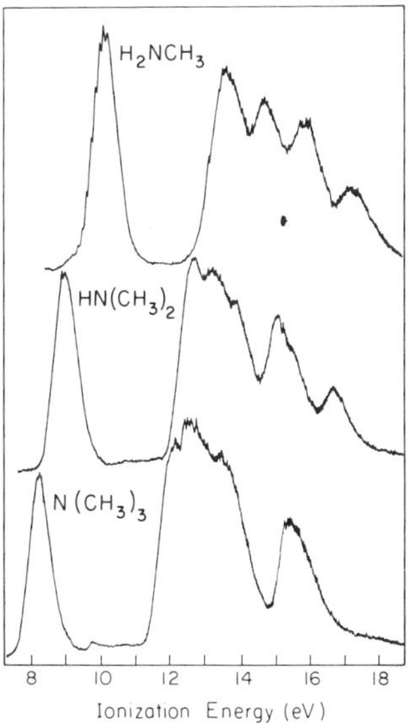

H2NCH3
HN(CH3)2
N(CH3)3
8
10
12
14
16
18
Ionization Energy (eV)

Fig. VIII.17. He I spectra of methylamine, dimethlamine, and trimethylamine.

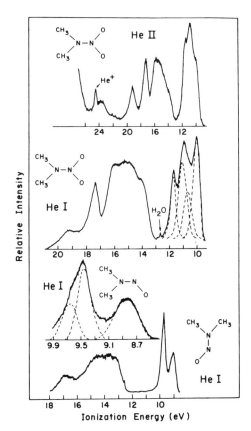

Fig. VIII.18. PE spectra of *N,N*-dimethylnitramine and *N,N*-dimethylnitrosamine. The 8.7- to 10.1-eV region of the $(CH_3)_2NNO$ spectrum is expanded in the insert. Approximate deconvolutions are indicated for the outer bands of both molecules. Reproduced with permission from Ref. 800.

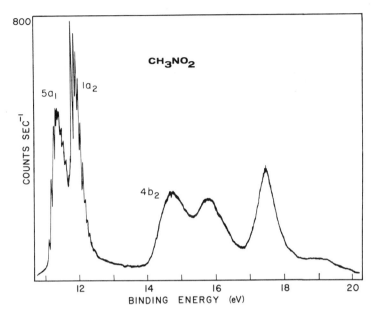

Fig. VIII.19. He I spectrum of nitromethane. Reproduced with permission from Ref. 647.

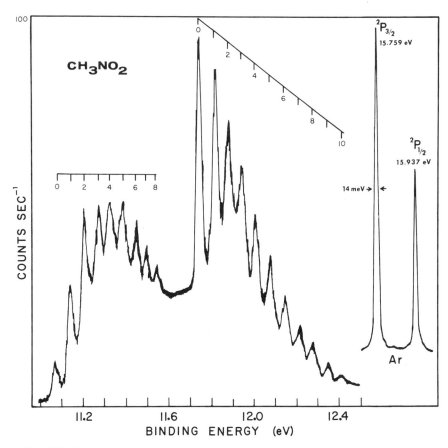

Fig. VIII.20. The two lowest energy ionization bands of nitromethane on an expanded scale. Observed vibrational progressions are labeled. The $^2P_{3/2}$ and $^2P_{1/2}$ lines of Ar are shown to the right of the spectrum. Reproduced with permission from Ref. 647.

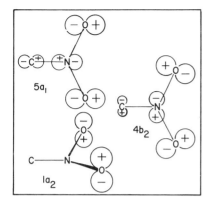

Fig. VIII.21. Schematic diagram of the three outer molecular orbitals of nitromethane. The hydrogen atoms are omitted because these orbitals are primarily localized on the CNO_2 group. Reproduced with permission from Ref. 647.

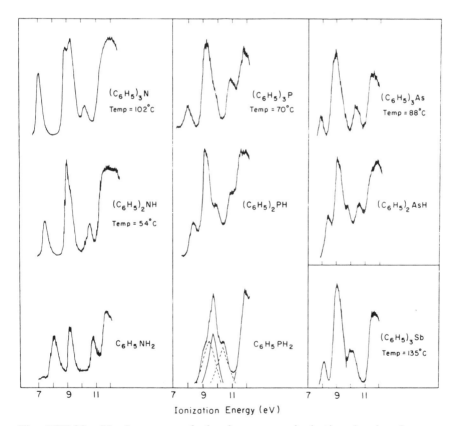

Fig. VIII.22. He I spectra of the low-energy ionization bands of some phenylamines, -phosphines, -arsines, and -stibine. Reproduced with permission from Ref. 264.

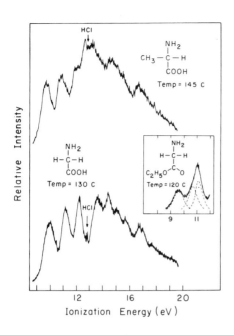

Fig. VIII.23. He I spectra of the amino acids glycine and alanine and the low energy region of glycine ethyl ester. Reproduced with permission from Ref. 263.

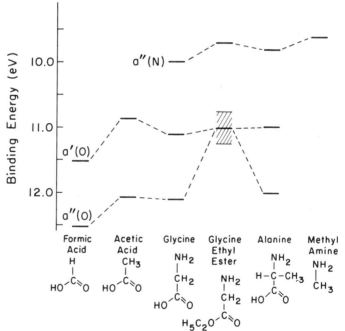

fig. VIII.24. Correlation of the outer electronic levels of glycine, alanine, and glycine ethyl ester with levels of the genealogically related molecules formic and acetic acid and methyl amine. Reproduced with permission from Ref. 263.

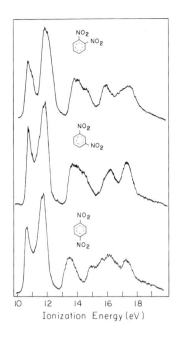

Fig. VIII.25. He I spectra of *ortho-*, *meta-*, and *para*-dinitrobenzene.

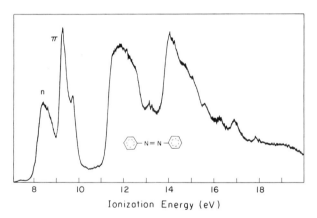

Fig. VIII.26. He I spectrum of azobenzene.

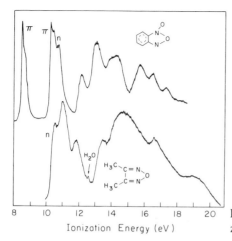

Fig. VIII.27. He I spectra of benzofuroxan and 3,4-dimethylfurazan.

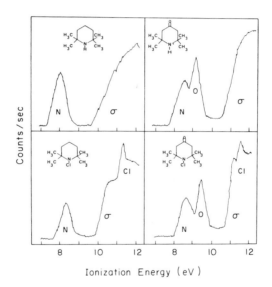

Fig. VIII.28. He I spectra of 2,2,6,6-tetramethylpiperidine, 2,2,6,6-tetramethyl-4-piperidone, 1-chloro-2,2,6,6-tetramethylpiperidine, and 1-chloro-2,2,6,6-tetramethyl-4-piperidone. Reproduced with permission from Ref. 115.

Appendix IX. Compilation of Atoms and Molecules with References to uv Photoelectron Spectral Data for the Period 1963 to early 1975

Atom or Molecule[a]	Reference	Figure in Text
Monatomic Gases		
Alkali atoms	569	—
H	217, 470, 472	—
Halogen atoms	470	—
Hg	69, 94, 266, 323, 367, 370, 583	—
Inert gases	2, 132, 135, 142, 188, 195, 261, 487, 488, 540, 556, 567, 568, 582, 594, 675, 677, 678, 681. 682, 684, 773	2.9, 4.2, VIII.20
N	193, 218, 267, 470, 472	—
O	196, 467, 470, 472, 685, 686	—
S	123	—
Zn, Cd	406, 782	—
Diatomic Molecules		
Alkali halides	73, 393, 643	—
CO	16, 20, 21, 22, 96, 188, 210, 215, 313, 373, 374, 376, 377, 456, 458, 480, 493, 609, 622, 641, 651, 652, 740, 750, 761, 763, 772	3.2, 7.1

Atom or Molecule[a]	Reference	Figure in Text
Diatomic Molecules (*continued*)		
CS	362, 468, 469, 622	—
CsX	59, 72, 756	—
Group VI diatomics	64	—
H_2, D_2	9, 74, 75, 139, 188, 227, 297, 456, 462, 486, 532, 616, 651, 652, 673, 674, 724, 726, 759, 760, 763	3.11, 10.2, VIII.1
HX	29, 62, 159, 261, 281, 368, 533, 711, 749, 750, 783, 796, 798	3.7, 4.2, 8.1
InX	67	—
KX	5	—
N_2	11, 20, 65, 96, 186, 188, 193, 210, 215, 313, 323, 373, 374, 376, 456, 458, 478, 480, 532, 540, 593, 609, 622, 641, 651, 652, 653, 675, 740, 750, 760, 763	7.1
NO	20, 22, 141, 158, 193, 211, 212, 214, 312, 316, 317, 375, 378, 484, 499, 528, 593, 595, 676, 751, 755, 761	5.1, 10.16
O_2	95, 136, 186, 194, 210, 215, 301, 303, 307, 313, 314, 323, 376, 471, 474, 496, 593, 605, 660, 748, 750, 757, 759, 761, 763	3.2, 3.12, 3.13, 5.3
SO	475	—
TlX	60, 63, 76, 268, 341a	—
X_2	7, 191, 194, 226, 273, 342, 368, 437, 440, 638, 749, 803	3.4, 8.2, 8.3, 8.4
Triatomic Molecules		
ClO_2	224, 225, 437	—
CO_2	22, 23, 96, 174, 189, 213, 321, 324, 376, 610, 641, 679, 680, 683, 740, 750, 764	VIII.2
COS	174, 189, 213, 610, 641, 764	—
COSe, CSSe	361	—
CS_2	174, 189, 213, 324, 493, 589, 610, 641, 750, 764	VIII.3

Atom or Molecule[a]	Reference	Figure in Text
Triatomic Molecules (*continued*)		
CSe_2	242, 361	—
Group VI hydrides	637	—
Group VI halides	635	—
HBS	345	—
HCN	37, 66, 364, 455, 519, 641, 750	10.19
HCP	364	—
HgX_2	61, 257, 319, 341a, 808	—
H_2O, D_2O	18, 32, 56, 128, 173, 175, 189, 261, 300a, 592, 645, 651, 652, 653, 715, 727, 763	1.1, 1.2, 4.1, 4.2, 7.2, 9.14–9.26, 10.27
H_2S, D_2S	189, 210, 279, 280, 300, 309, 360, 645, 651, 652, 715	7.2
H_2Se	210, 280	—
HOF	68	—
KrF_2	162, 163, 270	—
NF_2	222, 225	—
N_2O	22, 96, 158, 174, 189, 193, 452, 552, 553, 606, 641, 750, 764	VIII.4, VIII.9
NO_2	167, 315, 604, 608	5.4
NSX	228, 229, 275, 299, 669	—
OCl_2	199, 223	—
OF_2	173, 199, 223	10.27
O_3	160, 365, 765	—
PbX_2	63	—
SCl_2	216, 731	10.1
SiF_2	346, 799	—
SO_2	105, 324, 443, 549	—
S_2O	105, 363, 557, 669	—
XCN	429, 454, 522	—
XeF_2	140, 165, 171, 270, 714	—
ZnX_2, CdX_2	61, 118, 209, 257, 623	—
Tetraatomic Molecules		
AlX_3	525	—
AsX_3, SbX_3	531	8.7
ClF_3, BrF_3	273	—

Atom or Molecule[a]	Reference	Figure in Text
Tetraatomic Molecules (*continued*)		
Cl_2CO, Cl_2CS	199	—
CH_3	391, 466, 508a, 634	—
$(CN)_2$	455	—
F_2CS	517, 580	—
Group III halides	269, 525, 636, 566a	—
Group V hydrides	637	—
Group V halides and tetramethyls	326, 635	—
$HC\equiv CH$	34, 451, 453, 761	10.19
H_2CO	26, 450, 644	10.16, 10.26
H_2CS	399, 516, 517	—
HNCO, HNCS	322, 511	—
HN_3	322, 529	VIII.8, VIII.9
H_2O_2, H_2S_2	260, 625	—
NH_3, ND_3	22, 134, 261, 367, 654, 715, 774	4.2, 9.8, 9.9
N_2H_2	366	—
$(NPF_2)_n$	104, 199	—
NX_3	43, 46	—
PF_2H	250	—
PH_3	137, 265, 367, 561, 715	—
$POCl_2$, $PSCl_2$	198	—
P_4	166, 341	—
PX_3	46, 47, 58, 239, 323, 561	8.5
SO_3	274, 578	—
$SOCl_2$	—	VIII.7
S_2X_2	216, 781	VIII.12
X_2CO	753	—
X_2SO	104, 199	—
Pentaatomic Molecules		
C_3O_2	648, 762	3.5, 3.6, 3.8, VIII.13
CH_4	26, 57, 261, 298, 538, 600, 649, 651, 652, 653, 715	4.2, 9.2–9.7, 10.36
CH_nX_m	19, 169, 190, 241, 302, 304, 354, 483, 636, 661	10.36

Atom or Molecule[a]	Reference	Figure in Text
Pentaatomic Molecules (*continued*)		
C_3O_2	648	3.5, 3.6, 3.8, VIII.13
CrO_2Cl_2	530	VIII.10, VIII.11
Group IV B Tetrahalides	44, 237, 541, 628	—
Group IV hydrides, halides, and tetramethyls	42, 43, 120, 169, 177, 241, 243, 248, 254, 306, 311, 333, 354, 355, 356, 395, 403, 465, 565, 635, 714	9.12, 9.13, 10.36, 11.4, 11.5, VIII.5
HCOOH	176, 501, 752, 787	VIII.24
H_2CCO	35	—
H_2NCN	734	—
ONF_3	43, 46, 359	—
OsO_4, RuO_4	296, 337, 347	—
POX_3, PSX_3	46, 47, 58, 179, 239	8.6
$SnCl_4$	--	VIII.5
SO_2F_2, SO_2Cl_2	199, 276	VIII.7
SO_3F	225	—
$TiCl_4$	—	VIII.5
VCl_4	237, 628	—
$VOCl_3$	—	VIII.6
XeF_4	165, 271	—
Hexaatomic Molecules		
$H_2C{=}CH_2$	26, 102, 135, 161, 170, 173, 369, , 479, 574, 651, 652, 750, 801	7.2, 10.6, 10.16, 10.26
H_3CCN, $(CH_nX_m)CN$	357, 522	—
H_3CNC, D_3CNC	523	—
$H_3RSH(R{=}C, Si, Ge)$	252	—
IF_5, BrF_5	271, 272	—
PX_5	239, 394	—
N_2H_4	113	—
$XeOF_4$	164, 271	—
X_3CCN	667	—
Heptaatomic Molecules		
CH_3NO_2	647	VIII.19–VIII.21
$H_2C{=}C{=}CH_2$	763	—
$H_2C(CN)_2$ and derivatives	735	—

413

Atom or Molecule[a]	Reference	Figure in Text
Heptaatomic Molecules (*continued*)		
$(H_2C)_2O$	763	—
H_3CCHO, H_3CCFO	204	—
SeF_6, TeF_6, UF_6	635	—
SF_5Cl	272	—
SF_6	277, 278, 367, 635	—
XeF_6	165	—
Alkanes		
General saturated hydrocarbons	14, 30, 102, 149, 156, 185, 293, 308, 350, 435, 446, 551, 573, 599, 600, 640, 642, 645, 708, 763	—
$H_3C—CH_3$	26, 526, 538, 600, 655	9.10, 9.11
Cyclic Alkanes and Alkenes		
Adamantane and derivatives	114, 126, 577, 699, 700, 710, 794, 810, 811	—
Barrelene	89, 117, 416	—
Bullvalene and derivatives	85	—
Cyclopropane, cyclopropene, and derivatives	39, 40, 341, 405, 407, 408, 538, 665, 671, 695, 713	—
Cyclobutane, cyclobutadiene, and derivatives	87, 112, 291, 423, 672, 809	—
Cyclopentane, cyclopentene, cyclopentadiene, and derivatives	55, 127, 143, 148, 265, 288, 289, 351, 423, 449, 460, 630, 721	—
Cyclohexane, cyclohexenes, and derivatives	89, 127, 148, 448, 457, 460	—
Cyclooctane, cyclooctenes, and derivatives	49, 82, 86, 90, 383	10.35
Cyclononane, cyclononenes, and derivatives	89, 145, 155, 390, 410, 416, 663	—

414

Atom or Molecule[a]	Reference	Figure in Text
Cyclic Alkanes and Alkenes (*continued*)		
Cyclodecane, cyclodecenes, and derivatives	709, 725	—
General cyclic alkanes and alkenes	92, 146, 284, 424	—
Norbornane, norbornadiene, and derivatives	84, 89, 93, 151, 236, 284, 416, 422, 433, 538, 575, 627	10.34
Spiro-conjugated hydrocarbons	52, 53, 83, 282, 447, 719, 720, 815	—
Unsaturated Hydrocarbons and Derivatives		
Acetylene and derivatives	37, 80, 150, 157, 327, 357, 401, 426, 427, 428, 562, 706	—
Allene and derivatives	35, 144, 408, 538, 754	—
Butadiene and derivatives	50, 55, 168, 292, 412, 479, 743, 802	10.6, 11.1
Butatriene	152	—
Butenyne	775	—
Diacetylene	34	—
Dimethyldivinylmethane and tetravinylmethane	718	—
Ethylene and derivatives	88, 106, 110, 111, 197, 201, 283, 369, 371, 372, 473, 480, 491, 498, 521, 566, 581, 617, 687, 714, 802, 804, 805, 806	7.1, 10.21, 10.31, 10.33, 11.3
Hexatriene	55	—
$H_2C{=}CH{-}(CH_2)_n{-}CH{=}CH_2$	178	—
Propylene and derivatives (allyl compounds)	201, 481, 581, 701, 722, 801, 812	10.22–10.25, 10.33, 11.3
Tetracyanoethylene and tetracyanoquino-dimethane	459	—
Unsaturated fluorocarbon derivatives	254	10.26, 10.28, 10.29, 10.30, 11.1

Atom or Molecule[a]	Reference	Figure in Text
Benzene and Substituted Benzenes		
Allylanisol	310	—
Aniline and derivatives	235, 262, 388, 489, 505, 508, 563	10.11, 10.12, 10.13
Azabenzene and derivatives	430, 602, 692, 771	—
Azobenzene and derivatives	409, 508	VIII.26
Benzaldehyde	650	10.15, 10.16, 10.17
Benzene	6, 10, 17, 32, 36, 172, 187, 205, 381, 477, 479, 495, 570, 591, 592, 607, 639, 645, 672, 729	10.3, 10.4, 10.5, 10.8, 10.9, 10.13, 10.16, 10.19, 10.28, 11.7
Benzonitrile	398, 650	10.18, 10.19
Benzoquinone	172, 419, 584, 758	—
Biphenyl and derivatives	50, 564, 631	10.7
$(C_6H_5)_n RH_m$ (R = N, P, As, Sb)	51, 264	VIII.22
Ethynylbenzene	398	—
General substituted benzenes	33, 97, 108, 183, 184, 262, 387, 400, 461, 500, 503, 504, 586, 650, 739, 763, 777, 802	10.10–10.20, VIII.22, VIII.25, VIII.26, VIII.27
Halogenobenzenes	29, 262, 379, 591, 601, 603, 607, 639, 729, 738	10.11, 10.12, 10.13 10.14
Hexamethyl-Dewar-benzene	129, 571	—
Mesitylene and derivatives	510	—
Nitrobenzene	647	10.20, VIII.25
Nitrosobenzene	650	10.15, 10.16
Phenol and derivatives	262, 554, 563	10.11, 10.12, 10.13
Phenylacetylene	650	10.18, 10.19
Phenylsilane	558	—
Phosphabenzene	51, 692	—
Styrene and derivatives	507, 538, 562, 650	10.15, 10.16
Toluene	15, 262	10.11, 10.12, 10.13
Polycyclic Aromatic Hydrocarbons		
Anthracene	3, 206, 445	10.8, 10.9
Coronene, pyrene	124	—
Fluorenone	344	—

Atom or Molecule[a]	Reference	Figure in Text
Polycyclic Aromatic Hydrocarbons (*continued*)		
General delocalized π systems	117, 119, 121, 147, 290, 318, 320, 380, 464, 497, 586, 587, 763	10.8, 10.9
(α-Naphthyl)-(CH$_2$)$_4$-(α-naphthyl)	70, 71	—
Naphthalene and derivatives	91, 153, 172, 206, 294, 325, 404, 430, 464, 538, 559, 697, 698, 723, 733, 741, 768, 770	10.8, 10.9, 10.29
Perylene	723	—
Polyunsaturated propelanes	384	—
Quadricyclanes	572	—
Quinoline, isoquinoline	733, 767, 769	—
Boron Compounds		
Aminoboranes	100	—
B$_2$H$_6$	102, 170, 548, 670	—
B$_2$X$_4$, B$_4$X$_4$	545, 555	—
Borazine (B$_3$N$_3$H$_6$) and derivatives	99, 102, 358, 515, 544, 548	—
BX$_3$	41, 45, 131, 492, 542, 645, 784	—
(CH$_3$)$_{3-n}$BX$_n$	512	—
Complexes of the type (H$_3$B:NH$_3$, H$_3$B:PX$_3$, X$_3$B:NH$_3$, etc.)	442, 520, 543, 546	—
HBS	345, 518	—
H$_3$BCO	543	—
Organoboranes	371, 372	—
Pentaborane	476	—
Trivinylboron	463	—
Heterocyclic Molecules		
Azines	742, 816	—
Furan	28, 36, 265, 285, 289, 702, 758	—
Furazan and Furoxan derivatives		VIII.27
Isoxazole	28	—

Atom or Molecule[a]	Reference	Figure in Text
Heterocyclic Molecules (*continued*)		
Piperidine derivatives	—	VIII.28
Pyrazine	6, 36, 349, 381, 417, 602, 732	—
Pyrazole	28	—
Pyrene and derivatives	116	—
Pyridazine	6, 13, 36, 381	—
Pyridine and derivatives	6, 36, 172, 382, 389, 425, 430, 494, 506, 509, 560, 601, 618, 729, 795	10.30
Pyrrole and derivatives	265, 287, 289, 400	—
s-Tetrazine	6, 382, 732	—
s-Triazine	6, 348, 382, 732	—
Hetero-Organic Molecules		
Alcohols	208, 482, 527, 620, 621, 666	10.23, 10.31
Aldehydes	208, 491, 747, 758	10.24, 10.31
Amides	176	
Amines	4, 40, 90, 113, 221, 231, 247, 263, 482, 527, 620, 621, 642, 689, 779, 800, 802a	10.23, VIII.17, VIII.18, VIII.22
Amino acids	263	VIII.23, VIII.24
Aza compounds	1, 40, 90, 130, 411, 412, 413, 418, 419, 434, 513, 514, 612, 705	10.35
Azo compounds	414, 432	VIII.26
Azoles	249	—
Carbonates	259, 814	—
Carbonyl compounds	8, 25, 185, 230, 436, 576, 807	—
Esters	744, 746	—
Ethers	24, 40, 101, 203, 208, 258, 502, 633, 642, 713, 716, 746, 793	—
Halogeno-organic compounds	27, 115, 122, 148, 305, 308, 420, 421, 446, 490, 581, 688, 701, 703, 728, 766, 804, 805, 807, 813	3.3, 7.1, 8.8, 10.11, 10.14, 10.21, 10.25, 10.31, 10.33, 10.36, 11.2, 11.3
Hydrazines	611, 613, 614, 615, 624, 658, 659	—
Hydrazones	817	—

Atom or Molecule[a]	Reference	Figure in Text
Hetero-Organic Molecules (*continued*)		
Ketones	181, 202, 203, 208, 263, 457, 485, 491, 619, 717, 747, 793	
Organic nitro and nitroso compounds	647, 800	VIII.14, VIII.18–21. VIII.25, VIII.27
Organic acids	176, 263, 491, 501, 746, 752, 785, 786	10.23, VIII.23, VIII.24
Organic P and As compounds	79, 231, 232, 233, 234, 247, 599, 667, 689, 696, 712, 788, 789, 791, 792	VIII.22
Organic pseudohalides	54, 244, 253	—
Organic Si compounds	98, 101, 256, 397, 632, 662, 736	—
Organic S compounds	8, 57, 77, 78, 82, 103, 107, 109, 180, 200, 207, 216, 255, 258, 259, 263, 286, 289, 352, 353, 385, 386, 393, 399, 464, 480, 516, 579, 588, 597, 598, 620, 621, 633, 657, 690, 691, 693 704, 716, 730, 737, 745, 780, 807, 812	3.3, 7.1, 10.24, 11.2 VIII.12, VIII.15
Peroxides	48	—
S_8	664	—
Tropones	3a	—
Urea and thiourea	263	VIII.15, VIII.16
Transition Metal Complexes		
Allylmercuric chloride	707	—
Al_2X_6 and Ga_2X_6	525	—
π-arene complexes	329	—
Bis-(π-allyl)-nickel	668	—
π-$(C_5H_5)_2M$	238, 331, 332, 343, 437, 547, 656	5.5, 5.6, 5.7
Cyclic ethers co-ordinated with dimethyl zinc	534	—
π-Cyclopentadienylnickel nitrosyl	336	—
Cyclopropylcarbinyltri-methyltin and allyl-trimethyltin	154	—

Atom or Molecule[a]	Reference	Figure in Text
Transition Metal Complexes (*continued*)		
$K_4Fe(CN)_6$	38	—
$K[PtCl_3(C_2H_4)] \cdot H_2O$ (Zeise's salt)	161	—
M(hfa)$_n$	338, 339, 340, 539	—
M-trifluorophosphine complexes	596	—
M-carbonyl complexes and derivatives	138, 192, 220, 240, 246, 248, 295, 323, 328, 334, 335, 396, 402, 438, 439, 441, 536, 537, 550, 809	—
M-complexes with *N,N'*-ethylene-bis(acetyl-acetone-iminato)	219	—
M-complexes with tri-methylsilylmethyl and neopentyl	330	—
Nitrosyl-M-complexes	441, 550	—
Silylmethyl and neo-pentyl derivatives of Group IVA metals	524	—
Tetrakis(trifluoro-phosphine)M	396, 444	—
Triethylaluminum and dimethylzinc complexes	535	—

[a] Ac = acyl, hfa = enolate anion of hexafluoroacetylacetone, M = transition metal, R = hydrocarbon group, and X = halogen.

References

1. L. J. Aarons, J. A. Connor, I. H. Hillier, M. Schwarz, and D. R. Lloyd, *J. C. S. Faraday Transactions 2*, **70,** 1106 (1974).
2. P. Agostini, G. Barjot, G. Mainfray, and C. Manns, *IEEE J. Quantum Electronics*, **6,** 782 (1970).
3. J. Aihara and H. Inokuchi, *Chem. Letters*, 421 (1973).
3. a. M. Allan, E. Heilbronner, and E. Kloster-Jensen, *J. Electron Spectrosc.* **6,** 181 (1975).
4. M. E. Akopyan and Yu. V. Loginov, *Opt. Spektrosk.*, **37,** 442 (1974).
5. J. D. Allen, Jr., G. W. Boggess, T. D. Goodman, A. S. Wachtel, Jr., and G. K. Schweitzer, *J. Electron Spectrosc.*, **2,** 289 (1973).

6. J. Almlöf, B. Roos, U. Wahlgren, and H. Johansen, *J. Electron Spectrosc.*, **2**, 51 (1973).

7. C. P. Anderson, G. Mamantov, W. E. Bull, F. A. Grimm, J. Carver, and T. A. Carlson, *Chem. Phys. Letters*, **12**, 137 (1971).

8. M. Arbelot, J. Metzger, M. Chanon, C. Guimon, and G. Pfister-Guillouzo, *J. Amer. Chem. Soc.*, **96**, 6217 (1974).

9. L. Åsbrink, *Chem. Phys. Letters*, **7**, 549 (1970).

10. L. Åsbrink, O. Edqvist, E. Lindholm, and L. E. Selin, *Chem. Phys. Letters*, **5**, 192 (1970).

11. L. Åsbrink and C. Fridh, *Phys. Scripta*, **9**, 338 (1974).

12. L. Åsbrink, C. Fridh, B. O. Jonsson, and E. Lindholm, *Int. J. Mass Spectrom. Ion Phys.*, **8**, 215 (1972).

13. L. Åsbrink, C. Fridh, B. O. Jonsson, and E. Lindholm, *Int. J. Mass Spectrom. Ion Phys.*, **8**, 229 (1972).

14. L. Åsbrink, C. Fridh, and E. Lindholm, *J. Amer. Chem. Soc.*, **94**, 5501 (1972).

15. L. Åsbrink, C. Fridh, and E. Lindholm, *Chem. Phys. Letters*, **15**, 567 (1972).

16. L. Åsbrink, C. Fridh, E. Lindholm, and K. Codling, *Phys. Scripta*, **10**, 183 (1974).

17. L. Åsbrink, E. Lindholm, and O. Edqvist, *Chem. Phys. Letters*, **5**, 609 (1970).

18. L. Åsbrink and J. W. Rabalais, *Chem. Phys. Letters*, **12**, 182 (1971).

19. T. Baer and B. P. Tsai, *J. Electron Spectrosc.*, **2**, 25 (1973).

19. a. E. J. Baerends, C. Oudshoorn, and A. Oskam, *J. Electron Spectrosc.*, **6**, 259 (1975).

20. J. L. Bahr, A. J. Blake, J. H. Carver, J. L. Gardner and V. Kumar, *J. Quant. Spectrosc. Radiat. Transfer*, **11**, 1839 (1971).

21. J. L. Bahr, A. J. Blake, J. H. Carver, J. L. Gardner, and V. Kumar, *J. Quant. Spectrosc. Radiat. Transfer*, **11**, 1853 (1971).

22. J. L. Bahr, A. J. Blake, J. H. Carver, J. L. Gardner, and V. Kumar, *J. Quant. Spectrosc. Radiat. Transfer*, **12**, 59 (1972).

23. J. L. Bahr, A. J. Blake, J. H. Carver, and V. Kumar, *J. Quant. Spectrosc. Radiat. Transfer*, **9**, 1359 (1969).

24. A. D. Bain, J. C. Bunzli, D. C. Frost, and L. Weiler, *J. Amer. Chem. Soc.*, **95**, 291 (1973).

25. A. D. Bain and D. C. Frost, *Can. J. Chem.*, **51**, 1245 (1973).

26. A. D. Baker, C. Baker, C. R. Brundle, and D. W. Turner, *Int. J. Mass Spectrom. Ion Phys.*, **1**, 285 (1968).

27. A. D. Baker, D. Betteridge, N. R. Kemp, and R. E. Kirby, *Anal. Chem.*, **43**, 375 (1971).

28. A. D. Baker, D. Betteridge, N. R. Kemp, and R. E. Kirby, *J. Chem. Soc. D*, 286 (1970).

29. A. D. Baker, D. Betteridge, N. R. Kemp, and R. E. Kirby, *Int. J. Mass Spectrosc. Ion Phys.*, **4**, 90 (1970).

30. A. D. Baker, D. Betteridge, N. R. Kemp, and R. E. Kirby, *J. Mol. Struct.*, **8**, 75 (1971).

31. A. D. Baker, M. Brisk, and M. Gellender, *J. Electron Spectrosc.*, **3**, 227 (1974).
32. A. D. Baker, C. R. Brundle, and D. W. Turner, *Int. J. Mass Spectrom. Ion Phys.*, **1**, 443 (1968).
33. A. D. Baker, D. P. May, and D. W. Turner, *J. Chem. Soc. B*, 22 (1968).
34. C. Baker and D. W. Turner, *Chem. Commun.*, 797 (1967).
35. C. Baker and D. W. Turner, *J. Chem. Soc. D*, 480 (1969).
36. A. D. Baker and D. W. Turner, *Phil Trans. Roy. Soc. London*, **A268**, 131 (1970).
37. C. Baker and D. W. Turner, *Proc. Roy. Soc. London,*, **A308**, 19 (1968).
38. R. E. Ballard and G. A. Griffiths, *J. Chem. Soc. D*, 1472 (1971).
39. H. Basch, *Mol. Phys.*, **23**, 683 (1972).
40. H. Basch, M. B. Robin, N. A. Kuebler, C. Baker, and D. W. Turner, *J. Chem. Phys.*, **51**, 52 (1969).
40. a. P. J. Bassett, B. R. Higginson, D. R. Lloyd, N. Lynaugh, and P. J. Roberts, *J. C. S. Dalton Transactions*, 2316 (1974).
41. P. J. Bassett and D. R. Lloyd, *Chem. Commun.*, 36 (1970).
42. P. J. Bassett and D. R. Lloyd, *Chem. Phys. Letters*, **3**, 22 (1969).
43. P. J. Bassett and D. R. Lloyd, *Chem. Phys. Letters*, **6**, 166 (1970).
44. P. J. Bassett and D. R. Lloyd, *J. Chem. Soc. A*, 641 (1971).
45. P. J. Bassett and D. R. Lloyd, *J. Chem. Soc. A*, 1551 (1971).
46. P. J. Bassett and D. R. Lloyd, *J. C. S. Dalton Transactions*, 248 (1972).
47. P. J. Bassett, D. R. Lloyd, I. H. Hillier, and V. R. Saunders, *Chem. Phys. Letters*, **6**, 253 (1970).
48. C. Batich and W. Adam, *Tetrahedron Lett.*, 1467 (1974).
49. C. Batich, P. Bischof, and E. Heilbronner, *J. Electron Spectrosc.*, **1**, 333 (1972/1973).
50. C. Batich, O. Ermer, E. Heilbronner, and J. R. Wiseman, *Angew. Chem. (Int. Edit.)*, **12**, 312 (1973).
51. C. Batich, E. Heilbronner, V. Hornung, A. J. Ashe, D. T. Clark, U. T. Cobley, D. Kilcast, and I. Scanlan, *J. Amer. Chem. Soc.*, **95**, 928 (1973).
52. C. Batich, E. Heilbronner, and M. F. Martin, *Helv. Chim. Acta*, **56**, 2110 (1973).
53. C. Batich, E. Heilbronner, E. Rommel, M. F. Semmelhack, and J. S. Foos, *J. Amer. Chem. Soc.*, **96**, 7662 (1974).
54. P. Baybutt, M. F. Guest, and I. H. Hillier, *Mol. Phys.*, **25**, 1025 (1973).
55. M. Beez, G. Bieri, H. Bock, and E. Heilbronner, *Helv. Chim. Acta*, **56**, 1028 (1973).
56. T. Bergmark, L. Karlsson, R. Jadrny, L. Mattsson, R. G. Albridge, and K. Siegbahn, *J. Electron Spectrosc.*, **4**, 85 (1974).
57. T. Bergmark, J. W. Rabalais, L. O. Werme, L. Karlsson, and K. Siegbahn, *Electron Spectroscopy*, D. A. Shirley (Ed.), p. 413, North-Holland, Amsterdam, 1972.
58. J. L. Berkosky, F. O. Ellison, T. H. Lee, and J. W. Rabalais, *J. Chem. Phys.*, **59**, 5342 (1973).
59. J. Berkowitz, *J. Chem. Phys.*, **50**, 3503 (1969).

60. J. Berkowitz, *J. Chem. Phys.*, **56**, 2766 (1972).
61. J. Berkowitz, *J. Chem. Phys.*, **61**, 407 (1974).
62. J. Berkowitz, *Chem. Phys. Letters*, **11**, 21 (1971).
63. J. Berkowitz, *Electron Spectroscopy*, D. A. Shirley (Ed.), p. 391, North-Holland, Amsterdam, 1972.
64. J. Berkowitz and W. A. Chupka, *J. Chem. Phys.*, **50**, 4245 (1969).
65. J. Berkowitz and W. A. Chupka, *J. Chem. Phys.*, **51**, 2341 (1969).
66. J. Berkowitz, W. A. Chupka, and T. A. Walter, *J. Chem. Phys.*, **50**, 1497 (1969).
67. J. Berkowitz and J. L. Dehmer, *J. Chem. Phys.*, **57**, 3194 (1972).
68. J. Berkowitz, J. L. Dehmer, and E. H. Appelman, *Chem. Phys. Letters*, **19**, 334 (1973).
69. J. Berkowitz, J. L. Dehmer, Y. K. Kim, and J. P. Desclaux, *J. Chem. Phys.*, **61**, 2556 (1974).
70. J. Berkowitz, J. L. Dehmer, K. Shimada, and M. Szwarc, *J. Electron Spectrosc.*, **2**, 211 (1973).
71. J. Berkowitz, J. L. Dehmer, K. Shimada, and M. Szwarc, *J. Electron Spectrosc.*, **3**, 164 (1974).
72. J. Berkowitz, J. L. Dehmer, and T. E. H. Walker, *J. Chem. Phys.*, **59**, 3645 (1973).
73. J. Berkowitz, J. L. Dehmer, and T. E. H. Walker, *J. Electron Spectrosc.*, **3**, 323, (1974).
74. J. Berkowitz and P. M. Guyon, *Int. J. Mass Spectrosc. Ion Phys.*, **6**, 301 (1971).
75. J. Berkowitz and R. Spohr *J. Electron Spectrosc.*, **2**, 143 (1973).
76. J. Berkowitz and T. A. Walter, *J. Chem. Phys.*, **49**, 1184 (1968).
77. A. J. Berlinsky, J. F. Carolan, and L. Weiler, *Can. J. Chem.*, **52**, 3373 (1974).
78. D. Betteridge, A. D. Baker, P. Bye, S. K. Hasanuddin, N. R. Kemp, D. I. Rees, M. A. Stevens, M. A. Thompson, and B. J. Wright, *Z. Anal. Chem.*, **263**, 286 (1973).
79. D. Betteridge, M. Thompson, A. D. Baker, and N. R. Kemp, *Anal. Chem.*, **44**, 2005 (1972).
79. a. D. Betteridge, M. A. Williams, and G. G. Chandler, *J. Electron Spectrosc.*, **6**, 327 (1975).
80. G. Bieri, F. Brogli, E. Heilbronner, and E. Kloster-Jensen, *J. Electron Spectrosc.*, **1**, 67 (1972).
81. G. Bieri and E. Heilbronner, *Helv. Chim. Acta*, **57**, 546 (1974).
82. G. Bieri, E. Heilbronner, E. Kloster-Jensen, A. Schmelzer, and J. Wirz, *Helv. Chim. Acta*, **57**, 1265 (1974).
83. P. Bischof, R. Gleiter, A. DeMeijere, and L. Meyer, *Helv. Chim. Acta*, **57**, 1519 (1974).
84. P. Bischof, R. Gleiter, and E. Heilbronner, *Helv. Chim. Acta*, **53**, 1425 (1970).
85. P. Bischof, R. Gleiter, E. Heilbronner, V. Hornung, G. Schröder, *Helv. Chim. Acta*, **53**, 1645 (1970).

86. P. Bischof, R. Gleiter, M. J. Kukla, and L. A. Paquette, *J. Electron Spectrosc.*, **4,** 177 (1974).

87. P. Bischof, E. Haselbach, and E. Heilbronner, *Angew. Chem. (Int. Edit.,)* **9,** 953 (1970).

88. P. Bischof, J. A. Hashmall, E. Heilbronner, and V. Hornung, *Angew. Chem. (Int. Edit.)*, **8,** 878 (1969).

89. P. Bischof, J. A. Hashmall, E. Heilbronner, and V. Hornung, *Helv. Chim. Acta,* **52,** 1745 (1969).

90. P. Bischof, J. A. Hashmall, E. Heilbronner, and V. Hornung, *Tetrahedron Lett.,* 4025 (1969).

91. P. Bischof, J. A. Hashmall, E. Heilbronner, and V. Hornung, *Tetrahedron Lett.,* 1033 (1970).

92. P. Bischof, and E. Heilbronner, *Helv. Chim. Acta,* **53,** 1677 (1970).

93. P. Bischof, E. Heilbronner, H. Prinzbach, and H. D. Martin, *Helv. Chim. Acta,* **54,** 1072 (1971).

94. A. J. Blake, *Proc. Roy. Soc. London,* **A325,** 555 (1971).

95. A. J. Blake, J. L. Bahr, J. H. Carver, and V. Kumar, *Phil. Trans. Roy. Soc. London,* **A268,** 159 (1970).

96. M. Block, and D. W. Turner, *Chem. Phys. Letters,* **30,** 344 (1975).

97. G. Bobykina, *Z. Fiz. Khim.,* **45,** 2953 (1971).

98. H. Bock and W. Enblin, *Angew. Chem. (Int. Edit.),* **10,** 404 (1971).

99. H. Bock and W. Fuss, *Angew. Chem. (Int. Edit.),* **10,** 182 (1971).

100. H. Bock and W. Fuss, *Chem. Ber.,* **104,** 1687 (1971).

101. H. Bock, P. Moller, G. Becker, and G. Fritz, *J. Organometal. Chem.,* **61,** 113 (1973).

102. H. Bock and B. G. Ramsey, *Angew. Chem.,* **12,** 734 (1973).

103. H. Bock and B. Solouki, *Angew. Chem.,* **11,** 436 (1972).

104. H. Bock and B. Solouki, *Chem. Ber.,* **107,** 2299 (1974).

105. H. Bock, B. Solouki, P. Rosmus, R. Steudel, and W. Schultheis, *Angew. Chem.,* **85,** 987 (1973).

106. H. Bock and G. Stafast, *Chem. Ber.,* **105,** 1158 (1972).

107. H. Bock and G. Wagner, *Angew. Chem. (Int. Edit.),* **11.** 150 (1972).

108. H. Bock, G. Wagner, and J. Kroner, *Chem. Ber.,* **105,** 3850 (1972).

109. H. Bock, G. Wagner, and J. Kroner, *Tetrahedron Lett.,* 3713 (1971).

110. H. Bock, G. Wagner, K. Wittel, J. Sauer, and D. Seebach, *Chem. Ber.,* **107,** 1869 (1974).

111. H. Bock and K. Wittel, *J. C. S. Chem. Commun.,* 602 (1972).

112. N. Bodor, B. H. Chen, and S. D. Worley, *J. Electron Spectrosc.,* **4,** 65 (1974).

113. N. Bodor, M. J. S. Dewar, W. B. Jennings, and S. D. Worley, *Tetrahedron,* **26,** 4109 (1970).

114. N. Bodor, M. J. S. Dewar, and S. D. Worley, *J. Amer. Chem. Soc.,* **92,** 19 (1970).

115. N. Bodor, J. J. Kaminski, S. D. Worley, R. J. Colton, T. H. Lee, and J. W. Rabalais, *J. Pharm. Sci.,* **63,** 1387 (1974).

116. V. Boekalheide, J. N. Murrell, and W. Schmidt, *Tetrahedron Lett.,* 575 (1972).

117. V. Boekalheide and W. Schmidt, *Chem. Phys. Letters*, **17**, 410 (1972).
118. G. W. Boggess, J. D. Allen, and G. K. Schweitzer, *J. Electron Spectrosc.*, **2**, 467 (1973).
119. R. Boschi, E. Clar, and W. Schmidt, *J. Chem. Phys.*, **60**, 4406 (1974).
120. R. Boschi, M. F. Lappert, J. B. Pedley, W. Schmidt, and B. T. Wilkins, *J. Organometal Chem.*, **50**, 69 (1973).
121. R. Boschi, J. N. Murrell, and W. Schmidt, *Faraday Discussions Chem. Soc.* **54**, 116 (1972).
122. R. A. Boschi and D. R. Salahub, *Can. J. Chem.*, **52**, 1217 (1974).
123. R. Boschi and W. Schmidt, *Inorg. Nucl. Chem. Lett.*, **9**, 643 (1973).
124. R. Boschi and W. Schmidt, *Tetrahedron Lett.*, 2577 (1972).
125. R. Boschi, W. Schmidt, and J. C. Gfeller, *Tetrahedron Lett.*, 4107 (1972).
126. R. Boschi, W. Schmidt, R. J. Suffolk, B. T. Wilkins, H. J. Lempka, and J. N. A. Ridyard, *J. Electron Spectrosc.*, **2**, 377 (1973).
127. R. Botter, F. Menes, Y. Gounelle, J. M. Pechine, and D. Solgadi, *Int. J. Mass Spectrom. Ion Phys.*, **12**, 188 (1973).
128. R. Botter and H. M. Rosenstock, *J. Res. Nat. Bur. Stand. A*, **73**, 313 (1969).
129. D. Bougeard, B. Schrader, P. Bleckmann, and T. Plesser, *Justus Liebigs Ann. Chem.*, 137 (1974).
130. R. J. Boyd, J. C. Buenzli, J. P. Snyder, and M. L. Heyman, *J. Amer. Chem. Soc.*, **95**, 6478 (1973).
131. R. J. Boyd and D. C. Frost, *Chem. Phys. Letters*, **1**, 649 (1968).
132. G. R. Branton and C. E. Brion, *J. Electron Spectrosc.*, **3**, 129 (1974).
133. G. R. Branton, C. E. Brion, D. C. Frost, K. A. R. Mitchell, and N. L. Paddock, *J. Chem. Soc. A*, 151 (1970).
134. G. R. Branton, D. C. Frost, F. G. Herring, C. A. McDowell, and I. A. Stenhouse, *Chem. Phys. Letters*, **3**, 581 (1969).
135. G. R. Branton, D. C. Frost, T. Makita, C. A. McDowell, and I. A. Stenhouse, *J. Chem. Phys.*, **52**, 802 (1970).
136. G. R. Branton, D. C. Frost, T. Makita, C. A. McDowell, and I. A. Stenhouse, *Phil. Trans. Roy. Soc. London*, **A268**, 77 (1970).
137. G. R. Branton, D. C. Frost, C. A. McDowell, and I. A. Stenhouse, *Chem. Phys. Letters*, **5**, 1 (1970).
138. P. S. Braterman and A. P. Walter, *Discussions Faraday Soc.*, **47**, 121 (1969).
139. B. Brehm and R. Frey, *Z. Naturforsch., A*, **26**, 523 (1971).
140. B. Brehm, M. Menziner, and C. Zorn, *Can. J. Chem.*, **48**, 3193 (1970).
141. C. E. Brion, C. A. McDowell, and W. B. Stewart, *Chem. Phys. Letters*, **13**, 79 (1972).
142. C. E. Brion, C. A. McDowell, and W. B. Stewart, *J. Electron Spectrosc.*, **1**, 113 (1972/73).
143. F. Brogli, P. A. Clark, E. Heilbronner, and M. Neuenschwander, *Angew. Chem. (Int. Edit.)*, **12**, 422 (1973).
144. E. Brogli, J. K. Crandall, E. Heilbronner, E. Kloster-Jensen, and S. A. Sojka, *J. Electron Spectrosc.*, **2**, 455 (1973).

145. F. Brogli, W. Eberbach, E. Haselbach, E. Heilbronner, V. Hornung, and D. M. Lemal, *Helv. Chim. Acta*, **56**, 1933 (1973).

146. F. Brogli, E. Giovanni, E. Heilbronner, and R. Schurter, *Chem. Ber.*, **106**, 961 (1973).

147. F. Brogli and E. Heilbronner, *Angew. Chem. (Int. Edit.)*, **11**, 538 (1972).

148. F. Brogli and E. Heilbronner, *Helv. Chim. Acta*, **54**, 1423 (1971).

149. F. Brogli and E. Heilbronner, *Theor. Chim. Acta*, **26**, 289 (1972).

150. F. Brogli, E. Heilbronner, V. Hornung, and E. Kloster-Jensen, *Helv. Chim. Acta*, **56**, 2171 (1973).

151. F. Brogli, E. Heilbronner, and J. Ipaktsch, *Helv. Chim. Acta*, **55**, 2447 (1972).

152. F. Brogli, E. Heilbronner, E. Kloster-Jensen, A. Schmelzer, A. S. Manocha, J. A. Pople, and L. Radom, *Chem. Phys.*, **4**, 107 (1974).

153. F. Brogli, E. Heilbronner, and T. Kobayashi, *Helv. Chim. Acta*, **55**, 274 (1972).

154. R. S. Brown, D. F. Eaton, A. Hosomi, T. G. Taylor, and J. M. Wright, *J. Organometal. Chem.*, **66**, 249 (1974).

155. P. Bruckmann and M. Klessinger, *Angew Chem. (Int. Edit.)*, **11**, 524 (1972).

156. P. Bruckmann and M. Klessinger, *Chem. Ber.*, **107**, 1108 (1974).

157. P. Bruckmann and M. Klessinger, *J. Electron Spectrosc.*, **2**, 341 (1973).

158. C. R. Brundle, *Chem. Phys. Letters*, **5**, 410 (1970).

159. C. R. Brundle, *Chem. Phys. Letters*, **7**, 317 (1970).

160. C. R. Brundle, *Chem. Phys. Letters*, **26**, 25 (1974).

161. C. R. Brundle and D. B. Brown, *Spectrochim. Acta*, **A27**, 2491 (1971).

162. C. R. Brundle and G. R. Jones, *J. Chem. Soc. D*, 1198 (1971).

163. C. R. Brundle and G. R. Jones, *J. C. S. Faraday Trans. 2*, **68**, 959 (1972).

164. C. R. Brundle and G. R. Jones, *J. Electron Spectrosc.*, **1**, 403 (1973).

165. C. R. Brundle, G. R. Jones, and H. Basch, *J. Chem. Phys.*, **55**, 1098 (1971).

166. C. R. Brundle, N. A. Kuebler, M. B. Robin, and H. Basch, *Inorg. Chem.*, **11**, 20 (1972).

167. C. R. Brundle, D. P. Neumann, W. C. Price, D. Evans, A. W. Potts, and D. G. Streets, *J. Chem. Phys.*, **53**, 705 (1970).

168. C. R. Brundle and M. B. Robin, *J. Amer. Chem. Soc.*, **92**, 5550 (1970).

169. C. R. Brundle, M. B. Robin, and H. Basch, *J. Chem. Phys.*, **53**, 2196 (1970).

170. C. R. Brundle, M. B. Robin, H. Basch, M. Pinsky, and A. Bond, *J. Amer. Chem. Soc.*, **92**, 3863 (1970).

171. C. R. Brundle, M. B. Robin, and G. R. Jones, *J. Chem. Phys.*, **52**, 3383 (1970).

172. C. R. Brundle, M. B. Robin, and N. A. Kuebler, *J. Amer. Chem. Soc.*, **94**, 1466 (1972).

173. C. R. Brundle, M. B. Robin, N. A. Kuebler, and H. Basch, *J. Amer. Chem. Soc.*, **94**, 1451 (1972).

174. C. R. Brundle and D. W. Turner, *Int. J. Mass Spectrom. Ion Phys.*, **2**, 195 (1969).

175. C. R. Brundle and D. W. Turner, *Proc. Roy. Soc. London*, **A307**, 27 (1968).
176. C. R. Brundle, D. W. Turner, M. B. Robin, and H. Basch, *Chem. Phys. Letters*, **3**, 292 (1969).
177. W. E. Bull, B. P. Pullen, F. A. Grimm, W. E. Moddeman, G. K. Schweitzer, and T. A. Carlson, *Inorg. Chem.*, **9**, 2474 (1970).
178. J. C. Bünzli, A. J. Burak, and D. C. Frost, *Tetrahedron*, **29**, 3735 (1973).
179. J. C. Bünzli, D. C. Frost, and C. A. McDowell, *J. Electron Spectrosc.*, **1**, 481 (1972/1973).
180. J. C. Bünzli, D. C. Frost, and L. Weiler, *J. Amer. Chem. Soc.*, **95**, 7880 (1973).
181. J. C. Bünzli, D. C. Frost, and L. Weiler, *J. Amer. Chem. Soc.*, **96**, 1952 (1974).
182. J. C. Bünzli, D. C. Frost, and L. Weiler, *Tetrahedron Lett.*, 1159 (1973).
182. a. P. Burroughs, S. Evans, A. Hamnett, A. F. Orchard, and N. V. Richardson, *J. C. S. Faraday Trans. 2*, **70**, 1895 (1974).
    b. P. Burroughs, S. Evans, A. Hamnett, A. F. Orchard, and N. V. Richardson, *J. C. S. Chem. Commun.*, 921 (1974).
183. G. L. Caldow, *Chem. Phys. Letters*, **2**, 88 (1968).
184. G. L. Caldow, and G. F. S. Harrison, *Tetrahedron*, **25**, 3429 (1969).
185. P. Carlier, R. Hernandez, P. Masclet, and G. Mouvier, *Electron Spectroscopy*, R. Caudano and J. Verbist (Eds.), p. 1103, Elsevier, Amsterdam, 1974.
186. T. A. Carlson, *Chem. Phys. Letters*, **9**, 23 (1971).
187. T. A. Carlson and C. P. Anderson, *Chem. Phys. Letters*, **10**, 561 (1971).
188. T. A. Carlson and A. E. Jonas, *J. Chem. Phys.*, **55**, 4913 (1971).
189. T. A. Carlson and G. E. McGuire, *J. Electron Spectrosc.*, **1**, 209 (1973).
190. T. A. Carlson and R. M. White, *J. C. S. Faraday Discussions*, **54**, 285 (1972).
191. J. H. Carver and J. L. Gardner, *J. Quant. Spectrosc. Radiat. Transfer*, **12**, 207 (1972).
192. G. P. Ceasar, P. Milazzo, J. L. Cihonski, and R. L. Levenson, *Inorg. Chem.*, **13**, 3035 (1974).
193. V. Cermak, *J. Electron Spectrosc*, **3**, 329 (1974).
194. V. Cermak, *J. Electron Spectrosc.*, **6**, 135 (1975).
195. V. Cermak, M. Smutek, and J. Sramek, *J. Electron Spectrosc.*, **2**, 1 (1973).
196. V. Cermak and J. J. Sramek, *J. Electron Spectrosc.*, **2**, 97 (1973).
197. B. Cetinkaya, G. H. King, S. S. Krishnamurthy, M. F. Lappert, and J. B. Pedley, *J. Chem. Soc. D*, 1370 (1971).
198. D. Chadwick, *Can. J. Chem.*, **50**, 737 (1972).
199. D. Chadwick, A. B. Cornford, D. C. Frost, F. G. Herring, A. Katrib, C. A. McDowell, and R. A. N. McLean, *Electron Spectroscopy*, D. A. Shirley (Ed.), p. 453, North-Holland, Amsterdam, 1972.
200. D. Chadwick, D. C. Frost, F. G. Herring, A. Katrib, C. A. McDowell, and R. A. N. McLean, *Can. J. Chem.*, **51**, 1893 (1973).

201. D. Chadwick, D. C. Frost, A. Katrib, C. A. McDowell, and R. A. N. McLean, *Can. J. Chem.*, **50,** 2642 (1972).

202. D. Chadwick, D. C. Frost, and L. Weiler, *J. Amer. Chem. Soc.*, **93,** 4320 (1971).

203. D. Chadwick, D. C. Frost, and L. Weiler, *Tetrahedron Lett.*, 4543 (1971).

204. D. Chadwick and A. Katrib, *J. Electron Spectrosc.*, **3,** 39 (1974).

204. a. F. T. Chau and C. A. McDowell, *J. Electron Spectrosc.*, **6,** 357 (1975).

      b. F. T. Chau and C. A. McDowell, *J. Electron Spectrosc.*, **6,** 365 (1975).

205. Yu. V. Chizhov, V. I. Kleimenov, G. S. Mednskii, and F. I. Vilesov, *Opt. Spektrosk.*, **33,** 661 (1972).

206. P. A. Clark, F. Brogli, and E. Heilbronner, *Helv. Chim. Acta*, **55,** 1415 (1972).

207. P. A. Clark, R. Gleiter, and E. Heilbronner, *Tetrahedron*, **29,** 3085 (1973).

208. B. J. Cocksey, J. H. D. Eland, and C. J. Danby, *J. Chem. Soc. B*, 790 (1971).

209. B. G. Cocksey, J. H. D. Eland, and C. J. Danby, *J. C. S. Faraday Trans. 2*, **69,** 1558 (1973).

210. J. E. Collin, J. Delwiche, and P. Natalis, *Electron Spectroscopy*, D. A. Shirley (Ed.), p. 401, North-Holland, Amsterdam, 1972.

211. J. E. Collin, J. Delwiche, and P. Natalis, *Int. J. Mass Spectrom. Ion Phys*, **7,** 19 (1971).

212. J. E. Collin and P. Natalis, *Chem. Phys. Letters*, **2,** 194 (1968).

213. J. E. Collin and P. Natalis, *Int. J. Mass Spectrom. Ion Phys.*, **1,** 121 (1968).

214. J. E. Collin and P. Natalis, *Int. J. Mass Spectrom. Ion Phys.* **1,** 483 (1968).

215. J. E. Collin and P. Natalis, *Int. J. Mass Spectrom. Ion Phys.*, **2,** 231 (1969).

216. R. J. Colton and J. W. Rabalais, *J. Electron Spectrosc.*, **3,** 345 (1974).

217. F. J. E. Comes, *Z. Naturforsch*, **23,** 114 (1968).

218. F. J. E. Comes, *Z. Naturforsch*, **23,** 133 (1968).

219. G. Condorelli, I. Fragala, G. Centineo, and E. Tondello, *Inorg. Chim. Acta*, **7,** 725 (1973).

220. J. A. Connor, M. B. Hall, I. H. Hillier, W. N. E. Meredith, M. Barber, and Q. Herd, *J. C. S. Faraday Trans. 2*, **69,** 1677 (1973).

221. A. B. Cornford, D. C. Frost, F. G. Herring, and C. A. McDowell, *Can. J. Chem.*, **49,** 1135 (1971).

222. A. B. Cornford, D. C. Frost, F. G. Herring, and C. A. McDowell, *J. Chem. Phys.*, **54,** 1872 (1971).

223. A. B. Cornford, D. C. Frost, F. G. Herring, and C. A. McDowell, *J. Chem. Phys.*, **55,** 2820 (1971).

224. A. B. Cornford, D. C. Frost, F. G. Herring, and C. A. McDowell, *Chem. Phys. Letters*, **10,** 345 (1971).

225. A. B. Cornford, D. C. Frost, F. G. Herring, and C. A. McDowell, *Faraday Discussions Chem. Soc.*, **54,** 56 (1972).

226. A. B. Cornford, D. C. Frost, C. A. McDowell, J. L. Ragle, and I. A. Stenhouse, *J. Chem. Phys.*, **54,** 2651 (1971).

227. A. B. Cornford, D. C. Frost, C. A. McDowell, J. L. Ragle, and I. A. Stenhouse, *Chem. Phys. Letters*, **5,** 486 (1970).

228. D. O. Cowan, R. Gleiter, O. Glemser, and E. Heilbronner, *Helv. Chim. Acta*, **55**, 2418 (1972).

229. D. O. Cowan, R. Gleiter, O. Glemser, E. Heilbronner, and J. Schaeublin, *Helv. Chim. Acta*, **54**, 1559 (1971).

230. D. O. Cowan, R. Gleiter, J. A. Hashmall, E. Heilbronner, and V. Hornung, *Angew. Chem.*, **83**, 405 (1971).

231. A. H. Cowley, M. J. S. Dewar, J. W. Gilje, D. W. Goodman, and J. R. Schweiger, *J. C. S. Chem. Commun.*, 304 (1974).

232. A. H. Cowley, M. J. S. Dewar, D. W. Goodman, and M. C. Padolina, *J. Amer. Chem. Soc.*, **96**, 2648 (1974).

233. A. H. Cowley, M. J. S. Dewar, D. W. Goodman, and M. C. Padolina, *J. Amer. Chem. Soc.*, **96**, 3666 (1974).

234. A. H. Cowley, M. J. S. Dewar, D. W. Goodman, and J. R. Schweige, *J. Amer. Chem. Soc.*, **95**, 6506 (1973).

235. S. A. Cowling and R. A. W. Johnstone, *J. Electron Spectrosc.*, **2**, 161 (1973).

236. S. A. Cowling, R. A. W. Johnstone, A. A. Gorman, and P. G. Smith, *J. C. S. Chem. Commun.*, 627 (1973).

237. P. A. Cox, S. Evans, A. Hamnett, and A. F. Orchard, *Chem. Phys. Letters*, **7**, 414 (1970).

238. P. A. Cox, S. Evans, and A. F. Orchard *Chem. Phys. Letters*, **13**, 386 (1972).

239. P. A. Cox, S. Evans, A. F. Orchard, N. V. Richardson, and P. J. Roberts, *Faraday Discussions Chem. Soc.*, **54**, 26 (1972).

240. P. A. Cox and A. F. Orchard, *Chem. Phys. Letters*, **7**, 273 (1970).

241. S. Cradock, *Chem. Phys. Letters*, **10**, 291 (1971).

242. S. Cradock and W. Duncan, *Mol. Phys.*, **27**, 837 (1974).

243. S. Cradock and E. A. V. Ebsworth, *J. Chem. Soc. D*, 57 (1971).

244. S. Cradock, E. A. V. Ebsworth, and J. D. Murdock, *J. C. S. Faraday Trans. 2*, **68**, 86 (1972).

245. S. Cradock, E. A. V. Ebsworth, and A. Robertson, *Chem. Phys. Letters* **30**, 413 (1975).

246. S. Cradock, E. A. V. Ebsworth, and A. Robertson, *J. C. S. Dalton Trans.*, 22 (1973).

247. S. Cradock, E. A. V. Ebsworth, W. J. Savage, and R. A. Whiteford, *J. C. S. Faraday Trans. 2*, **68**, 934 (1972).

248. S. Cradock, E. A. V. Ebsworth, and R. A. Whiteford, *J. C. S. Dalton Trans.*, 2401 (1973).

249. S. Cradock, R. H. Findlay, and M. H. Palmer, *Tetrahedron*, **29**, 2173 (1973).

250. S. Cradock and D. W. H. Rankin, *J. C. S. Faraday Trans. 2*, **68**, 940 (1972).

251. S. Cradock and W. Savage, *Inorg. Nucl. Chem. Lett.*, **8**, 753 (1972).

252. S. Cradock and R. A. Whiteford, *J. C. S. Faraday Trans 2*, **68**, 281 (1972).

253. S. Cradock and R. A. Whiteford, *Trans. Faraday Soc.*, **67**, 3425 (1971).

254. W. R. Cullen, D. C. Frost, and W. R. Leeder, *J. Fluorine Chem.*, **1**, 227 (1971).

255. W. R. Cullen, D. C. Frost, and D. A. Vroom, *Inorg. Chem.*, **8**, 1803 (1969).
256. C. S. Cundy, M. F. Lappert, J. B. Pedley, W. Schmidt, and B. T. Wilkins, *J. Organometal Chem.*, **51**, 99 (1973).
257. L. C. Cusachs, F. A. Grimm, and G. K. Schweitzer, *J. Electron Spectrosc.*, **3**, 229 (1974).
258. J. Daintith, R. Dinsdale, J. P. Maier, D. A. Sweigart, and D. W. Turner, *Molecular Spectroscopy*, p. 16, Inst. of Petroleum, London, 1971.
259. J. Daintith, J. P. Maier, D. A. Sweigart, and D. W. Turner, *Electron Spectroscopy*, D. A. Shirley (Ed.), p. 289, North-Holland, Amsterdam, 1972.
260. D. W. Davies, *Chem. Phys. Letters*, **28**, 520 (1974).
261. T. P. Debies and J. W. Rabalais, *J. Amer. Chem. Soc.*, **97**, 487 (1975).
262. T. P. Debies and J. W. Rabalais, *J. Electron Spectrosc.*, **1**, 355 (1973).
263. T. P. Debies and J. W. Rabalais, *J. Electron Spectrosc.*, **3**, 315 (1974).
264. T. P. Debies and J. W. Rabalais, *Inorg. Chem.*, **13**, 308 (1974).
265. P. Dechant, A. Schweig, and W. Thiel, *Angew. Chem.*, **85**, 358 (1973).
266. J. L. Dehmer and J. Berkowitz, *Phys. Rev. A*, **10**, 484 (1974).
267. P. M. Dehmer, J. Berkowitz, and W. A. Chupka, *J. Chem. Phys.*, **60**, 2676 (1974).
268. J. L. Dehmer, J. Berkowitz, and L. C. Cusachs, *J. Chem. Phys.*, **58**, 5681 (1973).
269. J. L. Dehmer, J. Berkowitz, L. C. Cusachs, and H. S. Aldrich, *J. Chem. Phys.*, **61**, 594 (1974).
270. R. L. DeKock, *J. Chem. Phys.*, **58**, 1267 (1973).
271. R. L. DeKock, *J. Electron Spectrosc.*, **4**, 155 (1974).
272. R. L. DeKock, B. R. Higginson, and D. R. Lloyd, *Faraday Discussions Chem. Soc.*, **54**, 84 (1972).
273. R. L. DeKock, B. R. Higginson, D. R. Lloyd, A. Breeze, D. W. J. Cruickshank, and D. R. Armstrong, *Mol. Phys.*, **24**, 1059 (1972).
274. R. L. DeKock and D. R. Lloyd, *J. Chem. Soc. D*, 526 (1973).
275. R. L. DeKock, D. R. Lloyd, A. Breeze, G. A. D. Collins, D. W. J. Cruickshank, and H. J. Lempka, *Chem. Phys. Letters*, **14**, 525 (1972).
276. R. L. DeKock, D. R. Lloyd, I. H. Hillier, and V. R. Saunders, *Proc. Roy. Soc. London*, **A328**, 401 (1972).
277. J. Delwiche, *Bull. Cl. Sci. Acad. Roy. Belg.*, **55**, 215 (1969).
278. J. Delwiche, *Dyn. Mass Spectrom.*, **1**, 71 (1970).
279. J. Delwiche and P. Natalis, *Chem. Phys. Letters*, **5**, 564 (1970).
280. J. Delwiche, P. Natalis, and J. E. Collin, *Int. J. Mass Spectrom. Ion Phys.*, **5**, 443 (1970).
281. J. Delwiche, P. Natalis, J. Momigny, and J. E. Collin, *J. Electron Spectrosc.*, **1**, 219 (1973).
282. A. DeMeÿere, *Chem. Ber.*, **107**, 1684 (1974).
283. D. A. Demeo and M. A. ElSayed, *J. Chem. Phys.*, **52**, 2622 (1970).
284. D. A. Demeo and A. J. Yencha, *J. Chem. Phys.*, **53**, 4536 (1970).
285. P. J. Derrick, L. Åsbrink, O. Edqvist, B. O. Jonsson, and E. Lindholm, *Int. J. Mass Spectrom. Ion Phys.*, **6**, 161 (1971).

286. P. J. Derrick, L. Åsbrink, O. Edqvist, B. O. Jonsson, and E. Lindholm, *Int. J. Mass Spectrom. Ion Phys.*, **6,** 177 (1971).
287. P. J. Derrick, L. Åsbrink, O. Edqvist, B. O. Jonsson, and E. Lindholm, *Int. J. Mass Spectrom. Ion Phys.*, **6,** 191 (1971).
288. P. J. Derrick, L. Åsbrink, O. Edqvist, B. O. Jonsson, and E. Lindholm, *Int. J. Mass Spectrom. Ion Phys.*, **6.** 203 (1971).
289. P. J. Derrick, L. Åsbrink, O. Edqvist, and E. Lindholm, *Spectrochim. Acta,* **A27,** 2525 (1971).
290. M. J. S. Dewar, and D. W. Goodman, *J. C. S. Faraday Trans. 2,* **68,** 1784 (1972).
291. M. J. S. Dewar, M. C. Kohn, and N. Trinajstic, *J. Amer. Chem. Soc.,* **93,** 3437 (1971).
292. M. J. S. Dewar and S. D. Worley, *J. Chem. Phys.,* **49,** 2454 (1968).
293. M. J. S. Dewar and S. D. Worley, *J. Chem. Phys.,* **50,** 654 (1969).
294. M. J. S. Dewar and S. D. Worley, *J. Chem. Phys.,* **51,** 263 (1969).
295. M. J. S. Dewar and S. D. Worley, *J. Chem. Phys.,* **51,** 1672 (1969).
296. E. Diemann and A. Mueller, *Chem. Phys. Letters,* **19,** 538 (1973).
297. D. Dill, *Electron Spectroscopy,* D. A. Shirley (Ed.), p. 277, North-Holland, Amsterdam, 1972.
298. R. N. Dixon, *Mol. Phys.,* **20,** 113 (1971).
299. R. N. Dixon, G. Duxbury, G. R. Fleming, and J. M. V. Hugo, *Chem. Phys. Letters,* **14,** 60 (1972).
300. R. N. Dixon, G. Duxbury, M. Horani, and J. Rostas, *Mol. Phys.* **22,** 977 (1971).
300. a. R. N, Dixon, G. Duxbury, J. W. Rabalais, and L. Åsbrink, *Mol. Phys.,* **31,** 423 (1976).
301. R. N. Dixon and S. E. Hull, *Chem. Phys. Letters,* **3,** 367 (1969).
302. R. N. Dixon, J. N. Murrell, and B. Narayan, *Mol. Phys.,* **20,** 611 (1971).
303. P. H. Doolittle, R. I. Schoen, and K. E. Schubert, *J. Chem. Phys.,* **49,** 5108 (1968).
304. J. Doucet, P. Sauvageau, and C. Sandorpfy, *J. Chem. Phys.,* **58,** 3708 (1973).
305. J. Doucet, P. Sauvageau, and C. Sandorfy, *J. Chem. Phys.,* **62,** 355 (1975).
306. J. Doucet, P. Sauvageau, and C. Sandorfy, *J. Chem. Phys.,* **62,** 366 (1975).
307. R. G. Dromey, J. D. Morrison, and J. B. Peel, *Chem. Phys. Letters,* **23,** 30 (1973).
308. R. G. Dromey and J. B. Peel, *J. Mol. Struct.,* **23,** 53 (1974).
309. S. Durmaz, G. H. King, and R. J. Suffolk, *Chem. Phys. Letters,* **13,** 304 (1972).
309. a. J. M. Dyke, L. Golob, N. Jonathan, A. Morris, M. Okuda, and D. F. Smith, *J. C. S. Faraday Trans. 2,* **70,** 1818 (1974).
310. D. F. Eaton and T. G. Traylor, *J. Amer. Chem. Soc.,* **96,** 7109 (1974).
311. E. A. V. Ebsworth, *Kem. Kozl.,* **42,** 1 (1974).
312. O. Edqvist, L. Åsbrink, and E. Lindholm, *Z. Naturforsch., A,* **26,** 1407 (1971).
313. O. Edqvist, E. Lindholm, L. E. Selin, and L. Åsbrink, *Phys. Lett.,* **A31,** 292 (1970).

314. O. Edqvist, E. Lindholm, L. E. Selin, and L. Åsbrink, *Phys. Scripta*, **1**, 25 (1970).

315. O. Edqvist, E. Lindholm, L. E. Selin, and L. Åsbrink, *Phys. Scripta*, **1**, 172 (1970).

316. O. Edqvist, E. Lindholm, L. E. Selin, L. Åsbrink, C. E. Kuyatt, S. R. Mielczarek, and J. A. Simpson, *Phys. Scripta*, **1**, 172 (1970).

317. O. Edqvist, E. Lindholm, L. E. Selin, H. Sjogren, and L. Åsbrink, *Ark. Fys.*, **40**, 439 (1970).

318. J. H. D. Eland, *Int. J. Mass Spectrom. Ion Phys.*, **2**, 471 (1969)..

319. J. H. D. Eland, *Int. J. Mass Spectrom. Ion Phys.*, **4**, 37 (1970).

320. J. H. D. Eland, *Int. J. Mass Spectrom. Ion Phys.*, **9**, 214 (1972).

321. J. H. D. Eland, *Int. J. Mass Spectrom. Ion Phys.*, **9**, 397 (1972).

322. J. H. D. Eland, *Phil. Trans. Roy. Soc. London*, **A268**, 87 (1970).

323. J. H. D. Eland, *Photoelectron Spectroscopy*, Halsted, New York 1974.

324. J. H. D. Eland and C. J. Danby, *Int. J. Mass Spectrom. Ion Phys.*, **1**, 111 (1968).

325. J. H. D. Eland and C. J. Danby, *Z. Naturforsch., A*, **23**, 355 (1968).

326. S. Elbel, H. Bergmann, and W. Ensslin, *J. C. S. Faraday Trans. 2*, **70**, 555 (1974).

327. W. Ensslin, H. Bock, and G. Becker, *J. Amer. Chem. Soc.*, **96**, 2757 (1974).

328. S. Evans, J. C. Green, M. L. H. Green, A. F. Orchard, and D. W. Turner, *Discussions Faraday Soc.*, **47**, 112 (1969).

329. S. Evans, J. C. Green, and S. E. Jackson, *J. C. S. Faraday Trans. 2*, **68**, 249 (1972).

330. S. Evans, J. C. Green, and S. E. Jackson, *J. C. S. Faraday Trans. 2*, **69**, 191 (1973).

331. S. Evans, M. L. Green, B. Jewitt, G. H. King, and A. F. Orchard, *J. C. S. Faraday Trans. 2*, **70**, 356 (1974).

332. S. Evans, M. L. H. Green, B. Jewitt, A. F. Orchard, and C. F. Pygall, *J. C. S. Faraday Trans. 2*, **68**, 1847 (1972).

333. S. Evans, J. C. Green, P. J. Joachim. A. F. Orchard, D. W. Turner, and J. P. Maier, *J. C. S. Faraday Trans. 2*, **68**, 905 (1972).

334. S. Evans, J. C. Green, A. F. Orchard, T. Saito, and D. W. Turner, *Chem. Phys. Letters*, **4**, 361 (1969).

335. S. Evans, J. C. Green, A. F. Orchard, and D. W. Turner, *Discussions Faraday Soc.*, **47**, 112 (1969).

336. S. Evans, M. F. Guest, I. H. Hillier, and A. F. Orchard, *J. C. S. Faraday Trans 2*, **70**, 417 (1974).

337. S. Evans, A. Hamnett, and A. F. Orchard, *J. Amer. Chem. Soc.*, **96**, 6221 (1974).

338. S. Evans, A. Hamnett, and A. F. Orchard, *J. Chem. Soc. D*, 1282 (1970).

339. S. Evans, A. Hamnett, and A. F. Orchard, *J. Coord. Chem.*, **2**, 57 (1972).

340. S. Evans, A. Hamnett, A. F. Orchard, and D. R. Lloyd, *Faraday Discussions Chem. Soc.*, **54**, 227 (1972).

341. S. Evans, P. J. Joachim, A. F. Orchard, and D. W. Turner, *Int. J. Mass Spectrom. Ion Phys.*, **9**, 41 (1972).

References

341. a. S. Evans and A. F. Orchard, *J. Electron Spectrosc.*, **6**, 207 (1975).
342. S. Evans and A. F. Orchard, *Inorg. Chem. Acta.*, **5**, 81 (1971).
343. S. Evans, A. F. Orchard, and D. W. Turner, *Int. J. Mass Spectrom. Ion Phys.*, **7**, 261 (1971).
344. G. Favini, G. Buemi, D. Grasso, and G. Capietti, *J. Electron Spectrosc.* **2**, 239 (1973).
345. T. P. Fehlner and D. W. Turner, *J. Amer. Chem. Soc.*, **95**, 7175 (1973).
346. T. P. Fehlner, and D. W. Turner, *Inorg. Chem.*, **13**, 754 (1974).
347. S. Foster, S. Felps, L. C. Cusachs, and S. P. McGlynn, *J. Amer. Chem. Soc.*, **95**, 5521 (1973).
348. C. Fridh, L. Åsbrink, B. O. Jonsson, and E. Lindholm, *Int. J. Mass Spectrom. Ion Phys.*, **8**, 85 (1972).
349. C. Fridh, L. Åsbrink, B. O. Jonsson, and E. Lindholm, *Int. J. Mass Spectrom. Ion Phys.*, **8**, 101 (1972).
350. C. Fridh, L. Åsbrink, and E. Lindholm, *Chem. Phys. Letters*, **15**, 282 (1972).
351. C. Fridh, L. Åsbrink, and E. Lindholm, *Chem. Phys. Letters*, **15**, 408 (1972).
352. D. C. Frost, F. G. Herring, A. Katrib, and C. A. McDowell, *Chem. Phys. Letters*, **20**, 401 (1973).
353. D. C. Frost, F. G. Herring, A. Katrib, C. A. McDowell, and R. A. N. McLean, *J. Phys. Chem.*, **76**, 1030 (1972).
354. D. C. Frost, F. G. Herring, A. Katrib, R. A. N. McLean, J. E. Drake, and N. P. C. Westwood, *Can. J. Chem.*, **49**, 4033 (1971).
355. D. C. Frost, F. G. Herring, A. Katrib, R. A. N. McLean, J. Drake, and N. P. C. Westwood, *Chem. Phys. Letters*, **10**, 347 (1971).
356. D. C. Frost, F. G. Herring, C. A. McDowell, M. R. Mustafa, and J. S. Sandhu, *Chem. Phys. Letters*, **2**, 663 (1968).
357. D. C. Frost, F. G. Herring, C. A. McDowell, and I. A. Stenhouse, *Chem. Phys. Letters*, **4**, 533 (1970).
358. D. C. Frost, F. G. Herring, C. A. McDowell, and I. A. Stenhouse, *Chem. Phys. Letters*, **5**, 291 (1970).
359. D. C. Frost, F. G. Herring, K. A. R. Mitchell, and I. A. Stenhouse, *J. Amer. Chem. Soc.*, **93**, 1596 (1971).
360. D. C. Frost, A. Katrib, C. A. McDowell, and R. A. N. McLean, *Int. J. Mass Spectrom. Ion Phys.*, **7**, 485 (1971).
361. D. C. Frost, S. T. Lee, and C. A. McDowell, *J. Chem. Phys.*, **59**, 5484 (1973).
362. D. C. Frost, S. T. Lee, and C. A. McDowell, *Chem. Phys. Letters*, **17**, 153 (1972).
363. D. C. Frost, S. T. Lee, and C. A. McDowell, *Chem. Phys. Letters*, **22**, 243 (1973).
364. D. C. Frost, S. T. Lee, and C. A. McDowell, *Chem. Phys. Letters*, **23**, 472 (1973).
365. D. C. Frost, S. T. Lee, and C. A. McDowell, *Chem. Phys. Letters*, **24**, 149 (1974).

366. D. C. Frost, S. T. Lee, C. A. McDowell, and N. P. C. Westwood, *Chem. Phys. Letters*, **30**, 26 (1975).

367. D. C. Frost, C. A. McDowell, J. S. Sandhu, and D. A. Vroom, *Advan. Mass Spectrom.*, **4**, 781 (1968).

368. D. C. Frost, C. A. McDowell, and D. A. Vroom, *J. Chem. Phys.*, **46**, 4255 (1967).

369. D. C. Frost and J. S. Sandhu, *Indian J. Chem.*, **9**, 1105 (1971).

370. V. Fuchs, and H. Hotop, *Chem. Phys. Letters*, **4**, 71 (1969).

371. W. Fuss, and H. Bock, *J. Chem. Phys.*, **61**, 1613 (1974).

372. W. Fuss, and H. Bock, *Chemical Spectroscopy and Photochemistry in the Vacuum-Ultraviolet*, C. Sandorpfy, P. J. Ausloos, and M. B. Robin (Eds.), p. 223, Reidel, Dordrecht, Holland, 1974.

373. J. L. Gardner and J. A. R. Samson, *J. Chem. Phys.*, **60**, 3711 (1974).

374. J. L. Gardner and J. A. R. Samson, *Chem. Phys. Letters*, **26**, 240 (1974).

375. J. L. Gardner and J. A. R. Samson, *J. Electron Spectrosc.*, **2**, 153 (1973).

376. J. L. Gardner and J. A. R. Samson, *J. Electron Spectrosc.*, **2**, 259 (1973).

377. J. L. Gardner and J. A. R. Samson, *J. Electron Spectrosc.*, **2**, 267 (1973).

378. J. L. Gardner and J. A. R. Samson, *J. Opt. Soc.*, **63**, 511 (1973).

379. R. Gilbert, P. Sauvageau, and C. Sandorfy, *Chem. Phys. Letters*, **17**, 465 (1972).

380. R. Gleiter, E. Heilbronner, and A. DeMeijere, *Helv. Chim. Acta*, **54**, 1029 (1971).

381. R. Gleiter, E. Heilbronner, and V. Hornung, *Angew. Chem. (Int. Edit.)*, **9**, 901 (1970).

382. R. Gleiter, E. Heilbronner, and V. Hornung, *Helv. Chim. Acta*, **55**, 255 (1972).

383. R. Gleiter, E. Heilbronner, M. Kekman, and H. D. Martin, *Chem. Ber.*, **106**, 28 (1973).

384. R. Gleiter, E. Heilbronner, L. A. Paquette, G. L. Thompson, and R. E. Wingard, *Tetrahedron*, **29**, 565 (1973).

385. R. Gleiter, V. Hornung, B. Lindberg, S. Hogberg, and N. Lozach, *Chem. Phys. Letters*, **11**, 401 (1971).

386. R. Gleiter, E. Schmidt, D. O. Cowan, and J. P. Ferraris, *J. Electron Spectrosc.*, **2**, 207 (1973).

387. M. Godfrey, *J. Chem. Soc. B*, **7**, 1537 (1970).

388. H. Goetz, F. Marschner, and H. Juds, *Tetrahedron*, **30**, 1133 (1974).

389. C. Goffart, J. Momigny, and P. Natalis, *Int. J. Mass Spectrom. Ion Phys.*, **3**, 371 (1969).

390. M. J. Goldstein, S. Natowsky, E. Heilbronner, and V. Hornung, *Helv. Chim. Acta*, **56**, 294 (1973).

391. L. Golob, N. Jonathan, A. Morris, M. Okuda, and K. J. Ross, *J. Electron Spectrosc.*, **1**, 506 (1973).

392. D. Gonbeau, C. Guimon, J. Deschamps, and G. Pfister-Guillouzo, *J. Electron Spectrosc.*, **6**, 99 (1975).

393. T. D. Goodman, J. D. Allen, Jr., L. C. Cusachs, and G. K. Schweitzer, *J. Electron Spectrosc.*, **3**, 289 (1974).

394. D. W. Goodman, M. J. S. Dewar, J. R. Schweiger, and A. H. Cowley, *Chem. Phys. Letters*, **21**, 474 (1973).
395. J. C. Green, M. L. Green, P. J. Joachim, A. F. Orchard, and D. W. Turner, *Phil. Trans. Roy. Soc. London*, **A268**, 111 (1970).
396. J. C. Green, D. I. King, and J. H. D. Eland, *J. Chem. Soc. D*, 1121 (1970).
397. M. C. Green, M. F. Lappert, J. B. Pedley, W. Schmidt, and B. Wilkins, *J. Organometal. Chem.*, **31**, C55 (1971).
398. R. Griebel, G. Hohlneicher, and F. Dörr, *J. Electron Spectrosc.*, **4**, 185 (1974).
399. C. Guimon, D. Gonbeau, G. Pfister-Guillouzo, L. Åsbrink, and J. Sandström, *J. Electron Spectrosc.*, **4**, 49 (1974).
400. H. J. Haink, J. E. Adams, and J. R. Huber, *Ber. Bunsenges. Phys. Chem.*, **78**, 436 (1974).
401. H. J. Haink, E. Heilbronner, V. Hornung, and E. Kloster-Jensen, *Helv. Chim. Acta*, **53**, 1073 (1970).
402. M. B. Hall, M. F. Guest, and I. H. Hillier, *Chem. Phys. Letters*, **15**, 592 (1972).
403. M. B. Hall, M. F. Guest, I. H. Hillier, D. R. Lloyd, A. F. Orchard, and A. W. Potts, *J. Electron Spectros.*, **1**, 497 (1973).
404. Y. Harada, K. Ohno, and H. Inokuchi, *Bull. Chem. Soc. Japan*, **47**, 1608 (1974).
405. Y. Harada, K. Seki, A. Suzuki, and H. Inokuchi, *Chem. Letters*, 893 (1973).
406. H. Harrison, *J. Chem. Phys.*, **52**, 901 (1970).
407. W. R. Harshbarger, N. A. Kuebler, and M. B. Robin, *J. Chem. Phys.*, **60**, 345 (1974).
408. E. Haselbach, *Chem. Phys. Letters*, **7**, 428 (1970).
409. E. Haselbach, *Helv. Chim. Acta*, **53**, 1526 (1970).
410. E. Haselbach and W. Eberbach, *Helv. Chim. Acta*, **56**, 1944 (1973).
411. E. Haselbach, J. A. Hashmall, E. Heilbronner, and V. Hornung, *Angew. Chem. (Int. Edit.)*, **8**, 878 (1969).
412. E. Haselbach and E. Heilbronner, *Helv. Chim. Acta*, **53**, 684 (1970).
413. E. Haselbach, E. Heilbronner, A. Mannschreck, and S. Albrecht, *Angew. Chem. (Int. Edit.)*, **9**, 902 (1970).
414. E. Haselbach, E. Heilbronner, A. Mannschreck, and W. Seitz, *Angew. Chem.*, **82**, 879 (1970).
415. E. Haselbach, E. Heilbronner, H. Musso, and A. Schmelzer, *Helv. Chim. Acta*, **55**, 302 (1972).
416. E. Haselbach, E. Heilbronner, and G. Schroder, *Helv. Chim. Acta*, **54**, 153 (1971).
417. E. Haselbach, Z. Lanyiova, and M. Rossi, *Helv. Chim. Acta*, **56**, 2889 (1973).
418. E. Haselbach and A. Schmelzer, *Helv. Chim. Acta*, **54**, 1575 (1971).
419. E. Haselbach and A. Schmelzer, *Helv. Chim. Acta*, **55**, 1745 (1972).
420. J. A. Hashmall and E. Heilbronner, *Angew. Chem. (Int. Edit.)*, **9**, 305 (1970).

421. J. A. Hashmall, B. E. Mills, D. A. Shirley, and A. Streitweiser, Jr., *J. Amer. Chem. Soc.*, **94,** 4445 (1972).
422. E. Heilbronner, *Israel J. Chem.*, **10,** 143 (1972).
423. E. Heilbronner, R. Gleiter, H. Hopf, V. Hornung, A. DeMeijere, *Helv. Chim. Acta*, **54,** 783 (1971).
424. E. Heilbronner, R. Gleiter, T. Hoshi, and A. DeMeijere, *Helv. Chim. Acta*, **56,** 1594 (1973).
425. E. Heilbronner, V. Hornung, H. Bock, and H. Alt, *Angew. Chem. (Int. Edit.)*, **8,** 524 (1969).
426. E. Heilbronner, V. Hornung, and E. Kloster–Jensen, *Helv. Chim. Acta*, **53,** 331 (1970).
427. E. Heilbronner, *First world Quantum Chem. Proc. Int. Congr. Quantum Chem.*, p. 211, 1974.
428. E. Heilbronner, V. Hornung, J. P. Maier, and E. Kloster–Jensen, *J. Amer. Chem. Soc.*, **96,** 4252 (1974).
429. E. Heilbronner, V. Hornung, and K. A. Muszkat, *Helv. Chim. Acta*, **53,** 347 (1970).
430. E. Heilbronner, V. Hornung, F. H. Pinkerton, and S. F. Thames, *Helv. Chim. Acta*, **55,** 289 (1972).
431. E. Heilbronner and J. P. Maier, *Helv. Chim. Acta*, **57,** 151 (1974).
432. E. Heilbronner and H. D. Martin, *Chem. Ber.*, **106,** 3376 (1973).
433. E. Heilbronner and H. D. Martin, *Helv. Chim. Acta*, **55,** 1490 (1972).
434. E. Heilbronner and K. A. Muszkat, *J. Amer. Chem. Soc.* **92,** 3818 (1970).
435. C. R. Helms and W. E. Spicer, *Appl. Phys. Letters*, **21,** 237 (1972).
436. G. Hentrich, E. Gunkel, and M. Klessinger, *J. Mol. Struct.*, **21,** 231 (1974).
437. F. G. Herring and R. A. N. McLean, *Inorg. Chem.*, **11,** 1667 (1972).
438. B. R. Higginson, D. R. Lloyd, P. Burroughs, D. M. Gibson, and A. F. Orchard, *J. C. S. Faraday Trans. 2*, **69,** 1659 (1973).
439. B. R. Higginson, R. D. Lloyd, J. A. Connor, and I. H. Hillier, *J. C. S. Faraday Trans.*, **70,** 1418 (1974).
440. B. R. Higginson, D. R. Lloyd, and P. J. Roberts, *Chem. Phys. Letters*, **19,** 480 (1973).
441. I. H. Hillier, M. F. Guest, B. R. Higginson, and D. R. Lloyd, *Mol. Phys.*, **27,** 215 (1974).
442. I. H. Hillier, J. C. Marriott, V. R. Saunders, M. J. Ware, D. R. Lloyd, and N. Lynaugh, *Chem. Commun.*, **23,** 1586 (1970).
443. I. H. Hillier and V. R. Saunders, *Mol. Phys.*, **22,** 193 (1971).
444. I. H. Hillier, V. R. Saunders, M. J. Ware, P. J. Bassett, D. R. Lloyd, and N. Lynaugh, *J. Chem. Soc. D*, 1316 (1970).
445. S. Hino and H. Inokuchi, *Chem Letters*, 363 (1974).
446. R. J. Hodges, D. E. Webster, and P. B. Wells, *J. Chem. Soc. D*, 2577 (1972).
447. G. Hoehne, F. Marschner, and K. Praefcke, *Z. Naturforsch.*, **29,** 546 (1974).
448. R. Hoffmann, P. D. Mollere, and E. Heilbronner, *J. Amer. Chem. Soc.*, **95,** 4860 (1973).

449. R. W. Hoffmann, R. Schuttler, W. Schafer, and A. Schweig, *Angew. Chem.* (*Int. Edit.*), **11,** 512 (1972).
450. G. Hohlneicher, L. S. Cedarbaum, and S. Peyerimhoff, *Chem. Phys. Letters,* **11,** 421 (1971).
451. J. M. S. Hollas, *Mol. Phys.,* **27,** 1001 (1974).
452. J. M. Hollas and T. A. Sutherley, *Chem. Phys. Letters,* **21,** 167 (1973).
453. J. M. Hollas and T. A. Sutherley, *Mol. Phys.,* **21,** 183 (1971).
454. J. M. Hollas and T. A. Sutherley, *Mol. Phys.,* **22,** 213 (1971).
455. J. M. Hollas and T. A. Sutherley, *Mol. Phys.,* **24,** 1123 (1972).
456. H. Hotop and A. Niehaus, *Int. J. Mass Spectrom. Ion Phys.,* **5,** 415 (1970).
457. K. N. Houk, L. P. Davis, G. R. Newkome, R. E. Duke, Jr., and R. V. Nauman, *J. Amer. Chem. Soc.,* **95,** 8364 (1973).
458. J. T. J. Huang, F. O. Ellison, and J. W. Rabalais, *J. Electron Spectrosc.,* **3,** 339 (1974).
459. I. Ikemoto, K. Samizo, T. Fujikawa, K. Ishio, T. Ohta, and H. Kuroda, *Chem. Letters,* **4,** 785 (1974).
460. S. Ikuta, K. Yoshihara, T. Shiokawa, M. Jinno, Yu. Yokoyama, and S. Ikeda, *Chem. Letters,* 1237 (1973).
461. G. Innorta, S. Torroni, and S. Pignataro, *Org. Mass Spectrom.* **6,** 113 (1972).
462. Y. Itikawa, *J. Electron Spectrosc.,* **2,** 125 (1973).
463. R. A. W. Johnstone A. K. Holliday, W. Reade, A. Neville, *J. Chem. Soc. D,* 51 (1971).
464. R. A. W. Johnstone and F. A. Mellon, *J. C. S. Faraday Trans.* 2, **69,** 1155 (1973).
465. A. E. Jonas, G. K. Schweitzer, F. A. Grimm, and T. A. Carlson, *J. Electron Spectrosc.,* **1,** 29 (1972).
466. N. Jonathan, *Discussions Faraday Soc.,* **54,** 64 (1973).
466. a. N. Jonathan, A. Morris, M. Okuda, K. J. Ross, and D. J. Smith, *J. C. S. Faraday Trans.,* **70,** 1810 (1974).
467. N. Jonathan, A. Morris, M. Okuda, K. J. Ross, and D. J. Smith, *Discussions Faraday Soc.,* **54,** 48 (1973).
468. N. Jonathan, A. Morris, M. Okuda, K. J. Ross, and D. J. Smith, *Faraday Discussions Chem. Soc.,* 48 (1972).
469. N. Jonathan, A. Morris, M. Okuda, D. J. Smith, and K. J. Ross, *Chem. Phys. Letters,* **13,** 334 (1972).
470. N. Jonathan, A. Morris, M. Okuda, D. J. Smith, and K. Ross, *Electron Spectroscopy,* D. A. Shirley (Ed.), p. 345, North-Holland, Amsterdam, 1972.
471. N. Jonathan, A. Morris, K. J. Ross, and D. J. Smith, *J. Chem. Phys.,* **54,** 4954 (1971).
472. N. Jonathan, A. Morris, D. J. Smith, and K. J. Ross, *Chem. Phys. Letters,* **7,** 497 (1970).
473. N. Jonathan, K. Ross, and V. Tomlinson, *Int. J. Mass. Spectrom. Ion Phys.,* **4,** 51 (1970).
474. N. Jonathan, D. J. Smith, and K. J. Ross, *J. Chem. Phys.,* **53,** 3758 (1970).

475. N. Jonathan, D. J. Smith, and K. J. Ross, *Chem. Phys. Letters*, **9,** 217 (1971).

476. R. W. Jones and W. S. Koski, *J. Chem. Phys.*, **59,** 1228 (1973).

477. B. Ö. Jonsson and E. Lindholm, *Ark. Fys.*, **39,** 65 (1969).

478. G. Joyez, R. I. Hall, J. Reinhardt, and J. Mazeau, *J. Electron Spectrosc.*, **2,** 183 (1973).

479. I. G. Kaplan and A. P. Markin, *Opt. Spectrosk.*, **25,** 493 (1968).

480. A. Katrib, T. P. Debies, R. J. Colton, T. H. Lee, and J. W. Rabalais, *Chem. Phys. Letters*, **22,** 196 (1973).

481. A. Katrib and J. W. Rabalais, *J. Phys. Chem.*, **77,** 2358 (1973).

482. S. Katsumata, T. Iwai, and K. Kimura, *Bull. Chem. Soc. Jap.*, **46,** 3391 (1973).

482 a. S. Katsumata and K. Kimura, *J. Electron Spectrosc.*, **6,** 309 (1975).

483. S. Katsumata and K. Kimura, *Bull. Chem. Soc. Jap.*, **46,** 1342 (1973).

484. V. I. Keimenov, Yu. V. Chizhov, and F. I. Vilesov, *Opt. Spectrosc.* **32,** 371 (1972).

485. J. Kelder, H. Cerfontain, H. B. Higginson, and D. R. Lloyd, *Tetrahedron Lett.*, 739 (1974).

486. H. P. Kelly, *Chem. Phys. Letters*, **20,** 547 (1973).

487. P. C. Kemeny, R. T. Poole, J. G. Jenkin, and J. Lecky, *Phys. Rev.* A10, 190 (1974).

488. D. J. Kennedy and S. T. Manson, *Phys. Rev. A*, **35,** 227 (1972).

489. O. S. Khalil, J. L. Meeks, and S. P. McGlynn, *J. Amer. Chem. Soc.*, **95,** 5876 (1973).

490. K. Kimura, S. Katsumata, Y. Achiba, H. Matsumoto, and S. Nagakura, *Bull. Chem. Soc. Jap.*, **46,** 373 (1973).

491. K. Kimura, S. Katsumata, T. Yamazaki, and H. Wakabayashi, *J. Electron Spectrosc.*, **6,** 41 (1975).

491 a. K. Kimura, T. Yamazaki, and K. Osafune, *J. Electron Spectrosc.*, **6,** 391 (1975).

492. G. H. King, S. S. Krishnamurthy, M. F. Lappert, and J. B. Pedley, *Faraday Discussions Chem. Soc.*, **54,** 70 (1972).

493. G. H. King, H. W. Kroto, and R. J. Suffolk, *Chem. Phys. Letters*, **13,** 457 (1972).

494. G. H. King, J. N. Murrell, and R. J. Suffolk, *J. C. S. Dalton Trans.*, 564 (1972).

495. J. A. Kinsinger and J. W. Taylor, *Int. J. Mass Spectrom. Ion Phys.*, **10,** 445 (1972/1973).

496. J. A. Kinsinger and J. W. Taylor, *Int. J. Mass Spectrosc. Ion Phys.*, **11,** 461 (1973).

497. T. Kitagawa, *J. Mol. Spectrosc.*, **26,** 1 (1968).

498. M. Klasson and R. Manne, *Electron Spectroscopy*, D. A. Shirley (Ed.), p. 471, North-Holland, Amsterdam, 1972.

499. V. I. Kleimenov, Yu. V. Chizhov, and F. I. Vilesov, *Opt. Spektrosk.*, **32,** 702 (1972).

500. M. Klessinger, *Angew. Chem. (Int. Edit.)*, **11,** 525 (1972).

501. D. J. Knowles and A. J. C. Nicholson, *J. Chem. Phys.*, **60**, 1180 (1974).
502. T. Kobayashi and S. Nagakura, *Bull. Chem. Soc. Jap.*, **46**, 1558 (1973).
503. T. Kobayashi and S. Nagakura, *Bull. Chem. Soc. Jap.*, **47**, 2563 (1974).
504. T. Kobayashi and S. Nagakura, *Chem. Letters*, 903 (1972).
505. T. Kobayashi and S. Nagakura, *Chem. Letters*, 1013 (1972).
506. T. Kobayashi and S. Nagakura, *J. Electron Spectrosc.*, **4**, 207 (1974).
507. T. Kobayashi, K. Yokota, and S. Nagakura, *J. Electron Spectrosc.*, **2**, 449 (1973).
508. T. Kobayashi, K. Yokota, and S. Nagakura, *J. Electron Spectrosc.*, **6**, 167 (1975).
508. a. T. Koenig, T. Balle, and W. Snell, *J. Amer. Chem. Soc.* **97**, 662 (1975).
509. T. Koenig and H. Longmaid, *J. Org. Chem.*, **39**, 560 (1974).
510. T. Koenig and M. Tuttle, *J. Org. Chem.*, **39**, 1308 (1974).
511. W. Kosmus, B. M. Rode, and E. Nachbaur, *J. Electron Spectrosc.*, **1**, 409 (1972/1973).
512. J. Kroner, D. Noelle, and H. Noeth, *Z. Naturforsch.*, *B*, **28**, 416 (1973).
513. J. Kroner, D. Noelle, H. Noeth, and W. Winterstein, *Z. Naturforsch.*, *B*, **29**, 476 (1974).
514. J. Kroner, H. Noeth, and K. Niedenzu, *J. Organometal. Chem.*, **71**, 165 (1974).
515. J. Kroner, D. Proch, W. Fuss, and H. Bock, *Tetrahedron*, **28**, 1585 (1972).
516. H. W. Kroto and R. J. Suffolk, *Chem. Phys. Letters*, **15**, 545 (1972).
517. H. W. Kroto and R. J. Suffolk, *Chem. Phys. Letters*, **17**, 213 (1972).
518. H. W. Kroto, R. J. Suffolk, and N. P. C. Westwood, *Chem. Phys. Letters*, **22**, 495 (1973).
519. C. Lageot, *J. Chim. Phys. Physicochim. Biol.*, **68**, 214 (1971).
520. R. F. Lake, *Spectrochim. Acta*, **A27**, 1220 (1971).
521. R. F. Lake and H. Thompson, *Proc. Roy. Soc.*, *Ser. A*, **315**, 323 (1970).
522. R. F. Lake and H. Thompson, *Proc. Roy. Soc.*, *Ser. A*, **317**, 187 (1970).
523. R. F. Lake and H. W. Thompson, *Spectrochim. Acta*, **27A**, 783 (1971).
524. M. F. Lappert, J. B. Pedley, and G. Sharp, *J. Organometal. Chem.*, **66**, 271 (1974).
525. M. F. Lappert, J. B. Pedley, G. J. Sharp, and N. P. C. Westwood, *J. Electron Spectrosc.*, **3**, 237 (1974).
526. W. A. Lathan, L. A. Curtiss, and J. A. Pople, *Mol. Phys.*, **22**, 1081 (1971).
527. S. Leavell, J. Steichem, and J. L. Franklin, *J. Chem. Phys.*, **59**, 4343 (1973).
528. H. Lefebvre-Brion, *Chem. Phys. Letters*, **9**, 463 (1971).
529. T. H. Lee, R. J. Colton, M. G. White, and J. W. Rabalais, *J. Amer. Chem. Soc.*, **97**, 4845 (1975).
530. T. H. Lee and J. W. Rabalais, *Chem. Phys. Letters*, **34**, 135 (1975).
531. T. H. Lee and J. W. Rabalais, *J. Chem. Phys.*, **60**, 1172 (1974).
532. T. H. Lee and J. W. Rabalais, *J. Chem. Phys.*, **61**, 2747 (1974).
533. H. J. Lempka, T. P. Passmore, and W. C. Price, *Proc. Roy. Soc.*, *Ser. A*, **304**, 53 (1968).
534. G. Levy and P. DeLothe, *C. R. Acad. Sci.*, *Paris,* *Ser. C*, **279**, 331 (1974).

535. G. Levy, P. DeLoth, and F. Gallais, *C. R. Acad. Sci., Paris, Ser. C,* **278,** 1405 (1974).

536. D. L. Lichtenberger and R. F. Fenske, *Inorg. Chem.* **13,** 486 (1974).

537. D. L. Lichtenberger, A. C. Sarapu, and R. F. Fenske, *Inorg. Chem.,* **12,** 702 (1973).

538. E. Lindholm, C. Fridh, and L. Åsbrink, *Faraday Discussions Chem. Soc.,* **54,** 127 (1972).

539. D. R. Lloyd, *J. Chem. Soc. D,* 868 (1970).

540. D. R. Lloyd, *J. Phys. E,* **3,** 629 (1970).

541. D. R. Lloyd and P. J. Bassett, *J. Chem. Soc. A,* 641 (1971).

542. D. R. Lloyd and P. J. Bassett, *J. Chem. Soc. A,* 1551 (1971).

543. D. R. Lloyd and N. Lynaugh, *Chem Commun.,* **22,** 1545 (1970).

544. D. R. Lloyd and N. Lynaugh, *J. Chem. Soc. D,* 125 (1971).

545. D. R. Lloyd and N. Lynaugh, *J. Chem. Soc. D,* 627 (1971).

546. D. R. Lloyd, and N. Lynaugh, *J. C. S. Faraday Trans.* 2, **68,** 947 (1972).

547. D. R. Lloyd and N. Lynaugh, *Electron Spectroscopy,* D. A. Shirley (Ed.), p. 445, North-Holland Amsterdam, 1972.

548. D. R. Lloyd and N. Lynaugh, *Phil. Trans. Roy. Soc. London,* A268, 97 (1970).

549. D. R. Lloyd and P. J. Roberts, *Mol. Phys.,* **26,** 225 (1973).

550. D. R. Lloyd and E. W. Schlag, *Inorg. Chem.,* **8,** 2544 (1969).

551. L. L. Lohr, and M. B. Robin, *J. Amer. Chem. Soc.,* **92,** 7241 (1970).

552. J. C. Lorquet and C. Cadet, *Chem. Phys. Letters* **6,** 198 (1970).

553. J. C. Lorquet and C. Cadet, *Int. J. Mass Spectrom. Ion Phys.,* **7,** 245 (1971).

554. A. E. Lutskü and N. I. Gorokhova, *Theor. Exper. Chem. (USSR),* **6,** 490 (1970).

555. N. Lynaugh, D. R. Lloyd, M. F. Guest, M. B. Hall, and I. H. Hillier, *J. C. S. Faraday Trans.* 2, **68,** 2192 (1972).

556. M. J. Lynch, A. B. Gardner, K. Godling, and G. V. Marr, *Phys. Letters A,* **43,** 237 (1973).

557. C. A. McDowell, *Discussions Faraday Soc.,* **54,** 297 (1973).

557. a. S. P. McGlynn and J. L. Meeks, *J. Electron Spectrosc.* **6,** 269 (1975).

558. R. A. N. McLean, *Can. J. Chem.,* **51,** 2089 (1973).

559. J. P. Maier, *Helv. Chim. Acta,* **57,** 994 (1974).

560. J. P. Maier and J. F. Muller, *Tetrahedron Lett.,* 2987 (1974).

561. J. P. Maier and D. W. Turner, *J. C. S. Faraday Trans.* 2, **68,** 711 (1972).

562. J. P. Maier and D. W. Turner, *J. C. S. Faraday Trans.* 2, **69,** 196 (1973).

563. J. P. Maier and D. W. Turner *J. C. S. Faraday Trans.* 2, **69,** 531 (1973).

564. J. P. Maier and D. W. Turner, *Faraday Discussions Chem. Soc.,* **54,** 149 (1972).

565. R. Manne, *Chem. Phys. Letters,* **5,** 125 (1970).

566. R. Manne, *J. Electron Spectrosc.,* **3,** 327 (1974).

566. a. R. Manne, R. Wittel, and B. S. Mohanty, *Mol. Phys.,* **29,** 485 (1975).

567. S. T. Manson and J. W. Cooper, *Phys. Rev. A,* **32,** 2170 (1970).

568. S. T. Manson and D. J. Kennedy, *Chem. Phys. Letters,* **7,** 387 (1970).

569. G. V. Marr, *J. Phys. B,* **7,** L47 (1974).

570. F. G. Marschner and H. Goetz, *Tetrahedron Lett.*, **29**, 3105 (1973).

571. F. Marschner, H. Juds, and H. Goetz, *Tetrahedron Lett.*, 3983 (1973).

572. H. D. Martin, C. Heller, E. Haselbach, and Z. Lanyjova, *Helv. Chim. Acta*, **57**, 465 (1974).

573. P. Masclet, D. Grosjean, and G. Mouvier, *J. Electron Spectrosc.*, **2**, 225 (1973).

574. D. C. Mason, A. Kuppermann, and D. M. Mintz, *Electron Spectroscopy*, D. A. Shirley (Ed.), p. 269, North-Holland, Amsterdam, 1972.

575. G. D. Mateesca, *Tetrahedron Lett.*, 5285 (1972).

576. J. L. Meeks, J. F. Arnett, D. Larson, and S. P. McGlynn, *Chem. Phys. Letters*, **30**, 190 (1975).

577. L. L. Miller, V. R. Koch, T. Koenig, and M. Tuttle, *J. Amer. Chem. Soc.*, **95**, 5075 (1973).

578. G. W. Mines and R. K. Thomas, *Proc. Roy. Soc. London*, A336, 355 (1974).

579. G. W. Mines, R. K. Thomas, and H. Thompson, *Proc. Roy. Soc. London*, A329, 275 (1972).

580. G. W. Mines, R. K. Thomas, and H. Thompson, *Proc. Roy. Soc. London*, A333, 171 (1973).

581. G. W. Mines and H. W. Thompson, *Spectrochim. Acta*, **A29**, 1377 (1973).

582. P. Mitchell and K. Codling, *Phys. Letters A*, **38**, 31 (1972).

583. P. Mitchell and M. Wilson, *Chem. Phys. Letters*, **3**, 389 (1969).

584. R. P. Mitra and K. L. Kapoor, *Int. J. Quantum Chem.*, **6**, 387 (1972).

585. P. D. Mollere, *Tetrahedron Lett.*, 2791 (1973).

586. P. Mollere and H. Bock, *J. Chem. Educ.*, **51**, 506 (1974).

587. P. Mollere, H. Bock, G. Becker, and G. Fritz, *J. Organometal. Chem.* **46**, 89 (1972).

588. P. Mollere, H. Bock, G. Becker, and G. Fritz, *J. Organometal. Chem.*, **61**, 127 (1973).

589. J. Momigny and J. Delwiche, *J. Chim. Phys. Physiochim. Biol.*, **65**, 1213 (1968).

590. J. Momigny, C. Goffart, and P. Natalis, *Bull. Soc. Chim. Belges*, **77**, 533 (1968).

591. J. Momigny and J. C. Lorquet, *Chem. Phys. Letters*, **1**, 505 (1968).

592. J. Momigny and J. C. Lorquet, *Int. J. Mass Spectrom. Ion Phys.* **2**, 495 (1969).

593. K. Monahan and T. S. Wauchop, *J. Geophys. Res.*, **77**, 6262 (1972).

594. R. Morgenstern, A. Niehaus, and M. W. Ruf, *Chem. Phys. Letters*, **A268**, 147 (1970).

595. I. Morishim, K. Yoshikaw, T. Yonezawa, and H. Matsumot, *Chem. Phys. Letters*, **16**, 336 (1972).

596. J. Mueller, K. Fenderl, and B. Mertschenk, *Chem. Ber.*, **104**, 700 (1971).

597. C. Mueller and A. Schweig, *Tetrahedron*, **29**, 3973 (1973).

598. C. Mueller, A. Schweig, and W. L. Mock, *J. Amer. Chem. Soc.*, **96**, 280 (1974).

599. J. N. Murrell, *Chem. Phys. Letters*, **15**, 296 (1972).

600. J. N. Murrell and W. Schmidt, *J. C. S. Faraday Trans.* 2, **68,** 1709 (1972).
601. J. N. Murrell and R. J. Suffolk, *J. Electron Spectrosc.*, **1,** 471 (1973).
602. K. A. Muszkat and J. Schaublin, *Chem. Phys. Letters*, **13,** 301 (1972).
603. B. Narayan and J. N. Murrell, *Mol. Phys.*, **19,** 169 (1970).
604. P. Natalis and J. E. Collin, *Chem. Phys. Letters*, **2,** 79 (1968).
605. P. Natalis and J. E. Collin, *Chem. Phys. Letters*, **2,** 414 (1968).
606. P. Natalis and J. E. Collin, *Int. J. Mass Spectrom. Ion Phys.*, **2,** 221 (1969).
607. P. Natalis, J. E. Collin, and J. Momigny, *Int. J. Mass Spectrom. Ion Phys.*, **1,** 327 (1968).
608. P. Natalis, J. Delwiche, and J. E. Collin, *Chem. Phys. Letters*, **9,** 139 (1971).
609. P. Natalis, J. Delwiche, and J. E. Collin, *Chem. Phys. Letters*, **13,** 491 (1972).
610. P. Natalis, J. Delwiche, and J. E. Collin, *Faraday Discussions Chem. Soc.*, **54,** 98 (1972).
611. S. F. Nelson and J. M. Buschek, *J. Amer. Chem. Soc.* **96,** 2392 (1974).
612. S. F. Nelsen and J. M. Buschek, *J. Amer. Chem. Soc.* **96,** 6424 (1974).
613. S. F. Nelsen and J. M. Buschek, *J. Amer. Chem. Soc.* **96,** 6982 (1974).
614. S. F. Nelsen and J. M. Buschek, *J. Amer. Chem. Soc.*, **96,** 6987 (1974).
615. S. F. Nelsen, J. M. Buschek, and P. J. Hintz, *J. Amer. Chem. Soc.*, **95,** 2013 (1973).
616. A. Niehaus and M. W. Ruf, *Chem. Phys. Letters*, **11,** 55 (1971).
617. S. Nishida, I. Moritani, and T. Teraji, *J. C. S. Chem. Commun.* 1114 (1972).
618. H. Oehling, W. Schafer, and A. Schweig, *Angew. Chem. (Int. Edit.)*, **10,** 656 (1971).
619. H. Ogata, J. Kitayama, M. Koto, S. Kojima, Y. Nikei, and H. Kamada, *Bull. Chem. Soc. Jap.*, **47,** 958 (1974).
620. H. Ogata, H. Onizuka, Y. Nikei, and H. Kamada, *Bull. Chem. Soc. Jap.*, **46,** 3036 (1973).
621. H. Ogata, H. Onizuka, Y. Nikei, and H. Kamada, *Chem. Letters*, 895 (1972).
622. M. Okuda and N. Jonathan, *J. Electron Spectrosc.*, **3,** 19 (1974).
623. A. F. Orchard and N. V. Richardson, *J. Electron Spectrosc.*, **6,** 61 (1975).
624. K. Osafune, S. Katsumata, and K. Kimura, *Chem. Phys. Letters*, **19,** 369 (1973).
625. K. Osafune and K. Kimura, *Chem. Phys. Letters*, **25,** 47 (1974).
626. A. Padva, P. R. LeBreton, R. J. Dinerstein, and J. N. A. Ridyard, *Biochem. Biophys. Res. Commun.*, **60,** 1262 (1974).
627. M. H. Palmer and R. H. Findlay, *Chem. Phys. Letters*, **15,** 416 (1972).
628. T. Parameswaran and D. E. Ellis, *J. Chem. Phys.*, **58,** 2088 (1973).
629. S. Pignataro and G. Aloisi, *Z. Naturfosch., A*, **27,** 1165 (1972).
630. S. Pignataro, and G. Distefano, *Chem. Phys. Letters*, **26,** 356 (1974).
631. S. Pignataro, V. Mancini, J. N. A. Ridyard, and H. J. Lempka, *Chem. Commun.*, 142 (1971).
632. C. G. Pitt and H. Bock, *Chem. Commun.*, 28 (1972).

633. A. A. Planckaert, D. J. Doucet, and C. Sandorfy, *J. Chem. Phys.*, **60,** 4846 (1974).
634. A. W. Potts, K. G. Glenn, and W. C. Price, *Discussions Faraday Soc.*, **54,** 65 (1973).
635. A. W. Potts, H. J. Lempka, D. G. Streets, and W. C. Price, *Phil. Trans. Roy. Soc. London,* **A268,** 59 (1970).
636. A. W. Potts and W. C. Price, *Proc. Roy. Soc., Ser. A*, **326,** 165 (1972).
637. A. W. Potts and W. C. Price, *Proc. Roy. Soc., Ser. A*, **326,** 181 (1972).
638. A. W. Potts and W. C. Price, *Trans. Faraday Soc.*, **67,** 1242 (1971).
639. A. W. Potts, W. C. Price, D. G. Streets, and T. A. Williams, *Faraday Discussions Chem. Soc.*, **54,** 168 (1972).
640. A. W. Potts and D. G. Streets, *J. C. S. Faraday Trans. 2,* **70,** 875 (1974).
641. A. W. Potts and T. A. Williams, *J. Electron Spectrosc.*, **3,** 3 (1974).
642. A. W. Potts, T. A. Williams, and W. C. Price, *Faraday Discussions Chem. Soc.*, **54,** 104 (1972).
643. A. W. Potts, T. A. Williams, and W. C. Price, *Proc. Roy. Soc., Ser. A*, **341,** 147 (1974).
664. T. M. Praet and J. Delwich, *Int. J. Mass Spectrom. Ion Phys.*, **1,** 321 (1968).
645. W. C. Price, A. W. Potts, and D. G. Streets, *Electron Spectroscopy*, D. A. Shirley (Ed.), p. 187, North-Holland, Amsterdam, 1972.
646. B. P. Pullen T. A. Carlson, W. E. Moddeman, G. K. Schweitzer, W. E. Bull, and F. Grimm, *J. Chem. Phys.*, **53,** 768 (1970).
647. J. W. Rabalais, *J. Chem. Phys.* **57,** 960 (1972).
648. J. W. Rabalais, T. Bergmark, L. O. Werme, L. Karlsson, M. Hussian, and K. Siegbahn, *Electron Spectroscopy*, D. A. Shirley (Ed.), p. 425, North-Holland, Amsterdam, 1972.
649. J. W. Rabalais, T. Bergmark, L. O. Werme, L. Karlsson, and K. Siegbahn, *Phys. Scripta*, **3,** 13 (1971).
650. J. W. Rabalais and R. J. Colton, *J. Electron Spectrosc.*, **1,** 83 (1972).
651. J. W. Rabalais, and T. P. Debies, *Electron Spectroscopy*, R. Caudano and J. Verbist (Eds.), p. 847, Elsevier, Amsterdam, 1974.
652. J. W. Rabalais, T. P. Debies, J. L. Berkosky, J. T. J. Huang, and F. O. Ellison, *J. Chem. Phys.*, **61,** 516 (1974).
653. J. W. Rabalais, T. P. Debies, J. L. Berkosky, J. T. J. Huang, and F. O. Ellison, *J, Chem. Phys.*, **61,** 529 (1974).
654. J. W. Rabalais, L. Karlsson, L. O. Werme, T. Bergmark, and K. Siegbahn, *J. Chem. Phys.*, **58,** 3370 (1973).
655. J. W. Rabalais and A. Katrib, *Mol. Phys.*, **27,** 923 (1974).
656. J. W. Rabalais, L. O. Werme, T. Bergmark, L. Karlsson, M. Hussain, and K. Siegbahn, *J. Chem. Phys.*, **57,** 1185 (1972).
657. J. W. Rabalais, L. O. Werme, T. Bergmark, L. Karlsson, and K. Siegbahn, *Int. J. Mass Spectrom. Ion Phys.*, **9,** 185 (1972).
658. P. Rademacher, *Angew. Chem. (Int. Edit.)*, **12,** 408 (1973).
659. P. Rademacher, *Tetrahedron Lett.*, **83,** (1974).
660. J. Raftery and W. C. Richards, *Int. J. Mass Spectrom. Ion Phys.*, **6,** 269 (1971).

661. J. L. Ragle, I. A. Stenhouse, D. C. Frost, and C. A. McDowell, *J. Chem. Phys.*, **53**, 178 (1970).

662. B. G. Ramsey, A. Brook, A. R. Bassindale, and H. Bock, *J. Organometal. Chem.*, **74**, C41 (1974).

663. M. T. Reetz, R. W. Hoffmann, W. Schafer, and A. Schweig, *Angew. Chem.* (*Int. Edit.*), **12**, 81 (1973).

664. N. V. Richardson and P. Weinberger, *J. Electron Spectrosc.*, **6**, 109 (1975).

665. M. B. Robin, C. R. Brundle, N. A. Kuebler, G. B. Ellison, and K. B. Wiberg, *J. Chem. Phys.*, **57**, 1758 (1972).

666. M. B. Robin and N. A. Kuebler, *J. Electron Spectrosc.*, **1**, 13 (1972).

667. M. B. Robin, N. A. Kuebler, and C. R. Brundle, *Electron Spectroscopy*, D. A. Shirley (Ed.), p. 351, North-Holland, Amsterdam, 1972.

668. M. M. Rohmer and A. Weillard, *J. Chem. Soc.*, 250 (1973).

669. P. Romus, P. Dacre, B. Solouki, and H. Bock, *Theor. Chem. Acta*, **35**, 129 (1974).

670. T. Rose, R. Frey, and B. Brehm, *J. Chem. Soc. D*, 1518 (1969).

671. C. G. Rowland, *Chem. Phys. Letters*, **9**, 169 (1971).

672. J. A. R. Samson, *Chem. Phys. Letters*, **4**, 257 (1969).

673. J. A. R. Samson, *Chem. Phys. Letters*, **12**, 625 (1972).

674. J. A. R. Samson, *Electron Spectroscopy*, D. A. Shirley (Ed.), p. 441, North-Holland, Amsterdam, 1972.

675. J. A. R. Samson, *Phil. Trans. Roy. Soc. London*, **A268**, 141 (1970).

676. J. A. R. Samson, *Phys. Letters*, **A28**, 391 (1968).

677. J. A. R. Samson, *Phys. Rev. Letters*, **22**, 693 (1969).

678. J. A. R. Samson and R. B. Cairns, *Phys. Rev.*, **173**, 80 (1968).

679. J. A. R. Samson and J. L. Gardner, *J. Geophys. Res.*, **78**, 3663 (1973).

680. J. A. R. Samson and J. L. Gardner, *J. Opt. Soc. Amer.*, **62**, 856 (1972).

681. J. A. R. Samson and J. L. Gardner, *Phys. Rev. Letters*, **31**, 1327 (1973).

682. J. A. R. Samson and J. L. Gardner, *Phys. Rev. Letters*, **33**, 671 (1974).

683. J. A. R. Samson, J. L. Gardner, and J. E. Mentall, *J. Geophys Res.*, **77**, 5560 (1972).

684. J. A. R. Samson and G. N. Haddad, *Phys. Rev. Letters*, **33**, 875 (1974).

685. J. A. R. Samson and V. E. Petrosky, *J. Electron Spectrosc.*, **3**, 461 (1974).

686. J. A. R. Samson and V. E. Petrosky, *Phys. Rev. A*, **9**, 2449 (1974).

687. J. S. Sandu, *Indian J. Chem.*, **10**, 667 (1972).

688. P. Sauvageau, J. Doucet, R. Gilbert, and C. Sandorphy, *J. Chem. Phys.*, **61**, 391 (1974).

689. W. Schafer and A. Schweig, *Angew. Chem.* (*Int. Edit.*), **11**, 836 (1972).

690. W. Schaefer and A. Schweig, *J. C. S. Chem. Commun.*, 824 (1972).

691. W. Schafer and A. Schweig, *Tetrahedron Lett.*, 5205 (1972).

692. W. Schafer, A. Schweig, F. Bickelhaupt, and H. Vermeer, *Angew. Chem.*, **11**, 924 (1972).

693. W. Schafer, A. Schweig, S. Gronowit, A. Talicchi, and F. Fringuel, *J. Chem. Soc.* 541 (1973).

694. W. Schaefer, A. Schweig, G. Maier, and T. Sayrac, *J. Amer. Chem. Soc.*, **96**, 279 (1974).

695. W. Schafer, A. Schweig, G. Maier, T. Sayrac, and J. K. Crandall, *Tetrahedron Lett.*, 1213 (1974).
696. W. Schafer, A. Schweig, G. Markl, H. Haupman, and F. Mathey, *Angew. Chem.*, **12**, 145 (1973).
697. W. Schafer, A. Schweig, G. Markl, and K. H. Heier, *Tetrahedron Lett.*, 3743 (1973).
698. W. Schafer, A. Schweig, H. Vermeer, F. Bickelhaupt, and H. DeGraaf, *J. Electron Spectrosc.*, **6**, 91 (1975).
699. W. Schmidt, *J. Electron Spectrosc.*, **6**, 163 (1975).
700. W. Schmidt, *Tetrahedron*, **29**, 2129 (1973).
701. H. Schmidt and A. Schweig, *Angew. Chem. (Int. Edit.)*, **12**, 307 (1973).
702. H. Schmidt and A. Schweig, *Chem. Ber.*, **107**, 725 (1974).
703. H. Schmidt and A. Schweig, *Tetrahedron Lett.*, 981 (1973).
704. H. Schmidt and A. Schweig, *Tetrahedron Lett.*, 1437 (1973).
705. L. Schmidt, A. Schweig, A. G. Anastassiou, and H. Yamamoto, *J. C. S. Chem. Commun.*, 218 (1974).
706. H. Schmidt, A. Schweig, and A. Krebs, *Tetrahedron Lett.*, 1471 (1974).
707. H. Schmidt, A. Schweig, and G. Manuel, *J. Organometal. Chem.*, **55**, C1 (1973).
708. W. Schmidt and B. T. Wilkins, *Angew. Chem. (Int. Edit.)*, **11**, 221 (1972).
709. W. Schmidt and B. T. Wilkins, *Tetrahedron*, **28**, 5649 (1972).
710. W. Schmidt, B. T. Wilkins, G. Fritz, and R. Huber, *J. Organometal. Chem.* **59**, 109 (1973).
711. B. S. Schneider and A. L. Smith, *Electron Spectroscopy*, D. A. Shirley (Ed.), p. 335, North-Holland, Amsterdam, 1972.
711 a. W. H. E. Schwarz, *J. Electron Spectrosc.*, **6**, 337 (1975).
712. A. Schweig, W. Schafer, and K. Kimroth, *Angew. Chem. (Int. Edit.)*, **11**, 631 (1972).
713. A. Schweig and W. Thiel, *Chem. Phys. Letters*, **21**, 541 (1973).
714. A. Schweig and W. Thiel, *J. Electron Spectrosc.*, **2**, 199 (1973).
715. A. Schweig and W. Thiel, *J. Electron Spectrosc.*, **3**, 27 (1974).
716. A. Schweig and W. Thiel, *Mol. Phys.*, **27**, 265 (1974).
717. A. Schweig, H. Vermeer, and U. Weidner, *Chem. Phys. Letters*, **26**, 229 (1974).
718. A. Schweig and U. Weidner, J. G. Berger, and W. Grahn, *Tetrahedron Lett.*, 557 (1973).
719. A. Schweig, U. Weidener, D. Hellwinkel, and W. Krapp, *Angew. Chem. (Int. Edit.)*, **12**, 310 (1973).
720. A. Schweig, U. Weidner, R. K. Hill, and D. A. Cullison, *J. Amer. Chem. Soc.*, **95**, 5426 (1973).
721. A. Schweig, U. Weidner, and G. Manuel, *Angew. Chem. (Int. Edit.)*, **11**, 837 (1972).
722. A. Schweig, U. Weidner, and G. Manuel, *J. Organometal. Chem.*, **54**, 145 (1973).
723. K. Seki, H. Inokuchi, and Y. Harada, *Chem. Phys. Letters*, **20**, 197 (1973).
724. G. B. Shaw and R. S. Berry, *J. Chem. Phys.*, **56**, 5808 (1972).

725. K. Shen and N. A. Kuebler, *Tetrahedron Lett.*, 2145 (1973).
726. J. M. Sichel, *Mol. Phys.*, **18,** 95 (1970).
727. K. Siegbahn, *Electron Spectroscopy,* R. Caudano and J. Verbist, (Ed.), p. 3, Elsevier, Amsterdam, 1974.
728. I. G. Sim, C. J. Danby, J. H. D. Eland, *Int. J. Mass Spectrom. Ion Phys.*, **14,** 285 (1974).
729. C. Sluse-Goffart and P. Natalis, *J. Electron Spectrosc.*, **2,** 215 (1973).
729 a. S. P. So and W. Richards, *J. C. S. Faraday Trans.* 2, **71,** 62 (1975).
730. B. Solouki, H. Bock, and R. Appel, *Angew. Chem. (Int. Edit.)*, **11,** 927 (1972).
731. B. Solouki, P. Rosmus, and H. Bock, *Chem. Phys. Letters*, **26,** 20 (1974).
732. J. Spanget-Larsen, *J. Electron Spectrosc.*, **2,** 33 (1973).
733. J. Spanget-Larsen, *J. Electron Spectrosc.*, **3,** 369 (1974).
734. H. Stafast and H. Bock, *Chem. Ber.*, **107,** 1882 (1974).
735. H. Stafast and H. Bock, *Z. Naturforsch.*, *B*, **28,** 746 (1973).
736. K. H. O. Starzewski, H. T. Dieck, and H. Bock *J. Organometal. Chem.*, **65,** 311 (1973).
737. D. G. Streets, *Chem. Phys. Letters*, **28,** 555 (1974).
738. D. G. Streets and G. P. Ceasar, *Mol. Phys.*, **26,** 1037 (1973).
739. D. G. Streets, W. E. Hall, and G. P. Caesar, *Chem. Phys. Letters*, **17,** 90 (1972).
740. D. G. Streets, A. W. Potts, and W. C. Price, *Int. J. Mass Spectrom. Ion Phys.*, **10,** 123 (1972/73).
741. D. G. Streets and T. A. Williams, *J. Electron Spectrosc.*, **3,** 71 (1974).
742. R. J. Suffolk, *J. Electron Spectrosc.*, **3,** 53 (1974).
743. R. Sustmann and R. Schubert, *Tetrahedron Lett.*, 2739 (1972).
744. R. Sustmann and H. Trill, *Tetrahedron Lett.*, 4271 (1972).
745. D. A. Sweigart and J. Daintith, *Sci. Progr. Oxford*, **59,** 325 (1971).
746. D. A. Sweigart and D. W. Turner, *J. Amer. Chem. Soc.*, **94,** 5592 (1972).
747. W. C. Tam, D. Yee and C. E. Brion, *J. Electron Spectrosc.*, **4,** 77, (1974).
748. K. Tanaka and I. Tanaka, *J. Chem. Phys.*, **59,** 5042 (1973).
749. W. Thiel and A. Schweig, *Chem. Phys. Letters*, **12,** 49 (1971).
750. W. Thiel and A. Schweig, *Chem. Phys. Letters.*, **16,** 409 (1972).
751. E. W. Thielstrup and Y. Oehrn, *J. Chem. Phys.*, **57,** 3716 (1972).
752. R. K. Thomas, *Proc. Roy. Soc. London,* A331, 249 (1972).
753. R. K. Thomas and H. Thompson, *Proc. Roy. Soc., Ser. A,* **327,** 13 (1972).
754. R. K. Thomas and H. Thompson, *Proc. Roy. Soc., Ser. A,* **339,** 29 (1974).
755. E. W. Thulstru and Y. Ohrn, *J. Chem. Phys.*, **57,** 3716 (1972).
756. M. M. Timoshenko and M. E. Akopyan, *Khim. Vys. Energ.*, **8,** 211 (1974).
757. H. C. Tuckwell, *J. Quant. Spectrosc. Radiat. Transfer*, **10,** 653 (1970).
758. D. W. Turner, *Phil. Trans. Roy. Soc. London,* A268, 7 (1970).
759. D. W. Turner, *Proc. Roy. Soc., Ser. A,* **307,** 15 (1968).
760. D. W. Turner and M. I. Al-Joboury, *J. Chem. Soc.*, 5141 (1963).
761. D. W. Turner and M. I. Al-Joboury, *J. Chem. Soc.*, 616 (1965).
762. D. W. Turner and C. Baker, *Chem. Commun.*, 400 (1968).

· 763. D. W. Turner, C. Baker, A. D. Baker, and C. R. Brundle, *Molecular Photoelectron Spectroscopy*, Wiley-Interscience, London, 1970.

764. D. W. Turner, D. P. May, and M. I. Al-Joboury, *J. Chem. Soc.*, 6350 (1965).

765. D. W. Turner and T. N. Radwan, *J. Chem. Soc. A*, 85 (1966).

766. Y. Uehara, N. Saito, and T. Yonezawa, *Chem. Letters*, 495 (1973).

767. D. M. Vandenha and D. Vanderme, *Chem. Phys. Letters*, **15**, 549 (1972).

768. D. M. W. Van Den Ham and D. Van Der Meer, *Chem. Phys. Letters*, **12**, 447 (1972).

769. D. W. M. Van Den Ham and D. Van Der Meer, *Chem. Phys. Letters*, **15**, 549 (1972).

770. D. W. M. Van Den Ham and D. Van Der Meer, *J. Electron Spectrosc.*, **2**, 247 (1973).

771. D. W. M. Van Den Ham, D. Van Der Meer, and D. Feil, *J. Electron Spectrosc.*, **3**, 479 (1974).

772. M. J. Van Der Wiel and C. E. Brion, *J. Electron Spectrosc.*, **1**, 309 (1972/1973).

773. M. J. Van Der Wiel and C. E. Brion, *J. Electron Spectrosc.*, **1**, 439 (1972/1973).

774. M. J. Van Der Wiel and C. E. Brion, *J. Electron Spectrosc.*, **1**, 443 (1972/1973).

775. M. D. Van Hoorn, *J. Electron Spectrosc.*, **6**, 65 (1975).

776. C. J. Vesely, R. L. Hengehold, and D. W. Langer, *Phys. Rev. B.* **5**, 6 (1972).

777. Y. Vignollet, J. C. Maire, A. D. Baker, and D. W. Turner, *J. Organometal. Chem.*, **18**, 349 (1969).

778. V. I. Vovna, S. N. Lopatin, R. Petzold, F. I. Vilesov, and M. E. Akopyan, *Opt. Spektrosk.*, **36**, 173 (1974).

779. V. I. Vovna and F. I. Vilesov, *Opt. Spektrosk.*, **36**, 436 (1974).

780. G. Wagner and H. Bock, *Chem. Ber.* **107**, 68 (1974).

781. G. Wagner, H. Bock, R. Budenz, and F. Seel, *Chem. Ber.*, **106**, 1285 (1973).

782. T. E. H. Walker, J. Berkowitz, J. L. Dehmer, and J. T. Waber, *Phys. Rev. Letters*, **31**, 678 (1973).

783. T. E. H. Walker, P. M. Dehmer, and J. Berkowitz, *J. Chem. Phys.*, **59**, 4292 (1973).

784. T. E. H. Walker and J. A. Horsely, *Mol. Phys.*, **21**, 939 (1971).

785. I. Watanabe, Yu. Yokoyama, and S. Ikeda, *Bull. Chem. Soc. Jap.*, **46**, 1959 (1973).

786. I. Watanabe, Yu. Yokoyama, and S. Ikeda, *Bull. Chem. Soc. Jap.*, **47**, 627 (1974).

787. I. Watanabe, Yu. Yokoyama, and S. Ikeda, *Chem. Phys. Letters*, **19**, 406 (1973).

788. U. Weidner and A. Schweig, *Angew. Chem. (Int. Edit.)*, **11**, 146 (1972).

789. U. Weidner and A. Schweig, *Angew. Chem. (Int. Edit.)*, **11**, 536 (1972).

790. U. Weidner and A. Schweig, *Angew. Chem. (Int. Edit.)*, **11**, 537 (1972).

791. U. Weidner and A. Schweig, *J. Organometal. Chem.*, **37**, C29 (1972).

792. U. Weidner and A. Schweig, *J. Organometal. Chem.*, **39**, 261 (1972).

793. L. Weiler, D. Chadwick, and D. C. Frost, *J. Amer. Chem. Soc.*, **93**, 4320 (1971).

794. L. Weiler, D. Chadwick, and D. C. Frost, *J. Amer. Chem. Soc.*, **93**, 4962 (1971).

795. M. A. Weimer and M. Lattman, *Tetrahedron Lett.*, 1709 (1974).

796. M. J. Weiss and G. M. Lawrence, *J. Chem. Phys.*, **51**, 2876 (1970).

797. M. J. Weiss and G. M. Lawrence, *J. Chem. Phys.*, **53**, 214 (1970).

798. M. J. Weiss, G. M. Lawrence, and R. A. Young, *J. Chem. Phys.*, **52**, 2867 (1970).

799. N. P. C. Westwood, *Chem. Phys. Letters*, **25**, 558 (1974).

800. M. G. White, R. J. Colton, T. H. Lee, and J. W. Rabalais, *Chem. Phys.*, **8**, 391 (1975).

801. R. M. White, T. A. Carlson, and D. P. Spears, *J. Electron Spectrosc.*, **3**, 59 (1974).

802. R. A. Wielesek and T. Koenig, *Tetrahedron Lett.*, 2429 (1974).

802. a. J. S. Wieczorek, T. Koenig, and T. Balle, *J. Electron Spectrosc.*, **6**, 215 (1975).

803. K. Wittel, *Chem. Phys. Letters*, **15**, 555 (1972).

804. K. Wittel and H. Bock, *Chem. Ber.*, **107**, 317 (1974).

805. K. Wittel and R. Manne, *Theor. Chim. Acta*, **33**, 347 (1974).

806. K. Wittel, H. Bock, and R. Manne, *Tetrahedron*, **30**, 651 (1974).

807. K. Wittel, A. Haas, and H. Bock, *Chem. Ber.*, **105**, 3865 (1972).

808. K. Wittel, B. S. Mohanty, and R. Manne, *Electron Spectroscopy*, R. Caudano and J. Verbist, (Eds.), p. 1115, Elsevier, Amsterdam, 1974.

809. S. D. Worley, *J. Chem. Soc. D*, 980 (1970).

810. S. D. Worley, *J. Electron Spectrosc.*, **6**, 157 (1975).

811. S. D. Worley, G. D. Mateescu, C. W. McFarland, R. C. Fort, Jr., and C. F. Sheley, *J. Amer. Chem. Soc.*, **95**, 7580 (1973).

812. C. W. Worrell, *J. Electron Spectrosc.*, **3**, 359 (1974).

813. T. Yamazaki, S. Katsumota, and K. Kimura, *J. Electron Spectrosc.*, **2**, 335 (1973).

814. Y. Yokoyama, M. Jinno, I. Watanabe, and S. Ikeda, *Electron Spectroscopy*, R. Caudano and J. Verbist (Eds.), p. 1095, Elsevier, Amsterdam, 1974.

815. H. Yoshiya, K. Ohno, K. Seki, and H. Inokuchi, *Chem. Letters*, 1081 (1974).

816. V. V. Zverev, V. I. Vovna, M. S. El'man, Yu. P. Kitaev, and F. I. Vilesov, *Dokl. Akad. Nauk. SSSR*, **213**, 1117 (1973).

817. V. V. Zverev, V. I. Vovna, M. S. El'man, Yu. P. Kitaev, and F. I. Vilesov, *Dokl. Akad. Nauk. SSSR*, **213**, 1319 (1973).

Index